APPLICATION OF FREQUENCY AND RISK IN WATER RESOURCES

APPLICATION OF FREQUENCY AND RISK IN WATER RESOURCES

Proceedings of the International Symposium on Flood Frequency and Risk Analyses, 14–17 May 1986, Louisiana State University, Baton Rouge, U.S.A.

Edited by

VIJAY P. SINGH

Department of Civil Engineering, Louisiana State University, Baton Rouge, U.S.A.

D. REIDEL PUBLISHING COMPANY

A MEMBER OF THE KLUWER ACADEMIC PUBLISHERS GROUP

DORDRECHT / BOSTON / LANCASTER / TOKYO

Library of Congress Cataloging in Publication Data

CIP

International Symposium on Flood Frequency and Risk Analyses (1986:
Louisiana State University, Baton Rouge)
Application of frequency and risk in water resources.

Includes indexes.
1. Floods—Congresses. 2. Water resources development—Congresses.
I. Singh, V. P. (Vijay P.) II. Title.
GB1399.I58 1986 551.48′9 87–20630
ISBN 90–277–2573–X
ISBN 90–277–2576–4 (set)

Published by D. Reidel Publishing Company,
P.O. Box 17, 3300 AA Dordrecht, Holland.

Sold and distributed in the U.S.A. and Canada
by Kluwer Academic Publishers,
101 Philip Drive, Assinippi Park, Norwell, MA 02061, U.S.A.

In all other countries, sold and distributed
by Kluwer Academic Publishers Group,
P.O. Box 322, 3300 AH Dordrecht, Holland.

Printed in The Netherlands

PREFACE

Floods constitute a persistent and serious problem throughout the United States and many other parts of the world. They are responsible for losses amounting to billions of dollars and scores of deaths annually. Virtually all parts of the nation--coastal, mountainous and rural--are affected by them. Two aspects of the problem of flooding that have long been topics of scientific inquiry are flood frequency and risk analyses. Many new, even improved, techniques have recently been developed for performing these analyses. Nevertheless, actual experience points out that the frequency of say a 100-year flood, in lieu of being encountered on the average once in one hundred years, may be as little as once in 25 years. It is therefore appropriate to pause and ask where we are, where we are going and where we ought to be going with regard to the technology of flood frequency and risk analyses. One way to address these questions is to provide a forum where people from all quarters of the world can assemble, discuss and share their experience and expertise pertaining to flood frequency and risk analyses. This is what constituted the motivation for organizing the International Symposium on Flood Frequency and Risk Analyses held May 14-17, 1986, at Louisiana State University, Baton Rouge, Louisiana.

The objectives of this symposium were therefore (1) to assess the current state of the art of flood frequency and risk analyses, (2) to demonstrate applicability of flood frequency and risk models, (3) to establish complementary aspects of seemingly different models, (4) to enhance interdisciplinary interaction, (5) to discuss practice of flood frequency and risk analyses technology by federal agencies in the U.S., (6) to discuss research needs in frequency and risk analyses, and (7) to determine directions for further research.

We received an overwhelming response to our call for papers. It was indeed a difficult task to select amongst the many excellent papers that were submitted, and we regret that we could not include all of them. The sole criterion for selection of a paper was its merit in relation to Symposium objectives. The subject matter of the Symposium was divided into 20 major topics encompassing virtually all facets of flood frequency and risk analyses. Each topic comprised of an invited state-of-the-art paper and a number of contributed papers. These contributions blended naturally to evolve a synthesized body of knowledge on that topic. Extended abstracts of all the invited and contributed papers were assembled in a pre-Symposium proceedings volume. This helped stimulate discussion and exchange of ideas during the Symposium.

The papers presented at the Symposium were refereed in a manner similar to that employed for publishing a journal article. As a result, many papers did not pass the review and were therefore eliminated from inclusion in the final proceedings. The papers contained in this book, APPLICATION OF FREQUENCY AND RISK IN WATER RESOURCES, represent one part of the Symposium contributions. The other parts are embodied in three separate books, FLOOD HYDROLOGY, HYDROLOGIC FREQUENCY MODELING, and REGIONAL FLOOD FREQUENCY ANALYSIS, which are being published simultaneously. Arrangement of these books under four different titles was a natural consequence of the diversity of technical material discussed in the papers. These books can be treated almost independently, although some overlap does exist between them.

This book contains seven sections encompassing major hydrologic applications of frequency and risk analyses. Each section starts usually with an invited state-of-the-art paper, followed by contributed papers. Beginning with a discussion of flood characteristics, the papers go on to discuss design flood, errors and outliers, application in design of water resources projects, risk and reliability analyses, flood damage assessment, and practice of frequency and risk analysis by federal agencies in the United States.

The book will of of interest to researchers as well as those engaged in practice of Civil Engineering, Agricultural Engineering, Hydrology, Water Resources, Earth Resources, Forestry and Environmental Sciences. The graduate students as well as those wishing to conduct research in flood hydrology will find this book to be of particular value.

I wish to take this opportunity to express my sincere appreciation to all the members of the Organizing and Advisory Committees, and the Louisiana State University administration for their generous and timely help in the organization of the Symposium. A lack of space does not allow me to list all of them by name here. Numerous other people contributed to the Symposium in one way or another. The authors, including the invited keynote speakers, contributed to the Symposium technically and this book is a result of their efforts. The session chairmen administered the sessions in a positive and professional manner. The referees took time out from their busy schedules and reviewed the papers. Graduate students assisted in smooth conduct of the sessions. I owe my sincere gratitude to all of these individuals.

If the success of a Symposium is measured in terms of the quality of participants and presentations then most people would agree that this was a very successful Symposium. A very large number of internationally well-known people, who have long been recognized for their contributions and have long been at the forefront of hydrologic research came to participate in the Symposium. More than 35 countries, covering the five continents and most of the countries of the world active in hydrologic research, were represented. It is hoped that long and productive personal associations will develop as a result of this Symposium.

March 1987
Baton Rouge, Louisiana

Vijay P. Singh
Symposium Director

ACKNOWLEDGMENTS

The International Symposium on Flood Frequency and Risk Analyses was sponsored and co-sponsored by a number of organizations. The sponsors provided financial support without which the Symposium might not have come to fruition. Their financial support is gratefully acknowledged. The co-sponsors extended their help in announcing the Symposium through their journals, transactions, newsletters or magazines. This publicity helped with attendance at the Symposium, and is gratefully acknowledged. The following is a list of Symposium sponsors and co-sponsors.

SYMPOSIUM SPONSORS

Louisiana State University
- Department of Civil Engineering
- Louisiana Water Resources Research Institute

National Science Foundation
U.S. Army Research Office
U.S. Geological Survey, Louisiana District, Baton Rouge
Woodward-Clyde Consultants, Baton Rouge

SYMPOSIUM CO-SPONSORS

American Geographical Society
American Geophysical Union
American Meteorological Society
American Statistical Association
American Society of Agricultural Engineers
American Society of Civil Engineers
American Water Resources Association
Association of American Geographers
Association of State Floodplain Managers
International Association for Hydraulic Research
International Association of Hydrological Sciences
International Association of Theoretical and Applied Limnology
International Commission on Irrigation and Drainage
International Geographical Union
International Water Resources Association
National Wildlife Federation
North American Lake Management Society
Pan American Institute of Geography
Society for Risk Analysis
Soil Conservation Society of America

TABLE OF CONTENTS

RESEARCH NEEDS ON FLOOD CHARACTERISTICS

Vujica Yevjevich and Nilgun Bayraktar Harmancioglu
Research Professor and Director, and Visiting Scientist, respectively
School of Engineering and Applied Science
George Washington University
Washington, D.C. 20052

ABSTRACT. Flood risk is nature related. Flood uncertainty is investigation related. The former can be changed only by changing the characteristics of floods. The latter can be changed only by more observation and investigation. The question to be answered yet is whether or not there is a physical upper bound to flood magnitude. The concept of probable maximum precipitation (PMP) seems to imply that there is an upper bound to flood magnitude. Flood characteristics are estimated by one or more methods: frequency curves, transfer of information (joint probability), regional data, paleohydrology, Bayesian and PMF groups of methods. The contemporaneous aspects of analysis of flood characteristics are related to reliability of estimation of floods of small exceedence probabilities in the range 10^{-2} to 10^{-7}, whether there is an upper bound to floods and whether it is feasible to attach a probability value to probable maximum flood (PMF). To eventually answer these basic questions, three investigations are needed: (1) study of properties of the upper tail of probability distributions of floods; (2) use of regional data for drawing either the envelopes of largest floods or the curves of average largest flood characteristics for given sample sizes in the region, with probabilities attached to these envelopes or average curves; and (3) finding of the composite probability of PMF by studying the aggregated probabilities of random variables which are "maximized" in the process of computing PMP and PMF. Flood characteristics change with time due to changes in river basins. The need exists for methods of estimation of flood characteristics over periods of time of 25-100 years, particularly for planning flood mitigation measures over an extended future.

1. INTRODUCTION

1.1. Coping with floods

The general activities in coping with floods are: (1) procurement of reliable information on characteristics of floods, and (2) planning, implementation and operation of flood mitigation measures. The term "coping with floods" refers only to flood problems which affect the

V. P. Singh (ed.), Application of Frequency and Risk in Water Resources, 1–21.

safety and welfare of humans in direct or indirect way. This term encompasses two sets of measures, those which are implemented in the physical environment in order to change flood characteristics, and those which are related to humans and their communities in order to change their attitudes and flood-related activities. Knowledge on floods and their impacts, and technology of coping with floods are basically investigated in three aspects: physical-environmental, technological and socio-economic. This paper treats only the procurement of information on floods.

It is logical to assume that reliability and effects of flood mitigation measures cannot be better than the level of information available on flood characteristics. As a consequence, planning of measures for coping with floods must be in some way a function of the amount and reliability of this information. The more limited this information, the more flexible and adjustable should be the planned control measures, and vice versa. The effort in procuring information on floods should be commensurate with the importance of flood problems and the investment expected to be made in flood alleviation measures.

1.2 Basic Characteristics of Floods

Floods are a relatively rapidly evolving hazard. They can occur either several times per year or may not occur at all in some years. Floods are produced not only by rainfall, snowmelt or by the combination of both in various proportions, but also by special geophysical conditions, such as backwater floods of river ice jams and landslides into rivers, or as downstream floods due to breaks of ice jams, landslide dams, break of structures as dikes, levees, dams, gates, retention walls, etc., or by an inappropriate or erroneous operation of outlets and spillways. Therefore, floods may be nature-induced or human-produced, with many combinations of their causal factors.

Flood characteristics of practical importance are: peak, volume, duration, time-of-the year occurrence, produced erosion, velocity, sediment transported, water quality effects, full hydrograph and similar characteristics. The socio-economic flood characteristics of major interest are flooded area, damage incurred, loss of lives, sediment deposited, erosion-produced failures, various dislocations and destructions, post-flood health hazards, etc. As for the general character of flood hazards, the above flood related magnitudes are all random variables. Therefore, they are subject to risk and uncertainty.

1.3 Definition of Risk and Uncertainty

Different definitions can be found in literature on three concepts related to flood random variables: risk, uncertainty and reliability. For purposes of this paper these concepts are defined.

Risk is assumed to occur objectively in nature in the sense that it exists whether or not it is properly conceived or assessed. In case of flood variables risk is defined as probabilities of exceedence of variable values which are singled out for any reason or were selected by a decision making process in design as the critical values. If properties of random variables -- used as determinants of the risk -- were known, risk would be exactly assessed in any decision. Unfortunately, hydrologic random variables are never exactly known since data samples are always finite, while populations of these variables nearly always are conceived as infinite.

The lack of knowledge on risk-defining random variables, involving whatever sources of errors, is conceived as uncertainty. Basic errors come from sampling of nature of limited duration or size. However, included also must be the random and systematic observational errors, non-homogeneity in data (changes in nature), modeling errors, errors in methods for parameter estimation, etc. Uncertainty depends on investigations and may be a function of time and financial means available, methods applied, properties of instruments used in procuring and processing of data, and of similar factors. Risk is a nature-bound concept while uncertainty is an investigation-bound concept. Risk can be changed only by changing the nature, while uncertainty can be changed only by more and better observations and investigations. Figure 1 is a graphical representation of definitions of risk and uncertainty.

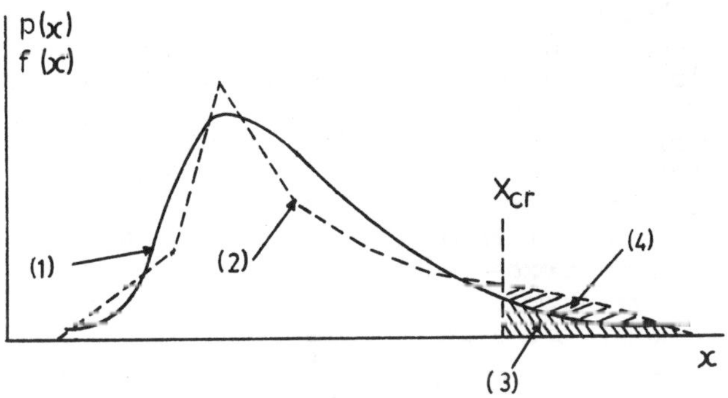

Figure 1. Definitions of risk (probability of exceedence) and uncertainty for floods: (1) unknown probability density function, p(x), of annual flood peaks; (2) estimated flood frequency density (histogram) curve, f(x); (3) area as unknown risk; (4) area as unknown uncertainty; area (3)+(4) as a joint estimate of "risk plus uncertainty."

Reliability, as the safety or assurance of performance, is defined here as a complementary value of risk to unity. If risk is 0.02, the reliability is 0.98. Different definitions or reliability are found in other areas related to random variables.

Risk and uncertainty cannot be separated because the exact magnitude of risks is never known. Therefore, the estimate of flood risk always contains uncertainty. The colloquial term of risk in flood analysis is always meant as basic risk plus uncertainty (Figure 1). Uncertainty is usually measured either by the variance of flood estimate when the probability distribution of an estimate is not known, or by the confidence limits or interval around the estimate when its probability distribution is known or could be estimated by the corresponding frequency distribution.

2. ESTIMATION OF FLOOD CHARACTERISTICS

2.1. Basic Concepts on Floods

Both statistical and deterministic concepts on floods have been developed. The variation in the magnitude of flood peak or flood volume, either as the largest event each year or as all the events above a threshold event during the period of observations, are easily shown to be random variables following frequency distributions which are then conceived as estimates of the hypothesized probability distributions. This concept does not assume the upper bound for distribution functions. It assumes that whatever maximum value was observed in the past, an incrementally larger value could still occur in the future. This implication for the boundless upper tail of fitted probability distribution functions means in practice that floods are chance-governed natural phenomenon which is not subject to a clear limitation in its extremes except by the probability related restrictions (the larger the flood, the smaller its probability of exceedence).

The concept of the boundless flood probability distribution conflicted (mainly in the United States) with two other prevailing concepts: (1) the nature cannot be boundless (say a small creek cannot produce precipitation and a flood equal to the largest flood of the Lower Mississippi River); and (2) the ocean-atmospheric processes must have a limitation in the supply of moisture to river basins through storms, or that an upper precipitation boundary exists. They resulted in the deterministic concept of maximum precipitation (MP) in the 1930's. Second thoughts among the originators of this concept on whether the MP could or could not be further exceeded resulted in a revised concept called the probable maximum precipitation (PMP), as it is used at present. However, the word "probable" in this name did not mean that a probability could be attached to MP. It should have been called "likely maximum precipitation." The word "likely" would then imply that it is an upper limit except for the uncertainty in its estimation. By the use of word "probable" two confusions resulted: (1) that PMP is a kind of maximum, assumed to exist in nature, except for the errors in its estimation; and (2) that the quantitative probability value cannot, or should not, be attached to the PMP concept.

The PMP was usually computed by maximizing some basic factors which produce the observed large storms. The concept of the storm is the first link between PMP and the resulting flood. The observed but maximized storm is transposed over a river basin in a position which produces the largest flood. This position may be conceived probabilistically as the possible event, however, with a probability of zero for its occurrence.

The assumed conditions of various river basin factors, especially the antecedent moisture condition index, are the next link between PMP and the resulting flood event. Because this index is conceived as a random variable, without an upper limit, the resulting flood should also be distributed without an upper limit regardless whether PMP has or has not an upper boundary. The use of the unit hydrograph for the estimated effective precipitation (rainfall minus all abstractions), or the use of any rainfall/runoff model, is the further link between PMP and the resulting flood, named during the 1930's as the probable maximum flood (PMF). In this case "probable" was meant as the possibility of rare occurrence of flood events which exceed PMF, however, no probability could be attached to PMF. The fact that a storm of rainfall can fall on the ground already covered by snow, with the rain melting it, leads to the conclusion that PMF cannot be conceived as an upper bound even if PMP could be accepted as having that bound.

The two concepts, frequency-based determination of floods and the PMF-based design of flood control measures, have been used in USA in parallel for already half a century. Several controversies still exist. Fitting frequency curves by probability distribution functions, and reliability of their extrapolations to determine probabilities of rare floods -- floods with the probabilities of exceedence of 10^{-3} to 10^{-5} (or even 10^{-5} to 10^{-7}) -- are still unsettled and controversial topics. This is the case even when uncertainty is taken into account by using the confidence limits around the estimated critical or design flood values.

The PMF concept did not go well with some practitioners, because: (1) it cannot be used in economic optimizations by searching for a design flood less than PMF; (2) if a coefficient $a<1.00$ is used as a ratio of the design flood to PMF, this design flood falls in the range of probabilities which should be subject to economic optimization, while there is no method of assigning the probability to (a.PMF); and (3)when PMF is used for design purposes the question nearly always arises whether it is economically affordable.

The present state of problems in coping with floods within the general problems of water resources development, conservation, control and protection seems to require new solutions, specifically:

(1) to find a bridge between the two opposite concepts of extreme large floods by resolving the basic controversy of whether the upper bound exists or not for storm precipitation for a given river basin;

(2) to determine whether the antecedent moisture conditions, including snow cover and melt during the occurrence of PMP, do or do not have an upper bound, with the resulting conclusion whether PMF has or has not an upper bound;

(3) to find the feasibility whether PMF can be located on the extrapolated tail of probability distribution functions fitted to flood frequency curves by any method available, incase the conclusion is made that floods do not have an upper bound, with this being equivalent of finding $P(Q_{max} > PMF)$ as the exceedence probability of PMF;

(4) to find out how to determine the fraction "a" for the design flood (a. PMF) by a rational method; and

(5) to sharpen the techniques of extrapolating the fitted flood probability distribution functions much beyond the return periods equal to observational lengths, say for the range of probabilities of 10^{-2} to 10^{-7}.

2.2 Methods of Estimation

The large number of methods for estimating flood characteristics, with their many variations, can be classified in several ways. Here, six groups of methods are selected to cover the techniques that exist for determining flood characteristics. Basically, and according to the source of data and to mathematical or conceptual approaches, these groups are:

(1) frequency curve methods, which use the directly observed flood data, by fitting the probability distribution functions to frequency curves, and by extrapolating these functions to ranges of large floods;

(2) transfer of information on floods from other variables, by such methods as the rainfall/runoff models, with freuqency analysis of the resulting floods by the established mathematical or physical relationships;

(3) regional data methods by estimating parameters of frequency curves or the curves themselves at given points by using the regional data on floods or flood-related variables;

(4) paleohydrology methods by using either the old chronicles on floods (historical markers or records), and/or the proxy chronological data either resulting from floods or related to past flood events;

(5) Bayesian methods by using the Bayes Theorem, in combining the information on floods from several sources; and

(6) Methods which lead to PMF through the estimation of PMP.

A large book manual (or manuals, such as the five volumes of United Kingdom Flood Study Report, 1975) are needed for a full description of all the methods which have been either used or proposed for the estimation of flood characteristics. The description of these methods is beyond the scope of this paper. However, some general remarks on the above groups of methods are in order if one has to project into the future the need for research that will give the largest payoffs.

2.3 Brief Description of Methods

Frequency curve methods depend on the accuracy of flood data and the length of record. Several problems are often encountered with these methods: (a) the reluctance to extrapolate the fitted probability distribution functions much beyond the flood of the probability 1/N, with N=the number of annual flood peaks; (b) largest floods with probabilities smaller than 1/N often occur in samples of N years, thus creating the problem of upper outliers, and largest floods with probabilities greater than 1/N also occur in samples of size N, thus creating the problems of inliers (lower outliers), Figure 2, with both cases pulling away the fitted probability distribution functions either much above or much below the expected population curves (overestimation or underestimation of floods for given probabilities); (c) by using only the largest annual flood values (annual flood series) instead of all the floods above a threshold, which are greater in number than N (partial flood series), some additional valuable information often is discarded; and (d) the fitted probability function is selected only by the goodness-of-fit criterion or by the convention for the uniform use of the same function (such as the Log-Pearson function), instead of searching what must be or should be the true distribution function, starting with the time series characteristics of instantaneous or daily runoff. A broad scope for research is further needed for these frequency methods, and also beyond the above four types of problems. In general, these methods are considered among the most reliable for floods of probability greater than 10^{-2}, or for return period of less than 100 years, if sufficient data are available (see references: Conover and Benson, 1963; Alexander, 1969; Ott and Linsley, 1972; Wall, 1974; Houghton, 1978; Brady, 1979; Landwehr et al, 1980; Hydrology Subcommittee, 1982; Boughton and Shirley, 1983; Kovacz, 1984; Lane, 1985).

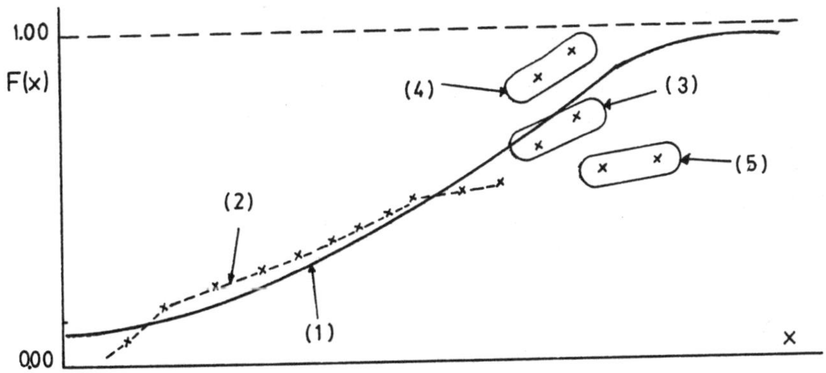

Figure 2. Outliers and inliers of flood frequency curves: (1) assumed population distribution, (2) frequency curve as estimation of flood distribution; (3) largest observed floods, if representative of the sample; (4) largest observed floods, as outliers (return period greater than the sample size); (5) largest observed floods, as inliers (lower outliers with return period smaller than the sample size).

Transfer methods for flood estimation are basically related to techniques of computing floods from precipitation (by the rainfall/runoff models) for given river basin moisture conditions, thus representing a joint distribution of rainfall and soil moisture measuring index. By searching for information in variables outside the runoff data, new sources of uncertainty are added, such as the sparse information on regional storm distributions, inaccuracy in computing various abstractions, deviations due to separation of hydrographs into surface and subsurface components, errors in models, errors in estimation of parameters, etc. The problem arises when one tries to measure how much information is contained in the resulting flood series, or how large is uncertainty in individual flood values. These methods are basically beneficial in areas with significant information on precipitation and limited information on direct runoff. They can also be used to fill in the missing records on floods or to extend the flood series based on runoff data to periods for which rainfall but no runoff is observed. Definitely, significant further contributions to estimation of flood characteristics can be made by improving the transfer techniques of hydrologic information (Eagleson, 1972; Wang and Revell, 1983;

Yankee Atomic Electric Co., 1984; Bras et al, 1985; Linsley and Kraeger, 1985).

Regional flood estimation methods combine all or most of the information available in a region in order to obtain flood characteristics either at a point or at all the points of interest in that region. This approach has led to methods which estimate the parameters of regional flood frequency distributions so that they could be plotted or their equations established. Often, only the characteristic floods (say the 20-year, 50-year, 100-year, and similar floods) are estimated by regional equations. The basic objection to regional methods may be the fact that even if storms are similar over a large region, the response hydrographs may be result of different unit hydrographs or they change from one basin to another. Some flood parameters may be quite different among these basins from those generalized by the regional information. A sound concept is that a region should contain much more information on floods than a point within it. However, the technology of generalization and transfer of that information are far from being reliable or satisfactory at present. The condensation of regional information and its transfer need some basic improvements. Future research is likely to produce many advancements. In general, these methods may be ranked by their accuracy immediately following the good frequency analysis methods (see Fuller, 1914; Clarke-Hafstad, 1942; Alexander, 1953, 1963; Linsley et al, 1958; Dalrymple, 1960; Matalas and Benson, 1961; Benson, 1962; Greis, 1963; Carrigan, 1971; Riggs, 1973; Rowe et al, 1975; Wallis, 1980, 1982; Hydrology Committee, 1981; Schaefer, 1982; Stedinger, 1983a; Buishand, 1984).

Paleohydrologic methods, whether they use the chronicle floods or paleoflood proxy data, must be considered as containing more uncertainty than the three previous groups because of large errors in individual values, in the first case, and various inaccuracies in relationships of past floods and proxy data, in the second case. Regardless of limited precision, these methods provide an insight into the past events, though they might lead to premature conclusions of "climatic changes," or that "floods have changed" significantly, due to significant inconsistencies (systematic errors) in basic data (see Helley and LaMarche, 1973; Costa, 1978, 1983; Baker et al, 1979, 1983; Patton, 1981; Baker, 1982; Foley et al, 1982; Kochel et al, 1982; etc).

Bayesian methods are a recent approach to flood hydrology, basically intended to combine information on floods from various sources. If the apriori probabilities could be reliably estimated by data-based frequencies, the aposteriori probabilities computed by Bayesian theorem would also be reliable. However, if subjective probabilities are used (therefore justifying the name of the "theorem of hypotheses"), the aposteriori probabilities can only tell how sensitive they are to variations in the apriori probabilities. The problems of objectivity or subjectivity in the estimation of the aposteriori probabilities have a very long history of controversies in probability theory and mathematical statistics. This history should not be neglected when applying the Bayesian methods to flood hydrology. Many approaches are used in the application of these methods (see Vicens et al, 1975; Wood, 1973; Wood and Rodriquez-Iturbe, 1975; Davis et al, 1976; Gentillon, 1981;

Kuczera, 1982; Russell, 1982; Stedinger, 1983b; etc.).

Methods of computation of PMP as well as synthesizing the storm hydrographs for the estimation of PMF need a review from time to time, as the new data on storms become available in each region. The objectivity or subjectivity in maximization of meteorological variables leading to PMP need a constant vigil, since the consequences of selecting these values have a significant impact on PMF and its affordability in design (for basic concepts see Bernard, 1944; Myers, 1969; WMO, 1973; Shipe and Riedel, 1976; Hansen et al, 1982; Newton, 1983).

2.4 Three Ranges of Flood Probabilities

Three ranges of flood probabilities can be singled out from the point of view of accuracy in their determination:

(1) floods of probabilities $2x10^{-1}$-10^{-2} (5-year to 100-year return period floods), or those used for design of many drainage or flood evacuation structures, which can be relatively reliably estimated by the first three groups of methods (frequency, transfer and regional methods);

(2) floods with probabilities 10^{-2} to 10^{-4} (100-year to 10,000-year return period flood), those that are usually estimated with the least accuracy by the first five groups of methods, and with the large uncertainty-based confidence intervals; this range of probabilities of floods require most attention in future research; and

(3) floods with probabilities equal or greater than the 10,000-year floods (including PMF and the very rare floods, with probabilities of exceedence of 10^{-4}-10^{-7}), which are crucial for survival of important and critical structures and systems; they represent the challenge to profession of nearly unprecedented scientific difficulties.

This division is made according to approaches needed to solve the problems of future increases in accuracies of flood estimates. For the lower range of floods, the multitude of techniques are already available, so that the problem may be the improvement and selection of the best existing method rather than the search for new methods. The main problem is the estimation of confidence intervals for computed floods once the method is selected and/or applied. For the middle range, with design floods of these magnitudes both critical and affordable, especially for its lower range of 10^{-2}-10^{-3} and important flood mitigation cases, both the improvement of existing methods and the design of new methods of estimation are needed. The probability of PMF would likely fall at the upper part of this range, say at or above 10^{-4}; however, this is still a hypothesis to be proven. Therefore, a location of PMF on the assumed unbounded tail of flood probability distribution functions might help the estimation of floods for both the middle and the upper ranges of probabilities.

2.5 Determination of Probability of Exceedence of PMF and Other Large Floods

To estimate the probability of exceedence of the probable maximum flood (PMF), as computed by using the probable maximum precipitation (PMP), it seems that future research should concentrate, if feasible also by parallel efforts, on the three basic lines of investigation:

(1) by searching for the shape (mathematical form) and location of the upper tail of probability distribution functions most likely to represent the population flood distributions, and by extrapolating this tail to low probability values to encompass the estimated PMF value; (2) by using both the envelope and the average curves of the largest observed floods in regions of relatively similar geophysical conditions, for a step-wise increase of periods of observation, by extrapolating these curves to curves of still smaller probabilities of exceedence, with the location of PMF value among these envelope or average curves; and (3) by studying the probability (frequency) distributions of all random variables than lead in their maximization to PMP and then to PMF, and by a composite aggregation of these variables in order to determine frequencies (as estimates of probabilities) of both PMP and PMF.

By finding the exceedence probabilities of PMF values, the other crucial problem, the extrapolation of flood frequency curves to probabilities in the range of 10^{-2} to 10^{-7} may be partially or fully resolved. Therefore, the problem is reduced to finding the most reliable upper tail of extreme floods, both in shape and in location.

2.6. Properties of Tails and Their Extrapolation

The results of studies on tails of flood frequency curves are controversial, namely whether the tails are light (exponential), normal (Gaussian) or very heavy (those of the stable probability distribution functions). Some studies (Tao et al, 1976) of daily flow series show tails of independent stochastic components to be nearly indistinguishable from the simple exponential tails. Then, sampling of extremes from such tails should also produce the exponential tails. Some other studies (Shen et al, 1980; Ochoa et al, 1980) have found the flood frequency tails to be heavy. Therefore, the problem must be resolved by finding what are the types of tails of most flood probability distributions by investigating their characteristics.

A hypothesis is advanced here, namely that tails of probability distributions of basic runoff processes (say of instantaneous discharge, often well approximated by daily runoff series for middle-sized and large rivers) can tell us as much if not more about the tails of distributions of flood peaks than the limited amount of data on annual flood peaks can tell about the tails of their distributions. Simply, much more data are available to assess accurately the character of tails of daily or instantaneous flows from which flood peaks are sampled than for flood distribution tails themselves. Distribution parameters of daily instantaneous flow change periodically over the year, even if one assumes that the probability function does not change with seasons. This latter hypothesis may or may not be valid. However, it may likely be valid for humid seasons in which floods usually occur than for the entire year composed of both the high flow and the low flow periods.

Figure 3 illustrates this problem of the tail. Equations of tails should have at least one parameter less (if not two parameters less) than the flood probability distribution function. Tails should be determined by its equation of the shape with parameters and its location (say by the initial tail coordinate values), such as Q_i and

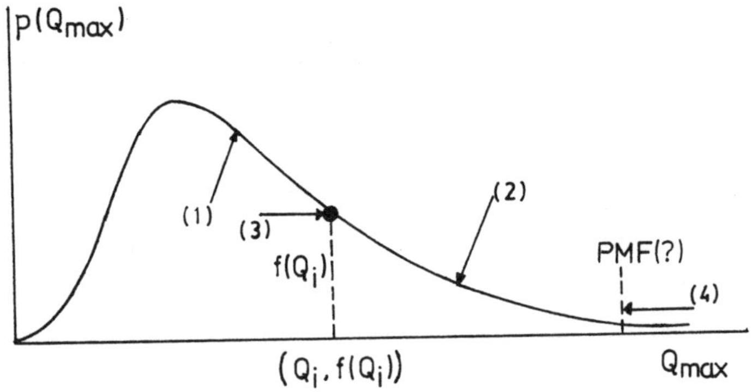

Figure 3. Location of the tail in flood probability graphs: (1) assumed probability density curve, $p(Q_{max})$; (2) upper tail function; (3) selected tail location point; (4) search for the location of PMF and its probability of exceedence.

$F(Q_i)$, but with no more than two parameters. An example may be

$$P(Q_{max} > Q) = \int_Q^{\infty} e^{-\alpha Q}\, dQ \qquad (1)$$

For $Q_{max}=Q_i$, Q_i and $F(Q_i)$ determine the tail location. If α= shape and β=scale parameters are those of the gamma distributions, β=1 may be often assumed for simple tails. In that case Q_i and $F(Q_i)$ uniquely determine the location of the tail. The larger Q_i, the closer the tail should be to an exponential tail. For small Q_i, $\beta \neq 1$ may be needed to take care of a more complex tail, with α and β estimates related (as both change with a change of Q_i). The location point Q_i should be selected so as to be to the right of the inflection point in Fig. 1, and α (and eventually β) to be determined by the full use of any information available on tails, both of the underlying process and flood frequency curve.

A concept is often advanced among flood specialists, namely that the superpopulation of floods is composed of different types of populations: those of frontal rainfall, convective rainfall, hurricane-related rainfall, rainfall/snowmelt combination, and similarly generated floods, often with various degrees of combination of these distinct general causes of extreme floods (Canfield et al, 1980). The question arises whether all these flood populations have the same type of tails, so that their combination in one "heterogeneous sample of floods" still preserves the same type of tail, but only with different parameters and location than is the case for a homogeneous population. Therefore, it will be necessary to study the tails in detail in order to investigate the problem posed by the concept of superpopulation of floods. This can be done both by assuming it to be valid and by contesting it on the basis of a myriad of flood producing causes, when each type of flood producing inputs is only one of those causal factors . In the cause-effect relationships with a very large number of causes, some causes may be absent or may have zero effects in some floods but not in others, and still the concept of a homogeneous population remains valid.

Once the tail equation is specified, both problems, that of finding $P(Q_{max}>PMF)$ and that of finding probabilities of floods with $P(Q_{max})<10^{-2}$ can be solved. Definitely, the estimation of Q_i, α and/or β, needs additional research efforts as well as finding the methods of drawing the confidence intervals to estimated Q_{max} of rare floods in general, and PMF in particular.

2.7 Methods of Drawing Envelope or Average Curves to Observed Floods in a Region

The longest reliable records on runoff and floods of medium or large rivers in the world is about 175 years. Many rivers have good records for 75-100 years. With thousands of river gauging stations available, various extreme flood events have occurred, and some of them are likely to be of much longer flood return period than the available sample size. If a region has 100 river gauging stations, and if extreme flood events are not excessively mutually dependent, say that 100 dependent flood stations are equivalent to 10 stations of independent flood events, each year the 10-year return period flood should be exceeded somewhere, 100-year flood should be exceeded each 10 years, and 1,000-year flood should be exceeded each 100 years, on the average. By using the effective number of stations in a region and the Poissonian law of computing the probabilities of exceedence of very rare floods which occur in relatively small samples, the expected (average) number of exceedence of various rare floods may be assessed. Similar considerations may be applied to precipitation of very large storms.

By drawing a graph of Q_{max} versus the river basin area A, or still better the ratio Q_{max}/A versus the area A (Figure 4), an envelope can be drawn for each effective length of years of observations. This envelope would then have a larger return period than the effective sample size (Crippen and Bue, 1977). The return period may then be estimated by using the Poissonian distribution of occurrence of rare floods in samples smaller than their return periods.

If, however, Q_{max}/A of all largest floods for all stations are plotted, an average curve can be fitted that will be close to the return period of the effective number of years of observation, computed by the number of stations, number of years and the dependence measure between flood peaks of these stations. By using a step-wise increase of observational length for regional data, both the envelope curves of the greater than sample size return periods and the average curves of the sample return period could be fitted, with the return period appropriately estimated. This should be done until data of the longest available samples are used for the region. As the length of samples increases, the incremental change between the two successive fitted curves will decrease. By studying the changes between the fitted envelope or average curves, their extrapolations can be designed (Figure 4), reaching much larger return periods than those

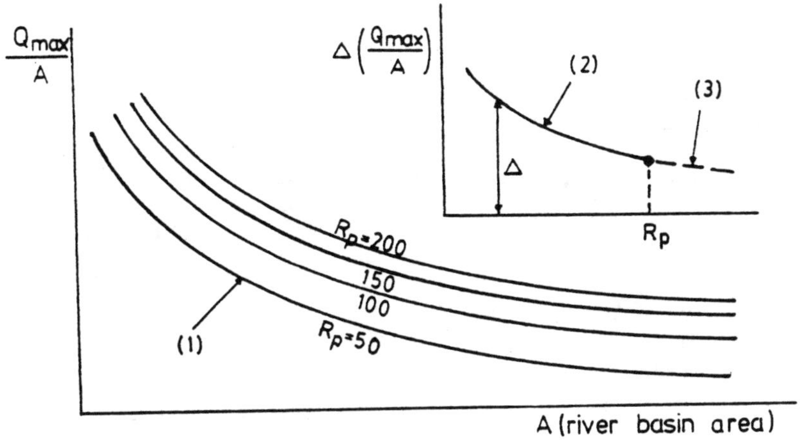

Figure 4. The envelope and average curves of Q_{max}/A versus area A for the largest floods in a region, which increase in their position on the graph as the return period R_p increases: (1) curves; (2) incremental increase in Q_{max}/A versus R_p; (3) extrapolation of envelope or average curves.

attached to these curves. Once a set of envelopes or a set of average curves have been drawn and extrapolated, the location of PMF enables the determination of its return period, and indirectly the location of the tail of probabilities of rare floods.

2.8 Probability Distribution of Variables That Lead to PMP

To estimate PMP one has to maximize several meteorologic random variables, such as the content of moisture in the air, temperature, dew point, eventually duration of flow of air mass of a storm over an area, etc., with all these variables depending on the method of computation used for PMP. Each of these variables has a probability distribution and it is safe to assume that they do not have the upper boundary. Marginal and joint distributions may be used to eventually assess what should be the probability (frequency) of exceedence of PMP and the resulting storm pattern. By using the PMP storm of given probability, with the antecedent moisture index conceived as a random variable, and probabilities of abstractions and effective precipitation also assessed as random variables, the composite probability of PMF could be eventually estimated.

Limitations in data may restrict the accuracy of this method. However, without trying this route even the need for future collection of data may not be clearly assessed and designed. This approach is based on the concept that PMP is a large value, although it can be exceeded provided precipitation is systematically observed for a long time. This concept is supported by the fact that the moisture content in an air mass can always be greater for a small increment, or the moisture-supplying ocean temperature can always be greater for a small increment, etc.; however, with smaller and smaller exceedence probabilities as these variables increase beyond the values that are used in computing PMP.

Definitely, to assess the probability of PMF, a multivariate probability distribution of PMP, the abstractions and moisture conditions index needs to be developed, in this case say as a trivariate or multivariate distribution. Then its integration could produce the marginal distribution of PMF by finding $P(Q_{max}>PMF)$. The transfer function of PMP minus abstractions would give the flood peaks with probabilities attached for given P(PMP). The P(PMF) would then be approximately estimated.

This third approach of determining the probability of exceedence of PMF would likely require the largest effort of the three approaches suggested above, both theoretically and practically. The reasons are the need for a good probabilistic framework of a relatively large number of random variables involved and the collection of significant amount of data on these variables.

3. EVOLUTION OF FLOOD CHARACTERISTICS WITH TIME

3.1 Natural Versus Human-Affected Floods

There is a very small number of important river basins in the world, which do not undergo significant changes in land, soil and water conditions by natural disruptions, slow natural or human-caused evolutions, and by various effects resulting from large human activities and built structures. Therefore, flood characteristics must also be changing either slowly or abruptly, depending on the type of change in river basins. Practical problems arise as the results of these changes, since most of the interest on floods is related to future occurrences, plans, projects and flood mitigation measures. Therefore, two aspects need to be studied: (1) what are the impacts of past measures and changes in river basins on present flood characteristics in comparison with flood characteristics of the years prior to those changes; and (2) future changes that will affect floods, since the flood control measures are designed for a relatively long future during which many changes in river basin could occur.

While the estimate of impacts of past measures or occurrences on present flood characteristics may be easily studied, provided sufficient amount of data is available to investigate both the changes themselves in river basins and their impacts on floods, the projection of potential or likely measures that will impact floods during long future (say for 25-75 years) is relatively difficult. This would be less difficult if good water resources plans existed for the corresponding river basins.

Four types of flood conditions may be distinguished in the study of changes in flood characteristics:

(1) natural floods, by assuming that precipitation regimes and river basin characteristics stay approximately unchanged with time;

(2) natural floods, assumed modified by natural abrupt disruptions (large landslides, large forest fires) and/or natural or human induced slow evolution (changes in forests, in land use, in soils);

(3) natural floods, assumed modified by large structures and projects which in their plans and operations do not have the flood control as the specific purpose, but with the accidental control (floods fill reservoirs and therefore become modified downstream); and

(4) natural floods, which are modified by both the structures for flood control purposes specifically or in conjunction with other water resources purposes, and by projects built only for the other water resources purposes but with accidental flood effects.

If the solution of flood problems requires the values of floods in the range of exceedence probabilities of 10^{-3}-10^{-6}, and if one has to take into account the impacts of future changes in a river basin, difficulties become compounded. Fortunately, the impacts of natural disruptions, human-made structures and slow evolutions in river basins, either by the natural or by human-induced factors, decrease percentage-wise as the magnitude of the flood increases. A classical example is the relative impact of forests on floods, which is the largest for small floods, moderate for the large floods, and nearly negligible for extremely large floods, for example the flood of the PMF type. For floods with the exceedence probabilities of 10^{-5}-10^{-7} the impact of all types of changes may eventually be neglected except for the case of very large reservoirs. In some cases these extremely high floods are increased by changes in river basin; forest land may retain large quantities of snow to increase floods by melting produced by large warm rainstorms; or floods can break dams, dikes, levees, and compound their magnitudes by the water released.

3.2 Methods of Determining Changes in Flood Characteristics

If the long runoff data are available, plotting of the largest annual flood versus time, and the tests of this discrete time series for jumps (slippages, shifts) and trends (tendency) represent the simplest statistical tools to be used. Similarly, a discrete time series of the number of flood peaks per year above the threshold flood level versus the time may show changes in this number due to an increase or a decrease of the average number of floods per year. However, any extrapolation of trends must be carefully treated since the future changes might be quite different from those of the past. This method may be conceived as a global (synthetic) approach.

When individual causal factors are studied for their incremental effects on floods, such as the effects of each reservoir, dike, levee, forest cut,urbanization, etc., and summarized in an aggregated effect, the method may be defined as analytical. The study of these individual effects through the cause/effect relationship may help the projection of future changes in flood characteristics, since it is easier to extrapolate the potential causes of changes (often contained in river basins or regional development plans) than to extrapolate the compounded effects themselves.

The study of future changes in flood characteristics is very important in present planning and decision making related to mitigation of floods. The large structural measures have an economic life of 50-100 years but their physical life with proper maintenance is approximately 150-200 years or more. If the optimization shows that a 100-year return flood is the most attractive design flood, the computation of this flood over 50-100 years (or for the physical life of 150-200 years or more) is highly speculative. However, when the data of observations of past 100-120 years are examined, significant conclusions can be drawn from them.

The easiest way of structural defense from floods is to build levees and to protect large flood plains and communities. This measure increases flood downstream because of the exclusion of flood attenuating plains. When the next phase starts, namely the construction of large reservoirs (with an accidental or a designed flood control), floods start to decrease downstream for a given probability of exceedence. When all major reservoirs are built upstream, then their mode of operation will decide whether there will be further changes in flood characteristics or not. In the case of constructing levees first and then reservoirs, the average 100-year flood during a prolonged period of changes will be greater than the 100-year flood estimated at the beginning of this period. This is equivalent of designing flood control to a lower than 100-year return period flood if the average 100-year flood over a long period is not used instead of its estimate under the initial flood producing conditions. Figure 5 illustrates this case.

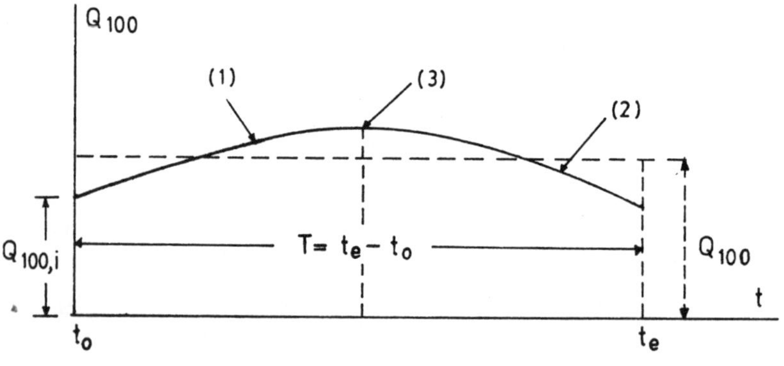

Figure 5. Hypothetical evolution of R_p=100-year flood peak during a long project life T: (1) phase of flood increases due to construction of levees; (2) phase of flood decreases due to construction of reservoirs; (3) design changeable 100-year flood curve with mean $Q_{100}>Q_{100,i}$, and maximum equivalent initial flood of $R_p>100$.

There are many rivers in the world with a history of observed flood of 50-175 years, and simultaneously there is a history of changes in river basins and in flood characteristics of 50-125 years (Danube, Rhine, Mississippi, Volga, Rhone and other rivers, etc.). Therefore, a need exists for the investigation in detail of what types and magnitudes of changes have occurred in the past, and how these changes have modified the flood characteristics. There must be some regularities in social and economic developments which have caused the changes in river basins with the corresponding modifications of flood characteristics. It is easier to apply regularities of the past to future projections by any acceleration of development into the future than to completely design what may be in store for future changes in floods.

While the study of characteristics of natural observed and/or changed but observed floods has been common in the past, the investigation of future flood characteristics based on anticipated changes have been sporadic and intuitive at best. Therefore, it seems logical to expect that future research on flood and flood mitigation measures will pay full attention to these problems in order to improve both planning of water resources schemes and of flood control measures. There is so much data accumulated (observed) on floods for the rivers of the world, that an extraction of information on relationships of changes in river basins and the resulting changes in flood characteristics seems very plausible and promising. Similarly, studying the state of floods, say at the times of 25-75 years ago with the projection of expected future changes at those times, and comparing the projected impacts on floods with the observations on flood in subsequent times, will enhance the accuracy in the projection on future flood characteristics.

ACKNOWLEDGMENT

The research leading to this paper is supported by the U.S. National Science Foundation, Grant No. 85-41631.

REFERENCES

Alexander, G.N., 1963. 'Using the Probability of Storm Transposition for Estimating the Frequency of Rare Floods'. Journal of Hydrology, 1, pp. 45-57.

Alexander, G.N., 1969. 'Application of Probability to Spillway Design Flood Estimation'. Proceedings of the Leningrad Symposium on Floods and Their Computation, August 1967, Vol. 1, International Association of Hydrologic Sciences, pp. 536-543.

Baker, V.R., 1982. 'Geology, Determinism and Risk Assessment'. Scientific Basis of Water-Resource Management, National Academy Press, Washington, DC, pp. 109-117.

Baker, V.R., Kochel, R.C., and Patton, P.C., 1979. 'Long-Term Flood Frequency Analysis Using Geological Data'. Proceedings of the Symposium on Hydrology of Areas of Low Precipitation, December 1979, Canberra, Australia, International Association of Hydrologic Sciences, No. 128, pp. 3-9.

Baker, V.R., Kochel, R.C., Patton, P.C., and Pickup, G., 1983. 'Paleohydrologic Analysis of Holocene Flood Slack-Water Sediments'. Special Publications International Association of Sedimentology, 6, pp. 229-239.

Benson, M.A., 1962. 'Evolution of Methods for Evaluating the Occurrence of Floods'. U.S. Geological Survey Water-Supply Paper 1580-A, 30pp.

Bernard, M., 1944. 'The Primary Role of Meteorology in Flood Flow Estimating'. Transactions of ASCE, 109, pp. 311-382.
Boughton, W.C., 1980. 'A Frequency Distribution for Annual Floods'. Water Resources Research, 16(2), pp. 347-354.
Boughton, W.C., and Shirley, E.D., 1983. 'Fitting Procedures for the Log-Boughton Distribution'. Journal of Hydraulic Engineering, ASCE, 109(4), pp. 579-589.
Brady, D.K., 1979. 'Flood Frequency Analysis: Logistic Method for Incorporating Probable Maximum Floods'. Research Report No. CE 1, Department of Civil Engineering, University of Queensland, St. Lucia, Australia.
Bras, R.L., Gaboury, D.R., Grossman, D.L., and Vicens, G.J., 1985. 'Spatially Varying Rainfall and Floodrisk Analysis'. Journal of Hydraulic Engineering, ASCE, 111(5), pp. 754-773.
Buishand, T.A., 1984. 'Bivariate Extreme-Value Data and the Station-Year Methodology'. Journal of Hydrology, 69, pp. 77-95.
Canfield, R.V., Olsen, D.R., Hawkins, R.H., and Chen, T.L., 1980. 'Use of Extreme Value Theory in Estimating Flood Peaks from Mixed Populations'. Hydraulics and Hydrology Series, Report No. 42, UWRL/H-80/01, Utah Water Research Laboratory, Logan, Utah, 42 pp.
Carrigan, PhH., Jr., 1971. 'A Flood-Frequency Relation Based on Regional Record Maxima'. U.S. Geological Survey Professional Paper 434-F, 22 pp.
Chan, S.O., and Brass, R. L. 1979. 'Urban Stormwater Management: Distribution of Flood Volumes'. Water Resources Research, 15(2), pp. 371-382.
Clarke-Hafstad, K., 1942. 'Reliability of Station-Year Rainfall-Frequency Determinations'. Transactions of ASCE, 107, pp. 633-683.
Conover, W.J., and Benson, M.A., 1963. 'Long-Term Flood Frequencies Based on Extremes of Short-Term Records'. U.S. Geological Survey Professional Paper 450-E, pp. 59-60.
Costa, J.E., 1978. 'Holocene Stratigraphy in Flood Frequency Analysis'. Water Resources Research, 14(4), pp. 626-632.
Costa, J.E., 1983. 'Paleohydraulic Reconstruction of Flashflood Peaks from Boulder Deposits in the Colorado Front Range'. Geological Society of America Bulletin, 94, pp. 986-1004.
Crippen, J.R., 1982. 'Envelope Curves for Extreme Flood Events'. Journal of the Hydraulics Division, ASCE, 108 (HY-10), pp. 1208-1210.
Crippen, J.R., and Bue, C.D., 1977. 'Maximum Floodflows in the Conterminous United States'. U.S. Geological Survey Water Supply Paper No. 1887, 52 pp.
Dalrymple, T., 1960. 'Flood-Frequency Analyses'. U.S. Geological Survey Water Supply Paper 1543-A, 80 pp.
Davis, D.R., Duckstein, L., and Fogel, M.M. 1976. 'Uncertainties in the Return Period of Maximum Hydrologic Events: A Bayesian Approach'. Journal of Hydrology, 31, pp. 81-95.
Eagleson, P.S., 1972. 'Dynamics of Flood Frequency'. Water Resources Research, 8(4), pp. 878-898.
Foley, M.G., Zimmerman, D.A., Doesburg, J.E., and Thorne, P.D., 1982. 'Review and Evaluation of Paleohydrologic Methodologies'. Battelle Pacific Northwest Laboratory, Richland, Washington (NUREG/CR 3055, PNL 4346, (NTIS)).

Fuller, W.E., 1914. 'Flood Flows'. Transactions of ASCE, 77, pp. 564-617.

Gentillon, C.D., 1981. 'FLOE: A Program for Bayesian Estimation of Flood Probabilities'. EG&G, Idaho, Inc., 123 (NUREG/CE-2259, (NTIS)).

Greis, N.P., 1983. 'Flood Frequency Analysis: A Review of 1979-1982.' Reviews of Geophysics and Space Physics, 21(3), pp. 699-706.

Hansen, E.M., J.F. Miller, L.C. Schreiner, 1982. 'Application of Probable Maximum Precipitation Estimates - United States East of the 105th Meridian'. NOAA Hydrometeorological Report No. 52, National Weather Service, U.S. Department of Commerce, Silver Spring, MD, 166 pp.

Helley, E.J., and LaMarche, V.C., 1973. 'Historic Flood Information for Northern California Streams From Geological and Botanical Evidence.' U.S. Geological Survey Professional Paper 485E, 16 pp.

Houghton, J.C., 1978. 'The Incomplete Means Estimation Procedure Applied to Flood Frequency Analysis'. Water Resources Research, 14(6), pp. 1111-1115.

Hydrology Committee, 1981. 'Estimating Peak Flow Frequencies for Natural Ungaged Watersheds: A Proposed Nationwide Test'. U.S. Water Resources Council, 346 pp.

Hydrology Subcommittee, 1982. 'Guidelines for Determining Flood Flow Frequency'. Hydrology Subcommittee Bulletin 17-B, with editorial corrections. Interagency Advisory Committee on Water Data, U.S. Geological Survey, 28 pp.

Kochel, R.C., and Baker, V.R., 1982. 'Paleoflood Hydrology.' Science, 215(4531), pp. 353-361.

Kochel, R.C., Baker, V.R., and Patton, P.C., 1982. 'Paleohydrology of Southwestern Texas'. Water Resources Research, 18(4), 1165-1183.

Kovacs, Z.P., 1984. 'Problems in the Estimation of Maximum Flood Peaks'. Proceedings of the South African National Hydrological Symposium, Department of Environmental Affairs, Technical Report No. T.R. 119, Pretoria, South Africa, 18 pp.

Kuczera, G., 1982. 'Combining Site-Specific and Regional Information: An Empirical Bayes Approach.' Water Resources Research, 18(2), pp. 306-314.

Landwehr, J.M., Matalas, N.C., and Wallis, J.R., 1980. 'Quantile Estimation with More or Less Floodlike Distributions'. Water Resources Research, 16(3), pp. 547-555.

Lane, W.L., 1985. 'Rare Flood Frequency Estimation - A Case Study of the Pecos River', (Abstract). EOS Transactions of AGU, 66(18), 267.

Lettenmaier, D.P., and Burges, S.J., 1978. 'A Nonparametric Approach to the Analysis of Extreme Hydrologic Events'. Proceedings of the International Symposium on Risk and Reliability in Water Resources, Vol. II, University of Waterloo, Ontario, Canada, June 26-28, 1978, pp. 799-818.

Linsley, R.K., Jr., and Kraeger, B., 1985. 'A System for Generating Long Streamflow Records for Study of Floods of Long Return Period.' Prepared for the Nuclear Regulatory Commission, Contract NRC-04-84-114, 115 pp.

Matalas, N.C., and Benson, M.A., 1961. 'Effect of Interstation Correlation on Regression Analysis'. Journal of Geophysical Research, 66

Myers, V.A., 1969. 'The Estimation of Extreme Precipitattion as the Basis for Design Floods - Resume of Practice in the United States'. Proceedings of the Leningrad Symposium on Floods and Their Computation, August 1967, Vol 1, International Association of Hydrologic Sciences, pp. 84-101.

Newton, D.W., 1983. 'Realistic Assessment of Maximum Flood Potential'. Journal of Hydraulic Engineering, ASCE, 109(6), 905-918.

Ochoa, I.D., Bryson, M.C., and Shen, H.W., 1980. 'On the Occurrence and Importance of Paretian-Tailed Distributions in Hydrology.' Journal of Hydrology, 48, pp. 53-62.

Ott, R.F., and Linsley, R.K., Jr., 1972. 'Streamflow Frequency, Using Stochastically Generated Hourly Rainfall.' Proceedings of the International Symposium on Uncertainties in Hydrologic and Water Resource Systems, Vol. 1, University of Arizona, Tucson, Arizona, pp. 230-244.

Patton, P.C., 1981. 'Geomorphic-Hydrologic Methods for Assessing Flood Potential in Southern New England.' Institute of Water Resources, University of Connecticut, 31 pp.

Riggs, H.C., 1973. 'Regional Analyses of Streamflow Characteristics.' U.S. Geological Survey Techniques for Water Resources Investigations, Book 4, Chapter B3, 15 pp.

Rowe, R.R., Long, G.L., and Royce, T.C., 1957. 'Flood Frequency by Regional Synthesis'. EOS Transactions of AGU, 38(6), pp. 879-884.

Russell, S.O., 1982. 'Flood Probability Estimation'. Journal of the Hydraulics Division, ASCE, 108(HYI), pp. 63-73.

Schaefer, M.G., 1982. 'Regional Risk Analysis for Spillway Design'. AWRA, International Symposium on Hydrometeorology, Denver, CO, 20 pp.

Shen, H.W., Bryson, M.C., and Ochoa, I.D., 1980. 'Effects of Tail Behavior Assumptions in Flood Predictions'. Water Resources Research, 16(2), pp. 361-364.

Shipe, A.P., and Riedel, J.T., 1976. 'Greatest Known Areal Storm Rainfall Depths for Contiguous United States'. NOAA Technical Memorandum NWS HYDRO-33, U.S. Department of Commerce, Silver Spring, MD, 174 pp.

Stedinger, J.R. 1983a. 'Estimating a Regional Flood Frequency Distribution'. Water Resources Research, 19(2), pp. 503-510.

Stedinger, J.R., 1983b. 'Design Events with Specific Flood Risk'. Water Resources Research, 19(2), pp. 511-522.

Tao Pen-Chich, Yevjevich, V., and Kottegoda N., 1976. 'Distributions of Hydrologic Independent Components'. Colorado State University Hydrology Papers, No. 82, January.

Vicens, G.J., Rodriguez-Iturbe, I., and Schaake, J.C., Jr., 1975. 'A Bayesian Framework for the Use of Regional Information in Hydrology'. Water Resources Research, 11(3), pp. 405-414.

Wall, I.B., 1974. 'Probabilistic Assessment of Flooding Hazard for Nuclear Power Plants'. Nuclear Safety, 15(4), pp. 399-408.

Wallis, J.R., 1980. 'Risk and Uncertainties in the Evaluation of Flood Events for the Design of Hydrologic Structures'. Seminar on Extreme Hydrologic Events: Floods and Droughts, Centro di Cultura Scientifica E. Majorana, Erice, Italy, March 22-25.

Wallis, J.R., 1982. 'Probable and Improbable Rainfall in California'. IBM Research Report RC9350 (No. 41172), Yorktown Heights, N.Y., 26 pp.

Wang, B.-H., and Revell, R., 1983. 'Conservatism of Probable Maximum

Flood Estimates.' Journal of Hydraulics, ASCE, 109(3), pp. 400-408.
Wood, E.F., 1973. 'Flood Control Design with Limited Data - A comparison of the Classical and Bayesian Approaches.' Symposium on the Design of Water Resource Projects with Inadequate Data, Madrid, Spain, June 1973. International Association of Hydrological Sciences, pp. 469-483.
Wood, E.F., and Rodriguez-Iturbe, I., 1975. 'Bayesian Inference and Decision Making for Extreme Hydrologic Events'. Water Resources Research, 11(4), pp. 533-542.
World Meteorological Organization, 1973. 'Manual for Estimation of Probable Maximum Precipitation'. Operational Hydrology Report No. 1, WMO No. 332, Geneva, Switzerland, 190 pp.
Yankee Atomic Electric Company, 1984. 'Probability of Extreme Rainfalls and the Effect of the Harriman Damage'. Framingham, MA, 16 pp.

APPLICATION OF PROBABLE MAXIMUM FLOOD ESTIMATES TO THE DESIGN OF WATER RESOURCE DEVELOPMENT PROJECTS

Bi-Huei Wang
Harza Engineering Company
150 South Wacker Drive
Chicago, Illinois 60606

ABSTRACT. Designs of most major dams and spillways are based on probable maximum floods (PMF). When properly estimated, PMFs provide the basis for confident sizing of spillway. There are, however, many factors which affect the estimation of a PMF. In analyzing each of these factors, certain assumptions have to be made and subjective judgements exercised. Depending on these assumptions and judgements, resulting PMF estimates may be quite different. This paper provides a brief summary of current practices in PMF estimation and application, and offers some guidelines for such practices. It also includes a comparison and discussion of PMF and risk-based analysis for spillway design.

1. INTRODUCTION

The main application of probable maximum flood (PMF) estimates to the design of water resource development projects is when the projects involve the construction of major dams and spillways. Any dam constructed on a natural stream requires a spillway to safely pass flood flows downstream. Unfortunately, it can never be certain how large a flood a dam will experience during its useful life. Therefore, any spillway will likely be designed either too large or too small.

If the spillway is designed to pass floods larger than those which could possibly occur, a large part of the investment may be wasted in providing excess capacity which will never be used. On the other hand, if it is designed to pass only floods smaller than those which could possibly occur, there will be a risk of failure of the dam and associated consequences.

Ideally, the selection of spillway capacity should be based on comprehensive economic, social, and political considerations, as well as technical criteria of hydrology, hydraulics and structural design. In practice, however, a criterion commonly used is to design the spillway for the PMF if it involves a high dam impounding a large amount of water, or where failure of the dam is likely to cause loss of human lives.

V. P. Singh (ed.), Application of Frequency and Risk in Water Resources, 23–33.

The PMF as defined by the U.S. Army Corps of Engineers (1979) is "the flood that may be expected from the most severe combination of critical meteorological and hydrologic conditions that are reasonably possible in the region". It is a hypothetical flood estimated by transforming a probable maximum precipitation (PMP).

In the following section, a brief summary of current practices in the estimation of PMP and its time and areal distributions is given. It is followed by a section on how an estimated PMP is transformed to a PMF which, in turn, is followed by a section on application of the PMF. The final section is devoted to a comparison and discussion of PMF and risk-based analysis for spillway design.

2. ESTIMATION OF PMP

PMP was once known as maximum possible precipitation (MPP). The main reason for the change from MPP to PMP was that MPP carried a stronger implication of a physical upper limit of precipitation than does PMP. PMP is preferred because of uncertainties surrounding any estimate of maximum precipitation.

In a publication by the World Meteorological Organization (1973), PMP is defined conceptually as "the theoretically greatest depth of precipitation for a given duration that is physically possible over a particular drainage basin at a particular time of year". Operationally, however, it is the estimated precipitation for a given duration, drainage area, and time of year for which there is virtually no risk of being exceeded, assuming that there is no long-term climatic trend.

2.1. Critical Duration

The magnitude and intensity of a PMP varies with duration. The duration of the PMP that causes the most critical flood at a damsite is the critical duration. In general, the critical duration is short for a small basin and long for a large basin. The flood regulating capacity of reservoirs located at or upstream of a project site also affects the length of critical duration.

To determine the critical duration, inflow hydrographs resulting from PMP of various durations should be derived and routed through the proposed reservoir. The duration of the PMP that causes the maximum flood surcharge and outflow is the critical duration. In practice, however, to avoid the need for estimating PMPs of various durations, the critical duration is often estimated based on the size of the drainage area and the flood storage capacity of all reservoirs located at and upstream from the proposed site.

For a small drainage area and a small reservoir storage capacity, a short duration of six hours or less may be critical. On the other hand, for a large drainage area and a large reservoir capacity, a long duration of three days or more may be critical.

The general guideline for determining the critical duration is that it should be at least equivalent to the time of concentration of the basin, and should be longer when a reservoir of substantial storage capacity is considered. In case of any doubt, a number of durations should be examined to determine the most critical one.

2.2. Type of Storm

For the purpose of estimating the PMP for a specific basin, rain storms can be classified as "general-storms" or "local-storms". A general-storm normally lasts more than 6 hours. It is associated with a major synoptic weather system and generally produces precipitation over an area in excess of 500 mi^2 (1,300 km^2). A "local-storm" seldom lasts more than 6 hours. It is commonly referred to as a thunder storm and produces precipitation over an area of less than 500 mi^2 (1,300 km^2).

The magnitude and time distribution of general- and local-storm PMPs are generally quite different. For a project basin with a critical duration of 6 hours or less, local-storm PMP generally would produce the PMF. For a critical duration of more than 24 hours, a general-storm normally would be more critical. For a critical duration of 6 to 24 hours, both types of storms would need to be analyzed to determine whether the local- or general-storm PMP would be more critical.

2.3. Generalized Estimates

For most of the United States, generalized PMPs have been estimated and published by the National Weather Service (previously the Weather Bureau) in collaboration with other agencies such as the Army Corps of Engineers, Bureau of Reclamation, the Tennessee Valley Authority, and the Soil Conservation Service. These publications are contained in a series of Hydrometeorological Reports, or Technical Papers, available from the Government Printing Office or the National Weather Service.

Using generalized PMP available in these publications, PMPs of various durations can be estimated for a storm area of any size and at any location. Based on these estimates, a synthetic PMP isohyetal map can be constructed and superimposed on the basin maps of the same scale. A step-by-step procedure for applying the generalized PMP estimates generally is included in the publication containing the estimates.

2.4. Site Specific Studies

There are basins of unique topographic characteristics where generalized PMP estimates cannot be properly applied. Under such conditions, or in areas where generalized PMP estimates are not available, a site specific study is needed (Riedel, Wang and Diebel, 1982). Basically, this involves the transposition of major historical storms from where they occurred to the project basin and the adjustment of the recorded rainfall for maximum moisture, topographic effects, and possibly other factors such as wind speed.

An important factor that affects the result of any site specific PMP study is the number of major storms considered in the study. If the region from which the recorded storms are transposed is too small or if the period of storm rainfall records is too short, the site specific study may result in an under-estimated PMP. Therefore, identifying the region which is considered to be meteorologically homogeneous with the project basin is extremely important. This may require an extensive study of meteorology of major rainstorms in a large region.

Another factor that greatly affects a site specific study is adjustment of the transposed storm for PMP. This is especially true for orographic regions where topographic effects are important. For such regions, an adjustment factor is determined to account for terrain differences. This is often accomplished by using an index map, such as the 100-year 24-hour precipitation map or seasonal isohyetal map, to compute the ratio of the index rainfalls in the project basin and in the area where the storm occurred. The resulting ratio is multiplied by the moisture maximization factor to obtain the combined adjustment factor (Wang, 1984).

The time distribution of PMP for a site specific study is also important. This can be determined in several ways depending on the degree of conservatism required for the PMF. The most conservative approach is to construct depth-duration curves of major storms in the region, draw an envelope curve, read incremental rainfall from the curve and arrange them in such a way so as to yield the most critical flood. A less critical but still conservative way is to construct rainfall mass curves of major storms in the region, and draw a near envelope curve which nearly envelopes the low values in the low end of the curves and high values in the high end of the curves (Wang, 1984). A less conservative approach would be to use the actual distribution of the storm which, after transposition and maximization, yielded the PMP estimate.

2.5. Snowmelt

For a project basin where snowmelt contributes significantly to major floods, critical combinations of rainfall and snowmelt must be considered. In general, the two combinations listed below are investigated:

(1) PMP during the snowmelt season plus snowmelt of a 50- to 100-year recurrence interval.

(2) Probable maximum snowmelt plus a 50- to 100-year rainfall during the snowmelt season.

These and possibly other combinations of rainfall and snowmelt are then compared with all-season PMP to select the most critical combination of events for estimating the PMF (Wang, 1986).

2.6. Antecedent and Subsequent Storms

A cyclonic storm could precede or follow another cyclonic storm within a 3- to 5-day period, depending on the season of the year and the meteorological characteristics of the region. Sequential thunderstorms could occur with even shorter periods between the storms. The antecedent and subsequent storms may or may not be important in the PMF analysis depending on their relative magnitudes with respect to the discharge capacity of the proposed reservoir at normal maximum pool.

In general, if the discharge capacity at the normal maximum pool is larger than the peak flow caused by the antecedent storm, it would not be necessary to consider the antecedent storm. Consideration of a subsequent storm also would not be necessary if the discharge capacity is large enough to return the high reservoir level caused by the PMP back to the normal maximum pool before the subsequent storm.

On the other hand, if the discharge capacity of the proposed reservoir is so small that the antecedent flood cannot be discharged before the PMP occurs, or if the subsequent storm would add to the maximum flood surcharge caused by the PMP, the antecedent and/or subsequent storms will have to be considered in determining the PMF.

3. TRANSFORMATION OF PMP TO PMF

Transformation of a PMP to a PMF is generally accomplished by the unit hydrograph method or by a simulation model (Wang and Jawed, 1986). The first method is more conventional, relatively simple, and generally requires less data and effort. The second method is more sophisticated and comprehensive but requires more data to justify its employment. In applying either method, an important step is to have a careful reconnaissance of the project basin and to divide the basin into a number of sub-basins so that areal variation of rainfall and basin characteristics can be properly accounted for. The number of sub-basins into which the project basin is divided depends on the size of the basin, the PMP pattern, and the basin characteristics such as topography, vegetative cover, and soil. The general guideline is to divide the basin in such a way that areal variation in rainfall, soil, vegetative cover, and topography in each sub-basin would be as small as practicable. For most cases, five to fifteen sub-basins are required.

3.1. Unit Hydrograph Method

Transformation of a PMP to a PMF by the unit hydrograph method requires estimation of retention losses, derivation of unit hydrographs, convolution of rainfall excesses with unit hydrographs, estimation of base flow, and routing of floods through the stream and reservoir system as discussed below.

3.1.1. <u>Retention losses</u>. Not all of the PMP runs off during the flood period. Some is intercepted by foliage before it reaches the ground, some is retained in surface depressions, some is lost due to evapotranspiration, and some infiltrates into the ground. These losses are generally grouped into a single term called retention losses.

Estimation of retention losses generally is accomplished by one of two methods; the infiltration approach and the runoff curve method. The infiltration approach follows the concept of representing the retention losses with a fixed amount of initial retention followed by a continuous retention at a constant rate. The initial retention may range from a few tenths of an inch (a few millimeters) during wet seasons to over 4 inches (100 millimeters) during dry seasons, depending primarily on the soil and vegetative cover of the basin. However, major storms generally occur in wet season during which the soil is wet or saturated. Therefore, initial retention is generally assumed to be negligible.

The continuous retention rate represents the infiltration capacity when the soil becomes saturated. It is controlled by the capability of lower soil horizons to transmit water downward, and may be determined by a trial and error process if rainfall and corresponding runoff data are available. In the absence of sufficient rainfall-runoff records, empirical retention rates are often used based on hydrologic soil groups (U.S. Bureau of Reclamation, 1973).

The runoff curve method uses soil and ground cover data for estimation of retention losses. Various tables and charts have been developed by the U.S. Soil Conservation Service to determine runoff curve numbers based on which direct runoff due to rainfall is estimated (U.S. Bureau of Reclamation, 1973).

3.1.2. Unit hydrographs. A unit hydrograph is the hydrograph resulting from one unit of rainfall excess over a specified duration. Generally accepted methods for deriving unit hydrographs include the direct method (Linsley, Jr., Kohler and Paulhus, 1975), the dimensionless graph-lag curve method (U.S. Bureau of Reclamation, 1973) and synthetic methods.

The direct method is applicable when there is a stream gaging station at or near the damsite and the drainage area upstream from the site is so small that areal variations of the storm rainfall, and basin characteristics such as topography, soils, and vegetative cover, can be ignored.

The dimensionless graph-lag curve method may be used to derive a unit hydrograph for any point of interest in a basin where flood hydrographs have been recorded at various gaging stations in the basin or its vicinity. The points of interest, including the damsite, may or may not coincide with the gaging sites.

For ungaged basins where no recorded flood hydrographs are available for analysis, synthetic methods such as the Snyder's method (Linsley, Jr., Koaler and Paulhus, 1975), the U.S. Soil Conservation Service's method (U.S. Soil Conservation Service, 1972), and Clark's method (Clark, 1945) can be used. The procedure for applying each of these method is described in the respective references.

3.1.3. Convolution. The direct runoff hydrograph resulting from a storm is derived by convolution of the incremental rainfall excesses with the appropriate unit hydrograph. Briefly, for each sub-basin, the direct runoff resulting from the first incremental rainfall excess is calculated by multiplying it with the unit hydrograph ordinates. The direct runoff resulting from the second incremental rainfall excess is then computed similarly and added to that resulting from the first incremental rainfall excess with the time lagged by one time interval. This procedure is continued until all incremental rainfall excesses are processed. The time increment for the rainfall excess is selected to be equal to the unit duration of the unit hydrograph.

3.1.4. Base flow. Base flow is that portion of streamflow derived from groundwater storage. Its contribution to the PMF peak discharge and volume is generally small, and a simplified estimate of base flow normally would not cause appreciable difference in the PMF estimate. A time-saving approach which is often adopted is to use the minimum daily flow during the month of highest flow in the period of record to approximate the base flow.

3.1.5. Stream and reservoir routing. When a basin upstream of a damsite is treated as one unit, the computed hydrograph is the PMF inflow hydrograph. When a basin is divided into a number of sub-basins, the hydrograph derived for the most upstream sub-basin is routed through the stream and reservoirs, if any, to the outlet of the second most upstream sub-basin. The routed hydrograph is then combined with that from the second most upstream sub-basin. This procedure is continued until all sub-basins are processed. The resulting hydrograph is the PMF inflow hydrograph.

3.2. Simulation Techniques

A simulation model generally consists of a number of mathematical equations each describing a component of the hydrologic cycle such as precipitation, interception, evaporation, transpiration, infiltration, percolation and runoff. These equations are linked by the mass continuity equation and sometimes also by the momentum equation, and are programmed on an electronic computer for rapid solution.

There are a large number of simulation models that have been developed by various organizations. Some of the models most commonly used include the HEC-1 and HEC-5 models developed by the U.S. Army Corps of Engineers, The Hydrologic Engineering Center (1981, 1979), the River Forecast System developed by the U.S. National Weather Service (1972), and the Stanford Watershed Model developed at Stanford University (Crawford and Linsley Jr. 1966). Any of these and other models can be used to transform a PMP to a PMF provided that sufficient data are available to properly calibrate the model.

On the other hand, for basins where sufficient data are not available, overly anxious employment of a simulation model can lead to erroneous results. Therefore, selection of the more conventional unit hydrograph method or the more recently developed simulation techniques should be based on a careful evaluation of available data.

4. APPLICATION OF PMF ESTIMATES

Before a PMF estimate is adopted for design purposes, its reasonableness is often investigated further by comparing the PMP and PMF estimates with recorded rainfalls and floods.

Observed rainfall data, such as those given in the "Manual of Probable Maximum Precipitation" by the World Meteorological Organization (1973, pp. 175-179), can be used to compare PMP estimates for their reasonableness. However, geographical, meteorological, and topographic similarities between the study area and the area used for comparison should be carefully evaluated and taken into consideration. For example, an extreme rainfall, such as the September 3-7, 1950 storm that occurred at Yankeetown, Florida, potentially could occur at other locations in southeastern United States but not in the Rocky Mountain region of Colorado.

Flood records are generally shorter than precipitation records, and extreme floods may or may not have occurred during the period of record. Therefore, a comparison of PMF with historical floods is more difficult and needs to be made with great caution. The procedure outlined below has been useful in making comparisons for some projects.

1. Conduct a thorough search of historical records on major floods in the region. These records may be of flood magnitudes, flood marks, and/or flood damages published in government or private documents and newspapers.
2. Perform a field investigation and interview of long-time local residents for their recollection of the floods. Questions to ask the residents include time, date, and year of occurrence of major floods, flood marks and their relative magnitudes with respect to the largest flood of known magnitude.
3. Estimate the magnitudes of these floods and determine the maximum flood of record for each location.
4. Plot the floods of record against their respective drainage areas and draw an envelope curve.
5. Compare the flood of record from the envelope curve with the PMF estimate. In general, the ratio between the PMF and the flood of record may be as small as two to three in tropical areas frequently invaded by typhoons or hurricanes but it may be very large in arid or semi-arid areas where heavy storms are scarce but do occur.

PMF estimates resulting from the analyses and comparisons described above often lead to large spillways which might be designed to discharge flows much larger than that which would occur under natural conditions. This, however, can be avoided by designing the dam and spillway to allow some flood surcharge in the reservoir. Generally, the PMF inflow hydrograph is routed through the proposed reservoir for a number of spillway configurations to determine the maximum flood surcharges in the reservoir and the maximum discharges through the spillway and other outlet facilities. Results of such routings are then used to select the optimum height of the dam and best design of the spillway based on economic and other considerations.

Under situations in which construction of a regular spillway to pass the entire PMF requires excessive investments, a smaller regular spillway supplemented by one or more fuse-plug spillways is occasionally used to reduce the total cost. Each fuse-plug spillway is designed to be activated when the reservoir level reaches a predetermined elevation. The regular and fuse-plug spillways, however, are designed such that, under no circumstance, will the total discharge from the reservoir exceeds the peak inflow. This often requires the construction of two or more fuse-plugs designed to be activated at different elevations.

In case of a pumped-storage project, water is released from the upper reservoir through the power house to the lower reservoir to generate electricity during the periods of peak demands. This water is pumped back to the upper reservoir during the off-peak hours. For such a project, special attention is required in applying the PMF estimate. This is because the PMF can arrive at any time in the pumping-generating cycle, and depending on when in the cycle the PMF arrives, the lower, or the upper, reservoir surface can rise rapidly and over-top the dam. In such cases, a flood operation plan has to be formulated to interrupt the normal operation cycle when the incoming flood exceeds a certain magnitude.

5. DISCUSSION AND CONCLUSION

There are two basic approaches for determining design floods for major dams and spillways. One approach is to use the PMF and the other approach is to use a flood of selected recurrence interval based on a risk analysis.

The first approach is based on the philosophy that dam failures should be prevented at any cost if such failures would cause loss of human lives. This approach does not directly consider optimum allocation of national resources but it offers the basis on which the engineering profession has the most confidence for sizing spillways. It appears to meet a standard of reasonable care as demonstrated by the performance of dams over the past several decades (U.S. National Research Council, 1985). It also minimizes dam owner's liability on damages due to failure of their dams. This is becoming increasingly important in view of recent court cases which indicate a trend of victim compensation for damage caused by dam failures.

The second approach is based on the theory that society cannot afford the cost of preventing all dams from failure and that the assignment of an infinite value to human life is wasteful and contrary to common practices on expenditures for safety in other phases of human society. This approach allows the incorporation of risk analysis in selecting the design floods and, thus, avoid overdesign of dams and spillways. The difficulty of a credible risk analysis, however, is tremendous. This is because most major dams and spillways would have to be designed for a flood having a recurrence interval of several thousand years or more. Any estimation of such a flood is subject to great uncertainty. The estimation of cost of risks also requires the determination of a flood damage function. Accurate determination of such a relationship between damage and flood magnitude is often difficult, especially considering the general reluctance in assigning monetary value to human life. Furthermore, a risk analysis considered valid today may be nullified in a short period of 10 or 20 years by changes in downstream development and other factors included in the risk analysis.

In a recent publication by the U.S. National Research Council (1985), an inventory of current practices on dam safety against extreme floods is given. This inventory shows considerable diversity in approach by various federal, state, and local government agencies and privately owned companies. However, there is a fair concensus on the spillway requirements for large, high hazard dams. This concensus is to design a new large, high hazard dam for the PMF. No similar concensus exists for existing dams of high hazard potential or new and existing dams of intermediate and low hazard potentials. For such dams, floods ranging from a PMF to a much smaller flood based on a risk or probability analysis are used for design purposes depending on the individual situation of the dam.

In any case, the utilization of PMFs for design of new major dams will likely continue. It is also likely that employment of a risk-based analysis in the process of selecting or evaluating the design flood will increase. Results of the risk-based analysis, however, will likely be used as supplementary information in the design of new large high hazard dams although they may be used as the sole or primary basis in the design of new, smaller, intermediate and low hazard dams. Risk-based analysis is also likely to find increased application in the selection of alternative measures for upgrading spillway capacities of existing dams of intermediate- and high- hazard potentials.

REFERENCES

(a) Books.

Linsley, Jr., R.K., Kohler M.A. and Paulhus, J.L.H. 1975. Hydrology for Engineers, 2nd ed., McGraw-Hill Book Co., New York, N.Y.

U.S. Bureau of Reclamation 1973. Design of Small Dams, 2nd ed., Washington, D.C.

(b) Journals and Periodicals.

Clark, C.O. 1945. 'Storage and Unit Hydrograph', ASCE Transactions, **Vol. 110**, pp. 1419-1488.

Wang, B.H. 1984. 'Estimation of Probable Maximum Precipitation: Case Studies', ASCE Journal of Hydraulic Engineering, **Vol. 110, No. 110**, pp. 1457-1472.

Wang, B.H. and Jawed, K., 1986. 'Transformation of Probable Maximum Precipitation to Probable Maximum Flood, Case Studies', ASCE Journal of Hydraulic Engineering, **Vol. 112, No. 7**, pp. 547-561.

(c) Edited Symposia.

Riedel, J.T., Wang, B.H. and Diebel, J.L. 1982. 'Site Specific Probable Maximum Precipitation Estimates, Upper South Platte River Basin Colorado', in: A.I. Johnson and R.A. Clark (Editors), Proceedings of the International Symposium on Hydrometeorology, American Water Resources Association, Bethesda, Md. pp. 517-522.

(d) Technical Reports.

Crawford, N.H. and Linsley Jr., R.K. 1966. 'Digital simulation in Hydrology: Stanford Watershed Model IV', Stanford University, Department of Civil Engineering, Technical Report 39, Palo Alto, Calif.

U.S. Army Corps of Engineers 1979. 'National Program for Inspection of Non-Federal Dams', ER 1110-106, Appendix D, Office of the Chief of Engineers, Washington, D.C.

U.S. Army Corps of Engineers, The Hydrologic Engineering Center 1979. 'HEC-5, Simulation of Flood Control and Conservation Systems, Users Manual, Generalized Computer Program 723-500', Davis, Calif.

U.S. Army >orp° of Engineers, The Hydrologic Engineering Center 1981. 'HEC-1, Flood Hydrograph Package, Users Manual, Generalized Computer Program 723-X6-L2010', Davis, Calif.

U.S. National Research Council 1985. 'Safety of Dams, Flood and Earthquake Criteria', National Academy Press, Washington, D.C.

U.S. National Weather Service 1972. 'National Weather Service River Forecast System Forecast Procedures', NOAA Technical Memorandum NWS Hydro-14, Silver Spring, Md.

Wang, B.H. 1986. 'Probable Maximum Flood and Its Application', Harza Engineering Company, Chicago, Il.

World Meteorological Organization 1973. 'Manual for Estimation of Probable Maximum Precipitation', WMO No. 332, Geneva, Switzerland.

HEAVY RAIN STORM FLOODS IN CHINA AND THEIR ESTIMATION

Zhan Daojiang
Department of Hydrology, Hohai University (East China Technical University of Water Resources), Nanjing, China
Zou Jinshang
Nanjing University, Nanjing, China

ABSTRACT. Four heavy rainstorm systems resulted in extraordinary floods in China are presented. Each has its synoptic feature which are different from those of ordinary heavy rainstorms. There is a reason to believe that extraordinary floods and frequent and minor floods do not come from the same parent population. In addition to this, it is difficult to estimate the recurrence interval of historic floods. And therefore the outliers could hardly match the ordinary floods even if the historic floods are involved. Closely scrutinizing the features of the heavy rain storms, we find that Probable Maximum Precipitation and Flood (PMP/PMF) estimation is favorable compared with that of frequency analysis.

1. SYNOPTIC ANALYSIS OF HEAVY STORMS IN CHINA

Floods on almost all rivers in China result from intensive and long duration rainstorms. Very heavy rains occur, not only in humid areas, but also in arid areas; and a few approach the world's record. See Fig. 1 (Luo Chenzheng, 1985) and Fig. 2 (Zhan et al., 1983).

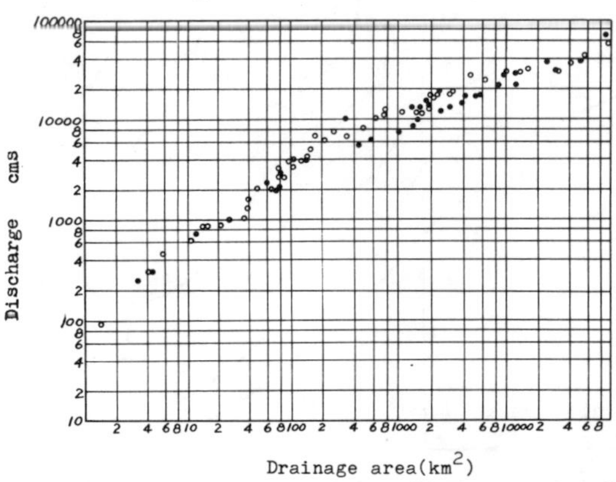

Fig. 1. The envelope of maximum discharge in China
o —— China • —— world

V. P. Singh (ed.), Application of Frequency and Risk in Water Resources, 35–44.

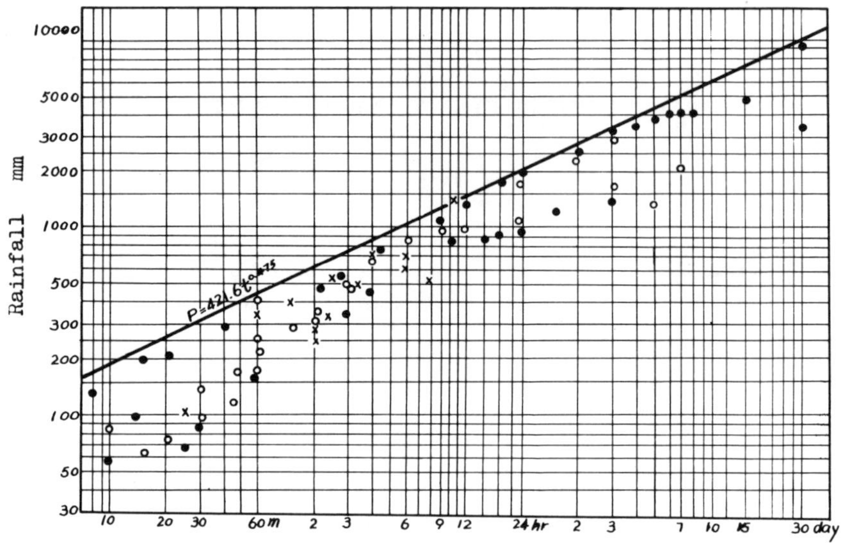

Fig. 2. The envelope of point rainfalls
o — China ● — World
x — Bucket survey in China

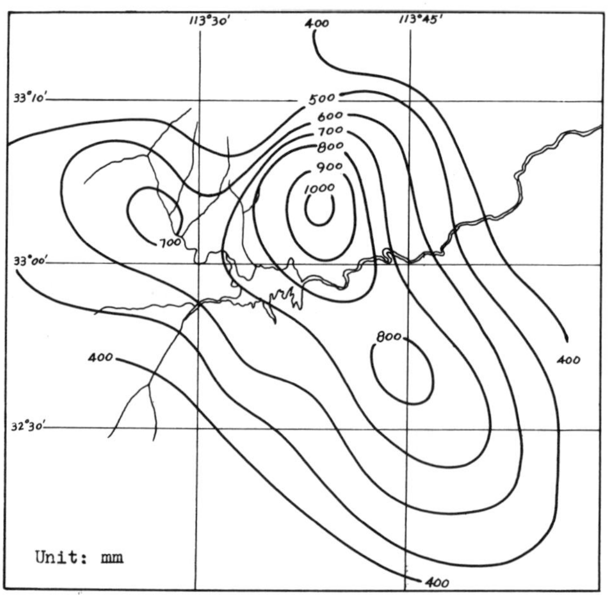

Fig. 3. Isohyetal map of typhoon-rainfall in 24-hr (from 0200 on Aug. 7 to 0200 on Aug. 8,1975)

According to synoptic meteorology, there are four important types of rainstorm in China.

(1) Typhoon-Storm Rainfalls: Typhoon storms cause intensive rainfalls and extremely high peak flows. As an example, in Aug. 1975, an unusual typhoon storm occurred in Henan province and resulted in an unprecedented catastrophe. There were three storm centers as shown in Fig. 3. The 24-hr total rainfall at Linzhung reached as high as 1060 mm and the 6-hr total was 830 mm.

Analytic results have indicated that the extraordinary rainfall occurs in the northeastern quadrant of typhoon, to the left of Jet stream in the lower layers, in front of the cold front and near the disturbance in the easterlies. Intensity and location of the typhoon rainfall are determined by the typhoon system, the direction of the jet in the lower layers, the position of westerly trough as well as the water vapour transfer, etc. This storm consisted of 7 synoptic systems.

(2) Storm Rainfall during the Period of Meiyu: In early summer, a quasi-stationary front accompanying the heavy rain belt in W-E direction stagnates over the valleys of the Yangtze and the Huaihe rivers. A case of extraordinary rainfall resulted from quasi-stationary front is given in Fig. 4. The determinative factors controlling the rainfall intensity are the persistence of the wind shear line with developing cyclonic vortices and the existense of the Jet stream in the lower layer. The heavy rain occurs generally in the region south of the wind shear line at the level of 850 hpa and north of the quasi-stationary front. The synoptic features of extraordinary rainfall resulted from quasi-stationary front appeared on July 21st, 1982 are shown in Fig. 5. The greatest values of rainfall occured in the left front of jet stream with abundant moisture from SW-direction, where the upward motion is very strong.

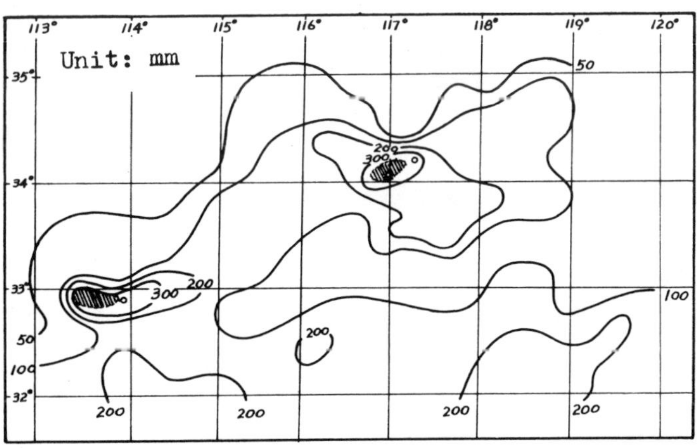

Fig. 4. Isohyetal map of the total rainfall occurred in the valleys of the Huaihe and Yellow rivers (from 0800 on July 19 to 2300 on July 23, 1982)

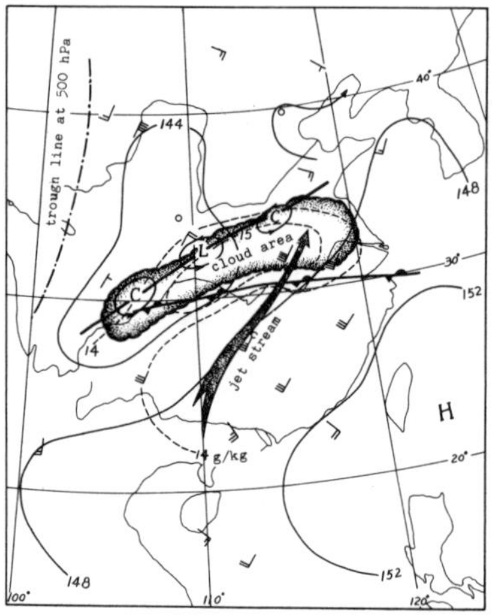

Fig. 5. Synoptic features of the heavy rain resulted from quasi-stationary front at 0800 on July 21, 1982

(3) Cyclonic Vortex Rainfalls: The cyclonic vortex is one of the synoptic scale systems releasing intensive rainfalls. There are two striking examples, one is the extraordinary heavy rain occurred in July 1935 at Wufeng in Hupeh Province. It was associated with two cyclonic vortices moving slowly from SW-direction, its total rainfall in 5 days being 1,282 mm. An other heavy rain occurred in Aug. 1963 at Changmo in Hebei Province, it also related to cyclonic vortices which came and moved slowly from SW-direction.

(4) Intensive Rainfall Occurred in the Warm Sector near a Cold Front in South China: This kind of storm rainfall occurs frequently in South China during early summer. The interaction of Jet streams at the lower level, cold front and topographical effect, resulted in intensive rainfall.

To sum up, it can be seen that various kinds of extraordinary heavy rain mentioned above have their own distinctive synoptic features. It is these peculiarities that distinguish the extraordinary heavy rainfalls (rare floos) from ordinary heavy rainfalls (frequent and minor floods) whose synoptic features are very simple. The occurrence of these super-events show that the phenomenon warrants sevious consideration and close scientific scruting and their importance can not be ignored.

2. FLOOD FREQUENCY ANALYSIS

Athough the underlying frequency distribution of floods is unknown, many types of distribution have been assumed and employed for plotting the frequency curve. Pearson type III distribution is widely used in China and has a probability density function

$$f(X)=\frac{\beta^{\alpha}}{\Gamma(\alpha)}(X-a_o)^{\alpha-1}e^{-\beta(X-a_o)}$$

where α, β, a_o are distribution parameters which may be expressed by another set of commonly used parameters, the mean ($\bar{x}$), coefficient of variation (c_v) and coefficient of skewness (c_s).

$$\alpha = 4/C_s^2$$
$$\beta = 2/\bar{X}\,C_v C_s$$
$$a_o = \bar{X}(1-2\,C_v/C_s)$$

Nonstationarity: Nearly all frequency distribution parameters such as C_v vary with space and time. For snowmelt streams in USSR, (Sokolovskee, 1963) C_v is more or less stable, being 0.25 to 0.40. However, for rain-fed rivers in China, C_v varies much more widely, from 0.24 to 1.90.

Historical Floods: China has a long history. The time series of flood may be extended as historic records are obtained from flood survey and literature search of archives. As an illustration (Northeast water survey and Design Institute, China, 1977), let us take the Hutuo River (a branch of the Haihe River) at Huangbizuang. This river has been gauged since 1918, but only 17 annual maximum flood were recorded up to 1954. Correlating with the stream data of a station upstream, an extended record of 20-yr annual flood series was obtained. A frequency curve was derived in accordance with the conventional method in 1954 (curve I in Fig. 6) for which, Q=1360 cms, C_v=0.80, C_s=3C_v. 100-yr and 1000-yr peak flows were computed to be 5490 and 8350 cms, respectively. However, floods with respective peak flows of 13100 and 12000 cms, occurred in 1956 and 1963, which are both greater than the 1000-yr flood. Obviously a period of 20 years is much too short for adequate frequency analysis of Huangbizuang floods.

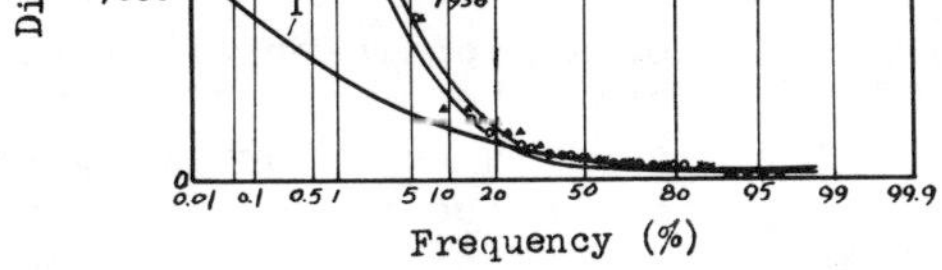

Fig. 6. Peak dischange frequency curve of the Hutao River at Huangbizuang
I–Curve deerved in 1954, historical floods not accounted for
II–Curve derived in 1954, historical floods taken into account
III–Curve derivaed in 1964, historical floods taken into account

A historic literature search reveals the following flood peaks at the station:

$$Q_{1794} = 20000\text{–}25000 \text{ cms}$$

$$Q_{1853} = 16000\text{–}18000 \text{ cms}$$

$$Q_{1917} = 13500 \text{ cms (roughly)}$$

$$Q_{1939} = 8300 \text{ cms}$$

Number of years during which Q_{1794} is known to be not exceeded =170-360.

Number of years during which Q_{1917} is known to be not exceeded =70 (approx).

Had the historical floods been taken into consideration in 1954, the following parameters would have been obtained.

$\bar{Q}$=1800 cms C_v=1.7, $C_s=3C_v$, and 1000-yr flood peak=16900 cms (Fig. 6, Curve II)

In 1964, after the occurrence of major floods in 1956 and 1964, the observed flood series was extrapolated up to 1963, and the observed flood peak in 1956 and 1963 are set apart for historical flood treatment. By the way, Q_{1853} is said to have a recurrence interval of 110 years according to a further study of local literatures. The final analysis gives the following results: $\bar{Q}$=2200 cm/s, C_v=1.5, $C_s=3C_v$. The 100-yr flood is estimated at 17000 cms (Fig. 6, Curve III).

Historical flood studies generally employ the following plotting position formula,

$$P = \frac{m}{N + 1} \tag{1}$$

where, m is the ordinal number of the flow events arranged in descending magnitude, N is the number of years of record, and P is the probability of such a magnitude being equalled or exceeded in any one year.

The analytical results in 1954 and 1964 are almost identical. A number of Chinese hydrologists presume that the historical floods stabilized the flood frequency curve, therefore the latter seems credible. Nevertheless, in locating the plotting position of historical floods, our common practice simply equates N the number of years during which the greatest discharge of the floods is known not to have been exceeded. However, there may be a prior period of unknown length, in which this historical flood event did not occur. Such treatment of flood data leads to too small N and higher probabilities. Baker (1985) mentioned a striking example of this problem. The Pecos River of West Texas has an unusually long systematic flood record (1900-present). But in this record, there exist two outliers. Untill 1953, the largest annual flood peak was 3280 cms, the 1954 flood reached as high as 26000 cms. As a historical flood, the 1954 flood would be the largest since 1746. However, some paleoflood information indicates that the 1954 event and 1974 flood of 16300 cms were the largest flood in 2000 years.

Another example is the Three-Gorge Project situated 40 km upstream Yichang, Hupeh Province, with a drainage area of 1,000,000 km^2. There have been 106-yr systmeatic record since 1877. The largest observed flood peak (1896) was only 71,000 cms and there were 6 historical floods as follows:

1870	105000 cms
1227	96300
1560	93600
1153	92800
1860	92000
1783	86000

The frequency curve of maximum peak discharges at Yichang is shown in Fig. 7. (Shih, 1985). Obviously, the outliers do not match the frequent and minor floods. Had there been a paleoflood information, all the plotting positions would be changed.

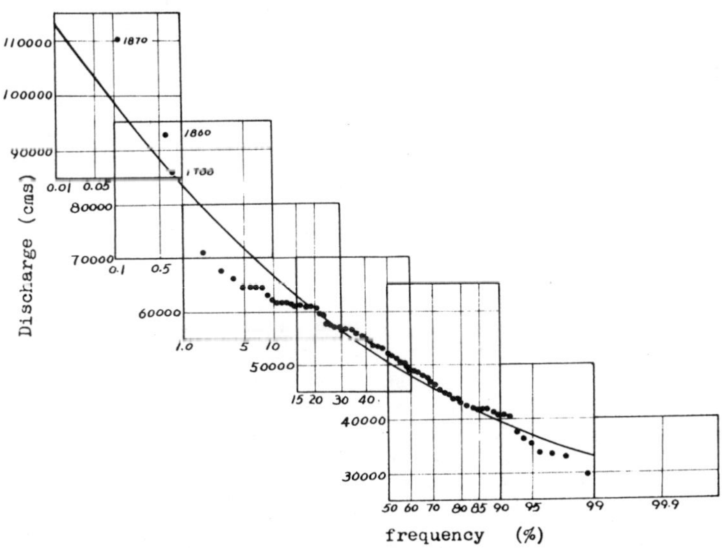

Fig. 7. The frequency curve of the flood peak at Yichang

These examples infer that the historical flood could hardly stabilize the flood frequency curve and may not be necessarily credible.

Moreover, it has been metioned above that the extraordinary floods are different from frequent and minor floods. They might not come from the same parent population and consequently, the outliers do not match the frequent and minor floods whatever probability density function is used.

In spite of the fact that frequency analysis can yield quantitative data for designers and decision makers, such data may still be open to criticism even if historic informations and treatment of outliers are involved.

3. PMP AND PMF – RESULTS OF CLOSE SCRUITING TO THE HEAVY RAIN STORM FLOOD

A PMP computation includes selection of a sequence of meteorological and hydrological events, i.e., the principal storm, antecedent and/or subsequent storms, time and areal distribution of rainfall, loss rates, initial reservoir level and the hydrograph determination that produce the probable maximum flood.

The probable maximum precipitation is PMP. According to Hansen (1982, 1985) the current definition of PMP is, "Theoretically the greatest depth of precipitation for a given duration that is physically possible over a given size storm area at a particular geographical location at a certain time of year."

Frequency analysis treats heavy rainstorm floods as outliers, PMP approach examines close to the extreme storms which cause flood outliers -synoptic and storm analysis, that is, close scrutinize to the heavy storm floods. Frequency analysis takes a sample including rare, frequent and minor floods and hardly cares about the cause of extreme floods.

The usage of PMP estimates as a basis for determining the PMF for use in hydrologic design has lasted about 50 years in the United States. The foundation for estimating PMP is the extreme storm and concepts of storm maximization transposition and envelopment. New concepts of within-storm precipitation, orientation of isohyetal pattern and concurrent precipitation outside the drainage area of the PMP storm have been developed during the last 10 years (Hansen, 1985).

About 150 PMP and PMF estimates have been completed for spillway design of major reservoirs and important small dams in China, and 24-hr point PMP generalized chart for China developed as well. Some hydrometeorologists in China have put forward their new ideas and methods (Zhan, 1984).

Considerable interests and attempts lay on the proper frequency of PMP and PMF, but there is no way to reliably solve this problem presently. (Zhan, 1980, National Academy of Science 1985)

Approaches in PMP and PMF are the boundary science between hydrology and meteorology. Modifications to the methods and theories have continued to evolve over this period and many countries have adopted PMP and PMF as general design standards for rainstorm floods(Zhan,1983).

4. PALEOFLOOD HYDROLOGY

Paleofloods may extend the flood series to as long as 10000 years and provide objective evidence of extraordinary floods. Such potential data should be used to demonstrate that calculated PMF and frequency floods values are credible and are neither unreasonably large nor small (National Research Council, 1985, p.278).

Obviously paleofloods would also be extremely useful in assessing the controversial estimates between PMF and frequency analysis.

5. CONCLUSION

In view of the hydrometeorological analysis, heavy-rain floods show marked difference from frequent and minor rain floods.

Current frequency analysis and techniques utilized in ordinary hydrology are inadequate for hazard assessment in regions of highly variable heavy rainfall behavior.

PMP and PMF approaches make close scientific scrutiny to the extraordinary rainstorm flood of catastrophes.

Paleofloods provide objective evidence of extraordinary floods. Such potential floods if available are of great value to be used in both frequency and PMF estimation.

ACKNOWLEDGMENT The authors are grateful to Professor Liu Kuangwen for consultation and thoughtful review during the preparation of this paper.

6. REFERENCES

[1] Baker, V. R. 1985, 'Paleoflood hydrology of extraordinary flood' events' in China-U. S. Bilateral Symposium on the analysis of extraordinary flood event, Nanjing, China.

[2] Hansen, E. M. 1985, 'Probable maximum precipitation for design flood in the United States' in China-U.S. Bilateral Symposium on the Analysis of Extraordinary Flood Events, Nanjing, China.

[3] Luo Chenzheng, 1985,'A survey of historical flood and its regionalization in China ' in China-U.S. Bilateral Symposium on the Analysis of Extraordinary Flood Events, Nanjing, China.

[4] National Academy of Science 1985, Safety of Dams, Flood and Earthquake Criteria. National Academy Press, Washington, DC. 1985. PP. 232-234.

[5] Northeast Water Survey and Design Institute China 1977, 'Field Investigation of Historical Floods,' Water Resources and Electric Power Press, Beijing, p. 10.

[6] Shih Winshing, 1985, 'Application of historic flood in the design of Three Gorge Project' in China-U.S. Bilateral Symposium on the

Analysis of Extraordinary Flood Events, Nanjing, China.

[7] Sokolovskee, 1963, Hydraulic Structure, No. **5**, pp. 33-36.

[8] Zhan Daojiang and Zhou Jinshang, 1980, ' Is PMP greater than 10000-yr flood? 'Water Power, 1980, **(2)** pp. 18-24.

[9] Zhan Daojiang and Zhou Jinshang, 1983, Probable maximum precipitation and flood. Water Resources and Electric Power Press, Beijing, China. p. 18.

[10] Zhan Daojing and Zhou Jinshang, 1984, 'Recent developments on the probable maximum procipitation (PMP) estimation in China,' Journal of Hydrology **68**(1984) pp. 285-293.

THE BAYESIAN FRAMEWORK FOR INFERENCE IN FLOOD FREQUENCY ANALYSIS

G. Kuczera
Department of Civil Engineering and Surveying
The University of Newcastle
Shortland N.S.W. 2308, Australia.

ABSTRACT. One of the major problems facing flood frequency analysis is that predictions typically require extrapolation beyond observed flood experience. Such extrapolations are very much affected by model and parameter uncertainty. This review reflects on the contributions that Bayesian theory has made, and can possibly make, in managing this uncertainty. Some of the issues examined from a Bayesian perspective include the choice of a flood distribution, the exploitation of gauged and historic site information possibly affected by measurement error, development of regional models and the pooling of site and regional information.

1. INTRODUCTION

The late sixties and early seventies saw the first applications of Bayesian ideas to flood frequency analysis. These works culminated with those of Wood and Rodriguez-Iturbe (1975a, b) and Vicens et al. (1975) who defined the principal elements of a Bayesian flood frequency analysis and also demonstrated the considerable benefits that could be realized. This initial flurry of activity was followed by a considerable lull. However, in recent years, interest in the Bayesian approach has been rekindled seeing wider applications and deeper probing of the original conceptual framework.

Flood frequency analysis displays the characteristics of a meta-physical problem in that it requires making possibly unverifiable assumptions and predicting beyond observed experience. The dominant theme is, thus, model and parameter uncertainty. Bayesian theory naturally handles this uncertainty, and moreover, offers a unified framework for using all available information to minimize it.

This paper aims to give an overview of the Bayesian framework that has evolved over the last decade in flood frequency analysis. It examines how Bayesian theory has been and may be used to look at problems such as the choice of a flood distribution, the use of all available site

V. P. Singh (ed.), Application of Frequency and Risk in Water Resources, 45–61.

information, the development of regional models and the pooling of site and regional information.

2. THE FLOOD FREQUENCY MODEL

As a preliminary, we briefly describe the flood frequency model and the data possibly associated with it.

We assume that the maximum annual flood magnitude q at some site is a random, independent realization from some probability distribution with probability density function (pdf) $f(q|\underset{\sim}{\theta})$ where $\underset{\sim}{\theta}$ is the parameter vector characterizing the flood distribution.

In the general case, the catchment is undergoing hydrologic change due to, say, changing land use or regulation. This implies that the parameter vector $\underset{\sim}{\theta}$ may be a function of time and/or catchment characteristics describing the change. For the present, we shall make the assumption that the catchment is hydrologically stationary implying $\underset{\sim}{\theta}$ is a constant.

The available data may consist of the following:

i. A gauged record of estimated maximum annual flood magnitudes denoted by w (note that because of measurement error, w does not necessarily equal q, the true maximum annual flood magnitude);

ii. Historic information typically on the occurrence of major floods before commencement of the continuous gauged record; and

iii. Catchment characteristics such as drainage area or degree of urbanization denoted by the vector $\underset{\sim}{x}$.

The goal of a flood frequency analysis is to infer the flood pdf $f(q|\underset{\sim}{\theta})$ using the available information at one or more sites.

3. BAYESIAN INFERENCE: SOME KEY IDEAS

The Bayesian theory of inference is conceptually very different to its classical counterpart, sampling theory which is typically used in flood frequency analysis. Therefore, it is worthwhile reviewing some of the key ideas in Bayesian inference. This will lay the foundation for outlining its role in flood frequency analysis.

3.1 Bayes' Theorem

Bayes' theorem is the cornerstone of Bayesian inference. It provides a mechanism for updating previous knowledge with new information.

Suppose we wish to make an inference about some parameter vector $\underset{\sim}{\theta}$. Prior to collection of new data D our knowledge about $\underset{\sim}{\theta}$ is summarized by the prior pdf $\xi(\underset{\sim}{\theta})$. All the information in D pertinent to the parameter vector $\underset{\sim}{\theta}$ is embodied in the likelihood function $\ell(\underset{\sim}{\theta}|D)$.

Bayes' theorem enables us to update our knowledge about $\underset{\sim}{\theta}$ as follows:

$$\xi(\underset{\sim}{\theta}|D) \quad \propto \quad \ell(\underset{\sim}{\theta}|D) \quad \xi(\underset{\sim}{\theta})$$

where $\xi(\underset{\sim}{\theta}|D)$ is the posterior pdf of $\underset{\sim}{\theta}$ summarizing our knowledge about $\underset{\sim}{\theta}$ after the new information in D has been pooled with previous or prior knowledge. Obtaining the posterior distribution of $\underset{\sim}{\theta}$ is the goal of Bayesian inference.

3.2 Subjective Probability and Prior Distributions

Unlike classical sampling theory which deals with distributions of statistics given fixed but unknown parameters, Bayesian theory deals with distributions of parameters given known data. The meaning of probability in this context has aroused much debate. The view taken here is that we "regard probability as a mathematical expression of our degree of belief with respect to a certain proposition" (Box and Tiao, 1973).

Taking a subjective view of probability should not restrict the generality of a Bayesian analysis. Provided a consensus is reached about the formulation of prior distributions, Bayesian results should be generally accepted. Box and Tiao (1973) argue that in scientific inference subjective prior beliefs are inappropriate. Rather they argue that a properly-formulated non-informative prior distribution is preferrable because the posterior distribution would then represent "what someone who a priori knew very little about $\underset{\sim}{\theta}$ should believe in the light of the data". In other words, a non-informative prior allows the "data to speak for themselves". It is my view that this is a sufficient condition for general acceptance of Bayesian results.

One should not conclude that only non-informative priors are acceptable. Suppose we analyze some data D_1. Our posterior pdf would be

$$\xi(\underset{\sim}{\theta}|D_1) \quad \propto \quad \ell(\underset{\sim}{\theta}|D_1) \quad \xi(\underset{\sim}{\theta})$$

where $\xi(\underset{\sim}{\theta})$ is a non-informative prior pdf. If we then analyze some more data D_2 our revised posterior pdf would be

$$\xi(\underset{\sim}{\theta}|D_1, D_2) \quad \propto \quad \ell(\underset{\sim}{\theta}|D_2) \quad \xi(\underset{\sim}{\theta}|D_1)$$

where $\xi(\underset{\sim}{\theta}|D_1)$ now assumes the role of a *data-based* informative prior. This example is important because it illustrates the Bayesian approach to using regional flood data - D_1 represents the regional flood data while D_2 represents site flood data.

4. CHOICE OF A FLOOD DISTRIBUTION

In a state-of-the-art review, Greis (1983) concluded that "as long as flood frequency analysis stays primarily within the statistical domain, the problem of making appropriate assumptions, for example the choice of a parent distribution, will remain". The root of the problem lies in the need to extrapolate well beyond observed flood experience. Statistical procedures based on goodness-of-fit principles lack the power to discriminate between candidate flood distributions which appear to fit observed flood data equally well but exhibit divergence in their tail behaviour. The problem is not with the statistical procedures themselves but rather with the minimal amount of information about tail behaviour that is contained in a typical flood record.

Despite this fundamental problem, there is a pressing need to adopt a parametric flood distribution with a minimal set of parameters. Without a parsimonious flood model, uncertainty would totally dominate the estimation of extreme floods. For example, Cooper and Clarke (1980) consider a non-parametric approach, which although distributionally robust, is not applicable for estimation of flood events far outside the range of the observed data.

In general flood data histograms exhibit a unimodal, positively skewed, bell-shaped form (Houghton, 1978). This implies that a flood distribution may need at least three parameters to exert the necessary control over its location, dispersion and shape. Regional empirical studies support this conclusion; for example, in Britain and New Zealand the general extreme value distribution was recommended (NERC, 1975; Beable and McKerchar, 1982), in the U.S., Beard (1977) recommended the log-Pearson III while in Queensland, Australia, Hadgraft (1982) opted for the power transformed normal. This is what Ashkanazy (1985) calls the "hard evidence".

The need for looking beyond three-parameter distributions arises because it may be that floods come from distributions with right-hand tails fatter than those of three-parameter distributions. However, the quest for more appropriate distributions is complicated by the fact that flood data may be affected by measurement error implying that virtually all past empirical studies have been fitting models to data coming from a mixed distribution rather than from the true flood distribution. The mixed distribution may differ from the underlying distribution. For example, Potter and Walker (1981) have shown that the presence of undetected measurement error inflates the true coefficient of variation, skew and kurtosis. Consequently, if these effects are not filtered out, there exists a distinct possibility that an over-parameterized flood distribution may be selected.

Even if the effects of measurement error are ignored, the empirical evidence in favour of fat-tailed distributions is equivocal.

For example, Houghton (1978) studied high quality U.S. flood records and concluded that a significant minority of flood records cannot be adequately modelled by a three-parameter log-normal distribution. He showed that the five-parameter Wakeby distribution could be made to fit all the records under study and thus proposed it as a grand parent flood distribution capable of mimicking conventional flood distributions and also spanning regions of distribution function space inaccessible to the conventional ones. In addition, he demonstrated that the Wakeby could mimic the separation effect discovered by Matalas et al. (1975). It should be noted, however, that other mechanisms are capable of explaining the separation effect - for example, that site skew within a region is distributed with a mean and non-zero variance.

In contrast, Hadgraft (1982) examined, amongst many others, the four-parameter symmetrical distribution based on Box and Tiao's (1973) exponential power distribution, which is capable of describing non-normal kurtotic behaviour. For the Queensland data he studied, he found no significant evidence favouring the symmetrical distribution over the three-parameter power transformed normal distribution.

It thus appears that the choice of a flood distribution remains unresolved, although strong arguments can be advanced that at least a three-parameter distribution is required. Because of the inherent stochasticity of flood data, carefully devised statistical procedures must play a major role in model selection. Nevertheless, we should not shrink away from the reality that judgements will have to be made - as Leamer (1978) observes, we should not believe "that real learning processes could be fully mathematized and therefore trivialized, or that actual learning should be altered to meet fully the mechanical features of any mathematical model".

The challenge lies in making the best use of gauged and historic information (Greis, 1983) whilst making allowance for data imperfections such as those due to measurement error. Historic information is particularly important as it can help "pin down those wagging distributional tails".

Wood and Rodriguez-Iturbe (1975b) demonstrated how Bayesian analysis could be used to compare alternative flood frequency models. They showed that the posterior probability that model M_i, $i = 1, \ldots, m$ is the true one is

$$p(M_i \text{ is true} \mid \text{data}) = \frac{K_i \, p(M_i \text{ is true})}{\sum_{i=1}^{m} K_i \, p(M_i \text{ is true})}$$

where $p(M_i$ is true) is the prior probability that M_i is the true model and K_i is the marginal likelihood function for model i defined as

$$K_i = \int \ell(\underset{\sim}{\theta} \mid \text{data}, M_i \text{ is true}) \; \xi(\underset{\sim}{\theta} \mid M_i \text{ is true}) \; d\underset{\sim}{\theta}$$

In the next section, we shall consider how gauged and historic information can be included in computing the marginal likelihood. If we assume that the m candidate flood distributions have flood-like properties, then it is reasonable to assume a diffuse prior model probability, namely

$$p(M_i \text{ is true}) = \frac{1}{m} \qquad i = 1, \ldots, m$$

Unless one finds a flood model with a dominant posterior probability of being true, it makes sense to incorporate explicitly model uncertainty into a flood frequency analysis. A Bayesian framework exists for this task (Wood and Rodriguez-Iturbe, 1975b; Bogardi et al., 1977; Tung and Mays, 1981a). Both model and parameter uncertainty can be allowed for using a composite Bayesian flood distribution defined as

$$f_*(q|\text{data}) = \sum_{i=1}^{m} p(M_i \text{ is true}|\text{data}) \quad f_i(q|\text{data})$$

where $f_i(q|\text{data})$ is the marginal or Bayesian pdf for the ith model given by

$$f_i(q|\text{data}) = \int f_i(q|\underset{\sim}{\theta}) \quad \xi(\underset{\sim}{\theta}|\text{data}, M_i \text{ is true}) \quad d\underset{\sim}{\theta}$$

Although this approach handles model uncertainty, judgements, nevertheless, have to be made as to which models to include in the composite set. Moreoever, the practical difficulties of maintaining several flood frequency models, should not be underestimated (Bodo and Unny, 1976).

Alternatively, and perhaps more practically, if a very general flood distribution like the Wakeby is adopted, it can then be argued that model uncertainty can be fully represented by parameter uncertainty.

5. BUILDING THE LIKELIHOOD FUNCTION

In a Bayesian analysis site information modifies prior information on $\underset{\sim}{\theta}$ solely through the likelihood function. The flood frequency literature is replete with likelihood functions, constructed largely in the context of the method of maximum likelihood.

Consider the simplest case of a gauged record with n annual maximum flood observations $\underset{\sim}{w}^T = \{w_1, \ldots, w_n\}$. Assuming the annual floods are distributed independently, the likelihood function of $\underset{\sim}{\theta}$ given $\underset{\sim}{w}$ is by definition

$$\ell(\underset{\sim}{\theta}|\underset{\sim}{w}) = \prod_{i=1}^{n} f(w_i|\underset{\sim}{\theta})$$

The implicit assumption has been made that the observation w is the true annual maximum flood magnitude, i.e., measurement error is negligible. Noting that measurement error can dilute the information in a flood record, distort inferences on the underlying distribution and induce outliers (Potter and Walker, 1981), it is important that it be allowed for in the likelihood function.

In principle, this can be done. Suppose the observed flood magnitude w is a random realization from the measurement error pdf $h(w|q)$ where q is the true flood magnitude. The likelihood function then becomes

$$\ell(\underset{\sim}{\theta}|\underset{\sim}{w}) = \prod_{i=1}^{n} g(w_i|\underset{\sim}{\theta})$$

where $g(w|\underset{\sim}{\theta})$ is the marginal pdf of w defined by

$$g(w|\underset{\sim}{\theta}) = \int h(w|q) \; f(q|\underset{\sim}{\theta}) \; dq$$

In all but the simplest cases the likelihood function will have to be evaluated numerically, as for example, done by Russell and Bonser (1980). However, the real challenge is in characterizing the measurement error distribution. Potter and Walker (1981) have shown that it is made up of several distributions depending on how the flood magnitude is computed (e.g., rating table extrapolation, slope-area estimate).

In contrast, historic information can augment the information in a gauged flood record bringing about significant reductions in parameter uncertainty (Leese, 1973). Sources of information ranging from newspaper articles to flood marks, can be represented in terms of a censored record where information is available only about floods above a threshold level. There are two cases to consider:

i. When the magnitude of floods above the censoring threshold w_0 can be estimated, the likelihood function can be expressed as

$$\ell(\underset{\sim}{\theta}|\text{gauged and historic censored data}) =$$

$$= \prod_{i=1}^{n+h} g(w_i|\underset{\sim}{\theta}) \binom{h+k}{k} G(w_0)^k$$

where h is the number of historic data above w_0, k is the number below w_0 and $G(w_0)$ is the probability that w is less than w_0. Leese (1973) has applied this function using the Gumbel distribution while Condie and Lee (1982) used the three-parameter log-normal.

ii. When the magnitude of the historic flood above the threshold cannot be reliably estimated the likelihood function takes the form (Russell and Bonser, 1980)

$$\ell(\underset{\sim}{\theta}|\text{gauged and historic censored data}) =$$

$$= \prod_{i=1}^{n} g(w_i|\underset{\sim}{\theta}) \begin{pmatrix} h+k \\ k \end{pmatrix} [1 - G(w_0)]^h \, G(w_0)^k$$

In summary, the likelihood function provides a framework for utilizing all available site information and also making allowance for measurement error.

6. DEFINING THE PRIOR DISTRIBUTION

The need to define a prior distribution is paradoxically one of the major strengths and also one of the major weaknesses of the Bayesian approach. The adoption of a subjective view of probability allows us to specify an informative prior which reflects our (perhaps strong) opinion about what the flood model parameters $\underset{\sim}{\theta}$ should be (Vicens et al., 1975; Russell and Bonser, 1980). This can lead to a potentially disturbing situation in the case of minimal site information because the subjective prior may dominate the likelihood function, and so if the prior is grossly mis-specified, the posterior distribution will similarly be mis-specified. Moreover, personal opinions differ between individuals (and even more so between 'experts') thereby thwarting the chance of developing a generally acceptable conclusion. As noted earlier, we can largely avoid such controversy if we restrict ourselves to data-based informative priors and relegate personal opinions to properly defined non-informative priors.

Provided we accept the legitimacy of data-based informative priors, then one of the major strengths of a Bayesian analysis can be exploited, namely, that in the flood frequency context, regional and site information can be pooled. The hope is that pooled information will lead to more precise inference about $\underset{\sim}{\theta}$ than would be obtained using site or regional information alone. The reality is that this hope is likely to be realized provided that site and regional information are, in a statistical sense, compatible.

6.1 The Empirical Bayes Model

Having imposed the requirement that informative priors must be data-based, one naturally gravitates to the empirical Bayes (EB) model. The conceptual framework was first articulated by Robbins (1964) who also suggested how it could be implemented. In flood frequency analysis it has seen several applications (for example, Vicens et al. (1975) and Kuczera (1982)).

Regional models exploit the notion of basin similarity by associating flood parameters $\underset{\sim}{\theta}$ with physiographic and climatic factors denoted by $\underset{\sim}{x}$. EB theory postulates that the flood parameter vector $\underset{\sim}{\theta}$ at each site in a region can be thought of as a random and independent realization from a pdf, called the regional prior, $\xi(\underset{\sim}{\theta}|\underset{\sim}{x}, \underset{\sim}{\beta})$ where $\underset{\sim}{\beta}$ is a parameter vector characterizing the regional prior. The notion of basin similarity is embodied in the regional prior in the sense that for a given site factor $\underset{\sim}{x}$, repeated sampling from the regional prior distribution would give values of $\underset{\sim}{\theta}$ that tend to cluster around some value. Note that this probability mechanism has a frequency interpretation.

Therefore, in a region with k sites, there are k realizations from the regional prior $\{\underset{\sim}{\theta}_i, i = 1, \ldots, k\}$. Given flood data at each site, inferences can be made about $\{\underset{\sim}{\theta}_i, i = 1, \ldots, k\}$ which in turn enables an inference to be made about the regional prior itself. Clearly in this context the regional prior assumes the status of an informative data-based prior.

6.2 Bayesian Inference of the Regional Prior Parameters

One common approach to inferring the regional prior parameter vector $\underset{\sim}{\beta}$ is to estimate $\underset{\sim}{\theta}$ at each site and then using method-of-moments or least squares, make an estimate of $\underset{\sim}{\beta}$ denoted by $\hat{\underset{\sim}{\beta}}$. Robbins (1964) specifies the necessary conditions for the estimator $\hat{\underset{\sim}{\beta}}$ to be consistent thereby ensuring that the estimated regional prior $\xi(\underset{\sim}{\theta}|\underset{\sim}{x}, \hat{\underset{\sim}{\beta}})$ converges to the true regional prior as the number of sites increases.

This approach is not necessarily sound in the flood frequency context where limited sites, short record lengths and spatial correlation tend to undermine the precision with which $\underset{\sim}{\beta}$ can be estimated. Thus there exists the possibility that the regional prior can be significantly mis-specified.

The Bayesian solution is to derive the marginal regional prior which explicitly allows for the uncertainty in $\underset{\sim}{\beta}$; that is:

$$\xi(\underset{\sim}{\theta}|\underset{\sim}{x}) = \int \xi(\underset{\sim}{\theta}|\underset{\sim}{x}, \underset{\sim}{\beta}) \; p(\underset{\sim}{\beta}|\text{regional data}) \; d\underset{\sim}{\beta}$$

where $p(\underset{\sim}{\beta}|\text{regional data})$ is the posterior pdf of $\underset{\sim}{\beta}$ derived from an analysis of regional flood data assuming a non-informative prior.

Kuczera (1983a) shows how the posterior pdf of $\underset{\sim}{\beta}$ can be obtained for the normal probability model which requires that:

i. The site estimate of θ be sufficient and have a normal sampling distribution;

ii. Inter-site correlation between estimates of θ can be estimated; and

iii. **The regional model follows the Matalas and Gilroy (1968) formulation where**

$$\theta \sim N(\underset{\sim}{x}^T \underset{\sim}{\beta}, A)$$

with A being the regional prior variance.

Inferring the marginal regional prior is not a trivial task, but the potentially significant effect of uncertainty on $\underset{\sim}{\beta}$ justifies the effort. For example, Kuczera (1983b) found that the regional variance of the power transformed normal shape parameter increased by 20% upon allowing for uncertainty - this was for a regional data base consisting of 42 sites each with records in excess of 40 years.

6.3 Regional Models of Dispersion and Shape Flood Parameters

Uncertainty in the dispersion and shape parameters of a flood distribution has a major effect on the precision of flood frequency predictions which typically involve extrapolation beyond observed experience. Wallis and Wood (1985) implicitly highlight the need to exploit regional information on both dispersion and shape parameters. In their sampling experiments with heterogeneous regions, they demonstrate the superiority of regional frequency curve methods over the log-Pearson III method using regional skew (WRC, 1977). One of the reasons for the poor performance of log-Pearson method was that it ignored the very significant regional information on the standard deviation, which the regional frequency curve methods exploited.

There have been few regionalization studies directed at inferring a regional prior distribution - Vicens et al. (1975) used regional regression relationships to get prior estimates of site mean and variance while Kuczera (1983b) used a Bayesian analysis to infer the regional prior for the power transformed normal shape parameter. The majority of the studies have employed one or both of the following approaches:

i. Regional Frequency Curve Methods

The basic approach involves scaling regional flood data by dividing observed flood magnitudes by the estimated mean annual flood and then fitting a regional frequency curve to smoothed scaled data. The method originated with Dalrymple (1960) and has seen several variations (for example, NERC, 1975 and Wallis, 1982).

ii. Parametric Methods

These methods associate some flood parameter such as mean annual flood or log skew with climatic, locational and physiographic factors (for example, Benson, 1962;

Thomas and Benson, 1970; NERC, 1975 and WRC, 1977).

For the regional frequency curve approach to be viable, the dispersion and shape parameters of the scaled flood distribution must come from regional priors that approximately take the form: site parameter equals constant plus random disturbance. In the British and New Zealand flood studies, (NERC, 1975 and Beable and McKerchar, 1982) regions were delineated in such a way that regional frequency curves could be meaningfully defined. In the U.S., this approach has been rejected because the coefficient of variation was found to be inversely proportional to drainage area (Stedinger et al., 1983).

The parametric approach offers more scope for regionalizing dispersion and shape parameters. This is because catchment characteristics can be incorporated in the analysis thereby offering a wider choice of regional mechanisms. However, the presence of spatial correlation and heteroscedasticity render traditional least squares methods inefficient and prone to inaccurate predictions of uncertainty (Stedinger and Tasker, 1985). Improved methods based on Matalas and Gilroy's (1968) regional hydrologic model have been devised (see Stedinger and Tasker (1985) and references therein).

7. POOLING OF SITE AND REGIONAL INFORMATION

It has been established conclusively that significant reductions in uncertainty can be realized when regional and site information are pooled (for example, Shane and Gaver (1970), Vicens et al., (1975), Tung and Mays (1981b), Kuczera (1982, 1983b)). The mechanism is best illustrated by considering the normal probability model (Kuczera, 1983a).

Assuming inference about some flood parameter θ is sought, let the regional prior be

$$\theta | \underset{\sim}{\beta}, A \sim N(\underset{\sim}{x}^T \underset{\sim}{\beta}, A)$$

and let site information be described by the sufficient statistic $\hat{\theta}$ with sampling distribution

$$\hat{\theta} | \theta, D \sim N(\theta, D)$$

Then the posterior distribution of θ is

$$\theta | \underset{\sim}{\beta}, A, D \sim N[\underset{\sim}{x}^T \underset{\sim}{\beta} + (1 - \eta)(\hat{\theta} - \underset{\sim}{x}^T \underset{\sim}{\beta}) , \frac{AD}{A + D}]$$

where η is the shrinkage factor defined as $D/(A + D)$. Observe that in this model pooling *always* reduces the posterior variance below both the regional and site variance, that the benefits of pooling depend on the relative precision of site and regional information and that the posterior mean shrinks away from $\hat{\theta}$ towards the regional mean $\underset{\sim}{x}^T \underset{\sim}{\beta}$.

7.1 The Issue of Compatability

When pooling different sources of information we have to guard against the possibility that the sources of information are not compatible, or equivalently, there is conflicting evidence. In such cases we have to study the causes of the conflict and select the most appropriate sources of information. In the flood frequency context a site may be atypical of the sites used to infer the regional prior because of, for example, regulation effects, gross errors in data and unusual climatic or physiographic features. It thus is important to test for incompatibility before pooling.

The normal probability model, discussed above, defies intuition in the sense that pooling is always advantageous irregardless of the difference between the site and regional means. This strong statement rests on the assumptions of normality and perfect knowledge of the regional parameters $\underset{\sim}{\beta}$ and A and the site variance D. However, once we admit uncertainty about A and D in particular, pooling is no longer always advantageous.

Box and Tiao (1973) demonstrate for the weighted mean problem, a special case of the normal probability model, that if the two sample means are sufficiently different the posterior distribution of the mean becomes bimodal indicating an obvious conflict of evidence.

Using first order analysis Kuczera (1983a) showed that for a given shrinkage factor η and level of uncertainty about A and D, there exists a positive value of Δ defined as

$$\Delta = \frac{(\hat{\theta} - \underset{\sim}{x}^T \hat{\underset{\sim}{\beta}})^2}{A + D} - 1$$

beyond which pooling becomes counterproductive in the sense that the posterior variance is greater than either the site variance D or the regional variance A. This result bears much resemblance to the limited translation rule proposed by Effron and Morris (1971) to protect EB estimators against the so-called "bad property" of Bayes estimators, namely poor performance at atypical sites.

The implications are clear. We cannot blindly pool different sources of information without running the risk of making poor inferences at some sites. The analyst must first satisfy himself that the site and regional flood information are compatible.

8. THE BAYESIAN FLOOD DISTRIBUTION

Once the posterior distribution of $\underset{\sim}{\theta}$ has been computed, it can be incorporated into analyses requiring knowledge of the flood distribution. Wood and Rodriguez-Iturbe (1975b) argued that the Bayesian pdf defined as

$$f(q|\text{data}) = \int f(q|\underset{\sim}{\theta}) \ \xi(\underset{\sim}{\theta}|\text{data}) \ d\underset{\sim}{\theta}$$

should be used in such analyses because it explicitly allows for parameter uncertainty in addition to the inherent or natural uncertainty about flood magnitudes; if a composite Bayesian pdf were computed then, depending on the range of models included, model uncertainty would also be allowed for.

Ignoring parameter uncertainty can lead to underdesign (Wood and Rodriguez-Iturbe, 1975b; Stedinger, 1983). The traditional approach of estimating the design flood by employing $f(q|\hat{\underset{\sim}{\theta}})$ where $\hat{\underset{\sim}{\theta}}$ is some "optimal" estimate, leads to design floods which will be exceeded with a probability in excess of that intended. The classical remedy is to compute design floods using an expected probability calculus (Beard, 1960) which gives similar design floods as the Bayesian pdf when only site information is used. However, unlike the expected probability calculus, the Bayesian pdf naturally allows for parameter uncertainty when both site and regional (or just regional) information is used (Stedinger, 1983).

The Bayesian pdf has a wider application in the context of decision theory (De Groot, 1970) where the preferred decision is the one which minimizes the expected loss

$$\int L(d,q) \ f(q|\text{data}) \ dq$$

where $L(d,q)$ is a loss function expressing the consequences of making a decision d given the flood magnitude q. See Davis et al. (1972), Wood and Rodriguez-Iturbe (1975b) and Stedinger (1983) (and references therein) for examples.

9. CONCLUDING REMARKS

The conceptual framework for Bayesian inference in flood frequency analysis is relatively well established. It offers a rational approach to the analysis of engineering works subject to extreme hydrologic events, where natural, parameter and even model uncertainty are significant. Moreover, it offers a unified approach to exploiting the many sources of site and regional information that may be available to the analyst.

However, its implementation remains far from complete. Analytically, considerable progress has been made with flood distributions and prior distributions which form conjugate families. In particular, flood distributions which after suitable transformation become normal (such as lognormal and power transformed normal) and the linear regional model permit extensive analytical development. But the complexities of utilizing historic information and incorporating measurement error remain to be tackled. Moreover, Bayesian flood studies, to date have been limited in scope, particularly in developing regional models. It is fair to say that the Bayesian methodology is still largely in the domain of the research hydrologist.

Needless to say, the data rather than analytical convenience should dictate the form of the analysis. The inclusion of historic site information and allowance for measurement error is likely to require greater use of numerical techniques to compute the posterior and Bayesian pdf's - with the advent of the (super) microcomputer this is no longer an impractical task. Moreover, the weight of empirical evidence may force the adoption of analytically less-tractable flood distributions and regional models thereby posing new challenges for the implementation of the Bayesian approach. Also the flood frequency model itself may need to undergo fundamental change to accommodate the effects of changing land use within catchments. What is certain, however, is that the critical analysis of flood data at both the site and regional scale will play a key role in guiding the future evolution of Bayesian flood frequency analysis.

10. REFERENCES

Ashkanasy, N.M., 1985, To Bayes or Not to Bayes - The Future Direction of Statistical Approaches in Hydrology, in Hydrology and Water Resources Symposium, The Institution of Engineers, Australia, Canberra, pp. 76 - 80.

Beable, M.E. and McKerchar, A.I., 1982, Regional Flood Estimation in New Zealand, Water and Soil Technical Publication No. 20, Water and Soil Division, Ministry of Works and Development, Wellington.

Beard, L.R., 1960, Probability Estimates Based on Small Normal Distribution Samples, J. Geophysical Research, Vol. 65, No. 7, pp. 2143 - 2146.

Beard, L.R., 1977, Appendix 14, Guidelines for Determining Flood Flow Frequency, Bulletin No. 17A of the Hydrology Committee, U.S. Water Resources Council, Washington, D.C.

Benson, M.A., 1962, Factors Influencing the Occurrence of Floods in Humid Regions of Diverse Terrain, U.S. Geological Survey Water Supply Paper, 1580-B.

Bodo, B. and Unny, T.E., 1976, Model Uncertainty in Flood Frequency Analysis and Frequency-Based Design, Water Resources Research, Vol. 12, No. 6, pp. 1109 - 1117.

Bogardi, I., Duckstein, L. and Castano, E., 1977, Effect of Stochastic Model Choice on Hydraulic Design, in: L. Gottschalk, G. Lindh and L. de Mare (Editors), Stochastic Processes in Water Resources Engineering, Water Resources Publications, Fort Collins, Colorado.

Box, G.E.P. and Tiao, G.C., 1973, Bayesian Inference in Statistical Analysis, Addison-Wesley, Reading, Massachusetts.

Condie, R. and Lee, K.A., 1982, Flood Frequency Analysis with Historic Information, J. Hydrology, Vol. 58, pp. 47 - 61.

Cooper, D.M. and Clarke, R.T., 1980, Distribution-Free Methods for Estimating Flood and Streamflow Exceedance Probabilities by Correlation, Water Resources Research, Vol. 16, No. 6, pp. 1121 - 1127.

Dalrymple, T., 1960, Flood Frequency Analysis, U.S. Geological Survey Water Supply Paper, 1543-A.

Davis, D.R., Kisiel, C.C. and Duckstein, L., 1972, Bayesian Decision Theory Applied to Design in Hydrology, Water Resources Research, Vol. 8, No. 1, pp. 33 - 41.

De Groot, M.H., 1970, Optimal Statistical Decisions, McGraw-Hill, New York.

Effron, B. and Morris, C., 1971, Limiting the Risk of Bayes and Empirical Bayes Estimators: Part 1 - The Bayes Case, J. American Statistical Association, Vol. 66, pp. 807 - 815.

Greis, N.P., 1983, Flood Frequency Analysis: A Review of 1979 - 1982 in: Reviews of Geophysics and Space Physics, Vol. 21, No. 3, pp. 699 - 706, U.S. National Report to International Union of Geodesy and Geophysics, 1979 - 1982.

Hadgraft, R.G., 1982, An Empirical Study of Flood Frequency Distributions, Queensland Water Resources Commission: Surface Water Branch, Hydrology Report 000716.PR.

Houghton, J.C., 1978, Birth of a Parent: The Wakeby Distribution for Modelling Flood Flows, Water Resources Research, Vol. 14, No. 5, pp. 745 - 752.

Kuczera, G., 1982, Combining Site-Specific and Regional Information: An Empirical Bayes Approach, Water Resources Research, Vol. 18, No. 2, pp. 306 - 314.

Kuczera, G., 1983a, Effect of Sampling Uncertainty and Spatial Correlation on an Empirical Bayes Procedure for Combining Site and Regional Information, J. Hydrology, Vol. 65, pp. 373 - 398.

Kuczera, G., 1983b, A Bayesian Surrogate for Regional Skew in Flood Frequency Analysis, Water Resources Research, Vol. 19, No. 3, pp. 821 - 832.

Leamer, E.E., 1978, Specification Searches: Ad Hoc Inference with Non-Experimental Data, Wiley, New York.

Leese, M.N., 1973, Use of Censored Data in the Estimation of Gumbel Distribution Parameters for Annual Maximum Flood Series, Water Resources Research, Vol. 9, No. 5, pp. 1534 - 1542.

Matalas, N.C. and Gilroy, E.J., 1968, Some Comments on Regionalization in Hydrologic Studies, Water Resources Research, Vol. 4, No. 6, pp. 1361 - 1369.

Matalas, N.C., Slack, J.R. and Wallis, J.R., 1975, Regional Skew in Search of a Parent, Water Resources Research, Vol. 11, No. 6, pp. 815 - 826.

NERC, 1975, Flood Studies Report Vol. 1 - 4, Natural Environment Research Council, London.

Potter, K.W. and Walker, J.F., 1981, A Model of Discontinuous Measurement Error and Its Effects on the Probability Distribution of Flood Discharge Measurements, Water Resources Research, Vol. 17, No. 5, pp. 1505 - 1509.

Robbins, H., 1964, The Empirical Bayes Approach to Statistical Decision Problems, Annals Mathematical Statistics, Vol. 35, pp. 1 - 20.

Russell, S.O. and Bonser, J.D., 1980, Peak Flood Estimation Program: A Program for Estimating Peak Flood Probabilities Using Bayesian Statistics, Department of Civil Engineering, University of British Columbia, Vancouver.

Shane, R.M. and Gaver, D.D., 1970, Statistical Decision Theory Techniques for the Revision Mean Annual Flood Flow Regression Estimates, Water Resources Research, Vol. 6, No. 6, pp. 1649 - 1654.

Stedinger, J.R., 1983, Design Events with Specified Flood Risk, Water Resources Research, Vol. 19, No. 2, pp. 511 - 522.

Stedinger, J.R., Potter, K., Kibler, D.F. and Tasker, G., 1983, Comment on "Regional Flood Frequency Estimation and Network Design" by N.P. Greis and E.F. Wood, Water Resources Research, Vol. 19, No. 5, pp. 1343 - 1345.

Stedinger, J.R. and Tasker, G.D., 1985, Regional Hydrologic Analysis 1. Ordinary, Weighted and Generalized Least Squares Compared, Water Resources Research, Vol. 21, No. 9, pp. 1421 - 1432.

Thomas, D.M. and Benson, M.A., 1970, Generalization of Streamflow Characteristics from Drainage-Basin Characteristics, U.S. Geological Survey Water Supply Paper, 1975.

Tung, Y.-K. and Mays, L.W., 1981a, Risk Models for Flood Levee Design, Water Resources Research, Vol. 17, No. 4, pp. 833 - 841.

Tung, Y.-K. and Mays, L.W., 1981b, Reducing Hydrologic Parameter Uncertainty, American Society of Civil Engineers, J. Water Resources, Planning and Management, Vol. 107, pp. 245 - 262.

Vicens, G.J., Rodriguez-Iturbe, I. and Schaake, J.C. Jnr., 1975, A Bayesian Framework for the Use of Regional Information in Hydrology, Water Resources Research, Vol. 11, No. 3, pp. 405 - 414.

Wallis, J.R., 1982, Risk and Uncertainties in the Evaluation of Flood Events for the Design of Hydrologic Structures, in: A. Giggino, G. Rossi and E. Todini (Editors), Proceedings of Eventi Hydrologici Extremi: Pieni e Siccita, Scuola de Gestione delle Risorse Idriche, Centro de Cultura Scientifica, "E. Majorana", Erice (Trapani).

Wallis, J.R. and Wood, E.F., 1985, Relative Accuracy of Log-Pearson III Procedures, American Society of Civil Engineers, J. Hydraulic Engineering, Vol. 111, No. 7, pp. 1043 - 1056.

Wood, E.F. and Rodriguez-Iturbe, I., 1975a, A Bayesian Approach to Analyzing Uncertainty Among Flood Frequency Models, Water Resources Research, Vol. 11, No. 6, pp. 839 - 843.

Wood, E.F. and Rodriguez-Iturbe, I., 1975b, Bayesian Inference and Decision Making for Extreme Hydrologic Events, Water Resources Research, Vol. 11, No. 4, pp. 533 - 542.

WRC, 1977, Guidelines for Determining Flood Flow Frequency, Bulletin 17A of the Hydrologic Committee, U.S. Water Resources Council, Washington, D.C.

DEVELOPMENT OF DESIGN PROCEDURES FOR EXTREME FLOODS IN AUSTRALIA

David H. Pilgrim, Ian A. Rowbottom and David G. Doran
School of Civil Engineering,
The University of New South Wales,
P.O. Box 1,
Kensington, N.S.W. 2033, Australia.

ABSTRACT. Two procedures are described that have been developed for the estimation of floods of very low probability as part of the revision of "Australian Rainfall and Runoff", the guide to flood estimation in Australia published by The Institution of Engineers, Australia. A relatively simple procedure for interpolation of floods between the 1 in 100 flood and the PMF was developed after other procedures were shown to be unsatisfactory over a wide range of Australian data. Choice between flood estimates based on design rainfalls and flood frequency analyses is an important design decision. Guidelines for this choice were developed from the statistical characteristics of Australian flood and design rainfall data.

1. INTRODUCTION

A program of development of design procedures has been carried out for a revised edition of "Australian Rainfall and Runoff" (ARR) to be published by The Institution of Engineers, Australia in mid 1987. ARR covers all aspects of flood estimation for Australia and is regarded as authoritative by all involved professionally with flood estimation and by the legal system. However, its recommendations are not prescriptive in the sense of a code, and use of new and innovative methods is encouraged.

There is an increasing need for the estimation of large and extreme design floods intermediate in magnitude between the flood with an annual exceedance probability (AEP) of 1 in 100 and the probable maximum flood (PMF). Two procedures that have been developed for use in the estimation of these floods are described in this paper. Probable maximum precipitation (PMP) and PMF are also required in the procedures, but will not be covered here. Their estimation in Australia is described by Pilgrim (1986). More detailed descriptions of the development of the two procedures in this paper are given by Rowbottom et al. (1986) and Doran and Pilgrim (1986).

V. P. Singh (ed.), Application of Frequency and Risk in Water Resources, 63–77.

2. ESTIMATION OF FLOODS BETWEEN THE PMF AND THE 1 IN 100 EVENT

2.1 The Need for a Recommended Procedure

The increasing need for estimation of floods of very low AEP arises from several sources. Design standards for the spillway provision for various classes of referable dams have been recommended by The Australian National Committee on Large Dams (ANCOLD, 1986). For most hazard classes these fall in the range of probabilities under consideration. The National Association of Australian State Road Authorities has prepared a draft standard for design of major bridges (NAASRA, 1983). The design flood for the ultimate failure limit state has an AEP of 1 in 2000. Increases in recent estimates of PMP by the Australian Bureau of Meteorology (Kennedy and Hart, 1984) have necessitated re-assessment of the probabilities of spillway capacity floods of existing dams so that the level of safety of these dams can be checked. These probabilities generally fall within the range under consideration.

Despite the difficulties of estimating floods of low AEP, both The Institution of Engineers, Australia and ANCOLD considered that a procedure for estimating these floods should be developed for recommendation in ARR. This will provide much-needed guidance for design practice, encourage a consistent approach by all Australian authorities, and help in setting priorities where existing structures require remedial work.

Recognizing that it is not possible to derive rigorously absolute or "true" flood magnitudes of very low probability, the procedure was developed to meet the criteria that the flood estimates should be consistent from site to site and reasonable in a subjective sense, and should provide some degree of conservatism. This provides reasonable and consistent values that are required for pragmatic design, rather than values that are intrinsically correct (but unattainable) from a theoretical viewpoint.

The approach adopted involves interpolation between the PMF and the 1 in 100 flood, which can be estimated with reasonable accuracy. This approach ensures compatibility of estimates of low probability floods with the PMF. It also reduces the effect of the choice of flood frequency distribution and fitting procedure. However, it does involve the problem of determining the probability of the PMF, which is not possible theoretically. Unfortunately, assumptions of some type are necessary in the practical estimation of floods of very low probability, and assigned probabilities of the PMF have been assumed in the approach that has been adopted as best meeting the design requirements. In individual Australian studies, the assigned AEP has ranged from 1 in 1000 to effectively zero. The Bureau of Reclamation (1981) recommends 1 in 10^4 as a conservative value for economic analysis only. The higher the AEP assigned to the PMF, the more conservative (that is, higher) will be the estimated flood discharge for a given intermediate probability when using a particular procedure. To maintain a reasonable degree of conservatism, the possible AEPs assigned to the PMF were restricted primarily to 1 in 10^4, 10^5 and 10^6, and to 1 in 10^3 and 10^7 in some circumstances.

2.2 Testing of Available Methods

Eight methods for estimating floods between the PMF and the 1 in 100 event were selected for testing. Several had been used by Australian water authorities. The methods of the Bureau of Reclamation (1981) and Bethlahmy (1984) were included. Six of the methods involve the use of annual flood data for the construction of a frequency curve, while two use rainfall data only. All require an estimate of the PMP, and the six using flood data require an estimate of the PMF. Where appropriate, those methods using flood data were tested assuming both log Normal and log Pearson III distributions. For methods requiring the assignment of a probability of the PMP or PMF, each of the selected values of 1 in 10^4, 10^5 and 10^6 was tested. In developing frequency curves, both conventional probabilities and expected probabilities corrected for small-sample bias were used. Where values were plotted on frequency diagrams, unbiased plotting positions (Cunnane, 1978) were used.

The methods were tested using data from 12 damsites around Australia for which recent estimates of PMP and the PMF had been made. Figure 1 shows the locations of the sites, which were selected to give a broad coverage of climatological and hydrological conditions across Australia. The standard deviations of the logarithms of the annual floods ranged from 0.15 to 0.80, and skews from -1.59 to 0.73. These ranges are lower than those for 172 Australian drainage basins reported by McMahon (1979) and Srikanthan and McMahon (1981), especially for skew, but the data represent a reasonable sample. Record lengths varied from 7 to 95 years.

In general, the six methods using frequency analysis of the annual flood series performed poorly over the range of conditions analyzed.

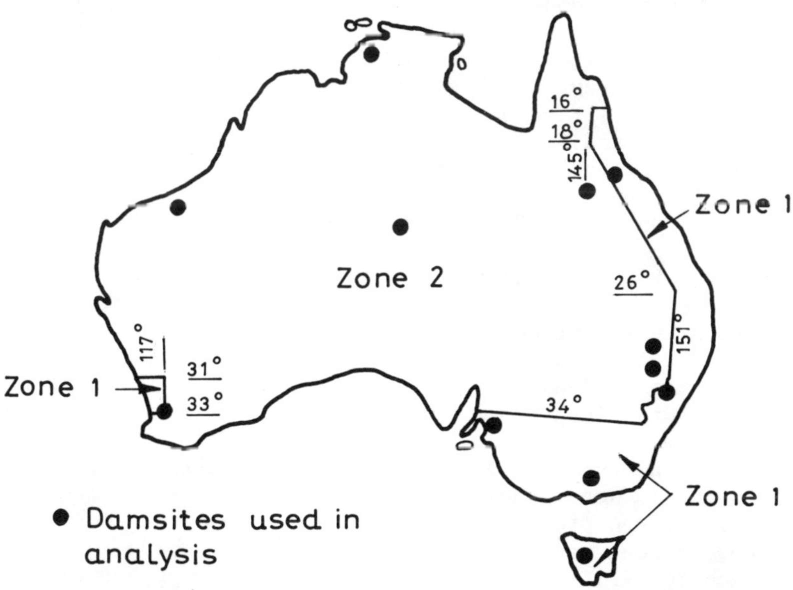

Figure 1 Locations of Damsites Used in Testing of Methods for Estimating Rare Floods, and Rainfall Zones for Basins Smaller than 100 Km^2.

Except for the Bethlahmy method which was unsatisfactory in all cases, each procedure was satisfactory at some sites but unsatisfactory at others. The method that was satisfactory at most sites was the extrapolation of the log Normal distribution from an AEP of 1 in 100 by means of a smooth curve drawn to the PMF plotted at its assigned probability. The upper bound on discharge implied by negative skew provided problems with the log Pearson III distribution at several sites. In the Bureau of Reclamation (1981) method, the extrapolated frequency curve is constrained to pass through a "gate" formed by horizontal lines at 0.4 and 0.6 of the PMF and vertical lines at AEPs of 1 in 200 and 1 in 500 on the log Normal plot. At most Australian locations, this "gate" was in an inappropriate position. This may reflect the greater variability of the statistics of Australian floods compared with those in the United States. The two methods using design rainfall performed better than all of the methods using flood frequency analysis, but were still not satisfactory in all cases.

Although none of the eight methods tested was satisfactory, the investigation provided three pointers that contributed to the development of the design procedure:

(i) The use of a smooth curve joining the 1 in 100 AEP event to the probable maximum event provides a simple procedure that is at least as effective as any other approach. Curves of this type in two methods tested gave better results than any other method.

(ii) Although the U.S. Bureau of Reclamation"gate" was found insufficiently flexible for Australian conditions, this type of approach for specifying the shape of the frequency curve would greatly assist the construction of the curve and the consistency of estimates of rare floods.

(iii) There are three essential elements of the data at a given site that affect the frequency curve between the 1 in 100 AEP and probable maximum events. These are the slope of the frequency curve at an AEP of 1 in 100, the ratio of the magnitudes of the probable maximum and 1 in 100 AEP events, and the assigned probability of the probable maximum event.

2.3 Development of the Basic Procedure for Determining Event Magnitude

The basic procedure for determining the magnitude of rare events and the assigned probability for the probable maximum event was developed from the three pointers in the preceding section. The term "events" will be used rather than "floods", as the procedure needs to be applied to rainfalls, peak flood magnitudes, and/or flood volumes for the solution of different types of practical design problems, as noted later. The probable maximum event will be denoted as the PME. It is assumed that logarithmic values of the events are used in the probability distributions adopted for design, in accordance with current Australian practice.

For a given assigned probability of the PME, the other two variables in (iii) of the previous section define the frequency curve in dimensionless form between an AEP of 1 in 100 and the PME on a log Normal plot. The slope of the plotted frequency curve at an AEP of 1 in 100 is closely determined by the difference between the logarithms of the 1 in 100 and

1 in 50 AEP events. The ratio of the actual values, denoted here as (X_{100}/X_{50}), is thus an index of this slope.

Consider frequency curves plotted on log Normal paper with a given assigned AEP of the PME. The complete range of possible frequency curves over the range of probabilities under consideration can then be defined in dimensionless form in terms of the difference in the logarithms of the PME and the 1 in 100 AEP event, divided by the slope of the frequency curve at an AEP of 1 in 100. Using the index of slope noted above, this measure can be replaced by the ratio of the differences of the logarithms of the PME and 1 in 100 AEP event, and of the 1 in 100 and 1 in 50 AEP events, expressed in the equality

$$\frac{\log X_{PM} - \log X_{100}}{\log X_{100} - \log X_{50}} = \frac{\log(X_{PM}/X_{100})}{\log(X_{100}/X_{50})} \qquad (2)$$

where X denotes event magnitude, the subscripts 50 and 100 refer to the 1 in 50 and 1 in 100 AEP events, and the subscript PM refers to the PME.

Development of the procedure for interpolating event magnitudes between the PME and 1 in 100 AEP event involved the drawing of frequency curves in this dimensionless form on a log Normal plot. The case with an assigned probability of 1 in 10^5 for the PME will be used as an example, and is illustrated in Figure 2. For an arbitrary slope of the

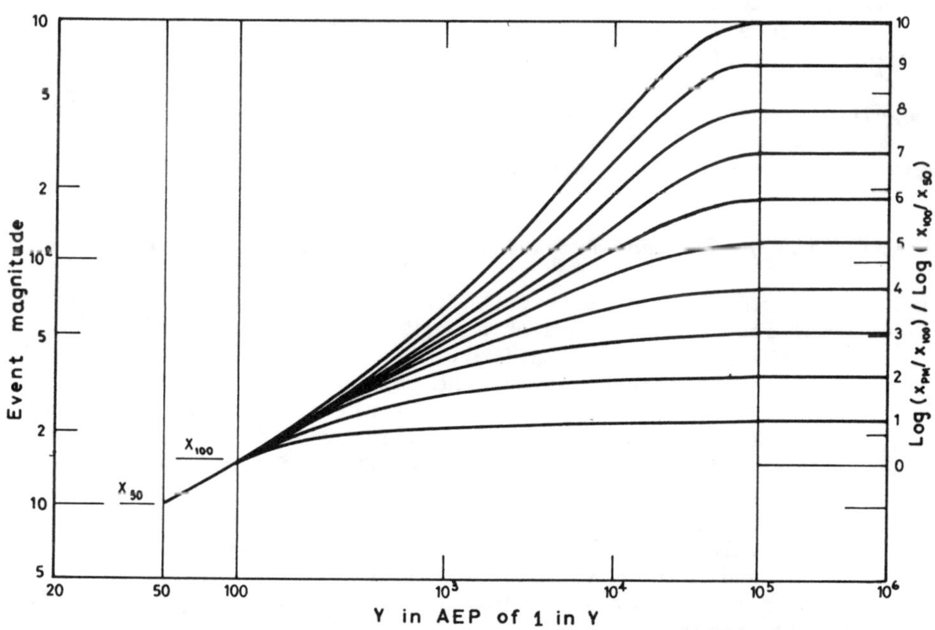

Figure 2 Set of Frequency Curves for Assigned AEP of 1 in 10^5 for Probable Maximum Event.

frequency curve between AEPs of 1 in 50 and 1 in 100, horizontal lines were drawn at an AEP of 1 in 10^5 for various values of the ratio in equality (2). Smooth curves were then sketched to be tangential to the frequency curve at an AEP of 1 in 100 (or at a lower probability), and to the horizontal lines at the AEP of 1 in 10^5. The curves have been drawn to meet the specified criteria of being of reasonable form in a subjective sense, of being consistent, and providing some degree of conservatism. Sets of curves were drawn for assigned probabilities of 1 in 10^3, 10^4, 10^5, 10^6, and 10^7 for the PME.

An attempt was made to fit a mathematical formula to the sets of curves. A third degree polynomial was tested, but gave unsatisfactory shapes as curvatures were too great at high probabilities. As no other suitable simple function was found, the sketched curves were adopted. In addition, the adopted approach is convenient for design, in harmony with the overall accuracy that is attainable, and has the advantage of forcing the designer to consider the form of the complete frequency curve. Care was taken to ensure that each curve was smooth, and that the curves for different AEPs of the PME were consistent with one another.

All combinations of data likely to be encountered throughout Australia should be covered by the sets of curves. By its nature, the procedure will provide consistent answers for the whole range of these data, in contrast to the eight methods tested.

In the recommended approach, the frequency curve is assumed to be a straight line on a log Normal plot between AEPs of 1 in 50 and 1 in 100. The effect of this assumption was tested by drawing sets of curves for cases with fairly extreme curvature between AEPs of 1 in 50 and 1 in 100. Both concave upward and concave downward curvatures were tested, corresponding to positive and negative skews. Although the resulting frequency curves were affected at relatively high probabilities, the changes in magnitude over the complete range of probabilities were considered to be negligible in relation to the overall accuracy of the procedure. A single set of curves was thus adopted for each assigned probability of the PME

2.4 Selection of Assigned Probability for the Probable Maximum Event

It was considered that large reversed curvature of the frequency curve on log Normal paper in the form of a marked S shape between an AEP of 1 in 100 and the PME would be unlikely. Consequently, a criterion adopted for selecting the probability of the PME was that the curve should have only simple curvature, or at most a small reversal of curvature if possible in this range. The tendency towards reversal of curvature increases as the assigned probability of the PME increases and as the value of the ratio $[\log(X_{PM}/X_{100})/\log(X_{100}/X_{50})]$ in equality (2) increases. The dimensionless sets of curves illustrated in Figure 2 were used to select maximum values of this ratio for which each assigned AEP of the PME should be used. These values are listed in Table I.

A second criterion for assigning an exceedance probability to the PME is adapted from recommendations of Kennedy and Hart (1984) of the Australian Bureau of Meteorology for AEPs of the PMP. The recommended probabilities in Table II are based on the ratio of the area of the drainage basin to the area of the region over which storm transposition can occur, the probabilities of occurrence of the most efficient storm and the extreme moisture content, and their joint probability of occurrence. As maximization in situ is now a little-used technique in most regions, and will be superseded by the further development of generalised PMP procedures, it is unlikely that an AEP of 1 in 10^3 will apply to many new estimates of PMP in the future.

TABLE I Assigned AEP of PME Based on Shape of Frequency Curves

Value of ratio $\log(X_{PM}/X_{100})/\log(X_{100}/X_{50})$	AEP of the PME
< 4.5	1 in 10^4
4.5-7.5	1 in 10^5
> 7.5	1 in 10^6

High adopted AEPs for the PME lead to high and thus conservative estimates at intermediate probabilities. To provide some degree of conservatism, the proposed procedure to be recommended in ARR is that probabilities should be determined from each of Tables I and II. The higher probability (e.g. 1 in 10^5 rather than 1 in 10^6) should then be adopted as the AEP of the PME.

TABLE II AEPs of Various Types of PMP Estimates Based on Meteorological Considerations - Adapted from Kennedy and Hart (1984)

PMP Method	Y in AEP of 1 in Y Catchment Area (km^2) 10^2	 10^3	 10^4	 10^5
Maximization in situ	10^3	10^3	10^3	10^3
Maximization and transposition (with a transposition area of 10^5 km^2)	10^6	10^5	10^4	10^3
Generalized method (effective transposition area 10^6 km^2)	10^7	10^6	10^5	10^4
Adjusted U.S. data (effective transposition area 10^7 km^2)	10^8	10^7	-	-

2.5 Shape of the Frequency Curves and Values at Intermediate Probabilities

To enable the designer to reconstruct the adopted frequency curves such as those illustrated in Figure 2, ordinates of the curves at two probabilities between 1 in 100 and the AEP of the PME are given in Table III. A more detailed table is given by Rowbottom et al. (1986). The ordinates for an intermediate probability of 1 in Y are given as values of the ratio $[\log(X_Y/X_{100})/\log(X_{PM}/X_{100})]$. This can be solved for the value of

X_Y. These values are then plotted on log Normal paper at the two relevant intermediate probabilities, together with the PME at its assigned probability, the 1 in 100 AEP event and the slope of the frequency curve at this point. These plotted values will enable the complete curve to be drawn in a consistent manner. Values for any intermediate frequency can then be read from the curve. For some probabilities, the value is given directly by the ordinates in Table III. It is still desirable that the complete curve be drawn in these cases to give an overall view of the assumed frequency relation.

Values are give in Table III for frequency curves incorporating considerable reverse curvature, even though this was considered undesirable in formulating the first criterion for selecting the AEP of the PME. This is necessary because the second criterion may require the use of a frequency curve with greater reversal of curvature. Also, large reversal of curvature will occur with the minimum AEP of the PME (1 in 10^6) for high values of the ratio in equality (2).

TABLE III Values of $[\log(X_Y/X_{100})/\log(X_{PM}/X_{100})]$ for Determining the Shape of the Flood Frequency Curve Between the 1 in 100 Event and the PME

AEP OF PME	1 in 10^4		1 in 10^5		1 in 10^6	
AEP of intermediate points (1 in Y) / $\frac{\log(X_{PM}/X_{100})}{\log(X_{100}/X_{50})}$	1 in 500	1 in 2000	1 in 1000	1 in 10 000	1 in 2000	1 in 50 000
1	0.877	0.976	0.928	0.989	0.948	0.994
2	0.711	0.936	0.802	0.970	0.850	0.982
3	0.585	0.880	0.690	0.944	0.752	0.967
4	0.489	0.816	0.600	0.911	0.667	0.945
5	0.419	0.783	0.527	0.862	0.597	0.915
6	0.372	0.766	0.467	0.811	0.539	0.874
8	0.328	0.749	0.385	0.770	0.447	0.797
10	0.306	0.742	0.342	0.757	0.385	0.771
12	0.291	0.738	0.321	0.751	0.347	0.762
15	0.276	0.735	0.301	0.746	0.320	0.754
20	0.260	0.732	0.279	0.743	0.294	0.751

2.6 Application of the Procedure to Practical Design Problems

The sets of dimensionless frequency curves presented in Table III provide a basic procedure for estimating rare floods, but only give event magnitudes. For many design problems, a complete flood hydrograph is required. An inflow hydrograph may be needed where duration of flooding is of interest, or possibly for erosion or water quality studies. More commonly, the inflow flood must be routed through storage to obtain an outflow hydrograph or peak outflow discharge. This is the case in the design of dam spillways and outlets of detention basins. Outflow floods with different critical durations of design rainfall will result from different configurations of the storage and spillways, and from different operating policies of spillway gates or outlet works.

These various types of problems require the basic procedure to be applied in different ways. In some cases peak discharges can be used, and in others, design rainfalls are the appropriate values to be used in the procedure. Use of flood volumes may also be helpful in some situations. In addition, other procedures for calibration, estimation and routing of flood hydrographs are often required to complement the application of the basic procedure to obtain the required design information. Detailed descriptions of these applications to practical design problems are beyond the scope of this paper, but will be given in the revised ARR as guides to designers.

3. CHOICE BETWEEN FLOOD ESTIMATES BASED ON DESIGN RAINFALLS AND FLOOD FREQUENCY ANALYSES

3.1 The Problem

As part of the procedure for estimation of large and extreme floods as well as for general flood estimation, flood discharges up to the 1 in 100 AEP event must be determined. For this, a choice may be needed between available methods. Often this choice is between rainfall-based methods and flood frequency analysis. Although this is commonly one of the most important decisions faced by the designer, little guidance has been available. Recommendations given in various manuals and books are rather arbitrary and generally give little consideration to many factors governing the relative accuracies of competing methods. Often, the choice of method has rested largely on the judgement and personal preference of the designer.

Guidelines have been developed for inclusion in ARR to assist the designer in this choice. Statistical theory has been applied to provide a basis for the guidelines, but assumptions based on judgement and experience have been necessary to quantify them. The guidelines are based on the statistical characteristics of Australian flood and design rainfall data, but the methodology employed could be applied elsewhere. The guidelines are based primarily on relative accuracies and hence the best method to use in a given situation. Less information is given on actual accuracies, and there is no reduction in the responsibility of the designer to eliminate bias and error from flood estimates, and to reconcile as far as possible the answers given by different methods.

The criterion adopted for the choice between methods was their relative accuracies as indicated by estimates of confidence limits or equivalent measures. In addition to the choice between flood frequency and rainfall-based methods calibrated from data at the site, guidelines have been developed for methods of both types based on regional relationships. Only a brief summary of the guidelines is given here. Doran and Pilgrim (1986) give a more complete description, including their development.

3.2 Factors Affecting Choice of Method

Considering first the case where the flood frequency analysis is based only on data from the site of interest, the factors affecting the choice between this approach and rainfall-based methods can be summarized as:

(i) The length (N years) of the available flood record at the gaged site.

(ii) The standard deviation (S) of the logarithms of the annual flood series. For Australian streams, S ranges from about 0.15 to 1.0, with a median of about 0.4 (McMahon, 1979; Srikanthan and McMahon, 1981). In accordance with current Australian practice, it is assumed that the logarithms of rainfalls and floods are used in the probability distributions adopted for design.

(iii) Coefficient of skewness (g) of the logarithms of the annual flood series (site or regional value). The range of values for Australian steams is -4 to 1.5 with most values between -1.25 and 0.75 (McMahon, 1979; Srikanthan and McMahon, 1981).

(iv) The assumed underlying probability distributions for both flood flows and rainfalls. The log Pearson III distribution was used for both floods and rainfalls in this analysis.

(v) The method of flood estimation, usually based on design rainfall, that is available as the alternative to flood frequency analysis. These rainfall-based methods ere divided into three general classes:
   - Class I - hydrograph models calibrated from data at the site.
   - Class II - methods calibrated by regional relations or transposition of values or parameters from an adjacent gaged basin.
   - Class III - generalized or arbitrary methods or formulae.

(vi) The length and quality of rainfall records from which the design rainfall data were derived. Development of design rainfall data for Australia is described by Canterford et al. (1987). As they are based on frequency analyses with the same characteristics as flood frequency analyses, the statistical properties of the design rainfall data will affect the choice of flood estimation method.

Where a regional flood frequency method is used, equivalent length of record and a representative standard deviation replace the first two factors listed above, and the skew would be a regional value. The statistical characteristics of the regional regression (Hardison, 1971) provide additional factors to (i) to (vi) listed above.

3.3 Approach Used in Developing Guidelines

The primary analysis involved an assessment of the relative accuracies of flood frequency and rainfall-based methods using design rainfalls developed from daily-read gages, where both types of methods are calibrated from observed data on the drainage basin of interest. Duration of rainfall on which the design rainfall data are based is a difficult variable to use as a design criterion, and basin size was used as a convenient surrogate. It was assumed that design rainfalls for basins larger than 150 km^2 would be based primarily on daily-read gage data. However, designers could use a different criterion in applying the guidelines if the critical duration of rainfall for a given basin was known.

The analysis was based on the estimated confidence limits of the floods obtained by the competing methods. Figure 3 shows schematically a flood frequency curve and its confidence limits, and the frequency curve of design rainfalls for the drainage basin and its confidence limits, which will generally be narrower than those for the flood frequency curve. Confidence limits of floods determined from the design rainfalls were estimated, as illustrated on Figure 3. For floods with lower exceedance probabilities than that corresponding to point A where these confidence limits become narrower than those for the frequency curve of annual floods, the methods based on design rainfall should be used in preference to flood frequency analysis.

The analysis to develop the guidelines used an error model for the rainfall-based methods. Various assumptions were required in calibrating the model, and these were based on available data and subjective judgement. From information supplied by the Bureau of Meteorology, a record length of 65 years was adopted for daily-read rain gages over the whole of Australia, and a constant skew of 0.24 was adopted for the logarithms of annual maximum rainfalls. Standard deviations of the logarithms of annual rainfalls range from 0.05 to 0.5, and a relationship between the standard deviations of the logarithms of the annual rainfalls and floods was assumed for use in the error model. The AEP of the point A on Figure 3 was then determined for a wide range of combinations of values of N, S and g for the flood sample.

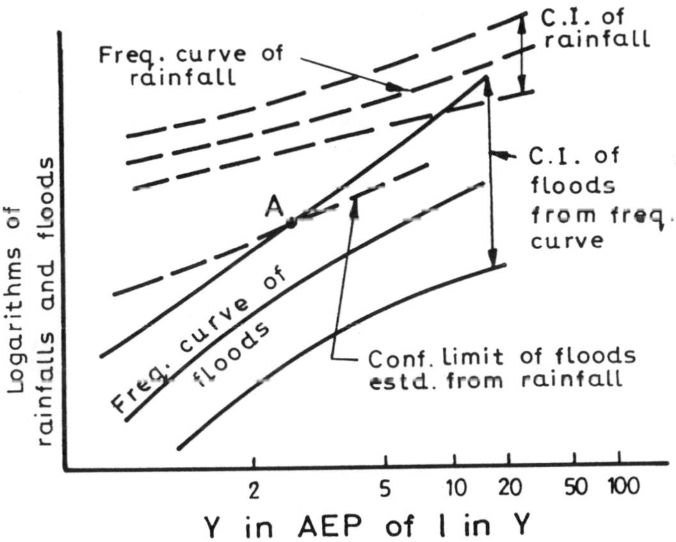

Figure 3 Frequency Curves and Confidence Limits of Design Rainfalls and Observed Floods

For drainage basins smaller than 100km^2 the design rainfalls were assumed to be derived primarily from recording rain gage data, although these design values are also based on regressions with daily data and topographic and locational variables (Canterford et al., 1987). Measures of the accuracy of the rainfall-based flood estimates for this case were scaled from those for the larger basins where daily-read data apply, using the statistical characteristics of the data. Information contents of the mean or estimated values (Matalas and Langbein, 1962) were used as the primary means of scaling. This was checked by estimation of confidence limits of the rainfall-based methods for small basins, in a similar manner to the analysis for the larger basins. For the analysis, Australia was divided into the two zones shown on Figure 1 on the basis of length of record and density of recording rain gages. In Zone 1, the effective average length of recording rain gage records used in determining the design rainfall data is 33 years, and the relative spatial density of recording to daily-read gages is approximately 1 to 20. In Zone 2, the corresponding figures are 16 years and 1 to 40.

Extension of the analysis to Class II and Class III methods based on design rainfalls involved comparison of the estimated errors in flood peaks resulting from errors in the hydrograph models of Class II and III methods with those for Class I methods. Only errors resulting from the hydrograph model are relevant, as the same design rainfalls, losses and baseflow are used for all classes of method. The errors in the hydrograph models, and their effects on hydrograph peaks, were determined from published data on the calibration of unit hydrograph and runoff routing hydrograph models, and from the statistical characteristics of regional relationships for Class II methods. In general, there is a need for more information on statistical characteristics to be published by developers of regional procedures.

For regional flood frequency methods or transposition of frequency parameters from an adjacent better-gaged basin, equivalent measures of N, S and g are required. These are beyond the scope of this paper, but are described in detail by Doran and Pilgrim (1986).

3.4 Recommended Guidelines

The primary analysis of the relative accuracies of flood frequency analyses and rainfall-based methods calibrated on observed data for basins larger than 150 km^2 produced a large number of values for the AEP of 1 in Y of point A on Figure 3. A value was obtained for each of the combinations of values of the variables, N, S and g. For application in design, a relatively simple formula is required. This was obtained by plotting the derived values of Y against N, S and g and deriving an empirical equation of best fit. This equation, which applies to Class I methods, is shown at the top left of Table IV. Values of F, which depend on the values of S and g for the sample of annual floods, are obtained from Table V.

The relative information contents were used to adjust the relation for large basins to obtain relations for basins smaller than 100 km^2 in the two zones in Figure 1. The resulting guideline relations for Class I methods are also shown in Table IV. Critical AEPs for areas between 100 and 150 km^2 are interpolated from the relations.

The remainder of Table IV shows the guideline relations for Class II and III rainfall-based methods. These were obtained by scaling the Class I relations to account for the relative model errors in the flood estimates. Based on Class II methods used and developed in Australia, the overall range of these errors was adopted as 25% to 70%, and three sets of guideline relations are given for different subsets within this error range. Rainfall-based methods with model error ε_{mII} greater than 70% were grouped with the Class III arbitrary methods.

TABLE IV Formulae for Finding Y for an AEP of 1 in Y, at which Rainfall Based Estimate is Preferable to Flood Frequency Estimate

Class of Method using Design Rainfall	Range of Hydrograph Model Error	Y for 1 in Y AEP Flood ($5 < Y < 500$) Area ≧ 150 km²	Area < 100 km² Zone	Formula
(I)	All	F N exp(0.02N)	1	$1.3\ F\ N^{1.1}\exp(0.03N)$
			2	$2\ F\ N^{1.2}\exp(0.04N)$
(II)	$\varepsilon_{mII}<40\%$	3 F N exp(0.02N)	1	$3.9\ F\ N^{1.1}\exp(0.03N)$
			2	$6\ F\ N^{1.2}\exp(0.04N)$
	$40\%\leqq\varepsilon_{mII}<55\%$	6 F N exp(0.02N)	1	$7.8\ F\ N^{1.1}\exp(0.03N)$
			2	$12\ F\ N^{1.2}\exp(0.04N)$
	$55\%\leqq\varepsilon_{mII}<70\%$	9 F N exp(0.02N)	1	$11.7\ F\ N^{1.1}\exp(0.03N)$
			2	$18\ F\ N^{1.2}\exp(0.04N)$
(III)	$\varepsilon_{mIII}\geqq 70\%$	12 F N exp(0.02N)	1	$15.6\ F\ N^{1.1}\exp(0.03N)$
			2	$24\ F\ N^{1.2}\exp(0.04N)$

Note:- Values of F found in Table V and N is actual or equivalent sample length.

TABLE V Values of F to be Used with Formulae in Table IV

S \ g	-1.5	-1.0	-.75	-0.5	0.0	0.5	.75	1.0	1.5
.05	2.35	5.68	8.85	5.68	2.35	0.97	0.62	0.40	0.16
0.1	0.68	1.64	2.56	1.64	0.68	0.28	0.18	0.12	0.05
0.2	0.37	0.88	1.38	0.88	0.37	0.15	0.10	0.06	0.03
0.3	0.30	0.72	1.12	0.72	0.30	0.12	0.08	0.05	0.02
≧0.4	0.27	0.65	1.00	0.65	0.27	0.11	0.07	0.05	0.02

Note:- S and g are sample or regression estimates.

4. CONCLUSION

There is an increasing need in Australia for the estimation of design floods of very low exceedance probability. Two procedures are described that have been developed to aid in the practical estimation of these floods, and will be included in a current revision of "Australian Rainfall and Runoff", the guide to flood estimation in Australia published by The Institution of Engineers, Australia. One of these is a relatively simple procedure for interpolation of floods between the 1 in 100 event and the PMF. While it is not possible to obtain theoretically correct estimates of these low probability floods, the procedure was developed to meet practical requirements after other procedures were shown to be unsatisfactory over a wide range of Australian data. The second procedure provides guidelines for the choice between flood estimates based on design rainfalls and flood frequency analyses. The guidelines were developed from the statistical characteristics of Australian flood and design rainfall data.

5. ACKNOWLEDGEMENTS

The revision of "Australian Rainfall and Runoff" is being carried out under the auspices of The Institution of Engineers, Australia. The financial assistance of the various sponsors who have given grants to the Institution for the revision is acknowledged. Reviews of various aspects of the procedures described have been provided by the Institution's National Committee on Hydrology and Water Resources, the Advisory Panel for the revision of "Australian Rainfall and Runoff", and the Australian National Committee on Large Dams Sub-Committee on Dam Safety.

6. REFERENCES

Australian National Committee on Large Dams. 1986. 'Guidelines on Design Floods for Dams'.

Bethlahmy, N. 1984. 'Long-term Hydrologic Events from Short Term Records'. Journal of Hydrology, Vol. 68, pp. 141-148.

Bureau of Reclamation. 1981. 'Criteria for Selecting and Accommodating Inflow Design Floods for Storage Dams and Guidelines for Applying Criteria to Existing Storage Dams'. U.S. Department of the Interior, Technical Memorandum No. 1, Assistant Commissioner - Engineering and Research, Denver, Colorado.

Canterford, R.P., Pescod, N.R., Pearce, H.J., Turner, L.H. and Atkinson, R.J. 1987. 'Frequency Analysis of Australian Rainfall Data as Used for Flood Analysis and Design'.

Cunnane, C. 1978. 'Unbiased Plotting Positions - A Review'. Journal of Hydrology, Vol. 37, pp. 205-222.

Doran, D.G. and Pilgrim, D.H. 1986. 'Choice Between Flood Estimates Based on Design Rainfalls and Flood Frequency Analyses'. Civil Engineering Transactions, Institution of Engineers Australia, Vol. CE28, pp. 74-91.

Hardison, C.H. 1971. ' Prediction Error of Regression Estimates of Streamflow Characteristics at Ungaged Sites'. U.S. Geological Survey Professional Paper 750-C, pp. C228-236.

Institution of Engineers, Australia. 1977. 'Australian Rainfall and Runoff'. The Institution, Canberra.

Kennedy, M.R. and Hart, T.L. 1984. 'The Estimation of Probable Maximum Precipitation in Australia'. Civil Engineering Transactions, Institution of Engineers Australia, Vol. **CE26**, pp. 29-36.

Matalas, N.C. and Langbein, W.B. 1962. 'Information Content of the Mean'. Journal of Geophysical Research, Vol. **67**, pp. 3441-3448.

McMahon, T.A. 1979. 'Hydrologic Characteristics of Australian Streams'. Civil Engineering Research Reports, Report No. 3/1979, 79 pp., Monash University, Clayton, Victoria.

National Association of Australian State Road Authorities. 1983. 'Bridge Design Code. Draft - Limit States Format'. Sydney.

Pilgrim, D.H. 1986. 'Estimation of Large and Extreme Design Floods'. Civil Engineering Transactions, Institution of Engineers Australia, Vol. **CE28**, pp. 62-73.

Rowbottom, I.A., Pilgrim, D.H. and Wright, G.L. 1986. 'Estimation of Rare Floods (Between the Probable Maximum Flood and the 1 in 100 Flood)'. Civil Engineering Transactions, Instituion of Engineers Australia, Vol. **CE28**, pp. 92-105.

Srikanthan, R. and McMahon, T.A. 1981. 'Log Pearson III Distribution - Effect of Dependence, Distribution Paramaters and Sample Size on Peak Annual Flood Estimates'. Journal of Hydrology, Vol. **52**, pp. 149-159.

DESIGN FLOOD FOR CAÑAFISTO DAM

F. Villegas
Technical Director Integral S.A.
Apartado aéreo 3036
Medellín, Colombia

J.L. Cadavid
Supervision Engineer, Integral
Apartado aéreo 3036
Medellín, Colombia

ABSTRACT. This paper presents the process of estimating the design flood for a large dam located in the equatorial region. The method applied was to estimate an exceptional synthetic storm and consequently, the corresponding hydrograph. Advantage was taken of techniques for evaluating storm potential, developed for temperate latitudes. The peak flow was computed as 28.500 m3/s. Because of limitations in the applied procedure it was not possible to assign probability level to this event.

1. INTRODUCTION

The Cauca river is one of the most important rivers in Central Colombia as several large cities and a significant portion of the population are located and live in the watershed. The basin has a total area of 58500 km2 and the river a length of 1230 km. The river has a predominant South-North direction between 2º and 9º N latitude discharging into the Magdalena River, the latter flowing into the Caribbean Sea. (See figure 1).

A chain of four large dams and powerplants with a total installed capacity of several million kilowatts is planned to be built along the middle portion of the river, called the Cauca Canyon, where it flows between elevations 900 m and 90 m in a 400 km reach. The first plant to be built in the near future will be Cañafisto which will have an installed capacity of 1500 MW. The second plant to be built will be Ituango, with and installed capacity of about 4000 MW which will be the largest hydroelectric project constructed in Colombia.

This paper describes the process which was followed in determining the storm producing the flood for the design of the Cañafisto spillway. The process has several significant features of interest given the lack of long-term river discharge records and the limited knowledge of storm producing mechanisms in the equatorial region where the watershed is located. It was necessary to analyze large historical storms, making full use of weather satellite pictures-ERTS (Earth Resources Satellite Technology), complemented by ATOLL (Analysis of Tropical Oceanic Lower

V. P. Singh (ed.), Application of Frequency and Risk in Water Resources, 79–90.

Layer) charts and ocean cruise records. After estimating the extreme storms, a rainfall-runoff mathematical model was calibrated and run for purpose of computing the corresponding flood hydrographs.

2. DESCRIPTION OF THE BASIN AND RIVER

The Cauca river basin is located between the Cordillera Central and the Cordillera Occidental of Colombia, which are two of the three mountain ranges which branch from the Andes in the southern part of the country. The basin may be subdivided in four physiographical zones (See figure 2).

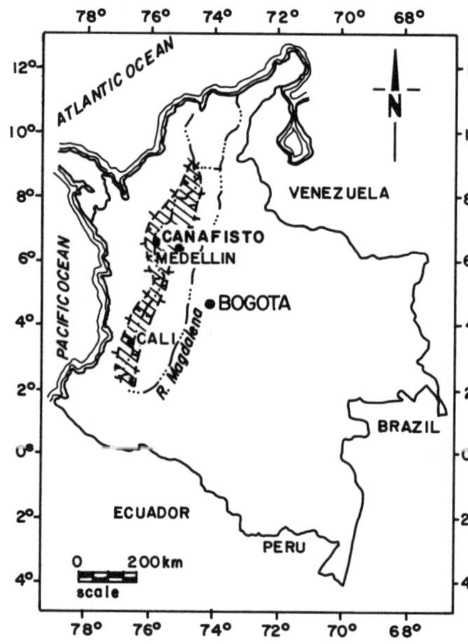

FIGURE 1. Location of Cauca basin in Colombia.

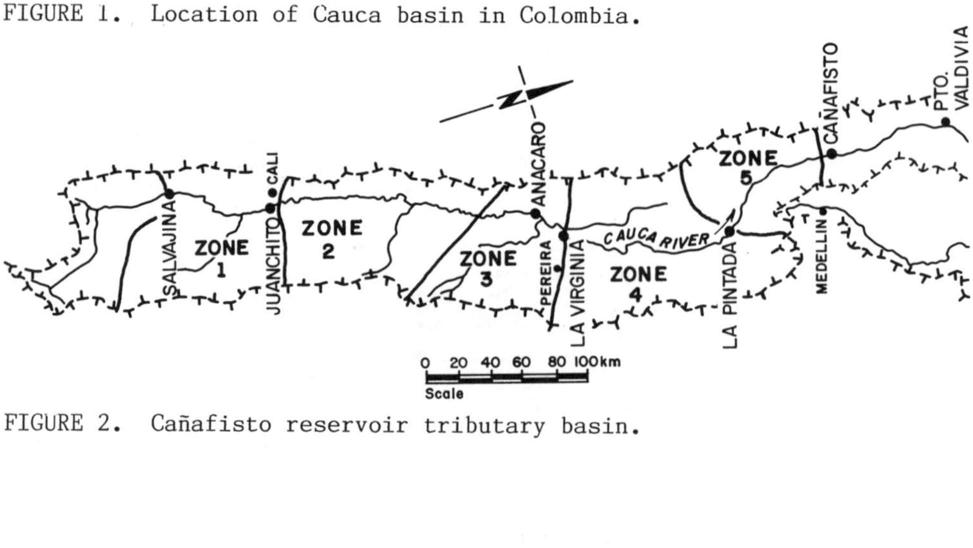

FIGURE 2. Cañafisto reservoir tributary basin.

The Upper Cauca, from the headwaters to Timba, with a 170 km river length, a 5450 km2 tributary area and a 198 m3/s average flow at the outlet. This region is characterized by steep slopes, where the river drops from an elevation of 4200 m to 1000 m above sea level. The Salvajina dam and power-plant which are located close to the outlet of the reach were recently completed.

The Cauca Valley, which is an extensive alluvial valley between the mountain ranges, goes from Timba to la Virginia. The river stretch is 404 km long and the partial tributary area, 17360 km2. The river falls from an elevation of 1000 m to 900 m.

The Cauca Canyon, a stretch where the river falls with steep slope along a deep, narrow canyon between La Virginia and Tarazá flowing between elevations 900 m and 90 m in a 400 km stretch. The partial watershed area is 15990 km2 and the average flow, 1210 m3/s at the outlet. This is the area of the watershed with the highest precipitation and where the significant storm centers are usually located. The chain of dams and powerplants is planned along this reach.

The Cauca Low Plain, which is the lower portion of the basin, where the river flows along a channel 260 km long, on a flat, deep alluvial plain. The partial tributary area is 19700 km2 and the average flow, 2900 m3/s at the confluence with the Magdalena river.

The longest, continuous discharge river records are available at the Juanchito and La Virginia stations located in the Cauca Valley zone. Records of stations within the Cauca Canyon are only available since 1965 at La Pintada and since 1959 at Puerto Valdivia.

3. HYDROMETEOROLOGY OF THE RIVER BASIN

The rainy seasons in the Cauca basin depend on the movement of the intertropical convergence zone (ICZ), which oscilates between latitudes 16º N and 5º S. The ICZ is associated with low barometric pressures, strong ascending winds and intense turbulence and mixing. Given the length and orientation of the Cauca River basin, the pattern of yearly dry and wet periods varies along the basin in relation to the movement of the ICZ. Hence in the northern portion of the basin there are two dry periods, December to April and July to August, the first of which is drier and the second, less defined. In the southern portion of the watershed, the period of June to September is the driest, while the period of December to March is not well defined. However, the movement of the ICZ is not always steady. Analysis of the ERST satellite pictures and meteorological records previous to and during important storms covering large portions of the basin has demonstrated several facts.

The ICZ, although showing a normally slow average cyclic displacement in latitude, may suffer significant short-term variations of several degrees, in such a way that it could cross the whole Cauca basin in a few days. This means that storms may simultaneously take place on the entire watershed.

Large storms are usually associated with inflows of humidity from several of the following sources: the Pacific Ocean across the Cordillera Occidental; the Panamá and Urabá gulfs of the Caribbean Sea entering through the Northwestern boundary of the basin; the Atlantic Ocean east of Brazil with the water-ladden air travelling across the Amazonas area and spilling above the southern portions of the Cordillera Central and the Cordillera Occidental; and the Atlantic Ocean east of Venezuela with clouds travelling counter-clockwise to the northern portion of the basin.

Storms covering large areas and producing long, continuous rainfall of several days duration are not the case in this region, as is normally observed in temperate areas of the World; instead, important storms are made of a patch-like aggregation of regional, isolated cells (usually less than 500 km2), occurring at different intervals in very complicated spatial and temporal patterns. However, the added effect of these regional storm cells is equivalent to a large storm with a duration of one to several days.

Regional storms are of the convective and orographic type; hence, besides the ICZ movement and other large-scale phenomena, orographic effects are essential in the formation of rainfall.

As in other humid equatorial regions, climatological parameters like precipitation, wind and humidity vary within relatively narrow ranges when compared with other areas of the World. Mean values of the parameters are relatively stable. This consistency in climatological variables in the Cauca is an advantage, because it allows one to place additional confidence on the relatively limited available information.

In general, average annual precipitation decreases from north to south and increases with increasing ground height. Maximum average precipitation is about 3800 mm/year, and the minimum 1000 mm/year. Foci of large storms are generally located at different places in the watershed.

Wind records show wind velocities to be normally less than 5 km/h, although short bursts could reach velocities of 60 km/h. Average wind velocity and direction do not vary much from year to year. Winds consistently blow from the oceans towards the mountains, thereby carrying humidity into the basin.

Temperature is mainly dominated by altitude with average annual temperatures ranging from 28º C in low areas to 0º C on top of the high mountains.

Daily variations are normally from 9º to 12ºC; average monthly variations are usually less than 2ºC, exceptionally 3ºC. The thermal gradient for mean temperatures decreases 0.60 degrees celsius for an increase of 100 m above sea level.

Average humidity of the air is between 75% and 80%. Monthly variation during the year is on the order of 12%.

4. FLOOD ESTIMATION APPROACH

The storage of the Cauca River reservoirs is not high when compared to the average annual runoff. In other words, the useful capacity of the Cañafisto reservoir, 1810 hm3, only amounts to 6.1% of the average annual runoff of 29860 hm3 at the same site. Hence, the impoundment will normally be full or almost full every year during the rainy season. This makes the spillway structure especially critical for dam safety.

Given the length of runoff records at the Cauca River gaging stations (longest of 38 years), statistical flood-frequency analyses are only reasonable for predicting with an adequate level of confidence, floods of low recurrence intervals, ie. in the range of 50 years or less. The flood-frequency analysis was improved by regional studies and the investigation of old historical records to help extend the range of past experience. These results were applied for designing the diversion works and construction activities at the river bed.

In order to design the spillway, a synthetic exceptional storm was estimated, which would generate the design flood, by applying a rainfall-runoff mathematical model. A reasonable degree of conservatism was followed in estimating the synthetic storm, although with the available information at this stage no precise frequency can be attached to such event.

5. HISTORICAL STORMS

Significant historical storms were selected and analyzed for purposes of estimating an extreme storm for synthesizing the design flood. Historical storms were considered when:

- Its average precipitation was high.
- It covered most of the watershed.
- It generated important floods.
- It had good coverage of meteorological information.

23 storms with durations from 5 to 40 days were selected between 1965 and 1982. Among them, 9 were analyzed in more detail (from years 1971 to 1976) following this approach:

- Areal and time distribution of precipitation were obtained.
- Magnitude and direction of winds, humidity, dew point, barometric pressure and cloud cover and their variation with time were studied based on station records, ATOLL charts and ERTS satellite pictures. Wherever possible these data were studied at the surface, at 600 mb and at 200 mb.
- ERTS satellite pictures were used to follow the sequence and location of cloud formation, maximum elevation of clouds, and movement and evolution of the same in order to appraise sources and zones of entry of humidity-ladden air into the basin.

6. ESTIMATION OF THE EXCEPTIONAL STORM

6.1 Methodology. In order to compute the exceptional storm, several procedures were applied: maximization of historical storms, transposition of the extreme storm calculated for the Panama Canal Zone, and adaptation of the Hershfield statistical approach (Hershfield, 1961). Results of the various procedures were compared in order to obtain a more reliable estimate of the extreme storm.

6.2 Maximization of Historical Storms. Among the large historical storms 4 were selected for which there was a good coverage of meteorological information and which caused high floods at La Pintada. For purposes of estimating inflow of humidity into the basin it was assumed a pseudo-adiabatic lapse rate of temperature of ascending, saturated air. Inflow of humidity was computed crossing above both the Cordillera Occidental and the Cordillera Central. The portion of the Cordillera Occidental divide across which inflow of humidity was considered to take place, was assumed to vary between elevations 2000 m and 2700 m on an added length of 443 km; for the Cordillera Central between elevations 2250 m and 3350 m on an added length of 275 km. Some 12-hour persistent dew point data were available at sea level on the windward side of the Cordillera Occidental, besides, some dew point records were known at varying elevations on the windward slopes of both cordilleras. Similarly, magnitude and direction of average 12-hour winds were also obtained from ATOLL charts. Efficiency of each historical storm (in percentage) was considered as the relationship between the average basin precipitation for a given time interval, and the average inflow of humidity divided by the basin area, for the same period. Maximum efficiencies for the 4 selected storms varied between 25% and 40% for 5-day periods, 26% and 49% for 4-day periods and 32% and 73% for 3-day periods.

The assumption was then made that the rainfall producing mechanism would be the same for the selected large historical as for the exceptional synthetic storms, so that the former could be maximized by maximizing the corresponding inflows of humidity. In other words the efficiencies were kept constant. The 12-hour persistent dew points and wind velocities were maximized and the most unfavorable wind directions were assumed using information on data recorded at stations in and around the basin, or at the Pacific Ocean water surface. In this way, the selected values of dew points and wind velocities would have a low probability of occurrence. A 12-hour persistent dew point was selected as 27ºC and wind velocitied of 47 km/h between 1000 mb and 600 mb and 61 km/h between 600 mb and 200 mb were applied for maximizations.

Higher values of maximized precipitations were found for the March, 1971 and November, 1973 storms. These precipitations were then used as one of the estimates of the exceptional storm by plotting the results on a graph relating average basin precipitation to rainfall duration, as shown in figure 3.

6.3 Storm Transposition. The U.S.Weather Bureau and the U.S. Corps of Engineers (1942) carried out a detailed and comprehensive study for calculating the maximized storm producing the probable maximum precipitation on the Panama Canal Zone. This Zone is close to the Cauca River basin and precipitation on the former is subject to and produced by similar meteorological influences and comparable phenomena as on the latter. Inflow of humidity into the Panama Canal Zone proceeds from similar sources as for the western and northwestern sides of the Cauca basin. Therefore, it was judged reasonable to transpose the Panama Canal Zone Maximum storm (area of 3424 km2, average ground elevation 304 m, maximum persistent dew point of 26.1ºC, and 5-day precipitation of 795 mm) to the Cauca River basin. Corrections were applied because of the topographical barrier height, dew point and area. The transposed storm data are shown in figure 3.

6.4 Hershfield Method. Hershfield (1961) proposed a statistical approach for estimating extreme precipitation at a given site, by adding the mean of the series of annual maxima plus k times the standard deviation of the same data. k is a statical variable depending on the applicable extreme value frequency distribution and the recurrence interval. The largest values of k were obtained from rainfall data at stations in the United States and were related to values of the mean of annual maximum rainfall. The method was initially studied for 24-hour precipitations and later extended to other durations less than 24 hours.

A search was made for maximum values of the variable k for precipitation durations of 1 to 5 days in the Cauca basin, complemented by a similar study for 1-day precipitations recorded in Colombia. It was decided to apply a modified Hershfield approach for durations of 1 to 5 days, using data obtained from 82 rainfall stations in Colombia. Values of k were found to increase with duration, a maximum of 8 was determined for 1 day and 10.4 for 5 days. For applying the method, the value of k was assumed as 10 for 1 and 2-day and 11 for 3, 4 and 5-day durations. These latter values are more in agreement with observations in equatorial areas where variability of meteorological phenomena is less pronounced than in higher latitudes.

Maximum precipitations were computed at 20 main storm centers, which showed Mistrató to be the station where the highest point rainfall would occur. It was assumed that the extreme storm would be centered at Mistrató distributed over the basin according to the precipitation-area curves for large storms which deduction is explained in the next section.

Precipitations in mm's for 1 to 5-day durations obtained by the Hershfield approach are shown in figure 3.

6.5 Precipitation-Area Curves. As was previously discussed, large storms on the Cauca River basin are an aggregation of local cells. No definite spatial or temporal pattern of occurrence of the cells has been found, at least with available historical records. Apparently they are mostly random arrangements, a suposition which is reasonable if one takes into account that inflows of humidity are multiple, from various directions and sources, and not necessarily inter-related. Besides, orographic effects exert a complex influence. Therefore, precipitation-area relationships were studied for the several historical storms under consideration without any distinction because of the main focus of the storm. The process was done for durations of 1 to 5 days, and upper envelope rainfall-area curves were obtained as shown in figure 4.

Although no definite or normal pattern of occurrence of rainfall cells within the main storm could be found, it was nevertheless consistently observed that spatial distribution of total storm precipitation followed a ranking similar to the average longterm annual precipitation on the zones in which the basin was divided. This division into zones was done through the course of the study to facilitate the analysis of the meteorological and hydrological behavior of the basin, taking into account the similarity of hydrologic regime and the homogeneity of geomorphological and other parameters regulating runoff. From upstream to downstream the zones are indicated in table 1, and shown in figure 2; ranking of average spatial distribution of total precipitation during large 3-day storms is also included, as well as average annual long-term precipitation.

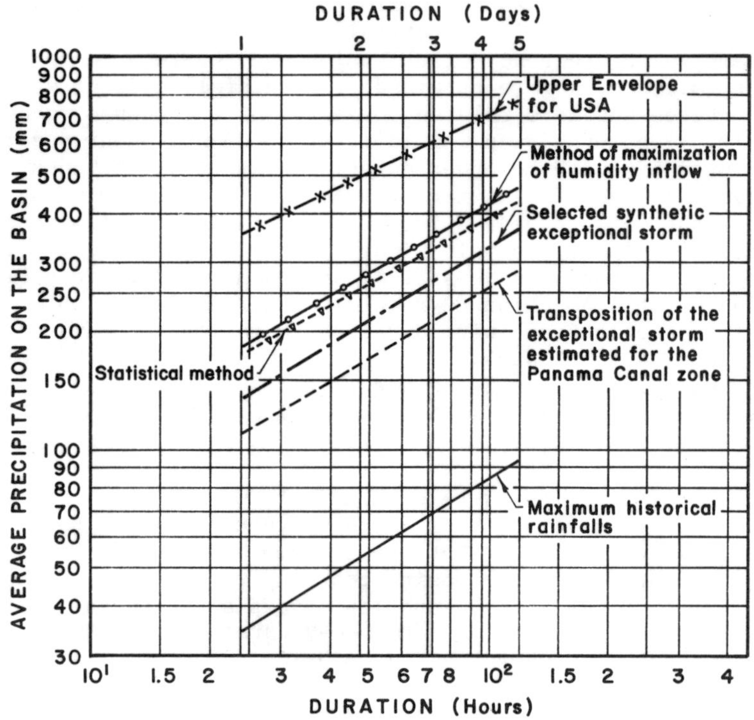

FIGURE 3. Design storm

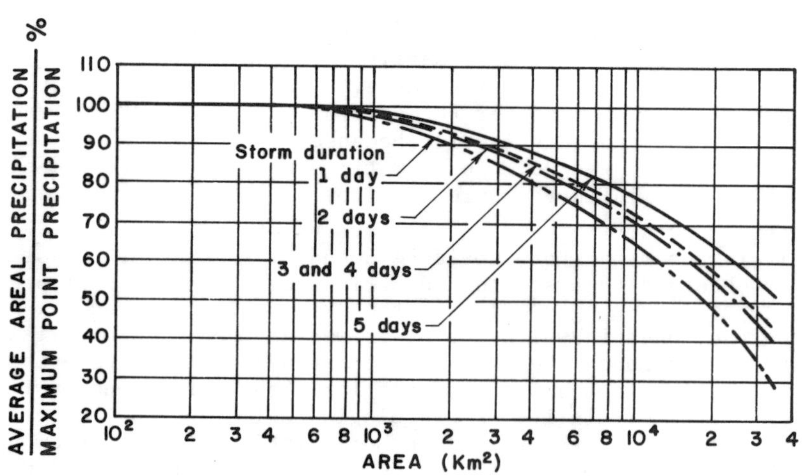

FIGURE 4. Rainfall-area curves for intense storms

Distribution of rainfall on the entire basin during the exceptional storm was adjusted to the precipitation area curve envelope for large historical storms; however, blocks of rainfall under such curves from the highest to the lowest were assigned to the various zones of the basin, following the ranking shown in table 1.

TABLE 1

Zone	Normal ranking of magnitude of 3-day storm precipitation	Average annual precipitation (mm)
1. Headwaters-Salvajina	4	2.020
2. Salvajina-Juanchito	5	1.890
3. Juanchito-Anacaro	6	1.470
4. Anacaro-La Virginia	3	2.100
5. La Virginia-La Pintada	2	2.340
6. La Pintada-Cañafisto	1	2.380

6.6 Storm Duration. Travel time of hydrographs between Salvajina and Cañafisto, normally varies between 2 and 2.5 days. Based on this, duration of the exceptional storm was assumed to be 3 days; although, as normal practice, a check was made by computing the peak flow caused by a 5-day storm in order to be sure that it would not be higher than the corresponding peak produced by the 3-day storm.

6.7 Exceptional Storm. It is clear from the previous sections that all methods have limitations in the procedures, in the values of the parameters, or require one to make some not-adequately substantiated assumptions. Therefore, in selecting the design storm it is necessary to be aware of the limitations of each method.

It could normally be expected that the results from the adapted statistical Hershfield approach would be on the conservative side, because of the very low probability that a maximum point rainfall coincides with an extreme storm covering the whole basin. The transposition of the Panama Canal Zone maximum storm would be of limited reliability, because of the limitations of extending an intense storm over an area several times greater, and the importance of orographic interferences. The method of maximization of inflow of humidity is physically reasonable but there are limitations because of the short length of records of the meteorological variables and the relative scarcity of stations.

Results of the various methods were drawn on figure 3, and a synthetic exceptional storm was selected giving more weight to the maximization of inflow of humidity approach. For purposes of comparison, the maximum historical World and Cauca basin precipitations are also shown on the same plot.

7. RAINFALL-RUNOFF MODEL

For this study the rainfall-runoff model developed for the U. S.

Geological Survey by Dawdy, Schaake and Alley (1978) was applied, to which several modifications and adjustments for obtaining better results were made. The model has four components: component of soil water content, component of excess precipitation, component of translation of surface runoff, and component of optimization of parameters.

Three representative rain gauges at each zone of the basin may be used for running the model. Because of this characteristic of the computer model, both the historical rainfall records of individual stations applied when calibrating the model, and the estimated average spatial rainfall on the various zones of the watershed corresponding to a synthetic storm, are processed in such a way that they correspond to idealized pluviograms of each of the three basic recording stations.

Large historical floods and the corresponding storms which produced them were selected for the calibration of the model. The values of the optimized infiltration and soil moisture parameters are shown in table 2. Symbols for the parameters are the same used by Dawdy, Schaake and Alley (1978).

TABLE 2 - Parameters of infiltration and soil moisture equations

Parameters	Salvajina-Juanchito	Juanchito-Anacaro	Anacaro-La Virginia	La Virginia-La Pintada	La Pintada Cañafisto
EVC	0.8	0.8	0.8	0.8	0.8
RR	0.7	0.7	0.7	0.7	0.7
DRN (cm/day)	0.6	0.6	0.15	0.17	0.6
PSP (cm)	19.40	25.15	25.00	29.93	25.10
KSAT (cm/s)	0.15×10^{-4}	0.17×10^{-4}	0.74×10^{-4}	0.14×10^{-4}	0.39×10^{-4}
RGF	0.07	3.82	3.05	4.91	19.68
BMSN (cm)	3.15	7.80	9.59	4.24	16.96

8. DESIGN FLOOD

Once the rainfall-runoff model was calibrated, the spillway design flood for Cañafisto dam was computed by introducing the exceptional 3-day storm into the model. For application of the model, the following concurrent events associated with the exceptional storm were supposed to take place.

The 3-day exceptional storm precipitation of 243 mm on the zone from the headwaters to Salvajina would produce a peak flow of 2750 m3/s at this site. The flood was routed through the Salvajina reservoir and spillway, based on the normal operational procedures for the project. The spillway discharge hydrograph with a peak of 2810 m3/s was used as input to the model.

The storm was assumed to occur during a very rainy period. In order to and for the sole purpose of estimating the soil-moisture previous to the design storm, a rainfall equal to the maximum recorded on the basin in a two month period was assumed.

Base flows at each zone were assumed as the two-month average flow recorded at the outlet of the zone coinciding with the highest historical

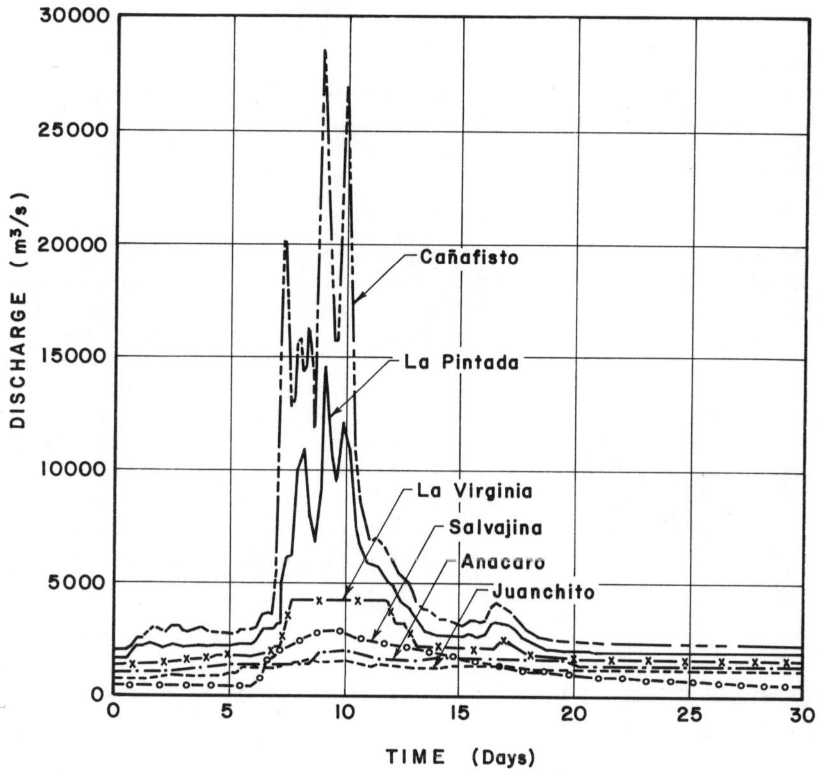

FIGURE 5. Cañafisto design flood

two-month average flow recorded at the La Pintada station.

Translation of the flood along the Cañafisto reservoir was assumed instantaneous.

The model was run using 6-hour time intervals. The peak flow at Cañafisto was computed to be 28500 m3/s.

9. CONCLUSIONS

This paper shows an application of the hydrometeorological approach for estimating an exceptional storm and the corresponding flood for purposes of designing the spillway of an important dam in a large equatorial basin. The availability of satellite images has helped determine inflows of humidity into the watershed during large storms and has provided additional confidence in the computation of synthetic extreme rainfalls.

The combination of the maximum values of several meteorological variables recorded during a 15 to 40 years span, plus the assumption of repetition of the physical processes associated with previous intense storms, led to the estimation of a flood of very low frequency, probably in the range of 10^{-7} as considered by Newton and Cripe (1973). A flood of this recurrence interval is reasonable for designing the spillway of such a high dam as Cañafisto, which useful life could conceivably be very long.

It would be convenient to continue research on the probability analysis of the various phenomena related to exceptional storms, which eventually would permit a more reliable estimate of the frequency of the corresponding floods.

10. REFERENCES

Dawdy, D.R., Schaake, J.C., and Alley, W.M., 1978, "User Guide for Distributed Routing Rainfall Runoff Model", U.S. Geological Survey. Water Resources Investigations 78-90, NSTL Station, Mississippi.

Hershfield, D.M., 1961, "Estimating the Probable Maximum Precipitation", Proc. ASCE, Journal of the Hydraulics Division, Vol. 88, No. HY6, September.

Leclerc, G., and Schaake, J.C., Jr. 1973, "Methodology for Assesing the Potentical Impact of Urban Development on Urban Runoff and the Relative Efficiency of Runoff Control Alternatives", Ralph M. Parsons Laboratory Report. No. 167, Massachusetts Institute of Technology.

Newton, D.W., and Cripe N.W., 1973, "Flood Studies for Safety of TVA Nuclear Plants, Hydrologic Embankment and Breaching Analysis". Tennessee Valley Authority, Knoxville, Tenn.

Rosembrock, H.H., 1960, "An Automatic Method of Finding the Greatest or Least Value of a Function", Computer Journal, Vol. 3, pp. 175-184.

THE EXPERIENCES OF DESIGN FLOOD OF RESERVOIRS IN MEDIUM AND SMALL BASINS BY THE METHOD OF DESIGN STORM

Hua Shiqian
The Nanjing Research Institute of Hydrology and Water Resources
Nanjing, China

ABSTRACT. This paper is based on the experiences accumulated from the actual design flood of more than eighty thousand reservoirs on main channel of medium and small basins. Most of these dams are earth dam, so that it is not permissible the water to flow over the dam. In China, the dam break is rarely due to the deviation error of design flood. Therefore, as a reference these experiences may be useful in actual design work. China is a developing country, most of the gaging stations were established from fifties, but, part of the raingage stations were established in early thirties, this is the reason why we like to use the design storm and then to transformed it into design flood.

This paper will introduce the method of design storm (including it's spatial and temporal distribution) and the method of transformation of design storm into design flood hydrograph. The conclusion of this paper is to introduce the crystalized experiences from many and many actual flood design works in medium and small basins. I'd very happy if something might be useful to the designers in other developing countries.

1. INTRODUCTION

In China, more than eighty thousand of reservoirs were established on the main channel of medium size (drainage area from 1,000 to 10,000 KM^2) and small size (drainage area less than 1,000 KM^2) basins. Most of them have no gaging stations at the outlet of the basin. A few gaging stations have a very short time of flood records in comparison with the raingage stations within these basin. Therefore, the design storm was generally used to derive the design flood by means of the rainfall-rainoff relation and unitgraph. In China, we call it the "indirect method" for design flood. For small basin, the spatial distribution of storm can be neglected, in accordance with the unitgraph of lumped system, the design point storm is generally expressed as the average rainfall depth of the basin. For medium basin, if there are several raingage stations within the basin, the longest raingage stations record is available to analyse the frequency curve of storms, because the statistical parameters of a long term storm series is more stable than that of a

V. P. Singh (ed.), Application of Frequency and Risk in Water Resources, 91–100.

short-term series. But, the spatial distribution must be considered, so that the method of computing spatial distribution will be introduced below.

The problem worthy of discussing in our pratice is how to prove the assumption that a certain frequency design storm will produce the design flood of same frequency. Surely, it is pretty well disproved. But, it was confirmed in our practice, though there may be some difference between the frequency of rainfall and flood, the deviation for a small basin would be very small if the soil moisture condition is considered reasonablely. The deviation for a medium basin could be controlled to a certain extent if the spatial distribution of rainfall within the basin is considered reasonably. Therefore, as an approach we can use the point rainfall to instead the average rainfall in small basin. But as for a medium basin, it is more complicate to design the spatial distribution of design storm, which will be described in more detail later.

The merits of the "indirect method" are not only more suitable for frequency analysis, but also there is little influence on it due to human activities.

2. THE DESIGN STORM

2.1. For small basin

In order that the spatial distribution of storm can be neglected, so that the parameters for frequency analysis (i.e. the mean, standard deviation and skewness) of a single gage station (point rainfall) is easy to be obtained. Therefore, the parameters of single gage stations for 1-hour, 3-hour, 6-hour and 24 hour duration were calculated all over the nation during the past years, now we have drawn isoline maps of $\bar{x}$ and C_V for different durations of point rainfall. (Fig. 1 and Fig. 2 is the mean and standard deviation C_V isoline maps of 6-hour) For small basin, we can readily obtain the frequency curve for different durations as shown in Fig. 3 , if a designed return period of 100 years is selected, we can easy to get the point rainfall of equal frequency of different durations.

The designed temporal distribution of a design point storm rainfall is determined by successive control of the design storm rainfall in different durations. A method of so called "envelope of equal frequency" is widely used for the temportal distribution of a certain frequency of design storm of the small basin. For example, to design a temporal pattern of 24 hours, it's maximum rainfall depth of 1-hours, 6-hours must be equal to the value obtained from 1 hour and 6-hour duration frequency curve as shown in Fig. 3, a designed storm temporal pattern is arranged for safety of the dam as shown in Fig. 4. Since the shortest duration of design storm (highest intensity of storm) located at the end of storm, which may cause the most severe flood at the outlet of the basin, it is designed on safe side.

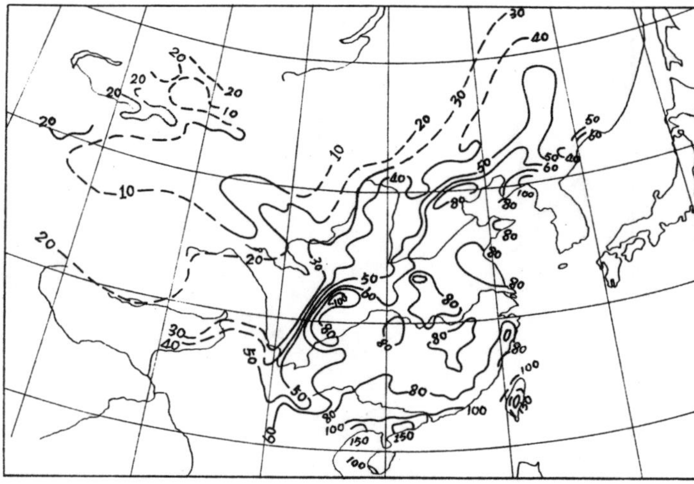

Fig. 1. Mean of maximum 6 hour point rainfall (m.m) of China

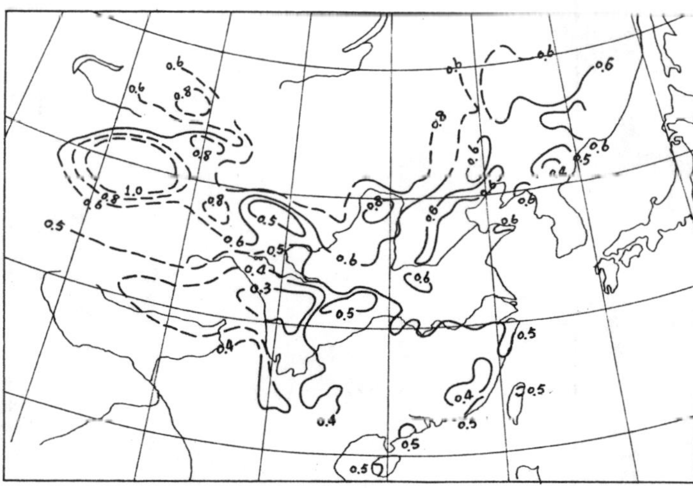

Fig. 2. C_V of maximum 6 hour point rainfall of Chian

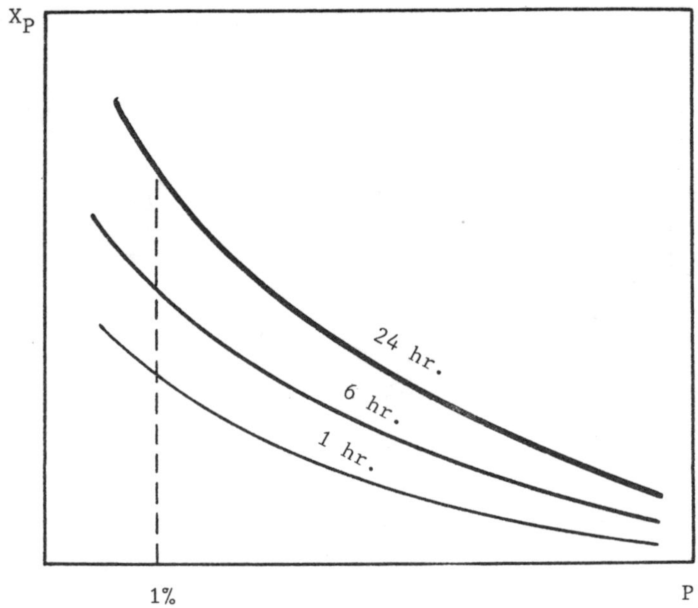

Fig. 3 Frequency curve of different durations for a single station

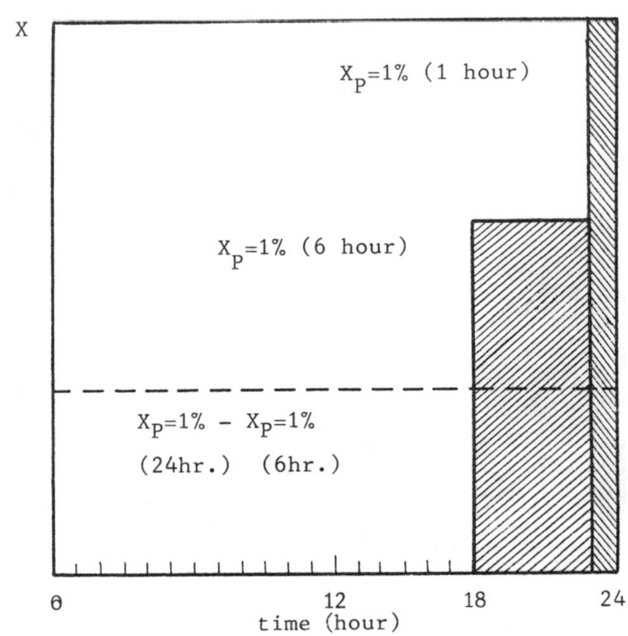

Fig. 4 Most severe time distribution of storm pattern

2.2. For medium basin

In medium basin the spatial distribution of rainfall can not be neglected. In general, the depth-area-duration curve is used. Base on 1,000 storm events, we had analyzed and synthesized some regional depth-area-duration curve over most areas of China, use this curve to determine the spatial and temporal distribution of design storm. If the length of rainfall records of raingage stations within a medium basin are approaching, we can use the longest rainfall series to prolong the shorter time series by interpolation or extrapolation. Then, it is possible to obtain a time series of average rainfall of different durations of the basin, it is also easy to obtained the frequency curve of average rainfall of different duration of the basin. But, in many cases, such as a longest raingage station have 50 years records, while other stations within the basin have only 10 years or less records, in such case, though the coefficient of correlation between longest and shorter station is relatively high, the error of interpolation and extrapolation may be still very large, therefore, we rather to use the relationship between the rainfall statistical parameters at a fixed point and the parameters of areal rainfall for fixed area which have define boundaries has been developed. It is so called the "fixed point-fixed area relationship", which can be transformed into "area reduction factor-area" relationship as shown in Fig. 5.

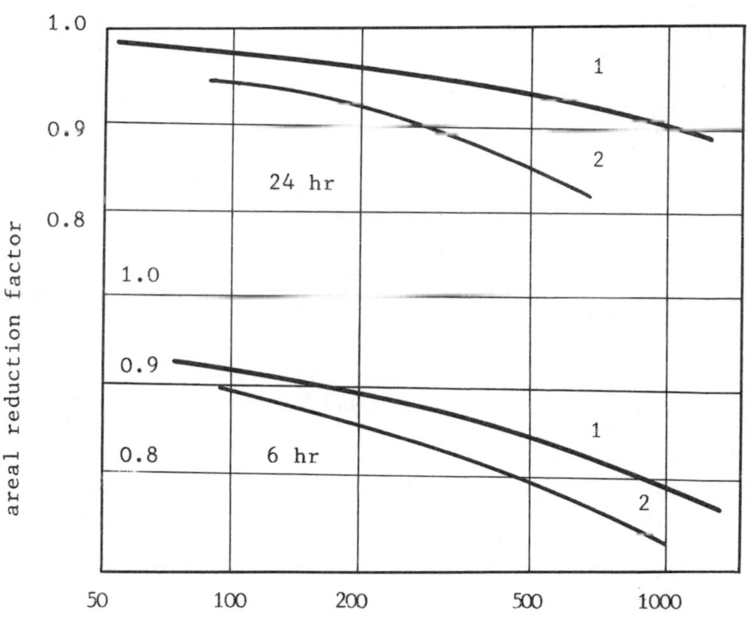

Fig. 5. Areal reduction factor for fixed point-fixed area (1 South China, 2 North West China)

From the statistical point of view, the former method may be more suitable for areal storm design. The design storm of any duration can be computed by maps of different duration and exponential formula.

If the density of raingage stations within the basin is enough to draw the isoline maps of rainfall of any storm event, we can develop a relationship between the average rainfall depth enclosed by an isohyet and the area enclosed, it is so called "moving point-moving area relationship".

The location of the point (center of storm) and the area enclosed relationship may vary for each storm, therefore, a large amount of heavy storms (more than thousands storm events) have been analysed and regionized, and the relationship of a reduction factor (i.e. a ratio of areal rainfall to point rainfall) and area is also given for most provinces in China.

The temporal pattern of the design storm (i.e. the rainfall depth arrangement of different durations) is subject to the same frequency and is often simplified by the hyetograph of a selected model times a coefficient of ratio to enlarge the model storm to design storm. To avoid the false, this selected model storm should be of an extensive rainfall event both in depth and intensity, that is being close to the magnitude of design storm. The temporal pattern should be so designed that the derived design hydrograph should be unfavourable for the operation of the project, this selection is made for the safety of the dam.

If the area of the design basin is relatively large, such as larger than 10,000 KM^2 , we still hope to find out the design storm of the basin, it is more complicate problem than the medium basin. Obviously, the spatial distribution of storm may be more uneven, so we must divide the total drainage basin (as a whole system) into several large tributaries (as the subsystem), then calculate the frequency curve of each subsystem separately, afterwards calculate a set of frequency curve of two, or three subsystems of different combinations, that is to say, we'll have a set of design storm of different storm center, if a combination cause the most severe flood hydrograph to the project, it may be the selected design storm of the drainage basin large than 10,000 KM^2. For illustration, an example is given from the planning of flood prevention of Yi-shu river basin, Si river basin and a large tributary of Pi-Chuang of the Huai drainage basin. The total drainage area (A) is about 50,000 KM^2 . (B)(C) and (D) are the subsystem of Yi-Shu river, Pi-Chuang tributary and Si river respectively. A simplified method far considering the spatial distribution of the system, the practical design procedures are introduced as follows:

(1) Take the annual maximum storm series of B, C, D, and A(equal to B+C+D) separately, calculate the frequency curve of storm depth of a certain duration (such as 5 day), then calculate the frequency curve of A, B, C, D,(B+C), and (C+D) independently.

(2) If the selected return period is 100 years and 5 days duration, we can find out it by the following calculations. By considering the different combinations of three subsystems, first we assume B+C have the same frequency of B+C+D, (i.e. the storm center is located either in Yi-Shu river or in Pi-Chuang fributary). Therefore the difference of $(B+C+D)_{1\%} - (B+C)_{1\%}$ is the coordinated design storm, of $D_{(B+C)1\%}$

Then, (a) if we assume the storm center is located on Yi-Shu River basin, then B will have the equal frequency of 1%, the difference of $(B+C)_{1\%}-(B)_{1\%}$ is the coordinated design storm depth of $C_{(B),1\%}$ (Pi-Chuang tributary). (b) If we assume C will have the equal frequency of 1%, the difference of $(B+C)_{1\%} - C_{1\%}$) is the coordinated design storm $B_{C(1\%)}$. In the same manner, assume D will have the same frequency of A (B+C+D), the difference of $(C+D)_{(1\%)}, -(D)_{(1\%)}$ is the coordinated design storm of $C_{(D),1\%}$ and $(B+C+D)_{1\%} - (C+D)_{1\%}$ will be the coordinated design storm depth of $B_{(C+D)1\%}$,

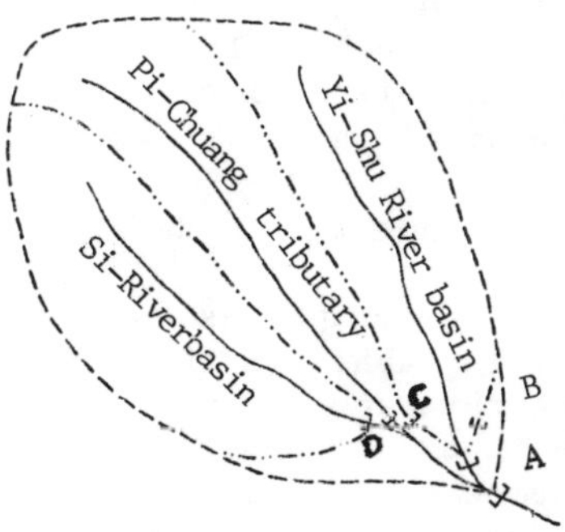

Fig. 6. The system of A and sub-system B, C, and D.

The design storm is determined that which combination may cause the most severe flood hydrograph at the project A.

3. THE FLOOD HYDROGRAPH OF THE DESIGN STORM

There are two steps to transform the design storm rainfall into design flood hydrograph.

3.1. To transform the design storm rainfall into design runoff

In the early fifties, the regional synthesized API model

$$R=f(P\ ,\ P_a\ ,\ T_e\ ,\ M\ \ldots\ldots) \qquad (1)$$

is general used to estimate the design runoff.

where

R —— Runoff in m.m.

P —— Rainfall in m.m.

P_a —— antecedent precipitation Index in m.m., it denotes the initial condition of soil moisture of the basin

T_e —— effective duration of rainfall, it indicates the factor of average intensity of storm rainfall.

M —— month, shows the factor of evapo-transpiration of the basin.

this model is a graphical representation of a co-axial correlation of parameters, in non-linear relations. In most cases, above relation can be simplified as the form

$$R = f\ (p + p_a) \tag{2}$$

the error of computation by this method is certificated always less than 20% in humid and semi-arid regions.

Afterwards, other methods such as the Φ - index method and the method of regional synthesized infiltration curve are also widely used in design flood of medium and small basins.

3.2. To transform the design runoff into design flood

In fifties, L. K. Sherman's unitgraph was commonly used to transform the storm runoff into flood hydrograph. In the begining of sixties, the J. E. Nash's instantaneous unithydraph (IUH) was widely used in China. A little promotion was made in actual design work. It is discovered that J. E. Nash's model is assumed to be a linear system, so that IUH's parameters will not change under the different rainfall intensity, therefore

$$t_p = (n - 1)\ K \tag{3}$$

where t_p- time to peak. But, the actual data tell us that the nonlinearity is obviously existed within the limit of i_f to i_L , and when i increases continously to exceed the threshold i_L , the IUH turns to a linear system again as shown in Fig.7.

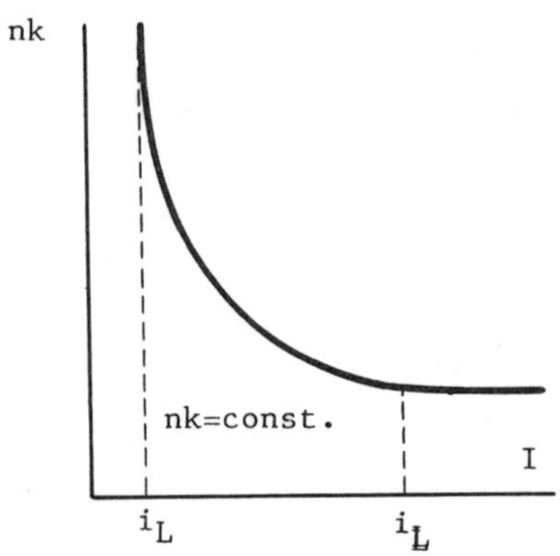

Fig. 7 The relationship of i and nk (or t_p)

This phenomena can be proved mathematically. According to the law of of unsteady flow in hydraulics following equations can be derived to illustrate it.

$$\begin{aligned} I-Q &= \frac{ds}{dt} \\ S &= KQ^b \end{aligned} \tag{4}$$

Putting imput I(t) into inpulse function $I\,\delta(t)$, then the response of this nonlinear system would be U(t), then,

$$I\ \delta(t) - U(t) = \frac{ds}{dt}$$
$$S = K U^{b}(t) \qquad (5)$$

It can be derived mathematically that above equations would be equivalent to the following equations:

$$-u(t) = \frac{d K U^{b}(t)}{dt}$$
$$u(+0) = \frac{S(+0)^{1/b}}{K} = \frac{I^{1/b}}{K} \qquad (6)$$

By solving the equations, we can obtain the non-linear IUH.

$$u(t) = \left\{ \frac{1-b}{b K I^{b-1}} \cdot t + \frac{1}{K^{(b-1)/b}\ I^{(b-1)^2/b}} \right\}^{\frac{1}{b-1}} \qquad (7)$$

according to the definition of moment, the first moment above origin is

$$m_1 = \frac{\int_0^{\infty} t \cdot u(t)\, dt}{\int_0^{\infty} u(t)\, dt} \qquad (8)$$

Considering the constrainted condition $\frac{2}{3} < b < 1$, the relationship between m and rainfall intensity I is obtained

$$m_1 = \frac{b K^{\frac{1}{b}}}{2b-1} \cdot I^{\frac{b-1}{b}} \qquad \text{or} \qquad (9)$$

$$m_1 = \alpha \cdot I^{\beta} \qquad (10)$$

from equation (10), if b=1, then $\beta = 0$, m_1 =constant. that is to say, in this case, m_1 do not depend upon the rainfall intensity; while $b \neq 1$, m_1 (or t_p) will be vary with the rainfall intensity I.

A general rule can be described on the basis of above analysis that the nonlinearity of unitgraph must be considered for lower frequency of design flood, in general,(the return period is less than 100 years). As for the high frequency design flood (more than 100 years), the linear system of unitgraph is more suitable than the non-linear system unitgraph.

4. CONCLUSION

4.1. The spatial and temporal distribution of design storm is much complicate for their uncertainties, so that the empirical method is prefered to solve the problem. By use of a large amount of storm events in each climate region of China, the synthetic analysis and regionalization of storm pattern is proved more effective and reliable for the safe of dam.

4.2. The type of frequency curve is chosen only by data fitting. We can't assess which type of frequency curve is better because the sampling series is too short in comparison with the population of storm events.

4.3. Does the transformed flood hydrograph have the same frequency as the design storm? it is still disproved. In China, we have studied this problem in past years. We have choicing a number of small and medium size basins which have long record raingage stations unitformly distributed within the basin and a long record stream gaging station at the outlet of the basin. We calculate the flood hydrograph by both "indirect" and "direct" method, then compare the result obtained by both methods. Afterwards, we have discovered that the deviation increases with increasing of drainage area; and that the deviation could be neglected if the drainage area is less than 1000 square kilometers.

4.4 The magnitude of design storm depth is always large, the error coming from computing runoff is relatively small, that is the reason why we like to use more simple method to compute the design runoff.

4.5. During the threat of design storm of high frequency, the time-to-peak of flood tends to be a linear system, so that it is more suitable to use the unitgraph of linear system to compute the design flood hydrograph.

4.6. If the area of drainage basin is large than 10,000 KM^2, better to divide the whole system to subsystems, and use relatively empirical method as mentioned above, it can get an acceptable result in general condition.

5. REFERENCES

[1] Hua Shiqian etc. " A general summary of dam break on medium and small basin of Huai River during the attack of extraordinary storm flood in Aug. 1975 of Huai River" Special Bulletin of the Commission of Harnessing Huai River, Ministry of Water Resources and Electric Power, (WRAEP) Dec. 1975.

[2] Wang Zulie, Hua Shiqian etc. " Revised water resources planning of Huai River Basin, Part I. Flood prevention planning" Special Bulletin of the Commission of Harnessing Huai River, Ministry of WRAEP. March 1976.

[3] Hua Shiqian, "The method of computation of runoff and flood hydrograph under the design storm of PMP" " Hydrology " No. 12, 1983.

ESTIMATION OF DESIGN FLOOD

Deep Kumar Gupta,
Executive Engr.,
I& P(WB)Dn. II,
IDUP,Roorkee,UP
INDIA

B.N.Asthana,
Executive Engr.,
I& P(WR)Unit VI,
IDUP,Lucknow,UP
INDIA

A.N.Bhargawa,
R.O., IRI,
IDUP,Roorkee,
INDIA

ABSTRACT. A new class of methods such as SMEMAX (Bethlahmy, 1977), Power Transformation (Chander et al., 1978), and Modified SMEMAX (Rasheed et al., 1982), has recently been developed for estimation of design flood using the approach of transforming the skewed distribution of observed data into normal distribution. It has been demonstrated that by these methods skewness practically **in all** cases reduces to a value very near to zero, but in many cases Kurtosis does not reduce to 3. To account for the effect of this residual Kurtosis,Chander et al. (1978) have recommended the use of a correction factor. These methods thus do not truly transform the skewed data into normal distribution. The complete transformation can be achieved by the proposed transformation called Two Step Power Transformation (TSPT), described in detail in the paper and it is demonstrated through its application in large number of cases that it is capable of converting flood data of any degrees of skewness and kurtosis to a normal distribution. A comparison of the results of TSPT with those obtained by other methods has been made.

1. INTRODUCTION

A designer of river structures, such as bridges, barrages, river training works etc, is generally interested in the estimation of design flood peak. Each of such structure is designed for flood peaks corresponding to different return periods such as 5,10,100 years or more depending upon the operational requirements and the importance of the structure. A large number of flood frequency methods are in vogue for the purpose of estimation of flood peaks of various return periods. These methods are based on standard probability distribution, log Pearson type III

V. P. Singh (ed.), Application of Frequency and Risk in Water Resources, 101–111.

distribution, etc. The assumption is that the annual maximum flood discharge at a certain location follows one of the standard probability distribution functions (PDF). This can be worked out by plotting the cummulative distribution function (CDF) of the observed flood peaks over a period of years on various probability papers. This trial procedure is difficult and time consuming. Furthermore the accuracy depends first on the availability of a large number of years of observations and secondly on the luck of finding a standard PDF close to the sample CDF. It is therefore, rarely applied in individual design problems. The Water Resources Council (1967) conducted a comparative study of different flood frequency methods at different locations in U.S.A. and recommended the use of Log Pearson type III distribution for analysing flood frequency unless the data warrant use of some other distribution. Another such study has been carried out by Natural Environmental Research Council (1975) for locations in U.K. concluding that three parameter distributions such as Log Pearson type III distribution fits well.

No such exhaustive and conclusive study has been made for Indian Conditions. As a sample study the CDFs of the observed flood peaks at five locations are shown with standard PDFs such as normal, log normal and Poissons in Fig. I. In all these cases it is seen that the CDF does not follow the normal distribution and is skewed to the left. In a few cases only it is found (on the basis of Chi Square test) some what close to the Poisson's distribution. Thus the selection of a reasonably reliable method of flood frequency analysis for a particular location is very difficult. The practice is, therefore to work out floods of different return periods by various methods of flood frequency analysis and the designer uses his judgement to decide the design flood.

It has been generally recognized by the researchers in this field that none of the common probability distributions can describe the true inherent variability of flood peak occurrences. It is believed that a probability model fitted through the observed data could provide a realistic estimate of the design flood of a given return period. But this is computationally a difficult task. Thus the use of common probability distribution models is in vogue because of ease of computations. However, there are continuous efforts by the researchers to reduce the element of luck in finding a standard probability distribution for a particular location.

A promising technique developed recently is to transform the observed data through a suitable transformation function such that the transformed series follows a particular distribution. Attempts have been made to transform the observed data to follow normal distribution which is

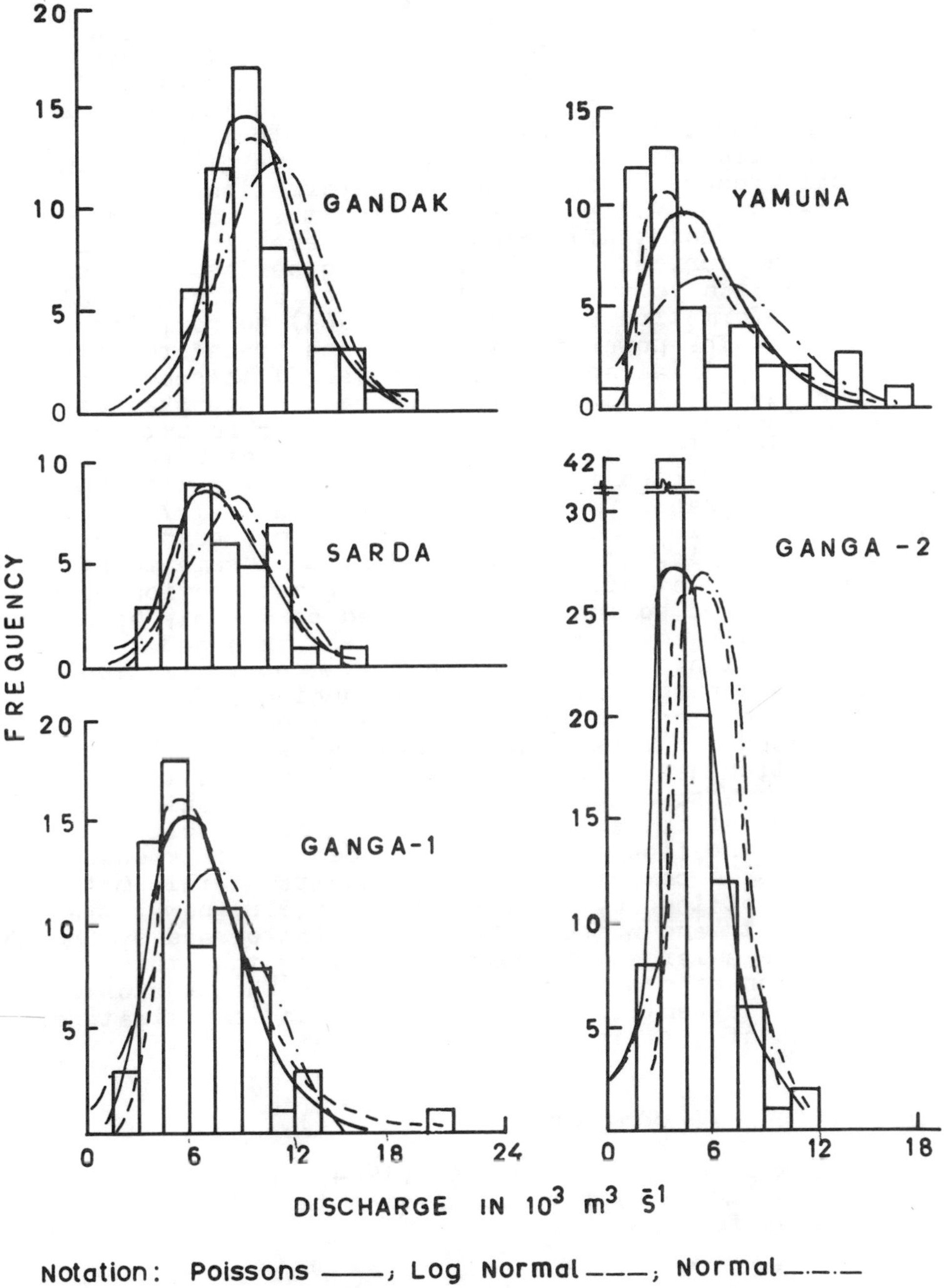

Fig. 1– Showing CDF with standard PDF

wholly defined by the mean and standard deviation. Bethlahmy (1977) has suggested SMEMAX transformation to normalize skewed data. This uses the smallest, median and the largest value of the given data, and makes the difference between smallest and median equal to that between median and the largest value. It is simple as well as graphical also. Rasheed et al. (1982) have given a simpler graphical procedure for SMEMAX transformation. But it is found that with SMEMAX transformation the resulting series are generally not truly normalised. There remains some skewness and kurtosis in the resulting series. Chander et al. (1978) have suggested the use of a more general power transformation (Box and Cox, 1964) to normalize the skewed data. The power function in this transformation is worked out by trial and error process. It is seen that with the use of this power transformation the skewness of the transformed series generally becomes zero but coefficient of kurtosis (the fourth moment) is not zero. Thus the resulting series is not truly normal even with the use of power transformation. Chander et al. (1978) have suggested the use of correction factor to account for the residual kurtosis. It is thus evident that some modification in Chander et al. (1978) power transformation procedure is required so that the observed series may be truly normalized with coefficient of skewness zero and kurtosis value of 3. It has been found that true normalization of skewed data can be achieved by introducing a further transformation in Chander et al. (1978) power transformation procedure. This is described as Two Step Power Transformation Method (TSPT) described in detail in this paper. It is also demonstrated through application in large number of cases that this method is capable of normalizing the flood series of any degree of skewness and kurtosis. A comparison of the results of this method with those obtained by Chander et al. (1978) power transformation with and without correction factor and by SMEMAX methods show a large variation indicating that true normalisation of observed data specially when kurtosis is large, is essential to obtain a realistic estimate of flood peaks of certain return periods.

2. TWO STEP POWER TRANSFORMATION (TSPT)

The Box-Cox power transformation (1964) suggested by Chander et al. (1978) for normalising the skewed flood series is as follows:

$$y_i = \frac{x_i^{\omega} - 1}{\omega} \quad , \quad \omega \neq 0 \qquad (1)$$

where: x_i = ith element of original series variate; y_i = ith element of power transformed series variate; and ω = parameter of power transformation.

With suitable value of ω the skewness of the observed data is reduced to zero or nearly zero and thus the histogram becomes symmetrical about the vertical axis passing through the mean. This transformed histogram may be truly normal only when kurtosis also becomes equal to 3. When kurtosisis not equal to 3 the histogram may be either peaky (Leptokurtic) or flat (Platykurtic). The correction for kurtosis is envisaged through another transformation in this two step approach. The two limbs of the transformed histogram obtained after Box Cox power transformation in the first step are equally stretched or compacted according to the following transformation:

$$t_i = (|y_i - \bar{y}|)^\gamma \text{ , } \gamma \text{ is positive and } t_i \text{ is given same sign as } (y_i - \bar{y}). \tag{2}$$

where: t_i = ith element of TSPT transformed series; $\bar{y}$ = mean of power transformed series; and γ = parameter of TSPT.

With suitable value of γ the condition of kurtosis equal to 3 may be achieved. In the above transformation it is evident that when $\gamma \rightarrow 0$, the kurtosis of the transformed 't' - series tends to be one and when $\gamma \rightarrow \infty$ the kurtosis of the transformed 't' - series tends to be infinite. For $\gamma = 1$ the kurtosis of 't' - series will be same as that of 'y' - series (obtained after Box-Cox power transformation). Therefore, if kurtosis of 'y' series is more than 3, γ will be between 1 and 0 and if kurtosis is less than 3, γ will be more than one. With the help of any simple iterative algorithm values of ω and γ are obtained such that both the skewness and kurtosis of the transformed series tends to be zero and three respectively. When a TSPT transformed 't'- series with normal distribution is obtained, the flood peak of a particular return period may be calculated by the following equation:

$$t_p = \bar{t} + k_p \cdot \sigma_t \tag{3}$$

where: t_p = value corresponding to pth return period in power transformed series; k_p = standard normal deviate corresponding to pth return period, and σ_t = standard deviation of TSPT transformed series.

The value corresponding to 't_p' in the original flood data series 'x_p' may be obtained by using the following equations:

$$x_p = (y_p \cdot \omega + 1)^{1/\omega} \tag{4}$$

where $y_p = y'_p + \bar{y}$ (5)

and $y'_p = (|t_p|)^{1/\gamma}$ (y'_p is given same sign as that of t_p) (6)

where: x_p = value corresponding to pth return period in original series; y_p = value corresponding to pth return period in power transformed series; and y'_p = intermediate dummy variable.

3. EFFICIENCY OF TSPT

Annual flood discharge data from 17 locations on Indian rivers were analysed to examine relative efficiency of SMEMAX, power transformation and the proposed TSPT methods in transforming the original data into normally distributed series. The results are given in Table I. It can be seen that SMEMAX reduces the skewness and kurtosis but there is no definite pattern of reduction. The values of skewness and kurtosis in the transformed series are quite high, and so the transformed series cannot be considered as a normally distributed one. In case of power transformation the skewness could be reduced to any desired degree of nearness to zero, but there was no definite pattern of reduction in kurtosis. The reduction in kurtosis was nearly of same order as that in SMEMAX. Thus the power transformed series also cannot be taken as truly normalized. In the proposed TSPT method both the skewness and kurtosis in all cases could be brought as near to the value of zero and three respectively as desired. In this method the value of ω obtained in power transformation was used as initial trial value and thereafter it was subsequently modified alongwith determination of γ iteratively. The value of ω was found slightly changing with γ because in majority of the cases 'y' - series was not truly symmetrical about the mean. The results show that TSPT can transform a flood series of any skewness and kurtosis into a truly normal distribution series.

4. ESTIMATION OF FLOODS OF DIFFERENT RETURN PERIODS

For all those cases given in Table I, the floods of different return periods were worked out by SMEMAX

TABLE - 1: Effect of Different Transformations on Coefficient of Skewness and Kurtosis

Sl. No.	Name of river	Length of flood data	Original Data		SMEMAX Transformed Data		Box-Cox Power Transformed data			TSPT Transformed data			
			Skewness	Kurtosis	Skewness	Kurtosis	OMEGA ω	Skewness	Kurtosis	OMEGA ω	GAMMA γ	Skewness	Kurtosis
1	2	3	4	5	6	7	8	9	10	11	12	13	14
1.	Ganga-1	68	1.64444	8.13113	-0.24566	3.24616	-0.082	-0.00010	3.06283	-0.082	0.985	-0.00014	2.99948
2.	Ganga-2	91	1.46564	5.61081	0.27466	3.65421	-0.299	0.00002	3.69413	-0.378	0.865	0.00047	2.99737
3.	Yamuna-1	55	1.67747	6.20009	-0.04422	2.86618	-0.188	-0.00040	2.73710	-0.176	1.086	0.00022	2.99862
4.	Yamuna-2	55	1.10431	4.77513	-0.03581	3.56419	0.277	-0.00073	3.64076	0.240	0.845	0.00029	3.00048
5.	Sarda	39	0.57468	3.28568	-0.11713	2.76251	0.278	0.00014	2.75276	0.279	1.090	0.00031	3.00106
6.	Gandak	58	0.91601	3.95175	-0.01148	2.94857	-0.414	0.00006	2.89777	-0.409	1.034	-0.00012	3.00042
7.	Sutlej-1	40	1.63574	6.37060	-0.16686	3.98283	-0.102	-0.00059	3.74065	-0.157	0.797	-0.00040	2.99875
8.	Sutlej-2	47	3.02419	14.45643	0.33020	6.12955	0.007	0.00028	6.91231	0.017	0.609	-0.00038	2.99838
9.	Sutlej-3	40	3.51497	17.14572	0.07906	5.37969	-0.507	0.00021	5.03060	-0.470	0.680	-0.00008	2.99944
10.	Rapti	54	2.18011	12.52728	-0.31801	3.16572	0.165	0.00076	3.62238	0.243	0.852	-0.00064	2.99929
11.	Betwa	39	1.56208	5.37728	0.09947	4.21584	-0.175	0.00078	3.76638	-0.235	0.779	0.00021	3.00144
12.	Ken	26	1.19678	6.67188	-0.23456	4.12383	0.343	0.00020	4.35026	0.508	0.657	0.00010	3.00019
13.	Panner	52	1.61711	4.93066	0.26839	2.60405	0.221	0.00014	2.73918	0.024	1.112	-0.00086	2.99937
14.	Mahanadi	32	-0.36669	2.58326	-0.12509	2.36111	1.677	0.00000	2.23814	1.930	1.771	-0.00023	3.00038
15.	Krishna	65	1.36180	5.06205	0.33195	3.23677	-0.624	0.00005	3.33545	-0.641	0.925	0.00034	3.00120
16.	Godawari	53	2.13643	9.52919	-0.24192	3.64149	-0.527	-0.00015	3.56931	-0.475	0.829	0.00031	2.99976
17.	Baitarni	84	1.48926	4.10403	0.28952	3.02049	0.105	0.00056	3.45285	0.069	0.889	0.00025	2.99836

transformation, power, power transformation (neglecting kurtosis), power transformation (adjusted for kurtosis) and TSPT methods. These are compared in Table II. A large variation in values obtained by different methods can be observed from the Table II. The values obtained by TSPT method in which the observed flood series could be truly normalised, are found much different from those obtained by power transformation after adjusting for kurtosis especially for higher return periods and in all those cases in which kurtosis after power transformation was far from 3 on either side. The results also show that by power transformation the assessment of flood peaks may be both either over or under estimated. If the kurtosis of the series after power transformation is more that 3 the estimated flood will be underestimated and if less than 3 the estimated flood value will be overestimated.

5. REVIEW OF RESULTS

A review of the results of the analysis of annual flood peak data at 17 locations on Indian rivers given in Table I and II have shown that:

1. the TSPT method is more efficient than SMEMAX and power transformation in normalising the skewed flood distribution,
2. the use of correction factor for incorporating the effect of kurtosis in power transformation method is not adequate to estimate floods of higher return periods (100 years or above) and in those cases where the kurtosis after transformation is far from 3 on either side,
3. the estimated values of floods of small return periods by all the three methods are quite close.

6. RECOMMENDATIONS

Since the proposed TSPT procedure has shown the capability of normalizing efficiently any kind of skewed flood distribution, its use is recommended for estimation of design flood for any major and important river structure. The design floods for purpose of project formulation and of small return periods for designing temporary or less important structures may be worked out by any of the three methods of transformation, but SMEMAX may be preferred due to ease of computation.

TABLE II - Estimated Floods of Different Return Periods in $m^3 s^{-1}$

S. No.	Name of River	Method	Estimated floods ($m^3 s^{-1}$) corresponding to return period (years) 4	10	20	40	100	200	1000
1	2	3	4	5	6	7	8	9	10
1.	Ganga-1	SMEMAX	6441	9125	10710	12118	13746	14846	17090
		PT	6347	8292	9736	11250	13324	14959	19007
		PTC	6320	8292	9736	11301	13448	15099	19371
		TSPT	6326	8286	9754	11302	13438	15130	19352
2.	Ganga-2	SMEMAX	5630	7156	8056	8857	9782	10407	11683
		PT	5414	6851	7936	9096	10729	12052	15483
		PTC	5294	6742	7970	9379	11552	13462	19217
		TSPT	5233	6792	8120	9676	12115	14318	21078
3.	Yamuna-1	SMEMAX	7690	11787	14205	16354	18839	20518	23943
		PT	6974	10515	13612	17312	23189	28522	44687
		PTC	7134	10664	13612	16789	21565	25798	37404
		TSPT	7175	10572	13356	16519	21263	25336	36711
4.	Yamuna-2	SMEMAX	3208	4180	4754	5263	5853	6251	7064
		PT	3129	4020	4624	5214	5962	6509	7737
		PTC	3057	3966	4642	5340	6277	7004	8771
		TSPT	2997	3968	4701	5465	6498	7297	9208
5.	Sarda	SMEMAX	9171	11326	12598	13729	15036	15920	17722
		PT	9123	11076	12361	13590	15119	16220	18644
		PTC	9222	11144	12361	13432	14778	15684	17712
		TSPT	9249	11104	12275	13368	14698	15637	17662
6.	Gandak	SMEMAX	11539	13756	15064	16226	17571	18479	20332
		PT	11374	13417	14873	16363	18362	19914	23709
		PTC	11404	13454	14873	16263	18187	19589	23050
		TSPT	11426	13433	14836	16252	18128	19566	23022
7.	Sutlej-1	SMEMAX	6691	9052	10445	11685	13117	14085	16060
		PT	6281	8305	9831	11449	13695	15487	19999
		PTC	6111	8156	9877	11846	14843	17435	24995
		TSPT	5895	8109	10088	12470	16270	19733	30294
8.	Sutlej-2	SMEMAX	9464	13248	15481	17466	19767	21312	24475
		PT	8878	11685	13739	15863	18728	20949	26323
		PTC			Correction factor not available				
		TSPT	7848	10970	14165	18348	25601	32686	56245
9.	Sutlej-3	SMEAX	9102	13880	16700	19207	22105	24063	28058
		PT	8236	10976	13291	16029	20405	24469	37695
		PTC	7753	10439	13290	17388	26441	38053	134600
		TSPT	7555	10512	13805	18752	29703	44586	164780
10.	Rapti	SMEMAX	3667	5684	6877	7938	9165	9994	11685
		PT	3680	4932	5824	6727	7913	8810	10989
		PTC	3584	4856	5851	6921	8417	9620	12702
		TSPT	3576	4882	5878	6927	8355	9466	12148
11.	Betwa	SMEMAX	7506	8501	9087	9609	10211	10619	11449
		PT	8755	11755	14091	16637	20296	23287	31192
		PTC	8552	11524	14163	17264	22273	26671	40896
		TSPT	8153	11439	14553	15508	25280	31929	54965

1	2	3	4	5	6	7	8	9	10
12.	Ken	SMEMAX	9632	12407	14056	15500	17164	18321	20641
		PT	9450	11764	13291	14752	16569	17876	20748
		PTC	9108	11498	13326	15254	17962	20024	25303
		TSPT	8825	11400	13464	15661	18654	20967	26467
13.	Pennar	SMEMAX	5359	8866	10937	12777	14904	16342	19275
		PT	4767	8149	11148	14703	20210	25026	38589
		PTC	4931	8292	11052	14203	18705	22393	32578
		TSPT	4997	8234	10895	13879	18242	21866	31403
14.	Mahanadi	SMEMAX	35209	39890	42710	45109	47948	49867	53780
		PT	35372	39712	42128	44200	46515	48037	51044
		PTC	36259	40153	41950	43304	44769	45634	47280
		TSPT	37298	39854	41061	42018	43022	43651	44835
15.	Krishna	SMEMAX	17329	21244	23554	25607	27982	29586	32858
		PT	16798	20545	23441	26618	31259	35192	45775
		PTC	16640	20412	23496	27024	32422	37288	52412
		TSPT	16587	20479	23658	27295	32887	37885	53233
16.	Godavari	SMEMAX	35717	48775	56481	63330	71250	76602	87518
		PT	34109	43142	50306	58323	70295	80651	110578
		PTC	33488	43127	50531	60076	75548	90426	142129
		TSPT	33027	42505	50872	61012	77623	93412	146771
17.	Baitarni	SMEMAX	4656	6174	7069	7865	8785	9407	10676
		PT	4484	6142	7220	8598	10540	12104	16067
		PTC	4395	5928	7121	8400	10179	11616	15277
		TSPT	4484	5999	7094	8214	9705	10845	13550

PT = Power Transformation, PTC = Power Transformation Corrected for Kurtosis and TSPT = Two Step Power Transformation.

7. REFERENCES

1. Bathlahmy, N. 1977. 'Flood analysis by SMEMAX Transformation'. Journal of the Hydraulics Division, ASCE, Vol. 103, No. Hy 1, Proc. paper 12701, 69-80.

2. Box, G.E.P. and D.R. Cox, 1964. 'An analysis of Transformation'. Journal of the Royal Statistical Society, Vol. B 26, 211-252.

3. Chander, S., S.K. Spolia and A. Kumar, 1978. 'Flood Frequency Analysis of Power Transformation'. Journal of the Hydraulics Division, ASCE, Vol. 104, No. HY 11, Proc. paper 14179, 1495-1503.

4. Rasheed, H.R., M.V. Ramamoorthy and A.S.Al-Dabbagh, 1982. 'Modified SMEMAX Transformation for Frequency Analysis'. Water Resources Bulletin, Vol. 18, No. 3, 509-515.

5. U.K. Natural Environmental Research Council, 1975. 'Flood Studies Report', London, England.

6. U.S. Water Resources Council, 1967. 'A Uniform Technique for Determining Flood Flow Frequencies'. Bulletin No. 15, Washington.

DESIGN FLOOD ESTIMATION FOR NARMADA SAGAR PROJECT USING PARTIAL DURATION SERIES - A CASE STUDY

N.K.Goel S.M.Seth Satish Chandra
National Institute of Hydrology,
Roorkee-247667
India

ABSTRACT. This paper highlights details of a study for estimation of flood magnitudes of different return periods for Narmada Sagar dam(India) using the partial duration series. The comparison of the efficiencies of annual and partial flood series has been given on the basis of exact theoretical approach and approximate theoretical approach. On the basis of exact theoretical approach it is seen that the partial flood series estimate of T year flood Q(T) (for any value of λ) always has a smaller sampling variance than that of the annual flood series for a return period T less than 11 years. For any return period the partial flood series estimate of Q(T) has a smaller sampling variance than annual flood series if the average number of peaks per year(λ) is at least 1.65. On the basis of approximate theoretical approach, it is observed that in the range of λ studied (1.0 to 2.437) the sampling variance of annual flood series is smaller than that of partial flood series.

1. INTRODUCTION

The basic requirement of flood frequency analysis is to estimate the probability for which a flow is equalled or exceeded during a stated design life of a particular project. The three approaches namely (i) time series approach (ii) partial duration series approach and (iii) the annual maximum series approach are generally used for frequency analysis. The first approach uses the whole flow hydrograph, the second uses only those peaks which exceed an arbitrary threshold and the third uses the highest peak in each year of record. The present study deals only with the annual maximum series and partial duration series approaches. A good description of time series models has been given by Hall and O'Connell (1972).

The classical dilemma in flood frequency analysis is whether to use annual maximum model or partial duration series model. The most frequent objection for the use of annual maximum model is regarding its use of only one largest flood for each year. Another shortcoming of annual flood series approach is that only a small

V. P. Singh (ed.), Application of Frequency and Risk in Water Resources, 113–122.

number of floods is considered. The estimate of higher moments like coefficient of skewness of historical flood series will not be reliable in case of annual flood series with small sample size. On the other hand, partial duration series model contains more floods than annual maximum model and as such the estimate of parameters of annual flood distribution from the partial flood series would be subjected to lesser uncertainty. Secondly, the theoretical expressions for annual flood distribution obtained through characteristics of partial floods have physical relevance and often are exact distributions rather than asymptotic.(Viraphol et.al,1978).

2. STUDY AREA

River Narmada is one of the major rivers of India. This rises in the Amarkantak plateau of Maikala range in the Shadol district of Madhya Pradesh at an elevation of 1057 meters above sea level. Narmada Sagar dam is a major project on river Narmada in Madhya Pradesh proposed to be located upstream of Mortakka site at about $22^{o}10'$ latitude and $76^{o}10'$ longitude. The total cost of project is expected to be Rs.1393 crores (about 1200 million U.S.Dollars). The dam is expected to be completed by 1994 A.D. and canal system by 2004 A.D. The catchment area of river Narmada up to Mortakka is 67170 km^2 and that up to Narmada Sagar is 61642 km^2.

3. AVAILABILITY OF DATA

Data for hourly stages and daily discharges is available from 1951-1982. The following stage-discharge relationship is developed and used to convert peak stages to corresponding discharges.

$$Q = 317.18\ (H-155.2)^{1.768} \qquad (1)$$

where, Q is discharge in cms and H is stage in meters above mean sea level.

4. STATEMENT OF THE PROBLEM

The objective of the present study is to estimate flood magnitudes for different return periods (100,500,1000 and 10000 years) using the partial duration series approach. The results of frequency analysis using annual peak discharge series have been presented earlier (Goel and Seth,1985). The objective is also to compare the efficiencies of annual and partial flood series models in the estimation of T year flood.

5. FREQUENCY ANALYSIS USING PARTIAL DURATION SERIES APPROACH

The models available for partial duration series differ from each other only in the way in which the number of peaks over the threshold each year is treated. These models vary from the simplest, in which a constant number of exceedances λ is assumed to occur each year to one in which the rate of occurrence of peaks and distribution of peak magnitudes vary with season in a year. The later was developed by Todorovic and Rousselle (1977). A good description of these models has been given in Flood Studies Report, Vol.1,NERC, 1975.

5.1. Model Used

The model used in this study acknowledges variation between years in the number of peaks but ignores variation between seasons. The number of peaks in a year is considered to be random variable with mean λ . The distribution of the number of exceedances in a year is assumed as Poissonian and the distribution of the magnitude of the exceedances as exponential.

The T year flood is given by (Viraphol et.al,1978)

$$\hat{Q}(T) = \hat{q}_o + \hat{\beta} \ln \lambda + \hat{\beta} y(T) \quad (2)$$

where, $Q(T)$ = annual flood magnitude for a given return period T

$\hat{q}_o$ = truncation level (estimated),

$\hat{\beta}$ = parameter estimated from partial duration series

λ = average number of peaks per year and

$y(T) = -\ln(-\ln(1-1/T))$

5.1. Preparation of Partial Duration Series

There are two methods of abstracting data from the discharge record depending upon whether the threshold is fixed first or number of floods to be extracted from the data is fixed first. The method in which the threshold is fixed first, the number and magnitudes of floods are used to estimate λ and β . If N is the number of years (also fixed) sufficient estimates of λ,β and Q(T) are

$$\hat{\lambda} = M/N \quad (3)$$

$$\hat{\beta} = \sum_{i=1}^{M} (q_i/M) - q_o \quad (4)$$

$$\hat{Q}(T) = q_o + \hat{\beta} \ln \lambda + \hat{\beta} \ln(T) \quad (5)$$

In the second method in which fixed number of floods (M) are extracted from the data, it is required to estimate jointly the location parameter q_o and scale parameter β of an Exponential distribution from a sample of fixed size M. The maximum likelihood estimate of q_o,β and Q(T) (after correction for bias) are:

$$\hat{q}_o = (Mq_{min}-\bar{q})/(M-1) \quad (6)$$

$$\hat{\beta} = M(\bar{q}-q_{min})/(M-1) \quad (7)$$

$$Q(T) = \hat{q}_o + \hat{\beta}[\ln \lambda + \ln(T)] \quad (8)$$

In the method adopted for this study the threshold level for peak stages was kept as 161.54 meters as first approximation, keeping in view the stage corresponding to the minimum peak discharge value in the annual maximum series of 162.68 meters. All the peak stages above 161.54 were compiled and converted to corresponding peak discharge values.

Two neighboring peaks were included only if (a) the flow between them dropped to less than two thirds of the earlier two and the time between the peaks exceeded $3T_p$ (231 hours) where T_p is the average time to peak of the first five hydrographs on the record. Average T_p in case of Narmada river upto Mortakka is 77 hours. A total of 78 peaks which satisfied above mentioned criterion were selected for further study. The peak discharges with date and time are given in Table 1.

Lag one serial correlation coefficient (r_1) was calculated for the series and it comes out to be 0.0441, which falls within the 95% tolerance limits to test the hypothesis of zero autocorrelation. For other values of λ the number of exceedances were fixed first.

5.2 Comparison of the Efficiency

To compare the efficiency of flood peaks of given return periods using annual and partial duration series, the sampling variance of Q(T) obtained from both flood series have been compared. It was shown by Zelenhasic (1970) that partial duration series gives double exponential or Gumbel distribution for annual floods when we assume Poisson distribution for number of exceedances and exponential distribution for the magnitude of exceedances. This theoretical finding is used in the comparison of sampling variances of Q(t) obtained from the two flood series, assuming the Gumbel distribution for annual maximum series.

For Gumbel distribution the estimate of T year flood, $Q(T)_a$, is given by

$$\hat{Q}(T)_a = \hat{u} + \hat{\alpha}\, y(T) \quad (9)$$

u and α are obtained from following equations

$$\mu = u + 0.5772\,\alpha \quad (10)$$

$$\sigma^2 = \pi^2\alpha^2/6 \quad (11)$$

Table 1 Partial duration series with date and time

S.N.	Date	Time (hrs)	Discharge (cms)	S.N.	Date	Time (hrs)	Discharge (cms)
1	9.8.51	8	11127	40	3.9.65	22	9110
2	4.8.52	8	13631	41	2.8.66	16	11338
3	28.8.52	8	9573	42	3.8.67	17	12753
4	1953		19521	43	22.8.67	17	10096
5	10.9.54	7	10761	44	31.8.67	24	19690
6	23.9.54	24	33915	45	5.8.68	22	31604
7	18.8.55	18	9475	46	5.8.69	7	27935
8	1.9.55	1	20746	47	18.8.69	8	9401
9	14.9.55	8	13345	48	23.9.69	24	8516
10	3.10.55	23	12392	49	4.7.70	8	14152
11	15.7.56	3	11982	50	19.8.70	9	9014
12	31.7.56	15	9038	51	6.9.70	4	41691
13	11.8.56	10	11258	52	5.7.71	15	13345
14	23.8.57	21	25023	53	30.7.71	8	10555
15	1.9.58	15	8586	54	7.9.71	12	18101
16	11.9.58	15	13005	55	18.8.72	12	47851
17	25.9.58	17	12503	56	2.9.72	8	21162
18	13.7.59	17	30372	57	16.7.73	8	29163
19	8.8.59	6	13717	58	23.7.73	8	27117
20	3.9.59	17	29644	59	31.8.73	2	54063
21	15.9.59	7	19521	60	20.8.74	16	36562
22	5.8.60	10	16822	61	14.8.75	19	24911
23	19.8.60	1	20540	62	25.8.75	8	14416
24	19.7.61	8	11022	63	11.9.75	9	33278
25	8.8.61	17	10324	64	6.8.76	5	12337
26	25.8.61	2	20437	65	29.8.76	2	17713
27	6.9.61	4	13890	66	14.9.76	11	9946
28	16.9.61	1	55323	67	8.8.77	5	22572
29	27.9.61	21	13005	68	31.8.77	20	17011
30	12.10.61	24	10761	69	15.9.77	1	24354
31	6.9.62	21	12669	70	14.7.78	2	11232
32	20.9.62	6	31604	71	17.8.78	5	22716
33	15.8.63	16	13890	72	29.8.78	23	29564
34	26.8.63	15	8376	73	10.8.79	12	26232
35	5.9.63	23	16446	74	6.8.80	24	13659
36	16.9.63	4	16135	75	30.8.80	16	22751
37	13.8.64	1	23438	76	9.7.81	7	9475
38	28.8.64	13	8751	77	10.8.81	10	25662
39	29.7.65	7	18591	78	1982		16602

The sampling variance of $Q(T)_a$ is given by (Kimball, 1946)

$$\mathrm{Var}\ (\hat{Q}(T)_a) = \frac{\alpha^2}{N} [1.11 + 0.52\ y\ (T) + 0.61 y^2(T)] \qquad (12)$$

Sampling variance of Q(T) for partial duration series is mainly based on the work by Cunnane (1973). The sampling variance of $Q(T)_p$ is given by (NERC, 1975, pp.195)

$$\mathrm{Var}\ (\hat{Q}(T)_p = \frac{\beta^2}{N\ \lambda - 1} \left[\frac{(1 - \ln\lambda - \ln T)^2}{N\ \lambda - 1} + (\ln \lambda + \ln T)^2 \right] \qquad (13)$$

5.2.1. <u>Comparison of efficiencies</u> . The following two approaches have been used to compare the sampling variance of Q(T) of annual and partial flood series:

(i) exact theoretical approach and

(ii) approximate theoretical approach.

In exact theoretical approach let R_1 be the ratio of the sampling variances of $Q(T)_a$ and $Q(T)_p$ obtained theoretically from annual and partial flood series respectively. R_1 is given by

$$R_1 = \frac{[\ 1.11 + 0.52\ y\ (T) + 0.61\ y^2(T)\ ]}{\left[\frac{(1 - \ln\lambda - \ln T)^2}{N\ \lambda - 1} + (\ln\ \lambda + \ln T)^2 \right]} \qquad (14)$$

In approximate theoretical approach, the parameters u and α are estimated from the annual flood series and parameters q_o and β from the corresponding partial flood series. Let R_2 denote the ratio of sampling variances of $Q(T)_a$ and $Q(T)_p$, then

$$R_2 = \frac{\lambda\alpha^2\ [1.11 + 0.52 y(T) + 0.61\ y^2(T)]}{\beta^2 \left[\frac{(1 - \ln\lambda - \ln T)^2}{N\ \lambda - 1} + (\ln \lambda + \ln T)^2 \right]} \qquad (15)$$

The difference between R_1 and R_2 is that the difference between α and β is taken into consideration in computing R_2.

6. RESULTS AND DISCUSSION

6.1. Annual Maximum Series Using Gumbel Distribution

The annual peak stage and discharge series is given in table 2 along with statistical parameters. Using equation (10) and (11) the values of location parameter (u) and scale parameter (α) for the Gumbel distribution are 20705.05 and 9061.48 respectively, 100,500, 1000 and 10000 years return period floods are 62389, 77009, 83295 & 104163 cms. The sampling variances corresponding to these

Table 2 Annual peak stage and discharge series at Mortakka
Stage discharge relationship: $Q=317.18*(H-155.2)^{1.768}$

Year	Peak stage above MSL (m)	Peak disch-arge(cms)	Year	Peak stage above MSL	Peak Discharge (cms)
1951	162.68	11127	1974	169.86	36562
1952	163.59	13631	1975	169.10	33278
1953	165.48	19521	1976	164.93	17713
1954	169.25	33915	1977	166.85	24354
1955	165.84	20746	1978	168.20	29564
1956	163.00	11982	1979	167.35	26232
1957	167.03	25023	1980	166.41	22751
1958	163.37	13005	1981	167.20	25662
1959	168.40	30372	1982	164.58	16602
1960	165.78	20540			
1961	173.73	55323	Mean		=25935.344 cms
1962	168.70	31604	Standard deviation		= 11615.904 cms
1963	164.43	16135	Coeff.of skewness		= 1.044
1964	166.60	23438			
1965	165.20	18591			
1966	162.76	11338			
1967	165.53	19690			
1968	168.70	31604			
1969	167.79	27935			
1970	170.99	41691			
1971	165.05	18101			
1972	172.27	47851			
1973	173.49	54063			

Table 3 Results of frequency analysis

	Recurrence Interval in Years			
	100	500	1000	10000
Q_T in cms from partial duration series	66487	83562	90903	115277
Sampling Variance using exact theoretical approach	$3.2x10^7$	$5.36x10^7$	$6.46x10^7$	$1.08x10^8$
Sampling variance using approximate theoretical approach	$4.37x10^7$	$7.32x10^7$	$8.82x10^7$	$1.48x10^8$
Q_T in cms from annual maximum series using Gumbel distribution	62389	77009	83295	104164
Sampling variance	$4.21x10^7$	$7.15x10^7$	$8.67x10^7$	$1.47x10^8$

are 4.21 $\times 10^7$, 7.15$\times 10^7$, 8.67$\times 10^7$ and 1.47 $\times 10^8$ respectively.

6.2. Partial Duration Series Using Poisson and Exponential Distributions

For q_o = 8376 cms the values of λ and β are 2.437 and 10583.098 respectively. Using equation (2) the 100, 500, 1000 and 10000 years return period floods are 66487, 83562, 90903 and 115277 cms respectively, which are slightly higher than the estimates of annual maximum flood series.

Using exact theoretical approach i.e. $\alpha = \beta$ the sampling variances of 100, 500, 1000 and 10000 years return period floods are 3.2$\times 10^7$, 5.36$\times 10^7$, 6.46$\times 10^7$ and 1.08$\times 10^8$ respectively.

Using approximate theoretical approach i.e. $\beta \neq \alpha$ and β =10583.098 the sampling variances are 4.37$\times 10^7$, 7.32$\times 10^7$, 8.82$\times 10^7$ and 1.48$\times 10^8$. These results have been tabulated in Table 3.

6.3. Comparison of the Efficiencies

6.3.1. Exact Theoretical approach. Equation 14 shows how the ratio of sampling variances obtained by the exact theoretical approach varies with the return period T. For a given value of λ , the relationship between the ratio R_1 and the return period T expressed as the Gumbel reduced variate y(T) can be derived. The results of these relationships for the range of λ =0.8 to 10.0 are shown in figure 1. It is evident from the figure that the partial flood series estimate of Q(T) always has a smaller sampling variance than that of the annual flood series for the return period T less than 11 years. For the whole range of return periods the partial flood series estimate of Q(T) has a smaller sampling variance than that of the annual flood series if the value of λ is atleast 1.65.

6.3.2. Approximate theoretical approach. By substituting the estimates of α , λ and β for each threshold level into equation (15) relationships between R_2 and T are obtained for various λ values. These have been plotted in figure 2. The value of R_2 increases with λ . For λ = 2.437 the ratio R_2 comes closer to unity. In the range of λ studied (1.0 to 2.437) the sampling variance of annual flood series is lesser than that of partial flood series.

7. CONCLUSION

For Narmada Sagar dam the 100, 500, 1000 and 10000 years return period floods are 66487, 83562, 90903 and 115277 cms respectively using partial duration series, which are slightly higher than corresponding estimates of annual flood series assuming that annual flood series follows Gumbel distribution.

On the basis of exact theoretical approach, it is seen that the partial flood series estimates of Q(T) always have a smaller

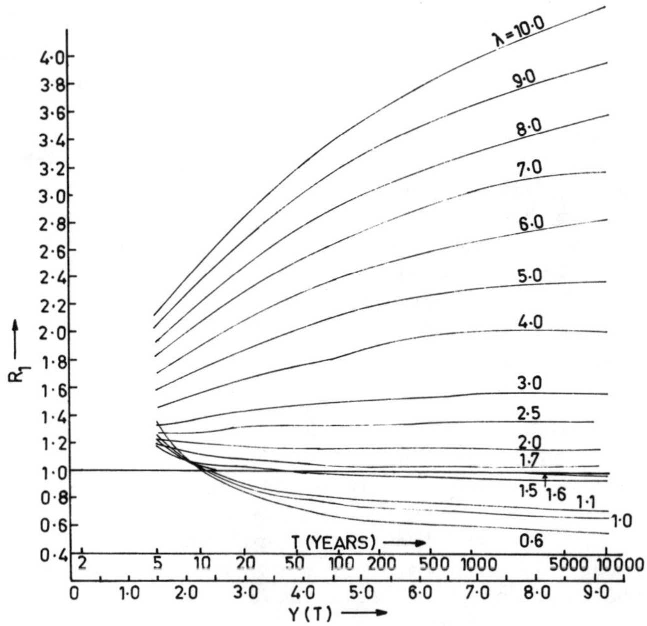

Figure 1 : RELATIONSHIP BETWEEN RATIO R_1 OF SAMPLING VARIANCES $Q(T)_a$ AND $Q(T)_p$ BASED ON EXACT THEORETICAL APPROACH AND THE RETURN PERIOD T FOR GIVEN λ

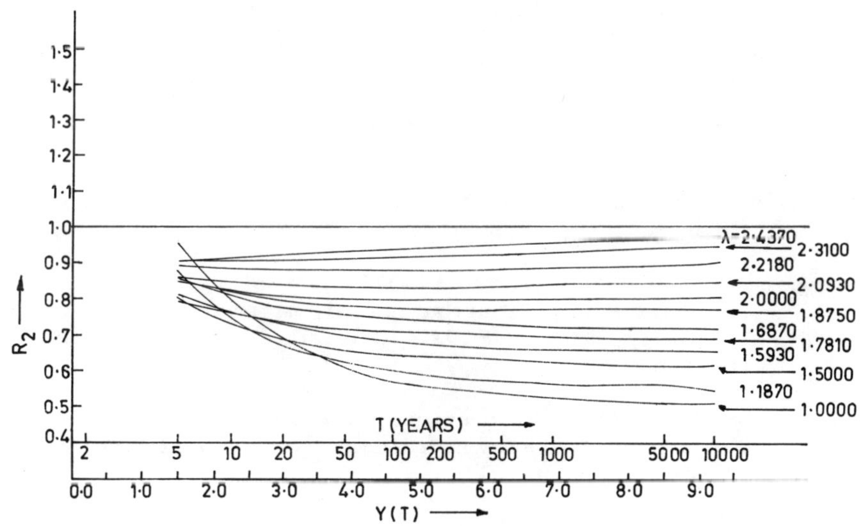

Figure 2 : RELATIONSHIP BETWEEN RATIO R_2 OF SAMPLING VARIANCES $Q(T)_a$ AND $Q(T)_p$ BASED ON APPROXIMATE THEORETICAL APPROACH AND THE RETURN PERIOD T FOR GIVEN λ

sampling variance than that of the annual flood series for the return period T less than 11 years. For the whole range of return periods the partial flood series estimate of Q(T) has a smaller sampling variance than that of the annual flood series if the value of λ is atleast 1.65.

On the basis of approximate theoretical approach(Computed estimates of α and β are taken) it is observed that in the range of λ studied (1.0 to 2.437) the sampling variance of partial flood series is more than the sampling variance of annual flood series. However, this difference reduces when λ increases. For λ =2.437 the sampling variances of the two series are comparable.

REFERENCES

Cunnane, C.1973.'A Particular Comparison of Annual Maxima and Partial Duration Series Methods of Flood Frequency Prediction', Journal of Hydrology, Vol.**18**,pp.257-271.

Goel,N.K. and Seth, S.M.1985.'Tentative Spillway Design Flood Estimation for Narmada Sagar Project-A Case Study', Seminar on Flood Frequency Analysis, pp.78-92, New Delhi.

Hall, M.J. and O'Connell, P.E.1972'Time Series Analysis of Mean Daily Flows' Water and Water Engineering, Vol.**76**, pp.125-133.

Kimball, B.F.1946,'Sufficient Statistical Estimation Functions for the Parameters of the Distribution of Maximum Values' Annals of Mathematical Statistics, Vol.**17**,pp.299-309.

Natural Environment Research Council, 1975, Flood Studies Report, Vol.**1**, pp.1-251.

Todorovic, P. and Rousselle, J.1971,'Some Problems of Flood Analysis', Water Resources Research, Vol.**7**, No.5, pp.1140-1150.

Viraphol, T. and Yevjevich, V.1978,'Use of Partial Flood Series for Estimating Distribution of Maximum Annual Flood Peak', Hydrology Paper No.**97**, Colorado State University, Fort Collins.

Zelenhasic, E.1970,' Theoretical Probability Distributions for Flood Peaks', Hydrology Paper No.**42**, Colorado State University, Fort Collins.

DESIGN FLOOD COMPUTATION WITH LIMITED DATA – CASE STUDY OF A RESERVOIR IN SUBERNREKHA RIVER BASIN

T. Prasad
Professor of Civil Engineering
Bihar College of Engineering
Patna University
Patna, India

K. Subramanya
Professor of Civil Engineering
Indian Institute of Technology
Kanpur, India

ABSTRACT. The paper deals with the various approaches, methods, techniques and steps adopted in estimating the design flood or design floods for a reservoir project having limitations of data in respect of length, reliability and instrumentation. Two approaches of defining design floods, the physical or genetic approach and the statistical approach have been indicated. The various steps in PMF estimation, the deviations from standard techniques, the need for making the deviations and their rationale have been described in some detail in the paper. The exercise, carried out with respect to a World Bank assisted project in eastern India, serves to demonstrate practically suitable and theoretically justifiable procedures to be adopted in applying standard techniques in design flood computation in a case where the available data and information have various limitations. This case is typical in developing countries and hence the study may be of interest to project planners and designers in these countries.

1. INTRODUCTION

A 38.5 m high dam is proposed to be constructed on the river Kharkai, which is a tributary of the river Subernrekha, at a site called Icha in order to create a gross reservoir storage of 1042 million m^3. The dominant purpose of the reservoir is to meet the irrigation needs of the states of Bihar and Orissa and is a part of the multipurpose Subernrekha project having another dam on the river Subernrekha, a barrage on Kharkai below Icha, another barrage at Galudih on Subernrekha below its confluence with Kharkai and a network of canal distribution systems (Fig.1). It is an interstate project with three states, Bihar, Orissa and West Bengal being involved and is being executed with the World Bank assistance.

The proposed dam at Icha intercepts a catchment of 2850 km. A gauge and discharge station near the dam site has been in operation by the Govt. of Bihar since 1974. Gauges and discharges are observed daily at this site and during high flood events, only hourly gauges are observed. The discharge is measured by float method. At another site,

V. P. Singh (ed.), Application of Frequency and Risk in Water Resources, 123–137.

Adityapur which intercepts a catchment 6050 sq.km. and is situated below a major tributary to Kharkai, gauge and discharge are observed daily by another agency, Central Water Commission of the Govt. of India. The data observed and collected by two different agencies at two different stations (Fig.2) sometimes do not reconcile with each other, putting the reliability of data of one or the other agency to question.

There are five raingauge stations in and around the basin; Rairangpur, Chaibasa, Potka, Champua and Manjhgaon out of which only the stations at Rairangpur and Champua have been equipped with self-recording gauges for the last few years (Fig.2). The other stations have ordinary raingauges with daily observations for many years. Incompatibility of rainfall data with discharge data is not infrequent.

The study reported in this paper is aimed at estimating the design flood or design floods for the proposed reservoir at Icha for various design purposes. Because of the limitations of data with respect to length, reliability and instrumentation, difficulties were experienced in applying the standard techniques for the purpose. Some compromise and deviation from theoretically valid procedures became inevitable in view of these difficulties. The rationale of such compromise and deviation and the appreciation of the results arrived at in that light have been indicated.

2. DESIGN FLOOD CRITERIA

A design flood at a given site on a river is the instantaneous peak discharge adopted for the design of a structure. Its value depends not only on the characteristics of the actual occurrences of floods but also on the economics of the project and the social acceptability of inherent risks resulting from the failure of the structure or of its anticipated performance. Thus, design flood is essentially a techno-socio-economic parameter which is a factor of the river, the site, its hydrology as well as socio-economic considerations.

Estimation of design flood depends upon, but is not solely determined by, the hydrologic behaviour of the river at the given site. Hence, the characteristics of the relevant hydrologic occurrences provide necessary, but seldom sufficient, basis for design flood estimation. For two reasons this may be so. One is that the length and the reliability of the record of the historical occurrences of the relevant hydrologic event may be so deficient that the characteristics derived on its basis may not yield true enough or good enough indication of the required hydrologic behaviour, no matter how sophisticated a technique is employed for the purpose. Secondly, for design flood estimation, this behaviour is to be interpreted in the socio-economic dimension where the time frame and other criteria are never the same as those valid for hydrologic behaviour, no matter how long is the record used for deriving this behaviour. So, necessarily the historical experience of hydrologic occurrences has to be expanded and transformed for design flood estimation.

There are various techniques for such expansion and transformation of the historical experience, depending upon the nature, length and reliability of this experience as well as the purpose of the project being designed. In general, there are two approaches on which these techniques are based. One is the genetic or physical approach in which the physical causative factors and processes in the occurrences of floods are considered and the expansion and transformation of the historical experience are sought through these factors and processes. Another is the statistical approach in which the flood occurrences are considered to represent the integral effects of various causative factors and processes and the historical record implicitly contains the probability laws governing these occurrences. The required expansion and transformation of the historical experience is done on the basis of this law which is derived through appropriate statistical analysis.

In India, both the approaches are followed either for designing different elements of the same project or for projects of various magnitudes and purposes. In the first approach, the prime genetic factor for floods is taken to be the precipitation which is transformed into floods through the various basin-bound hydrologic processes. The historical processes are suitably enlarged and convoluted to yield the design flood. If the structure is a major one, socio-economic considerations may require it to be designed for a flood which is somewhere near or related to the physical upper bound of the flood occurrences. Two floods are generally adopted in this category of design floods, probable maximum flood (PMF) and standard project flood (SPF). Probable maximum flood is a flood event that may be expected from the most severe combination of critical meteorologic and hydrologic considerations that are reasonably possible in the region, while a standard project flood is a flood event that may be expected from the most severe combination of meteorologic and hydrologic considerations that are considered reasonably characteristic of the given geographical region, excluding extremely rare combinations (Chow, 1964, pp. 25.25-25.26).

For expanding and transforming the historical experience of flood occurrences based on their underlying law of probability for design flood estimation, frequency analysis of extreme value series, generally the annual maximum series, is adopted. Flood events of different return periods are computed and selected as design floods for various types of structures. For a project or a structure of a given type or magnitude, one approach may be preferred to the other for design flood estimation, or both the approaches may be tried for different design purposes. In India, the following guidelines are used for design estimation for various types of structures (Subramanya, 1984, p. 226) (Table 1).

TABLE 1 : Guidelines for Selecting Design Floods

S.No.	Structure	Recommended design flood
1.	Spillways for major and medium projects with storages more than 60 Mm^3	(a) PMF determined by unit hydrograph and probable maximum precipitation (PMP) (b) If (a) is not applicable or possible, flood-frequency method with T = 1000 years
2.	Permanent barrage and minor dams with capacity less than 60 Mm^3	(a) SPF determined by unit hydrograph and standard project storm (SPS) which is usually the largest recorded storm in the region (b) Flood with a return period of 100 years (a) Or (b) whichever gives higher value
3.	Pickup weirs	Flood with a return period of 100 or 50 years depending on the importance of the project.
4.	Aqueducts	
	(a) Waterway	Flood with T = 50 years
	(b) Foundations and free board	Flood with T = 100 years
5.	Project with very scanty or inadequate data	Empirical formulae

3. DESIGN STORM

In the genetic or physical approach for expanding and transforming the historical experience of flood occurrences, the historical experience of storms, the genetic cause for floods, is first expanded. The storm event arrived at through this process carried out as per specified criteria is the design storm. If it is the probable maximum flood that is the design flood, the storm event must represent a realistic least upper bound for meteorologically feasible storms for the region. As per the practice followed in India, this storm, known as the probable maximum storm (PMS), is derived as follows.

3.1. Storm Transposition and Selection of Storms

The purpose of storm transposition is to increase the storm experience of the basin by considering not only storms which have occurred directly over it but also those which have occurred within the surrounding meteorologically homogeneous region. The records of all the rainfall stations in the region were examined to select the dates of occurrences of historical major rainfalls. The storm that occurred during July 27-30, 1927 in the vicinity of the basin under study was considered to be the most severe on record and was selected for further analysis (Dhar et al, 1977).

3.2. Depth-Area-Duration (DAD) Analysis

This analysis is carried out to determine the largest average depths of rainfall over varying areal sizes in the meteorologically homogeneous region under consideration for different passages of time. This analysis provides the basic data to estimate the PMS for the basin under study.

In India, most of the rainfall data of past storms consist of only daily amounts at ordinary raingauge stations measured at 8:30 a.m. each day. Thus the DAD analysis for the selected major storm was done by constructing maximum isohyetal maps of one-day, two-day and three-day durations, using incremental isohyetal method. From these maps, depth-area curves are drawn for different durations. Corresponding to the basin area of 2850 sq.km. the depths obtained for different durations are given in Table 2.

3.3. Storm Maximisation

The objective of storm maximisation is to ascertain the increase in the rainfall depth of any observed storm due to physically possible increases in the meteorological factors which produced the storm. These factors may be divided into two types, (a) the moisture content of the rain producing air mass involved in the storm and (b) the mechanical efficiency of the storm in changing water vapour and droplets in the atmosphere into rain.

It is rather difficult to evaluate the mechanical efficiency of any storm with any acceptable degree of accuracy. Also, it may be assumed that in severe and major storms, such efficiency is very near the maximum possible efficiency. Hence, it is the practice to maximise storms for moisture content only. This is done by multiplying the storm depth by a moisture maximisation factor which is determined by the dew point maximisation procedure. On the basis of previous studies (Dhar et al., 1977) this factor has been taken to be 1.3 in this case. Thus, the maximised storm depth values, or PMS values for different durations, are obtained as given in Table 2.

TABLE 2 : PMS Estimates over the Kharkai River Basin

Storm Period	Storm Depths obtained from DAD Analysis	PMS Depths
1 - day	42.8 cm	55.7 cm
2 - day	60.0 cm	78.0 cm
3 - day	61.2 cm	79.5 cm

3.4. Design Duration of PMS

On the basis of an examination of the depth duration characteristics of the DAD analysis, a 2-day duration for PMS is considered to be critical. Incremental depths of rainfall beyond this duration are found to be marginal.

4. UNIT HYDROGRAPH ANALYSIS

Unit hydrograph is the prevalent method of transforming the rainfall into the corresponding flood hydrograph. The unit hydrograph and the design storm, however, have to be modified and treated in certain ways so that their convolution may yield the design flood.

The steps adopted in derivation and modification of the unit hydrograph for its use in transforming the design storm into the design flood and their rationale in the present case of limited data are indicated below.

4.1. Selection of Hydrograph

Eleven years, 1974-84, of gauge and discharge data - hourly gauge values and daily discharge values at the G-D site near the dam - are available. First, the hydrograph based on daily discharge values were drawn for all the eleven years in the monsoon season, June - October. By examining these hydrographs, those found suitable for unit hydrograph derivation were identified, keeping in view the following points,

(a) Too high values of peaks, on account of unreliability of accuracy in measuring high discharges, and too low values, on account of the possibility of high percentage errors in measurement, were discarded.

(b) Those flood events which evinced simple and isolated peaks were chosen.

For the flood events so identified, hydrographs were redrawn with the hourly discharge values computed from the hourly gauge values by means of the following stage-discharge relationship developed for the G-D site on the basis of the available discharge and corresponding gauge observations.

$$Q = 21.88\ (G - 187.94)^{2.717}$$

where Q is the discharge in cumecs and G is the R.L. of water level at the G-D site,in m.

Several hydrographs which looked simple when drawn with daily computed values turned out to be multiple peaked when redrawn with computed hourly discharge values.

Corresponding rainfall data were examined and analysed for the hydrographs which looked prospective for unit hydrograph derivation. Some of such hydrographs could not be used for unit hydrograph derivation due to non availability of required rainfall data or due to incompatibility and indiscrepancy of the available data vis-a-vis the corresponding discharge data.

4.2. Derivation of Unit Hydrograph

A few hydrographs found suitable for the purpose were analysed by standard techniques for unit hydrograph derivation. The base-flow was estimated by drawing a tangent at the point of rise of the hydrograph upto the time to peak. This point was connected to a point on the recession limb N days after the peak where

$$N = 0.84\ A^{0.2}$$

Here, A is the catchment area in sq.km.

The ordinates of the unit hydrographs were obtained by dividing the ordinates of the direct runoff hydrographs by the direct runoff depth in cm.

The hyetographs of the storm events which produced the flood hydrographs used for unit hydrograph derivation were drawn on the basis of total areal rainfall depths determined from the rainfall measurements of all the five raingauge stations in and around the basin by Thiessen method and the temporal distribution determined by the rainfall record of the self-recording raingauge at Rairangpur. A uniform rate of abstraction, ϕ - index, was used to determine the unit duration of the unit hydrograph.

The unit durations of the 3 unit hydrographs thus derived were found to be 8 hours ± 2 hours. An 8-hour unit duration was assumed for all the unit hydrographs.

The salient points of the three flood events and the unit hydrographs derived therefrom are given in Table 3.

TABLE 3 : Salient Information of the Flood Events Analysed and the Unit Hydrographs Derived

Sr.No.	Dates of Flood Events	Peak Discharge of Total Runoff (Cumecs)	Peak of Unit Hydrograph (Cumecs)	Unit Duration of U.H. (Hours)
1	June 1981	3790	406	10
2	June 1984	2420	434	7
3	August 1984	2520	300	8

A graphical averaging process was used to derive a normal 8-hour unit hydrograph for the basin by plotting all the unit hydrographs on the same scale and on the same graph such that the peaks were on the same vertical axis.

4.3. Modification of the Unit Hydrograph

The unit hydrograph as derived above has to be modified in order to make it suitable for computing the design flood. Earlier studies and experience (US Army Corps of Engrs, 1959) have shown that the peak ordinates of unit hydrographs derived from major flood hydrographs having very large volumes are generally higher than those derived from minor floods. This probably results from differences in the areal distribution of precipitation and hydraulic conditions of channel conveyance. In this study, as only a very limited number of available hydrographs were suitable for unit hydrograph derivation and these represented depths of direct runoff much smaller than the design storm depth, the need for increasing the peak ordinate of the unit hydrograph to be used for design flood computation was indicated. There is no guideline for determining the quantum of the increase of the peak ordinate except that an increase of upto 50 percent may be applied on this count (Sokolov, et al., 1976, p. 200).

In this case, the peak ordinate of the unit hydrograph was increased by 20%. The widths of the unit hydrographs at 50% and 75% of the revised peak were also adjusted by standard synthetic unit hydrograph procedures in order to bring the volume under the unit hydrograph equal to 1 cm.

5. CONVOLUTION OF UH WITH PMS

The unit hydrograph as derived and modified suitably for design flood computation is to be convoluted with the design storm. For this purpose, the design storm was pre-treated as follows,

5.1. Temporal Distribution of Design Storm

The two-day PMS depth of 78.0 cm was distributed in blocks of 6 time intervals of 8 hours, the unit duration of the unit hydrograph. For this purpose, the temporal pattern of rainfall occurrences as recorded at the self-recording rain gauge station at Rairangpur, which is centrally located in the basin, commanding 69% of the basin in Thiessen weightage, is taken as a guide, leading to the distribution given in Table 4.

TABLE 4. Distribution of PMS Depth During the 2-day Storm Period

8 - Hour Period	% of Total Depth	Rainfall Depth (cm)
Ist 8-Hour	50	39.0
2nd 8-Hour	20	15.6
3rd 8-Hour	10	7.8
4th 8-Hour	8	6.2
5th 8-Hour	7	5.5
6th 8-Hour	5	3.9
Total	100%	78.0

5.2. Critical Sequencing

The rainfall depths in 8-hour time periods, as given in Table 4, were to be arranged in such a sequence that they corresponded with the storm characteristics of the area, tending if at all towards the conservative. In computing PMF, this sequence is usually the one which gives the most critical conditions of runoff, i.e. the maximum peak runoff. This was done by both trial and error as well as by following a procedure suggested for the purpose (Central Water & Power Commission, 1972, p. 85).

The unit hydrograph ordinates at 8 hour intervals were arranged chronologically. The 8-hour design storm depths were now so arranged that the maximum rainfall increment is against the maximum unit hydrograph ordinate, the second highest increment is against the second largest unit hydrograph ordinate, and so on. The arrangement of rainfall increments so obtained is reversed to obtain the critical sequence, given in Table 5.

TABLE 5. Critical Sequence of 8-Hr. Rainfall Depths in the Design Storm

Time	Rainfall Depths
0 - 8 hrs.	3.9 cm
8 - 16 hrs.	5.5 cm
16 - 24 hrs.	15.6 cm
24 - 32 hrs.	39.0 cm
32 - 40 hrs.	7.8 cm
40 - 48 hrs.	6.2 cm

5.3. Calculation of Effective Rainfall

Usually, the initial loss and infiltration index values are derived from the available hydrometeorological records. The effective rainfall from the adopted storm is then obtained by using the minimum initial loss and infiltration index values on the assumption that the basin would be saturated at the time of the design storm. However, unless these values are derived from the records of very severe storms that have occurred on wet soils, much reliance cannot be placed on its use in computing runoff from the design storm, particularly when the design storm happens to be PMS. Thus, it was considered reasonable to be conservative in adopting a uniform initial loss and infiltration index of 1 mm/hour throughout the storm, as suggested by the U.S. Army Corps of Engineers.

5.4. Convolution

The unit hydrograph - as derived and modified for the purpose - and the design storm with the time distribution, sequencing and abstraction done as indicated above, were convoluted linearly to produce the direct runoff hydrograph.

5.5. Base-Flow Addition

There does not seem to be any accepted or well defined criterion to estimate the base-flow during the design storm. However, this does not affect the design flood computation significantly as the base flow under all circumstances will be rather small percentage of the peak runoff of the design flood. A base-flow of 300 cumecs is added uniformly, which gives a peak value of total runoff as 21,100 cumecs.

6. 1000 – YEAR FLOOD COMPUTATION

Only 11 years (1974-84) of gauge-discharge data were available at a station near the proposed dam site on the river Kharkai. At the station Adityapur below this station 13 years (1972-84) of data were available with a different agency. A major tributary, however, meets the river between these two stations. Considering that these data were not readily available, they would have involved an unreliable procedure of transferance from the observed site to the desired site and they would have meant only a marginal increase in the time length of data, only 11 years of dam site data were taken into account for the statistical analysis. This is obviously too inadequate a data series for any reliable statistical analysis, particularly when such analysis is to be used for frequency analysis involving extrapolation for an event of 1000-year return period. In this study, however, frequency analysis was carried out to find out the order of magnitude of a 1000-year flood for the sake of comparison with the design flood estimated from the physical approach. Keeping this limited objective in view, the following methods and steps were adopted in this case,

(i) Eleven years of annual maximum series were analysed,

(ii) Chow's equation for hydrologic frequency analysis, i.e. the frequency factor method was used, as it possesses both validity in theory and simplicity in application.

(iii) Only one distribution, Log Pearson Type III was used as a probability model for the extreme flood occurrences. Based on an extensive study by a task force of U.S. Water Resources Council (Benson, 1968), this distribution was found to be the most general and most objective of several distributions tested and hence was recommended for general use. In another study on the data of six stations in the U.S. (Natural Environmental Research Council, 1975) also, this distribution was found more suitable than others, particularly for higher return periods.

(iv) Keeping in view the limited objective, the inherent limitations on account of data length, and preference for a particular distribution, no goodness-of-fit and confidence limit analysis were done in this case.

(v) The results of the frequency analysis for estimating the 1000-year flood are given in Table 6.

TABLE 6. Frequency Analysis with Log Pearson Type III Distribution

Computed mean of series	Computed Std. Dev. of logx series	Computed Co-eff.of skewness	Recurrence Interval in years	Frequency factor, obtd.from a table	T-year flood in cumecs (T=1000 yrs.)
3.232	0.382	-0.2697	1000	2.716	18,613

7. STANDARD PROJECT FLOOD

A standard project flood (SPF) is the flood discharge that may be expected from the most severe combination of meteorologic and hydrologic conditions that are considered reasonably characteristic of the region, excluding extremely rare combinations. Thus, SPF also is a design flood based on physical or genetic approach depending on both the causative factor of rainfall and the basin process of rainfall-runoff transformation. In general terms, it may be defined as the peak of a hydrograph representing runoff from standard project storm (SPS). SPS estimate of a drainage area which does not receive any substantial amount of runoff from snowmelt should represent the most severe flood producing rainfall depth-area-duration relationship and isohyetal pattern of any storm that is considered reasonably characteristic of the region in which the basin is located. As SPF excludes extremely rare combination of hydrometeorological factors, a reasonable estimate of SPS is the storm obtained by transposing the DAD pattern of the severest storm observed in a hydrometeorologically homogeneous region to the basin under study. No adjustment of the transposed storm needs to be done for moisture maximisation. Its duration is taken to be the same as that for PMS.

The hydrologic characteristics of the basin in transforming rainfall to runoff are taken to be the same those adopted for the PMF computation. Thus, the same unit hydrograph as derived and modified for PMF computation is adopted for SPF also. Time distribution, critical sequencing and abstraction of the 2-day SPS are done on the same basis and then convoluted with the unit hydrograph to yield ordinates of the direct runoff resulting from SPS. As this convolution is linear process, these ordinates are reduced from the corresponding ordinates of the direct runoff resulting from PMS in the same ratio as the SPS is reduced from the PMS, i.e. by the factor of moisture maximisation. Adding a base flow of 200 cumecs, SPF comes to 16,200 cumecs.

8. DESIGN IMPLICATIONS

The three design floods for the river Kharkai at the proposed dam site, computed from considerations and concepts indicated above, are given in Table 7.

TABLE 7 : Design Floods of the River Kharkai at Icha

Approach	Specification	Value
Physical	PMF	21,100 cumecs
	SPF	16,200 cumecs
Statistical	1000-Year Flood	18,600 cumecs

The concepts and assumptions on which these values have been obtained and the consequent limitations and deficiencies of this study have been indicated. In the extant state of the length and reliability of available data and information, these values would provide the best possible guide to the project designer in selecting particular values for designing various elements and components of the project. This is a significant improvement over the empirical approaches for design flood estimation which held sway in the past. Such approaches yielded only one design value of flood for a given basin with no relationship to the physical possibilities of future flood occurrences in the basin or with no specification for the inherent risks involved in its adoption for design. For example, in the present case, adopting the maximum permissible value of Dickens constant for the region under study, use of Dickens formula gives a value of 11,000 cumecs for the design flood. There is no basis of going beyond or below this value for the final selection of design flood.

In selecting the design flood based on the physical approach and/or statistical approach, the designer may and should keep in view the appropriate socio-economic considerations and criteria. As a matter of fact, such selection involves a policy decision and hence can be done at a higher decision-making level with a range of design flood values being available from these approaches. An economic analysis can be performed on the basis of these values, which will help in such decision making. Other things that will influence the final selection of the design flood are the sophistication in instrumentation in data collection, analytical techniques and computational facilities available, economic conditions and social awareness and acceptability of risks prevailing at a given time, thus making this exercise a veritable dynamic process.

9. ACKNOWLEDGEMENT

This paper is based on an ongoing study being carried out at the Centre of Water Resources Studies at Bihar College of Engineering, Patna. The authors are thankful to the authorities of Multipurpose Subernrekha Project for having provided the financial support and the data for carrying out this study. The discussions with Sri K.K. Srivastava, Superintending Engineer, Sri R.R.P. Sharma, Executive

MULTIPURPOSE SUBERNAREKHA PROJECT
- PROJECT ELEMENTS -

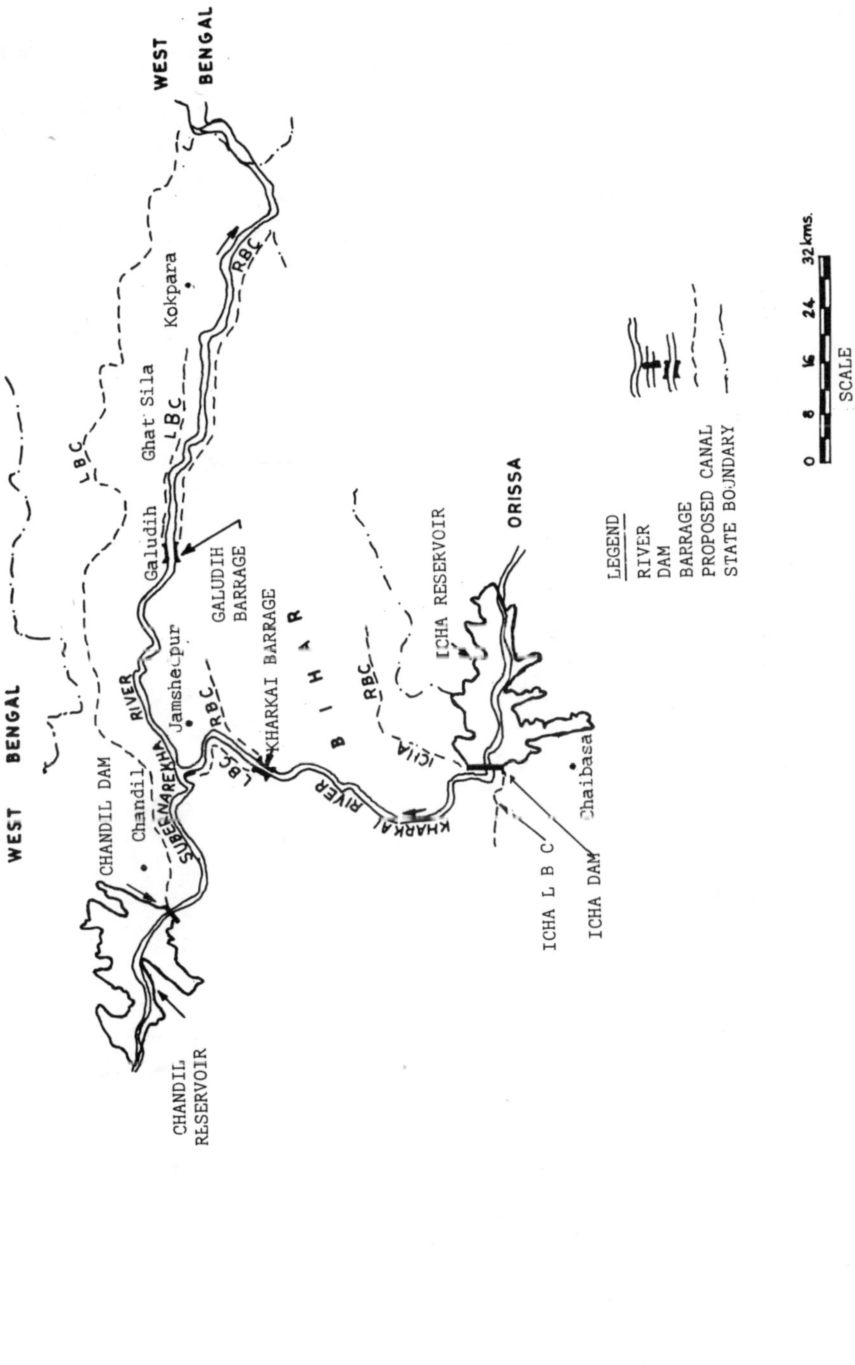

FIG. No.-1

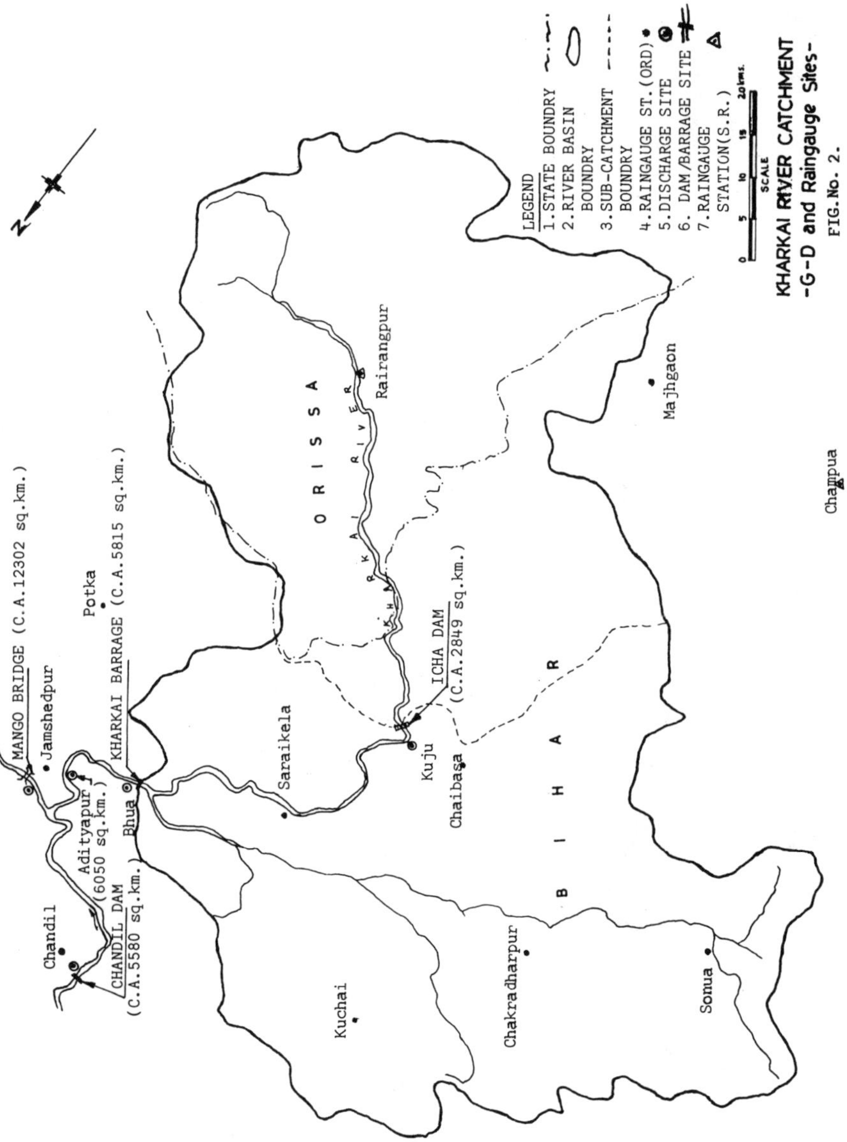

KHARKAI RIVER CATCHMENT
-G-D and Raingauge Sites-
FIG.No. 2.

Engineer and other officers of the Project from time to time during the progress of the study were very helpful and are gratefully acknowledged. Sri Naresh Kumar, Research Fellow at the Centre of Water Resources Studies worked diligently in data processing and computation for this study and the authors would like to express their appreciation for this help.

10. REFERENCES

United States Army Corps of Engineers, 1959. 'Flood-Hydrograph Analyses and Computations', Eng. Manual 1110-2-1405.

Chow, V.T. (Ed.), Handbook of Applied Hydrology, 'Flood Characteristics and Flow Determination' by Tate Dalrymple, pp. 25.25 - 25.26

Central Water & Power Commission, 1972. Estimation of Design Flood - Recommended Procedures, Govt. of India Publication, p. 85.

Natural Environment Research Council, 1975. Flood Studies Report, Vol. I, Hydrological Studies, 27 Charing Cross Road, London, pp. 155-158.

Sokolov, A.A., Rantz, S.E. and Roche, M. 1976. Floodflow Computation - Methods Compiled from World Experience; Studies and Reports in Hydrology No. 22, The Unesco Press, Paris.

Dhar, O.N., Rakhecha, R., and Mandal, B.N. 1977. 'Estimation of Design Storm for the Subernrekha Basin Upto Chandil and Ghatsila Dam Sites', Indian Journal of Power & River Valley Development, September, pp. 338-343.

Subramanya, K. 1984. Engineering Hydrology, Tata McGraw-Hill Pub. Co. Ltd., New Delhi, p. 226.

Benson, M.A. (1968) 'Uniform flood-frequency estimating methods for Federal agencies'. Water Resources Research, 4, 5, 891-908.

CONFIDENCE INTERVALS FOR DESIGN FLOOD EVENTS UNDER DIFFERENT STATISTICAL FLOOD MODELS

Fahim Ashkar, Bernard Bobée and Louise Fortier
Institut national de la recherche scientifique,
INRS-Eau, C.P. 7500
Sainte-Foy (Québec)
G1V 4C7 CANADA

ABSTRACT. In many engineering activities relating to floods one has to use a sample $x_1, \ldots x_n$ of flood data along with a statistical distribution $f(x)$ that approximately fits this sample, in order to estimate a particular extreme flood event x_p corresponding to a small probability of exceedance 1-p (return period $T = 1/(1-p)$). Although one can never tell with certainty what is the true form of $f(x)$ that represents the flood population, it is common practice to assume that one specific form applies, and to estimate its unknown parameters $\Theta_1, \ldots, \Theta_k$ from the sample values $x_1, \ldots, x_n$. The precision with which the event (or "quantile") x_p is estimated, depends on the accuracy with which the parameters of $f(x)$ have been obtained. This precision can be quantified by constructing confidence intervals for x_p. Since these confidence intervals can contain valuable information, they should be calculated with as much accuracy as possible. In hydrology, however, practitioners have often used simple but inaccurate methods for calculating confidence intervals for x_p. The purpose of the present study is to draw attention to some relatively simple techniques found in the mathematical literature for calculating confidence intervals for x_p that give more accurate and better informative results than those commonly used in hydrology at present. We present in particular some useful techniques for obtaining confidence intervals for the quantiles of the normal (log Normal), Pearson (log Pearson), Weibull (Gumbel), and Exponential distributions. We also discuss the problem of constructing confidence intervals for x_p when x_p is estimated via the partial-duration-series approach, under the assumption that the distribution of the number of flood events arriving in an arbitrary time interval $[0, t]$, is Poissonian, and that of the magnitude of flood events, is exponential.

1. INTRODUCTION

In such engineering activities as the design of a flood-control structure, or the estimation of flood damage, one has often to base one's decisions on a sample $x_1, \ldots x_n$ of flood data. It is common

V. P. Singh (ed.), Application of Frequency and Risk in Water Resources, 139–151.

practice to use a statistical distribution, or statistical "model", $f(x) = f(x; \theta_1, \ldots, \theta_k)$, to fit the sample, and to subsequently use $f(x; \theta_1, \ldots, \theta_k)$ to estimate a particular extreme flood event x_p corresponding to a small probability of exceedance 1-p (or return period $T = 1/(1-p)$). Two kinds of uncertainty are involved in this kind of estimation:

First kind : model uncertainty, or uncertainty in the choice of the parent distribution;

Second kind: parameter uncertainty, or uncertainty in the estimation of parameters of the chosen distribution.

Flood analysts have long been preoccupied with the study of these two kinds of uncertainty. New probability distributions suitable for modeling flood frequency are continuously being sought (model uncertainty) and new parameter-estimation techniques for these distributions are continuously being investigated (parameter uncertainty).

Although one can never tell with certainty what statistical distribution the flood data have come from, it is common practice to assume that a certain form of distribution applies, and to estimate its unknown parameters $\Theta_1, \ldots, \Theta_k$ from the sample values $x_1, \ldots, x_n$. Statistical tests such as the chi-square and Kolmogorov-Smirnov tests can be used to check whether a particular distribution is adequate, but unfortunately these tests are not powerful enough to discriminate between distributions.

The precision with which the event $x_p = g(p; \Theta_1, \ldots, \Theta_k)$ is obtained, depends on the accuracy with which the parameters $\Theta_1,\ldots,\Theta_k$ of f(x) have been estimated. This precision can be quantified by constructing confidence intervals for x_p. These confidence intervals calculated at a specific 100 $(1-2\alpha)$ % confidence level are based on the assumption that the form of f(x) is <u>known with certainty</u> (f(x) is usually assumed to be Pearson, log Pearson, log Normal, Weibull, etc.).

In practice, however, one can never be completely sure what is the true form of f(x), but this does not prevent constructing confidence intervals for quantiles x_p assuming that the only source of uncertainty in the estimation of these quantiles is the unknown <u>value</u> of the parameters of f(x) and not the unknown <u>form</u> of f(x) itself. This approach of course has its deficiencies but it has been used for a long time and continues to be used in practice. The confidence intervals obtained for x_p by this approach are used not only in the design of flood-control structures but in many other fields not relating to floods, and for this reason, they should be calculated with as much accuracy as possible.

In many instances, the problem of constructing confidence intervals for x_p is complex, and simplifying assumptions have to be introduced in order to be able to obtain an approximate solution to the problem. In hydrology, practitioners have often used simple but inaccurate methods for calculating confidence intervals for x_p, although in certain cases, putting a little more effort can produce more precise results.

The purpose of the present study is to draw attention to some relatively simple techniques found in the mathematical literature for calculating confidence intervals for x_p. These give more accurate and better informative results than those commonly used in hydrology at present.

In both the annual flood series approach (composed of the largest discharge of each year observed over an N-year period of record) and the partial duration series approach (obtained by fixing a base discharge, or base level Q_b, and only observing those flood peaks that exceed Q_b), the authors feel that there is a need for hydrologists, especially practitioners, to be well informed of some of the simple and quite precise methods available for constructing confidence intervals for the quantiles of different distributions.

We intend to present some useful techniques for obtaining confidence intervals for the quantiles of the normal (lognormal), Pearson (log Pearson), Weibull (Gumbel), and Exponential distributions. We intend also to discuss the problem of constructing confidence intervals for x_p when x_p is estimated via the partial-duration-series approach, under the assumption that the distribution of the number of flood events arriving in an arbitrary time interval $[0, t]$ is Poissonian, and that of the magnitude of flood events is exponential.

2. CONFIDENCE INTERVALS FOR THE NORMAL AND LOG NORMAL DISTRIBUTIONS

The normal and lognormal distributions are often encountered in flood frequency analysis. Their probability density functions are given respectively by:

$$f(y) = \frac{1}{\sigma(2\pi)^{\frac{1}{2}}} \exp\left[- \frac{1}{2} \left(\frac{y - \mu}{\sigma}\right)^2 \right] , \quad -\infty < y < \infty \qquad (1)$$

$$f(y) = \frac{1}{\sigma\,(2\pi)^{\frac{1}{2}} y} \exp\left\{ - \frac{1}{2} \left[\frac{\ln y - m}{\sigma}\right]^2 \right\} , \quad y \geqq 0 \qquad (2)$$

The mathematical and statistical properties of the lognormal distribution have been studied in particular by Chow (1954) and by Aitchison and Brown (1957). The lognormal distribution is related to the normal by a logarithmic transformation, i.e. if Y is lognormally distributed then U = lnY will be normally distributed.

As indicated in Johnson and Welch (1940), and Stedinger (1983), let the normally distributed random variable of interest be denoted by x, and its p-th quantile by x_p. Here x has mean μ and standard deviation σ. To construct a $100(1-2\alpha)\%$ confidence interval for the unknown quantile x_p, given a set of independent observations $\{x_1, \ldots, x_n\}$ one needs to know the distribution of the random variable

$$\xi(p) = (x_p - \bar{x}) \,/\, s_x \qquad (3)$$

or a similar statistic where

$$\bar{x} = \frac{1}{n} \sum_{i=1}^{n} x_i \quad ; \quad s_x^2 = \frac{1}{n-1} \sum_{i=1}^{n} (x_i - \bar{x})^2$$

Let Z_p = the 100 p percentile of the standard normal distribution. Then $x_p = \mu + Z_p\sigma$, and equation (3) becomes

$$\sqrt{n}\,\xi(p) = \frac{\sqrt{n}[\mu + Z_p\sigma - \bar{x}]}{s_x} = \frac{\dfrac{\sqrt{n}(\mu - \bar{x})}{\sigma} + Z_p \sqrt{n}}{\{[(n-1)\, s_x^2 / \sigma^2]/ (n-1)\}^{\frac{1}{2}}} \tag{4}$$

Now $[\sqrt{n}(\mu - \bar{x}) /\sigma + Z_p \sqrt{n}]$ has a normal distribution with mean $Z_p\sqrt{n}$, and unit-variance. The term $[(n-1)s_x^2 / \sigma^2] /(n-1)$ is the ratio of a Chi-squared random variable $(n-1)\, s_x^2 / \sigma^2$ to ν where $\nu = n - 1$ is the number of degrees of freedom which characterizes this Chi-squared variable. Since $\bar{x}$ and s_x^2 are independent, $\sqrt{n}\,\xi(p)$ has a noncentral t-distribution with noncentrality parameter $\delta = Z_p\sqrt{n}$ and degrees of freedom ν. Letting $\xi_\alpha(p)$ and $\xi_{1-\alpha}(p)$ denote the 100α and $100(1-\alpha)$ percentiles of $\xi(p)$'s distribution, then:

$$\Pr \left[\xi_\alpha(p) < \frac{x_p - \bar{x}}{s_x} < \xi_{1-\alpha}(p) \right] = 1 - 2\alpha \tag{5}$$

As a result, a 100 $(1-2\alpha)$ % confidence interval for X_p is:

$$[\bar{x} + s_x\xi_\alpha(p) \;;\quad \bar{x} + s_x\, \xi_{1-\alpha}(p)] \tag{6}$$

Various quantiles of $\xi(p)$'s distribution have been computed and tables for p = 0.90, 0.98, 0.99, 0.998 are given in Stedinger (1983). Other tables are available in WRC (1977, 1981), Locks <u>et al.</u> (1973), Resnikoff and Lieberman (1957).

Since the lognormal distribution is deduced from the normal by a logarithmic transformation, confidence intervals for the lognormal variable Y can be easily obtained once we know the confidence intervals for the corresponding normal variable X.

If we consider for instance the neperian logarithm such that X = ln Y and let the confidence intervals for X be given by (6), then since $Y = e^X$ the confidence intervals for Y are given by:

$$[e^{\bar{x}+s_x\xi_\alpha(p)},\; e^{\bar{x}+s_x\, \xi_{1-\alpha}(p)}]$$

3. THE WEIBULL AND GUMBEL DISTRIBUTIONS

The density function of the Weibull distribution is given by

$$f(t) = \frac{\gamma}{\beta} \left(\frac{t}{\beta}\right)^{\gamma-1} \exp\left[-\left(\frac{t}{\beta}\right)^{\gamma}\right] , \quad t \geqq 0 \qquad (7)$$

where $\gamma > 0$, $\beta > 0$ are parameters and t is the variate. Equivalently, we consider the so - called first asymptotic distribution of extreme values (Gumbel) with density

$$f(x;u,b) = \frac{1}{b} \exp\left(\frac{x-u}{b}\right) \exp\left[-\exp\left(\frac{x-u}{b}\right)\right] , \quad -\infty < x < \infty \qquad (8)$$

This is equivalent to considering the Weibull distribution, since if T has the Weibull distribution, then $X = \ln T$ has the Gumbel distribution, with $u = \ln \beta$ and $b = 1/\gamma$ (Lawless, 1978).

The relationship between the Weibull and Gumbel distributions is therefore the same as that between the lognormal and normal. Confidence intervals for one distribution can easily be transformed into corresponding confidence intervals for the other via the logarithmic or anti-logarithmic transformation.

The Gumbel distribution has been introduced as a model for flood frequency analysis by Gumbel (1941). It has known an increasing popularity in hydrology because of its apparent theoretical justification but one should always be careful that its underlying hypotheses are not always respected (Bobée *et al.*, 1984).

The Gumbel distribution cannot, therefore, be preferred over other laws for theoretical reasons but it is still one of the distributions that are widely used in hydrology.

The Weibull distribution has also been often employed in flood frequency analysis (Matalas *et al.*, 1975; Rao, 1981; Kuczera, 1982, and others).

The problem of constructing confidence intervals for quantiles of the Weibull and Gumbel distributions with probability density functions (7) and (8) respectively, has been studied by several authors (Lawless, 1975, 1978; Thoman *et al.*, 1970; Mann and Fertig, 1977; Bain and Engelhardt, 1981, and others). Among the methods proposed, the one presented by Bain and Engelhardt (1981) is particularly simple to apply and quite accurate for most applied problems. It is based on the parameters of the Weibull distribution being estimated by the method of maximum likelihood. For the details of the approach the reader is referred to (Bain and Engelhardt, 1981), but the following example should be sufficient to demonstrate the kind of calculations involved.

Suppose that we have a flood sample of size n that is assumed to come from a Gumbel distribution Y and that we wish to find a $100(1-2\alpha)\%$ confidence interval for the pth quantile Y_p (flood with return period $T = 1/1-p$); the confidence interval is calculated as follows:

1° Calculate $d = [n / (n-1)]^{\frac{1}{2}} [\lambda_p + (\lambda_p^2 - a.b)^{\frac{1}{2}}] /a.$

where $\lambda_p = \ln(-\ln(1-p))$

$$a = 1-(Z_{1-\alpha})^2 / 2\ (n-1)$$

$$b = \lambda_p^2 - (Z_{1-\alpha})^2 / n$$

and $Z_{1-\alpha}$ is the $(1-\alpha)^{th}$ quantile of the standard normal distribution,

2° Calculate $c = [n / (n-1)]^{\frac{1}{2}} [-\lambda_p + (\lambda_p^2 - a.b)^{\frac{1}{2}}] / a$

3° Calculate $p_1 = 1- \exp(-\exp(-c))$ and $p_2 = 1- \exp(-\exp(d))$

4° Let $K(p_i)$, $i = 1, 2$ be the standardized Gumbel variate corresponding to a probability of non exceedance equal to p_i. The value of $K(p_i)$ is given by (Kite, 1978):

$$K(p_i) = -[0.45 + 0.7797 \ln(-\ln(p_i))]; i = 1,2$$

with p_1 and p_2 given above.

The interval $[K(p_1), K(p_2)]$ will now be a 100 $(1-2\alpha)$% confidence interval for the standardized flood event; a 100 $(1-2\alpha)$ % confidence interval for the observed event will be $[\tilde{\mu} + K(p_1)\tilde{\sigma}, \tilde{\mu} + K(p_2)\tilde{\sigma}]$ where $\tilde{\mu}$ and $\tilde{\sigma}$ are the mean and standard deviation estimated from the sample.

3.1 Example:

Taking $n = 40$, $\alpha = 0.05$, $p = .99$ $(T = 100)$ above, gives:

$d = 2.004$	$b = 2.2644$	$p2 = 0.9940$
$\lambda_p = 1.5272$	$c = -1.2007$	$K(p_1) = 2.1254$
$a = 0.9653$	$p_1 = 0.9639$	$K(p_2) = 3.5366$

The interval [2.1254, 3.5366] is therefore a 90% confidence interval for the standardized 100-year flood event. If the population mean μ and population standard deviation σ are estimated by $\tilde{\mu}=200$ and $\tilde{\sigma}=40$, for instance (obtained by the method of moments or ML), then the desired 90% C.I. for the observed (non standardized) 100-year event will be [285.02, 341.46].

If the flood sample is assumed to have come from a Weibull distribution X with parameters estimated by $\tilde{\gamma}$ and $\tilde{\beta}$ (eq. 7) rather than from a Gumbel distribution, then a 100 $(1-2\alpha)$ % C.I. for the p-th quantile X_p will be given by:

$$\{\tilde{\beta}\,[-\ln(1-p_1)]^{1/\tilde{\gamma}}\ ;\ \tilde{\beta}\,[-\ln(1-p_2)]^{1/\tilde{\gamma}}\}$$

where p_1 and p_2 are the same as those obtained for the Gumbel distribution.

4. EXPONENTIAL DISTRIBUTION AND WEIBULL DISTRIBUTION WITH KNOWN SKEW

The exponential distribution is a widely used distribution in the partial duration series approach. It is closely related to the Weibull distribution with known skew.

In fact if Y is a Weibull random variable such that its skewness is known, i.e. such that the parameter γ in (7) is known (because the skew of the Weibull distribution is only a function of γ), it can easily be shown (Johnson and Kotz, 1970) that the variable $U = Y^{\gamma}$ is exponentially distributed with p.d.f.:

$$f_U(u) = \frac{1}{a'} e^{-u/a'}; \quad u \geqq 0 \text{ with } a' = a^{\gamma} \tag{9}$$

Therefore, once the confidence intervals for the exponential distribution are known, the confidence intervals for the Weibull distribution with known skew can be easily derived (and vice versa).

If $\{U_1, \ldots, U_n\}$ is a random sample from the exponential distribution with p.d.f. (9) and a' is estimated by its maximum likelihood estimator (Johnson and Kotz, 1970):

$$a' \simeq \bar{U} = \frac{1}{n} \sum_{i=1}^{n} U_i \tag{10}$$

then the limits:

$$2n\,\bar{U} / \chi^2_{2n,\,1-\alpha} \text{ and } 2n\,\bar{U} / \chi^2_{2n,\,\alpha} \tag{11}$$

will form a 100 (1-2α) % confidence interval for a', where $\chi^2_{m,p}$ stands for the p-th quantile of a chi-squared distribution with m degrees of freedom.

Now for the exponential distribution with p.d.f. (9) we have:

$$P[U \leqq U_p] = 1 - e^{-U_p/a'} = p \tag{12}$$

which when solved for a' gives:

$$a' = -U_p / \ln(1-p) \tag{13}$$

From confidence limits (11) for a', and relationship (13) between a' and U_p, one can easily obtain the corresponding 100 (1-2α)% confidence limits for U_p, the p-th quantile of the exponential distribution:

Lower confidence limit = $-\, 2n\,\bar{U} \ln(1-p) / \chi^2_{2n,1-\alpha}$

Upper confidence limit = $-\, 2n\,\bar{U} \ln(1-p) / \chi^2_{2n,\alpha}$

5. TWO-PARAMETER EXPONENTIAL DISTRIBUTION

Let Y now be a 2-parameter exponential random variable with p.d.f.

$$f_Y(y) = \frac{1}{a} e^{-(y-m)/a} \; ; \; y \geq m$$

A method for constructing confidence intervals for the pth quantile Y_p is discussed by Lawless (1982). According to this method, a $100(1-2\alpha)\%$ C.I. for Y_p is given by:

$$[\tilde{a}\,(-\ln\,(1-p_1)) + \tilde{m}; \quad \tilde{a}\,(-\ln\,(1-p_2)) + \tilde{m}]$$

where $\tilde{a}$ and $\tilde{m}$ are estimates for a and m, and:

$$p_1 = 1 - e^{-nk_1}$$

$$p_2 = 1 - e^{-nk_2}$$

n is the size of the sample

$$k_1 \simeq \frac{1}{n} \left[- m(p) - Z_{1-\alpha} \left(\frac{m^2(p)}{n} + \frac{1}{n^2} \right)^{\frac{1}{2}} \right]$$

$$k_2 \simeq \frac{1}{n} \left[-m(p) - Z_{\alpha} \left(\frac{m^2(p)}{n} + \frac{1}{n^2} \right)^{\frac{1}{2}} \right]$$

$$m(p) = [1 + n \ln\,(1-p)] \; / \; (n - 5/2)$$

Z_α and $Z_{1-\alpha}$ are the αth and $(1-\alpha)^{th}$ quantiles of the standard normal distribution.

6. PEARSON TYPE 3 AND LOG-PEARSON TYPE 3 DISTRIBUTIONS

The log Pearson type 3 (LP) distribution is deduced from the Pearson type 3 (P) distribution by a logarithmic transformation: If $Y = \ln X$ follows a Pearson type 3 distribution, X follows a log-Pearson type 3 distribution. It can be shown (Bobée, 1975) that the density function of the variate X is given by:

$$f(x) = \frac{|\alpha|}{\Gamma(\lambda)} \frac{e^{\alpha m}}{x^{1+\alpha}} [\alpha\,(\ln x - m)]^{\lambda-1}$$

$$\alpha > 0, \; e^m \leq x \qquad\qquad (14)$$
$$\alpha < 0, \; 0 < x \leq e^m$$

Similarly, the p.d.f. of the P distribution Y is given by:

$$f(y) = \frac{|\alpha|}{\Gamma(\lambda)} e^{-\alpha (y-m)} \left[\alpha (y-m)\right]^{\lambda-1} \tag{15}$$

$$\alpha > 0, \quad m \leqq y < \infty$$

$$\alpha < 0, \quad -\infty < y \leqq m$$

where $\Gamma(\cdot)$ is the gamma function.

Until recently, methods for constructing confidence intervals for quantiles of the P and LP distributions have been very crude. Some of the early methods proposed were based on asymptotic theory (WRC 1977, 1981; Kite 1975). Stedinger (1983) was the first to introduce confidence intervals for P and LP quantiles in hydrology not completely based on asymptotic theory. His approach was based on the idea of adjusting confidence intervals for quantiles Y_p of the normal distribution to obtain approximate confidence intervals for P quantiles X_p. Stedinger's method gives good results, but is valid only when the skew coefficient of the P variable is assumed to be known, which is an assumption seldom justified in practice.

In fact, there is no theoretical reason why Y should be taken to be normal. This is why Ashkar et al. (1986) proposed that one should search for the best-fit distribution Y for producing confidence intervals for P quantiles. They considered the normal, lognormal, Weibull, Gumbel and exponential distributions. They came out with a method of adjusting confidence intervals for Weibull quantiles Y_p to produce approximate confidence intervals for P and LP quantiles X_p in the case of unknown skew. Their method is based on certain similarities between the Weibull and Pearson distributions.

The method proposed by Ashkar et al. (1986) is as follows: Suppose that a method exists for constructing confidence intervals for quantiles Y_p of the random variable Y (Y is Weibull in this case). In other words, for any 100(1-2α)% confidence level, upper and lower confidence limits $U_\alpha(p)$ and $L_\alpha(p)$ can be calculated such that:

$$P\left[L_\alpha(p) \leqq Y_p \leqq U_\alpha(p)\right] = 1-2\alpha$$

Let

$$p_1 = P\left[Y \leqq L_\alpha(p)\right], \quad \text{i.e. } L_\alpha(p) = Y_{p_1}$$

and

$$p_2 = P\left[Y \leqq U_\alpha(p)\right], \quad \text{i.e. } U_\alpha(p) = Y_{p_2}$$

Knowing $L_\alpha(p)$ and $U_\alpha(p)$, p_1 and p_2 can be calculated using the relationships: $p_1 = F_Y\left[L_\alpha(p)\right]$ and $P_2 = F_Y\left[U_\alpha(p)\right]$. Once p_1 and p_2 have been obtained, values of K_{p_1} and K_{p_2} (frequency factor) can be obtained for the P distribution X from the Harter tables (1964, 1969) or by using approximation formulas such as the Wilson-Hilferty transformation (Wilson & Hilferty, 1931), modified Wilson-Hilferty transformation (Kirby, 1972) or Cornish-Fisher transformation (Fisher and Cornish, 1960). Consequently the interval

$$[L'_\alpha(p) \; ; \; U'_\alpha \; (p)] = [\bar{X} + K_{p_1} \; S_x \; ; \; \bar{X} + K_{p_2} \; S_x]$$

will be the desired 100 $(1-2\alpha)$% C.I. for the P quantile X_p.

For a mathematical justification of the above method the reader can refer to Ashkar et al. (1986) but here we give a simple numerical example:

6.1 Example:

Let us assume that a sample $x_1, \ldots, x_n$ of flood data of size n=40 is available which can be assumed to come from a P distribution. Let the mean, standard deviation and coefficient of skewness of the sample be given respectively by:

$$\bar{x} = 180 \text{ m}^3/\text{sec.}$$
$$s_x = 35 \text{ m}^3/\text{sec.}$$
$$c_s = .8$$

Suppose that we wish to use this information to construct a 90% $(1 - 2\alpha = .90$, i.e. $\alpha = .05)$ C.I. for the 100-year event $x_{.99}$.

For $n = 40$, $\alpha = .05$ and $p = .99$ we have already obtained in the example for the Weibull (Gumbel) distribution the values: $p_1 = .9639$ and $p_2 = .9940$. In the Harter Tables (WRC, 1981) we find for $C_s = .8$:

$$K_{.9639} \simeq 2.075 \quad \text{and } K_{.9940} \simeq 3.228$$

According to the method of Ashkar et al. (1986), the desired 90% confidence limits for the 100-year event under a Pearson type 3 model will be:

lower confidence limit $\simeq 180 + 2.075 \times 35 \simeq 253 \text{ m}^3/\text{sec.}$
upper confidence limit $\simeq 180 + 3.228 \times 35 \simeq 293 \text{ m}^3/\text{sec.}$

7. PARTIAL DURATION SERIES APPROACH

As mentioned in the introduction, a second type of flood data used in flood frequency analysis are partial flood series (p.f.s.) which are also called exceedance series (e.s.). These series are obtained by fixing a base discharge, or base level Q_b, and only observing those flood peaks that exceed Q_b. The reader may refer to Todorovic and Zelenhasic (1970) for a more detailed description of the e.s. approach.

The exponential distribution (with parameter β) as model for flood magnitude, and the Poisson distribution (with parameter λ) as model for flood count, are among the most widely used distributions in the e.s. approach. They are not the only distributions used, however, but their widespread applicability seems to warrant special consideration. In this e.s. model based on the Poisson and exponential distributions, all exceedances within the year are taken to be independent and identically distributed according to the exponential distribution. The statistical uncertainty in the estimation of the T-year event X_T for this model has already been studied by Ashkar and Rousselle (1981) who derived the p.d.f. of the estimate $\tilde{X}_T$ for an arbitrary return period T:

$$f_{T,N}(x;\lambda,\beta) = \begin{cases} \dfrac{e^{-N\lambda}}{x} \sum_m^* \left[m\, N\, \lambda\, \beta\, x \exp(-\beta x/u)\ /u\right]^m / m!(m-1)! & \text{for } x > 0 \\ 1 - \sum_m^* \dfrac{e^{-N\lambda}(N\lambda)^m}{m!} & \text{for } x = 0 \end{cases} \qquad (16)$$

where the summation $\sum^*$ is over all positive integers m that are greater than ln $[T/(T-1)]^N$.

An advantage of this p.d.f. $f_{T,N}(x)$ of $\tilde{X}_T$ is that it is not asymptotic. Therefore, it can be used to calculate the confidence limits around X_T at any confidence level $(1-\alpha)$ and any sample size N. It is to be noted that calculation of the asymptotic variance of $\tilde{X}_T$ has already been done by Cunnane (1973) but this was based on the assumption that Cov $(\tilde{\lambda}, \tilde{\beta}) = 0$, which is only true for large N. Ashkar and Rousselle (1981) and also Tavares and Da Silva (1983) have shown that, in the case of small sample sizes encountered in hydrology, this assumption of independence of $\tilde{\lambda}$ and $\tilde{\beta}$ can lead to considerable error.

Ashkar and El-Jabi (1986) have recently constructed tables, based on equation (16), that give probability limits for $\tilde{X}_T$ for values of T, N, λ and β that are of most interest in flood frequency analysis. This table can be used to obtain 90, 95 and 99 % probability limits for $\tilde{X}_T$ for return periods T = 10, 20, 50, 100, 200, 500, 1000 and 10,000 in the case when $10 < N < 100$ and $1 < \lambda < 4$.

8. CONCLUSION

The approaches proposed in the present study can be very useful to hydrologists who have to deal with small sample sizes. It has already been shown that methods based on asymptotic theory do not provide very good results for small samples and for this reason it is hoped that methods summarized in the present study will be employed more and more by hydrologists in practice. In the design of flood-control structures, what is important is not only the estimation of design events X_T, but also upper confidence limits for these events, calculated with as much precision as possible. Methods developed specifically for small sample sizes should therefore be considered as very valuable by practitioners to achieve this kind of precision.

9. REFERENCES

Aitchison, J., and Brown, J.A.C.: 1957, The Lognormal Distribution with Special Reference to Its Uses in Economics. Cambridge University Press, London.

Ashkar, F., and El-Jabi, N: 1986, 'Tables for calculating probability limits for design flood events under a partial duration series model'. Submitted to Water Resources Research.

Ashkar, F., and Rousselle, J.: 1981, 'Design discharge as a random variable: A risk study'. Water Resources Research, **17(3)**: 577-591.

Ashkar, F., Bobée, B., and M. Paradis: 1986, 'Confidence intervals for Pearson III and Log Pearson III quantiles. Water Resources Research (submitted for publication).

Bain, L.J., and, Engelhardt, M.: 1981, 'Simple approximate distributional results for confidence and tolerance limits for the Weibull distribution based on maximum likelihood estimators'. Technometrics, **23(1)**: 15-20.

Bobée, B.: 1975, 'The log-Pearson type 3 distribution and its application in hydrology'. Water Resources Research, **11(5)**: 681-689.

Bobée, B., Cavadias, G., and Ashkar, F: 1984, 'Statistical analysis of floods as extreme hydrologic events'. Fifth International Conference on Water Resources Planning and Management, Athens, Greece, October 1-4: 5.158-5.184.

Chow, V.T.: 1954, 'The log-probability law and its engineering applications'. Proc. Am. Soc. Civ. Engineering. **80(5)**: 1-25.

Cunanne, C.: 1973, 'A particular comparison of annual maxima and partial duration series methods of flood frequency prediction'. Journal of Hydrology, **18**: 257-271.

Fisher, R.A. and, Cornish: 1960, 'The percentile points of distributions having known cumulants'. Technometrics, **2(2)**: 209-225.

Gumbel, E.J.: 1941, 'The return periods of flood flows'. Annals of Mathematical Statistics, **12**: 163-190.

Harter, H.L.: 1964, 'New Tables of the Incomplete Gamma-Function Ratio and of Percentage Points of the Chi-Square and Beta Distributions' Aerospace Research Laboratories, Office of Aerospace Research, U.S. Air Force, 245 p.

Harter, H.L.: 1969. 'A New Table of Percentage Points of the Pearson Type 3 Distribution'. Technometrics, **2(1)**: 177-187.

Johnson, W.L. and, S. Kotz: 1970, Distributions in Statistics: Continuous Univariate Distributions 1. Houghton-Mifflin, Boston, Mass.

Johnson, N.L. and, Welch, B.L.: 1940, 'Applications of the non-central t-distribution'. Biometrika, **31**: 362-389.

Kirby, W.: 1972, 'Computer-oriented Wilson-Hilferty transformation that preserves the first three moments and the lower bound of the Pearson type 3 distribution'. Water Resources Research, **8(5)**: 1251-1254.

Kite, G.W.: 1975, 'Confidence limits for design events'. Water Resources Research, **11(1)**: 48-53.

Kite, G.W.: 1978, 'Frequency and risk analyses in hydrology'. Water Resources Publications, Fort collins, Colorado, 224. p.

Kuczera: 1982, 'Robust flood-frequency models'. Water Resources Research, **18(2)**: 315-324.

Lawless, J.F.: 1975, 'Construction of tolerance bounds for the extreme value and Weibull distributions'. Technometrics, **17**: 225-261.

Lawless, J.F.: 1978, 'Confidence interval estimation for the Weibull and extreme value distributions'. Technometrics, **20**: 225-261.

Lawless, J.F.: 1982, Statistical Models and Methods for Lifetime Data, Wiley, New York, N.Y. 580 p.

Locks, M.O., Alexander, M.J. and, Byars, B.J.: 1973, 'New tables of the non-central t-distribution'. Aerospace Research Laboratories Report ARL 63-19. Aeronautical Research Laboratories, Wright-Patterson Air Force Base, Ohio.

Mann, N.R. and, Fertig, K.W.: 1977. 'Efficient unbiased quantile estimators for moderate-size complete samples from extreme value and Weibull distributions: confidence bounds and tolerance and prediction intervals'. Technometrics, **19**: 87-93.

Matalas, N.C., Slack, J.R. and, Wallis, J.R.: 1975, 'Regional skew in search for a parent'. Water Resources Research, **11(6)**: 815-826.

Rao, D.V.: 1981, 'Three-Parameter probability distributions'. Journal of the Hydraulics Division, ASCE **107(HY3)**: 339-358, March. Process Paper 16124.

Resnikoff, G.J. and, Lieberman, G.J.: 1957. Tables of the non-central t-distribution. Stanford University press, Stanford, Calif.

Stedinger, J.R.: 1983, 'Confidence intervals for design events'. Journal Hydraulics Division, Am. Soc. Civ. Eng., **109(HY1)**: 13-27.

Tavares, L.V. and, Da Silva, J.E.: 1983, 'Partial duration series method revisited'. Journal of Hydrology, **64**: 1-14.

Thoman, D.R., Bain, L.J. and, Antle, C.E.: 1970, 'Reliability and tolerance limits in the Weibull distribution'. Technometrics, **12**: 363-371.

Todorovic, P. and, Zelenhasic, E.: 1970, 'A Stochastic model for flood analysis'. Water Resources Research, **6(6)**: 1641-1648.

Water Resources Council: 1977, 'Guidelines for Determining Flood-Flow Frequency'. Hydrology Committee, Bulletin 17A, Washington, D.C.

Water Resources Council: 1981. 'Guidelines for Determining Flood-Flow Frequency'. Hydrology Committee, Bulletin 17B, Washington, D.C.

Wilson, E.B. and, Hilferty, M.M.: 1931, 'The distribution of chi-square'. proceedings, National Academy of Science, New York, **17(12)**: 684-688.

USE OF ADDITIONAL HISTORICAL INFORMATION FOR ESTIMATION AND GOODNESS OF FIT OF FLOOD FREQUENCY MODELS

J. BERNIER - J. MIQUEL - A. LEBOSSE - A. GRIFFET
LABORATOIRE NATIONAL D'HYDRAULIQUE
6 Quai Watier - 78400 CHATOU - FRANCE

ABSTRACT. For a more reliable estimation of design floods, it is necessary to take in account sampling and model uncertainties. A better control of the involved errors requires the use of all available informations, and especially we emphasize the importance of additional historical data. These data are essentially the more or less precise estimations of the few extreme events observed in the past before the regular observation of the river.

In the first part of the paper we recall the way this information is integrated in the estimation process of classical parametric model such as partial duration curves. The results are associated with their approximate confidence intervals especially in the case of design flood discharges estimation. The involved approximations are now in progress and will be discussed. Moreover these confidence intervals constitute a first element for judging the quality of the fits of the models.

In a more developped second part, we emphasize the importance of the non parametric estimation of probabilities (sometimes called empirical frequencies) taking in account additional historical information for the graphical validation of the model.

The confidence intervals of these non parametric estimations are as important as the previous ones calculated with the parametric models. Both intervals constitutes the basis of a better tool for choosing the adequate probability distribution for floods on one hand, and a better appraisal of sampling and model uncertainties on the other hand.

For illustration the described methods are applicated at two rivers.

V. P. Singh (ed.), Application of Frequency and Risk in Water Resources, 153–164.

1. INTRODUCTION

The main interest of a probabilistic assessment of design floods is the characterisation and control of the various uncertainties involved, especially sampling and model uncertainties. A way for reducing the effect of these uncertainties is the use of additional historical data, such as the few extreme events observed in the past before the regular and systematic recording of data at a given river site.

Previous works have presented estimation methods of classical parametric models used in partial duration series approach (Bernier, Miquel, 1977), (Miquel, 1984). The following section sums up the general methodology used and some results obtained, mainly the significant reduction of sampling errors connected with the estimation of design discharges corresponding to large return periods.

These more complete data should be interesting enough to reduce model uncertainties in the domain of small exceedances probabilities by allowing a better choice of distribution function model based on more complete information in the domain. But this choice can be altered if the empirical probabilities of the extreme events (plotting positions in graphical meaning) have bad estimates. The main part of this paper is devoted to the discussion of the estimation problem concerning these empirical probabilities and using the statistical concept of missing data. This concept have had significant developments in statistical fields since the classical work of Dempster and al (1977). It should be noted that the methods based on this concept have large possibilities of applications in hydrology. Indeed the available information is generally far from owning the regular, symetrical property required by the classical estimation methods.

2. USE OF HISTORICAL INFORMATION IN PARAMETRIC MODELS

Let us consider the classical partial duration series approach of statistical analysis of floods in which all peak discharges of floods above a given threshold qo are selected (see the following fig.I)

Figure I

CLASSICAL PARTIAL DURATION SERIES APPROACH

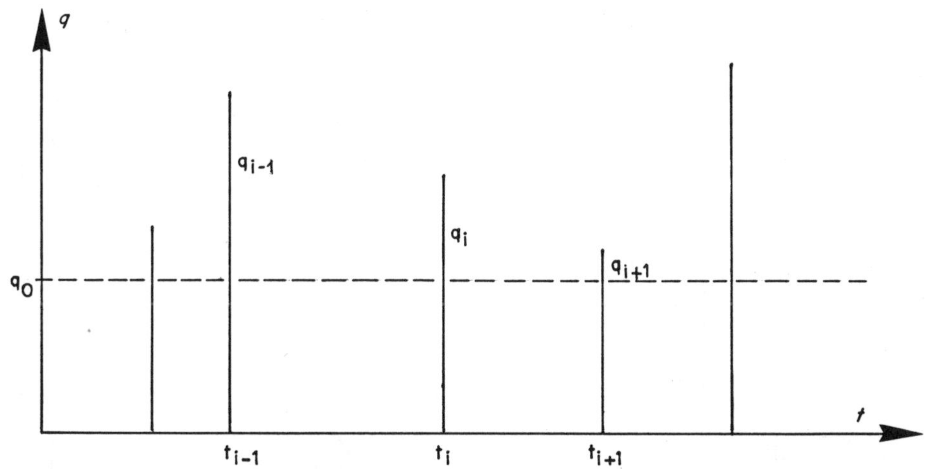

Based on the assumptions of identical distribution functions F of the q_i and the independence of successive selected q_i, we have the following expression for the distribution function of the maximum discharge Q_M over a period T (generally the year) :

$$H(q) = \text{prob}\left[Q_M \leqslant q\right] = G_T\ (F) \tag{1}$$

where G_T (u) is the generating function of the random number n of selected floods over T :

$$G_T\ (u) = \sum_{n=o}^{\infty} P_n\ (T)\ u^n$$

Classical models used in practice are :

- Poisson distribution of n : $P_n(T) = e^{-\mu T} \dfrac{(\mu T)^n}{n!}$

- Weibull distribution of q : $F(q) = 1 - e^{-\rho q^p}$

so that : $G_T(u) = e^{-\mu T(1-u)}$

and $$H(q) = e^{-\mu T e^{-\rho q^p}} \tag{2}$$

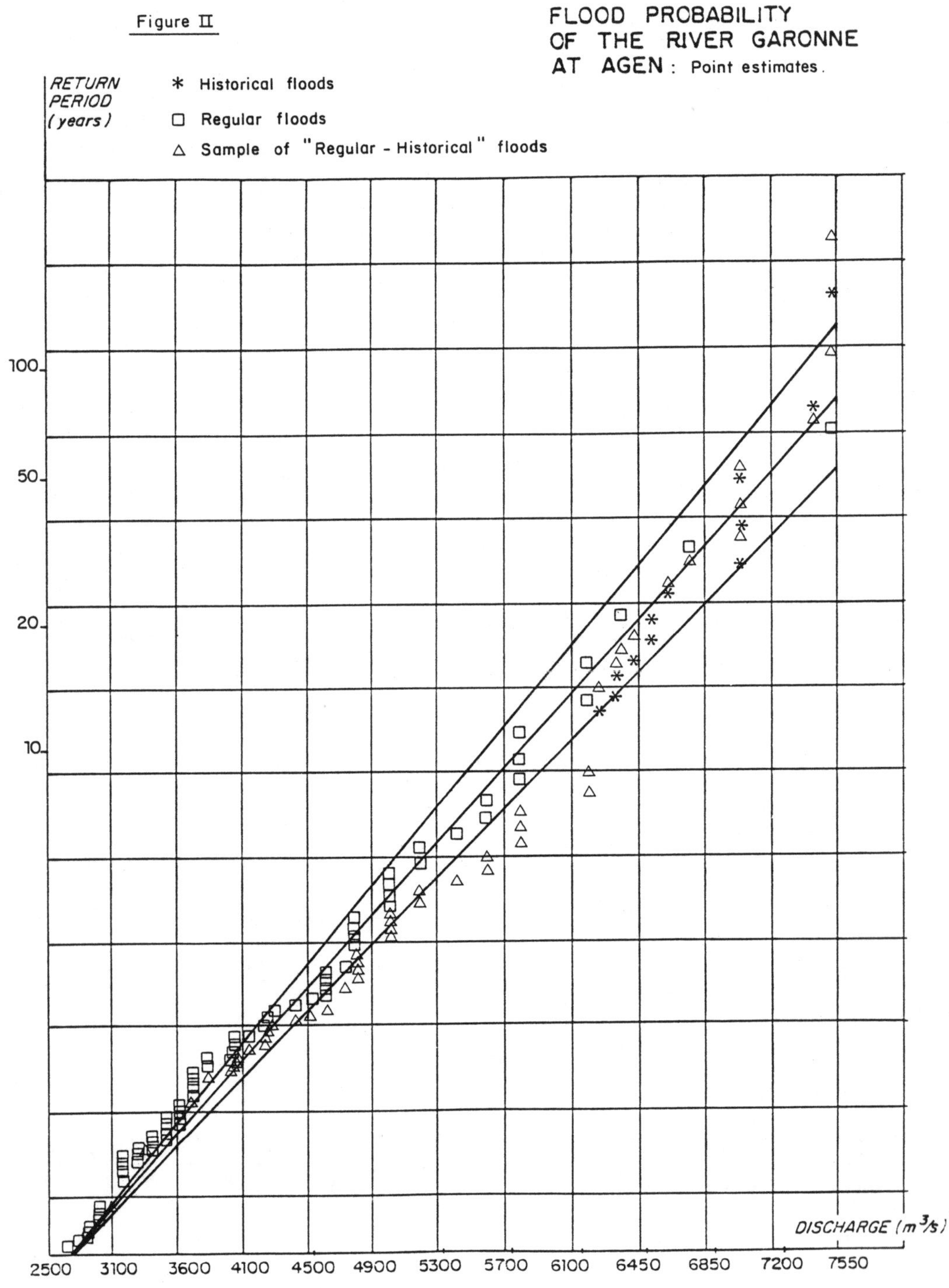
Figure II
FLOOD PROBABILITY
OF THE RIVER GARONNE
AT AGEN : Point estimates.
RETURN
PERIOD
(years)
* Historical floods
□ Regular floods
△ Sample of "Regular - Historical" floods
100
50
20
10
DISCHARGE (m³/s)
2500
3100
3600
4100
4500
4900
5300
5700
6100
6450
6850
7200
7550

The figure II shows an example of application for Garonne River at Agen in France, in which the probabilities F are directly graduated in annual probabilities H with respect to the formula (2).

Generally the fitting of this model uses information collected during systematic recording period. But often we can have additional historical data, as shown in figure II, which are not systematically observed but which are sparse notable extreme events in the past. We shall assume these historical floods be ranked in decreasing order of their maximum discharges :

$$Q'_1 > Q'_2 > \dots > Q'_{NP}$$

knowing they are the NP largest floods in N years before the systematic observation period has been recorded.

Parametric estimation methods have been developed, based on Bayes approach (Bernier, Miquel, 1977), (Miquel, 1984). This approach uses the posterior distribution of the three "parameters" μ , ρ, and p, and defines the posterior mode as point estimate. Furthermore there exists an approximate method for calculating sampling variances of these parameters. This method uses a trivariate normal distribution centered at the posterior mode and with variances and covariances calculated through a limited expansion around this mode.

For example in the Garonne River case where it has been used 65 years of systematic records, 12 extreme events observed on 143 years above the threshold q_o = 2500 m^3/s, we estimated the 1000 years flood associated with its 70% confidence interval ; the results are :

- without historical information : 8700 $m^3/s < Q_{1000} < 10500$ m^3/s

- with historical information : 9100 $m^3/s < Q_{1000} < 10100$ m^3/s

So the confidence interval goes from 1800 m^3/s to 1000 m^3/s which is a very significant decrease.

These historical data are often neglected because they are considered as unreliable. As a matter of fact they are of the uppermost interest even if they are marked with errors of measurements or recording, at which the estimated results are not so sensitive in many cases as shown by simulation analyses of the method. Other sensitivity analyses have been worked out for validation purposes. Namely the normality assumption for posterior distribution of parameters and quantiles has been verified using Monte-Carlo analyses.

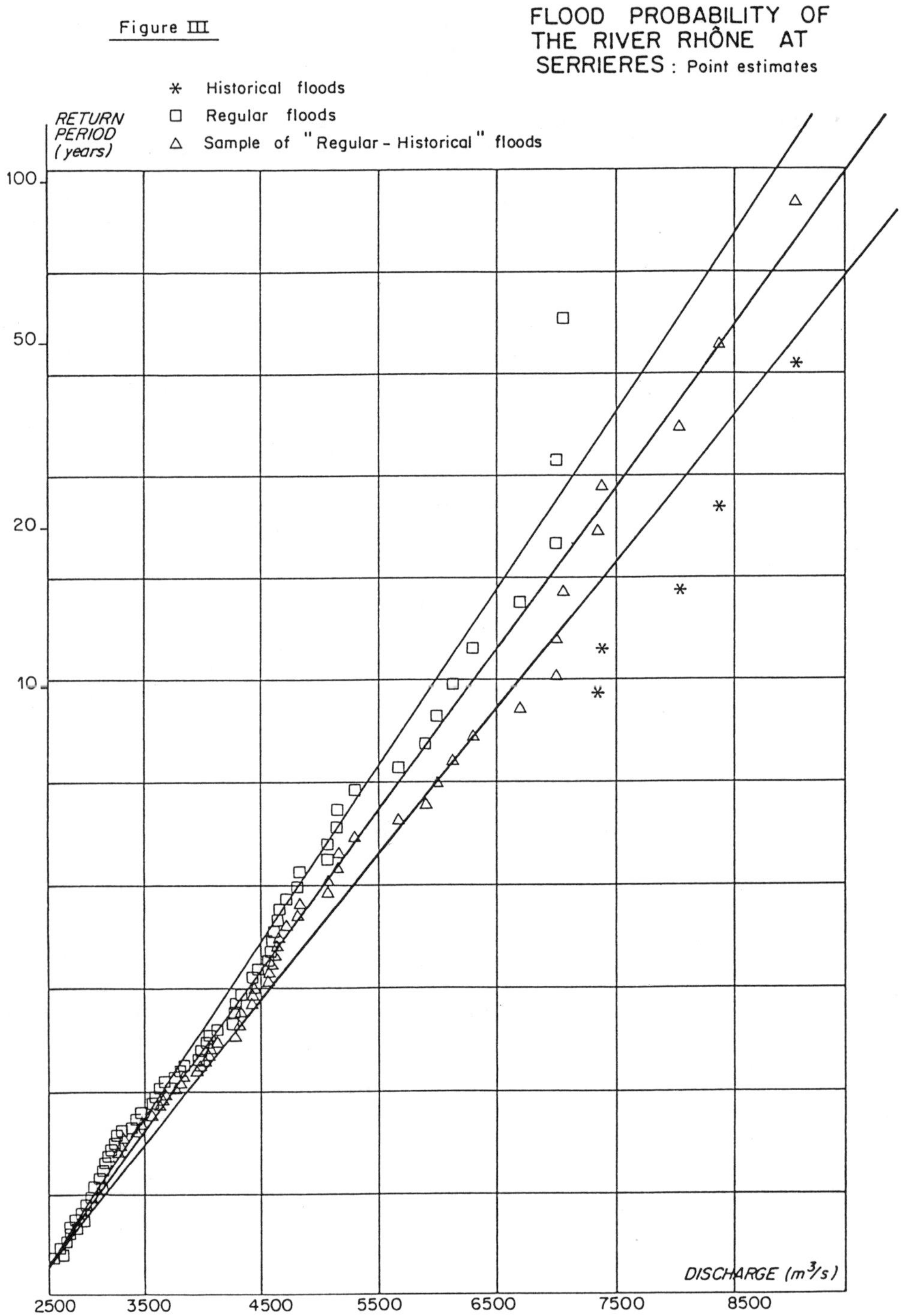
Figure III
FLOOD PROBABILITY OF THE RIVER RHÔNE AT SERRIERES : Point estimates
* Historical floods
□ Regular floods
△ Sample of "Regular - Historical" floods
RETURN PERIOD (years)
100
50
20
10
DISCHARGE (m³/s)
2500
3500
4500
5500
6500
7500
8500

3. DEALING WITH MODEL UNCERTAINTIES

3.1. Tests of Fit and Plotting Positions

An useful property of the statistical flood model described here is that the Poisson assumption is not necessary for large floods. Interpreting the parameter μT as the average number of selected floods above the threshold during the period T, this number can be distributed according any probability law, the result given by formula (2) is still asymptotically valid. On the contrary assumption like Weibull distribution regarding the intensities of floods is important and must be tested above all in the domain of large events.
Practically the model assumption is verified by comparing the theoretical parameter fitting using formula (2) with empirical estimation of probabilities of the observations ranked in increasing order as usual. For instance one of the most useful formula is the Weibull one :

$$f_i = \frac{R_i}{NC_1 + 1} \qquad (3)$$

where : f_i is the estimation of non-exceedance probability of observation q_i , R_i is the rank of q_i in the ordered sample of size NC_1.

Note that all the other formulae for empirical probalities are still function of rank R_i as the previous one.
In our previous works concerning historical sample, we considered this sample as independent of the systematic one, and estimated the non-exceedance probabilities of each historical observation Q'j by :

$$f'_j = \frac{R_j + NC_2 - NP}{NC_2 + 1}$$

NC_2, number of floods above the threshold during the historical period, is an unknown random variable distributed according Poisson law with expectation $\mu.NA_2$ (where NA_2 is the number of years in the historical period). We used the approximation :

$$f'_j \quad \frac{R_j + \mu\, NA_2 - NP}{\mu\, NA_2 + 1}$$

The plotting positions in figure II and also in figure III showing the case of Rhône River at Serrières have been performed according these principles. But the fact of dealing with independent estimations of probabilities can introduce breakdowns in the estimated probabilities when the theoretical distribution is continuous. This appears in the figure III concerning Serrières and we must ask the question :
is this behaviour due to model uncertainty by Weibull distribution or is it due to the plotting position ? The answer requires an analysis of the problem of plotting positions assessment.

3.2. Mixing of the Systematic and Historical Samples

Let us consider the union of the systematic and historical samples with total size equal to $NC_1 + NC_2$. A Weibull type estimation of non-exceedance probability for a given observation q_i in this sample is :

$$f_i^* = \frac{R_i}{NC_1 + NC_2 + 1} \tag{4}$$

where R_i is the rank of q_i in the total ordered sample and NC_2 is assumed known here.

Two cases are to be considered :

1 : q_i is larger then q'_{NP}, smallest known observation in the historical sample ; then R_i is perfectly determined and :

$$f_i^* = \frac{i^* + NC_2 - NP}{NC_1 + NC_2 + 1} \tag{5}$$

where i* is the known number of recorded observations lower or equal to q_i in the total sample (systematic + historical)

2 : q_i is lower or equal to q'_{NP} ; then R_i is random owing to the unknown missing historical observations lower then Q'_{NP}. It can be shown that the probability distribution of R_i is given by :

$$\text{Prob}\left[R_i = i + s \right] = \frac{i}{i+s} \; \frac{C^{s}_{NC_2-NP} \cdot C^{i}_{NC_1}}{C^{i+s}_{NC_1+NC_2-NP}} \tag{6}$$

where i is the known number of systematic observations lower or equal to q_i, and s has a value between 0 and $NC_2 - NP$

From formula (6) the expectation of R_i can be calculated :

$$E\left[R_i \right] = i \; \frac{NC_1 + NC_2 - NP + 1}{NC_1 + 1} \tag{7}$$

By using formulae (4) and (7) an estimation of the non-exceedance probability of q_i can be obtained :

$$\hat{f}_i = E\left[f_i^* \right] = \frac{i\left[NC_1 + NC_2 - NP + 1 \right]}{(NC_1 + 1)\ (NC_1 + NC_2 + 1)} \tag{8}$$

The methodology for obtaining formula (8) must be emphasized ; it is to consider the complete sample : unknown observations together with known observations, and to estimate the effect of unknown data by taking their conditional distribution given the known data. It is a part of the missing information principle used by Dempster and al. (1977) in the more general context of maximum likelihood method.

Figure IV

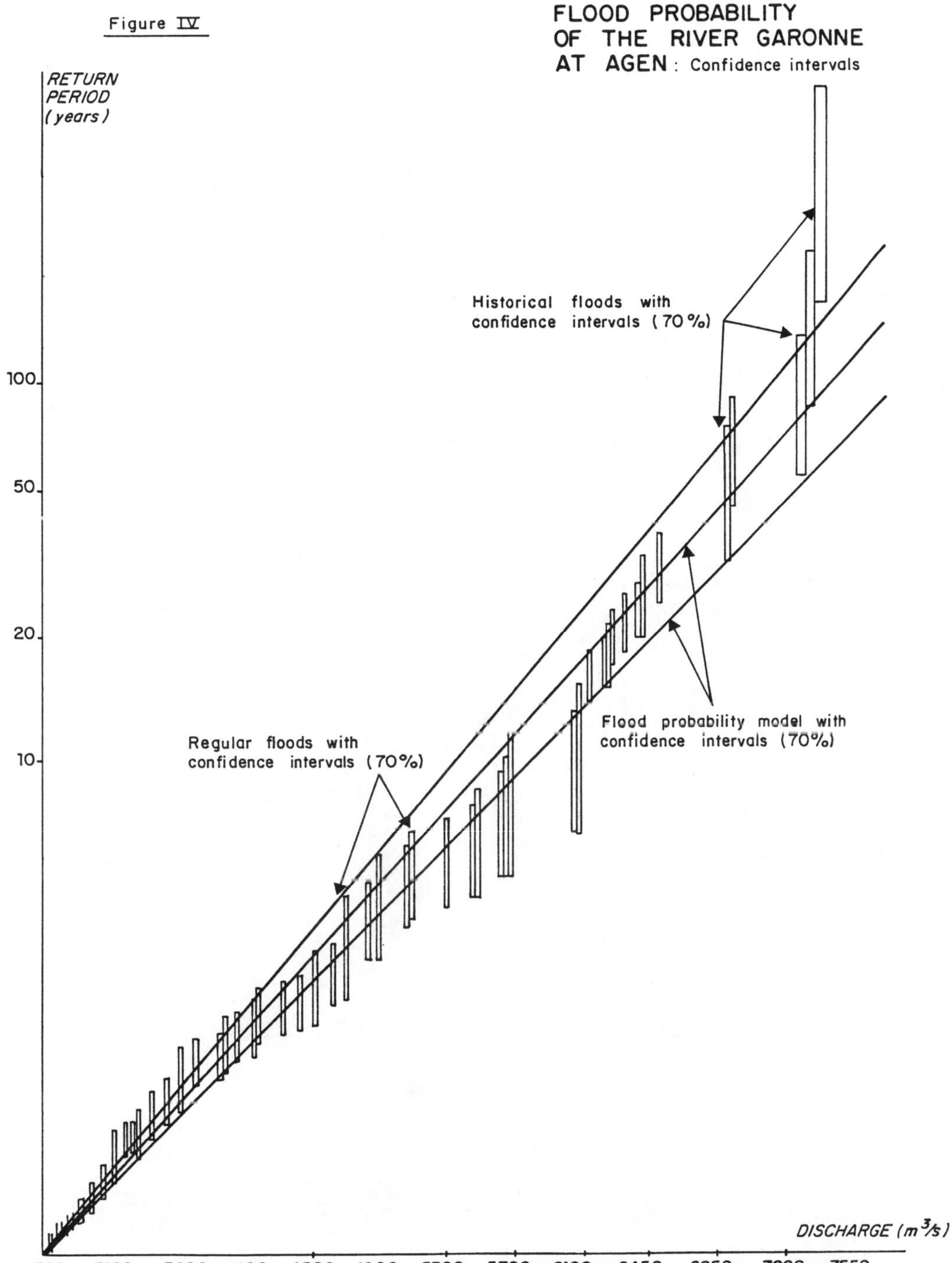
FLOOD PROBABILITY
OF THE RIVER GARONNE
AT AGEN : Confidence intervals
RETURN
PERIOD
(years)
100
50
20
10
Historical floods with
confidence intervals (70%)
Regular floods with
confidence intervals (70%)
Flood probability model with
confidence intervals (70%)
DISCHARGE (m³/s)
2500
3100
3600
4100
4500
4900
5300
5700
6100
6450
6850
7200
7550

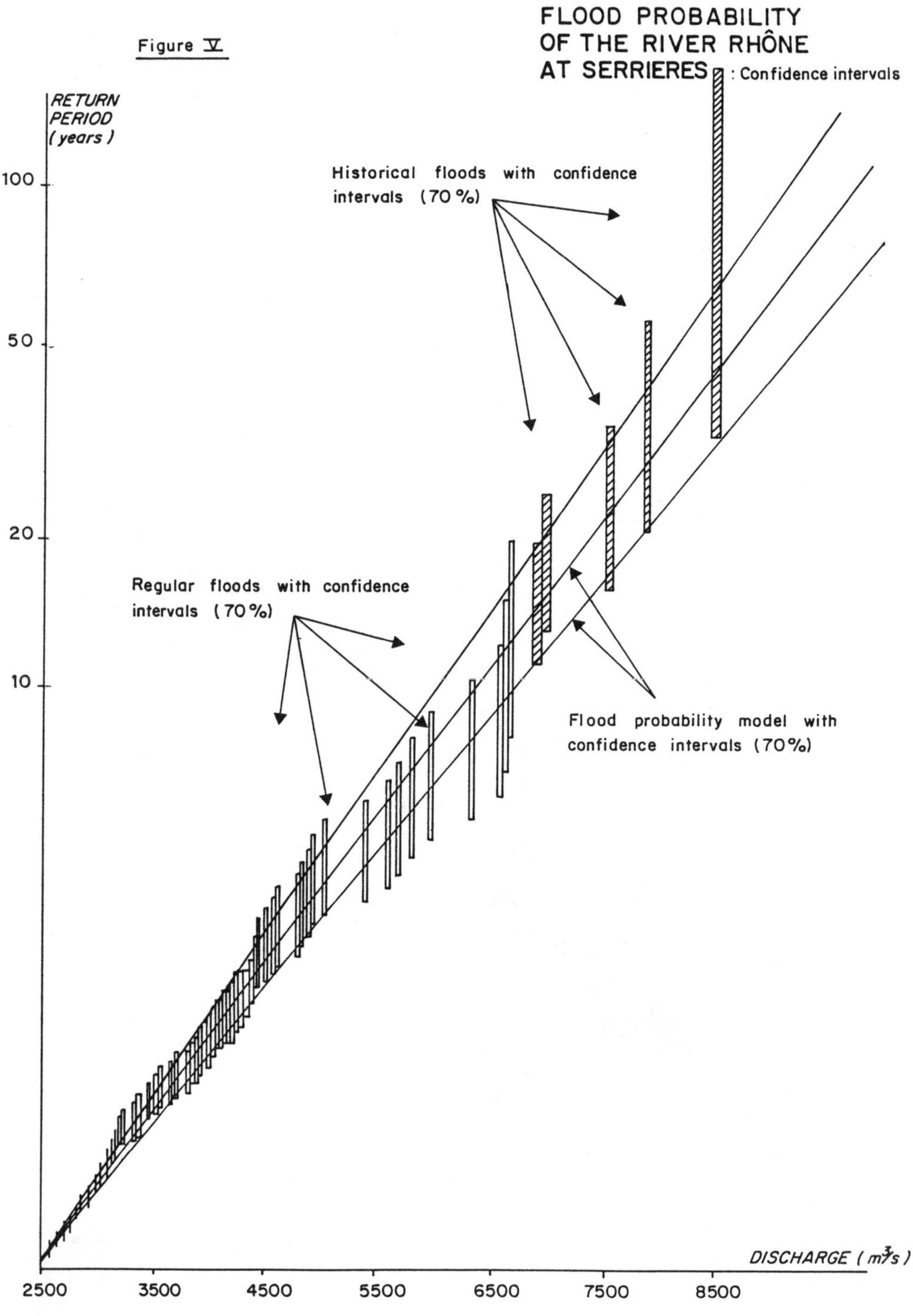
Figure V
FLOOD PROBABILITY
OF THE RIVER RHÔNE
AT SERRIERES
: Confidence intervals
RETURN
PERIOD
(years)
100
50
20
10
Historical floods with confidence
intervals (70 %)
Regular floods with confidence
intervals (70 %)
Flood probability model with
confidence intervals (70%)
DISCHARGE (m³/s)
2500
3500
4500
5500
6500
7500
8500

For using pratically the formula (8) it is necessary to estimate NC_2. An approximation is to put :

$$NC_1 \simeq \mu \, NA_2$$

with μ obtained by the parametric model. A theoretical analysis shows that the approximation is fairly accurate for not too small values of NA_2, as usual in pratice. Figures II and III show the application of the method to the Garonne and Rhône examples. For the last case the estimation of probabilities for historical data appears homogeneous with those of the systematic data and more comparable to the Weibull model than the previous ones.

3.3. Comparison Between Empirical and Theoretical Fitting

For comparing the non-parametric plotting positions with theoretical fitting like that one given by Weibull model associated with its confidence intervals, it is necessary to take in account sampling errors in the plotting position such as those given by formulae (5) an (8). As a matter of fact, i* and i are random variables, the distributions of which can be considered binomial with probability $F(q_i)$ and respective sizes equal to NC_1 + NP and NC_1 . This result is exact in the i case and only an approximation in the i* case. Therefore, using formulae (5) and (8) we can obtain approximate confidence intervals with size $1-\alpha$ with respect to $F(q_i)$:

Case : $q_i > q'_{NP}$:

$$f_i^* - \frac{u_{\alpha/2}\sqrt{f_i^*(1-f_i^*)(NC_1+NP)}}{NC_1 + NC_2 + 1} < F(q_i) < f_i^* + \frac{u_{\alpha/2}\sqrt{f_i^*(1-f_i^*)(NC_1+NP)}}{NC_1 + NC_2 + 1}$$

Case : $q_i < q'_{NP}$:

$$f_i - \frac{u_{\alpha/2}\sqrt{\hat{f}_i(1-\hat{f}_i).NC_1.(NC_1+NC_2-NP+1)}}{(NC_1 + 1)(NC_1 + NC_2 + 1)} < F(q_i) < f_i + \frac{u_{\alpha/2}\sqrt{\hat{f}_i(1-\hat{f}_i).NC_1(NC_1+NC_2-NP+1)}}{(NC_1 + 1)(NC_1 + NC_2 + 1)}$$

In these formulae $u_{\alpha/2}$ is the reduced normal variate with exceedance probability equal to $\alpha/2$.

Figures IV and V show these confidence intervals such as $1 - \alpha = 70\%$ calculated in the case of our two examples (Agen and Serrières). It appears that the plotting positions of our complete sample are coherent with the Weibull model, much more where than the previous plotting positions.

4. CONCLUSION

Use of fuller information is absolutely necessary for reducing both sampling and model uncertainties in the problem of design flood statistical estimation. The paper showed the interest of historical data in this respect. The method used allows not only the non - parametric estimation of probabilities for large values but also the appraisal of confidence intervals for these probabilities. Further works are needed for validation purpose of certain parts of the method, namely the normal assumption for the sampling errors of the estimated parameters in the case of historical data combined with systematic observations. It could be also interesting to generalize tests of goodness of fit like X^2 or Andersen-Darling tests which appears useful to the case of fuller information.

5. REFERENCES

BERNIER J., MIQUEL J. . 1977. 'Exemple d'application de la théorie de la décision statistique au dimensionnement d'ouvrages hydrauliques : prise en compte de l'information historique'. In proceedings XVIIth congress AIHR, Baden-Baden.

DEMPSTER A.R., LAIRD N.M., RUBIN D.B. . 1977. 'Maximum likelihood from incomplete Data via the E.M. Algorithm'. J.R Stat.Soc., Serie B, n° 39.1.

MIQUEL J. . 1984. 'Guide pratique d'estimation des probabilités de crue'. Eyrolles, Collection de la D.E.R. d'Electricité de France, n° 53.

EXTRAPOLATION ERROR IN FLOOD FREQUENCY ANALYSIS

A. Qadar and M. Muzzammil
Department of Civil Engineering
Z.H. College of Engineering & Technology
Aligarh Muslim University, Aligarh, India

ABSTRACT. The frequency analysis of annual peak flood data is considered to be one of the widely used methods for estimating frequency floods. The extra-apolation of frequency floods is generally avoided due to indadequate data available for such studies. In the present paper, an attempt has been made to study the amount of errors involved in extrapolation over sufficiently long periods by using short period series. In order to achive this aim, flood data available for six gauging stations with varying catchment areas and spread over different regions of the country (India) have been selected for studies and analysed by the most commonly used Gumbel's extreme value distribution using Least Square Method and by Method of Maximum Likelihood. It has been found that the extent of error in exrapolation does not increase appreciably with the extent of extrapolation but corresponds to the error in selection of sample series. In order to estimate the frequency floods within an error of ±10%, about 50 years flood record may be essential.

1. INTRODUCTION

Most of the hydraulic structures such as bridges, barrages, weirs, embankments, reservoirs etc. are affected by floods. While designing these structures, proper safeguards must be made for the safe passage of the maximum flood expected. The structure must be sound not for its own safety but also for the life and property which might be in danger by its failure. Thus, one could realize that the knowledge of magnitude and probability of occurrence of floods is not only essential for proper design of water resources projects but also for the flood control such as flood plain zoning and flood insurance. There are various methods available for estimating the magnitude of a flood peak. The use of

V. P. Singh (ed.), Application of Frequency and Risk in Water Resources, 165–174.

a particular method depends upon the desired objective, the available data and the importance of the project. One of the widely used methods is the flood frequency analysis of annual peak discharges. Frequency studies are made to predict the future probabilities of occurrence based on the past record of events. The historical data usually available for the frequency analysis is only a limited sample from the true population. Thus, it is clear that the data available should be for sufficiently long period so as to obtain satisfactory estimates. When the record of short period is used to estimate flood of long return periods, it may lead to inconsistent results as they are not representative of long-term trends. If extrapolation outside the observed range of records is made, it would reduce the accuracy of the estimates. The frequency analysis of flood data has been made to study the errors involved in extrapolation by many engineers and research workers. However, a little attention has been focussed on the extent of errors in extrapolation. With this view in mind, the present study has been undertaken.

2. RELEVANT LITERATURE REVIEW

The extrapolation of floods of high return periods has been a subject of controversy. Some engineers have set the limit to extrapolation of about the length of record available.

In U.S.G.S. Water Supply Paper, it has been pointed out that the tendency to use frequency graph for purposes of extrapolation can not be encouraged. It has been further argued that the linear distance from 25 to 200 years seems very short on most graphs, but extrapolation can only be justified when the phenomenon have been proven to conform to underlying law. It has also been added that the error of a curve fitted by whatever method may be extremely great at its outer end.

Benson (1952) demonstrated the variability inherent in short records. Synthetic 1000-year record of annual peak discharges was so distributed as to define exactly a straight line graph on extreme value graph paper. Individual peaks were then drawn by the author at random to form groups of samples equivalent to 10, 25, 50 and 100 years period of record. His studies showed that graphical variation in frequency curves on the short-term samples could, by chance, alone take the position of any of the curves. He, however, did not study the maximum likely percentage of variation using short period data.

During discussion in the Twelfth-International Congress on large Dams Mr. Benger, expert from USA(after Gole et al, 1978) emphasised that the extrapolation frequency curves much beyond a recurrence interval equal to the length of records is unrealistic. He has cited an example of results of frequency extrapolation for susquehanna River at Harrisburg, Pennsylvania (U.S.A.) to show the extent of errors. Based on these results, he concluded that the extrapolated curves gave widely different results and extended to recurrence intervals many times the respective lengths of records. However, he also did not study the percentage error in quantitative terms.

Gole et al (1978) studied the extent of errors involved in extrapolation for 100 yrs, 1000 yrs and 10000 yrs return period by using 20, 30, 40 and 50 years short period series for five gauging stations. Based on this study, they concluded that the extent of error in extrapolation of floods for longer return period does not appreciably increase with the extent of extrapolation but corresponds to the error in selection of sample series used for frequency analysis or the sample series available for study. They made their study only for five gauge stations.

3. PRESENT STUDY

The present study has been made to obtain the detailed informations about the extent of error in extrapolation of floods for longer return periods. In order to achieve this aim, six river basins in India spread over different regions and of varying catchment areas have been selected for study. The annual peak discharge data have been tabulated in Table I and details of the gauging stations are as follows :

Gauging Stations	Catchment areas (C.A.) (Sq.km)	Period of record available
1. Ganga at Hardwar	23388	1901-1977
2. A at Z	2530	1901-1972
3. Baiterni at Akhupada	10750	1905-1957
4. Yamuna at Tajewala		1924-1977
5. Krishna at Vijaywada	257340	1901-1950
6. Sutlej at Bhakra	57000	1912-1955

In the present study, each historical data has been split into 10, 25 and 50 years continuous series. Each of these series has been subjected to most commonly used method of Gumbel's extreme value distribution by Least Squares Method and by the Method of Maximum Likelihood. The variation in 100 years, 500 years and 1000 years flood in each of the series over the long-term historical series has been determined to obtain likely percentage errors due to extrapolation. The results obtained have been entered in Tables II. The variation due to sampling errors for 1000 years return period has been shown in Table III. The average standard error in estimates has also been worked out and tabulated in Table IV and its variation with the number of years of series considered has been plotted in Fig. 1 along with the data obtained by Gole et al.

4. RESULTS AND DISCUSSIONS

The results in Table II indicate that the extent of error in extrapolation of floods for longer return periods of 100 years, 500 years or even 1000 years does not appreciably increase with the extent of extrapolation but corresponds to the error in selection of sample series used for frequency analysis. It also reveals that the accuracy of estimates of peak flood by frequency analysis does not only depend on the length of series but also on the sample selected. The dry and wet years present in the historical data also affect the accuracy of the estimates. Whenever the highest annual peak flood is considered in any series, it results in higher estimates of frequency floods and consequently higher percentage deviation from long-term series. On the other hand, the lowest annual peak flood considered in any series results in lower percentage deviation from the long-term series. These results are in good agreement with those obtained by Gole et al.

It has been reported by Gole et al that the Method of maximum likelihood gives better results in comparision to the Method of Least Squares based on the sampling errors obtained by them for both the methods. This fact is not proved from the present analysis as shown in Table III. Both methods appear to be equally important.

It is clear from Fig. 1 that the present data are in good agreement with those of Gole et al although data points for the present analysis are few due to shorter length of the historical data available for some gauging stations. The Fig 1 shows that the length of the series required for estimates within an error of ±10% is of the order of 50 years. It also indicates that for obtaining estimates within an error of ±25%, about 15 years annual peak data may be essential.

5. CONCLUSIONS

From the present study, the following conclusions may be drawn:

(a) The extent of error in extrapolation of floods for longer return periods does not appreciably increase with the extent of extrapolation but corresponds to the error in selection of sample series.

(b) The length of series as well as the sample selected affects the accuracy of estimates of peak flood by frequency analysis.

(c) The accuracy of estimates of the peak floods by frequency analysis is also affected by the presence of the highest and the lowest peak floods in the series.

(d) About 50 years flood record may be used to estimate the frequency floods within an error of $\pm 10\%$.

TABLE I

Peak annual discharges in m^3/s

year	Ganga at Hard-war	A* at Z	Yamuna at Taje-wala	Baiter-ni at Akhur-pada	Krishna at Vijay-wala	Sutlej at Bhakra
1	2	3	4	5	6	7
1901	11400	11700			12602	
1902	9250	6860			15066	
1903	7400	8250			39346	
1904	8550	3480			14047	
1905	9070	9140		1586	12093	
1906	7070	7300		4870	12234	
1907	7530	4650		7787	20928	
1908	11500	8330		2860	26394	
1909	8320	4600		3469	15066	
1910	11880	6220		2718	13622	
1911	6940	8060		2322	8298	
1912	8320	8994		3540	27414	6700
1913	3510	7300		8212	12602	5660
1914	9230	5400		2619	34097	5660
1915	7400	18900		2350	16652	4120
1916	4700	9560		1844	33588	2120
1917	8410	3660		2860	14387	4530
1918	4640	7150		2407	6853	1985
1919	6280	3660		2860	13339	3740
1920	8150	4500		7447	10847	4100
1921	9070	5910		1841	17502	3845
1922	7400	3420		3596	13594	3110
1923	5480	6070		3964	19966	2640
1924	19130	5180	14200	1487	20136	6710
1925	9650	8920	2270	4899	19343	5410
1926	3680	5200	3370	3540	16624	2720
1927	7240	8280	5350	9253	18351	4000
1928	3680	3185	2380	4870	13140	2400
1929	4540	3470	8300	3540	13084	4580
1930	6160	2318	10260	2902	10960	4250
1931	3460	2400	2310	4191	14613	2035
1932	6150	3090	7530	4134	19314	3830
1933	5270	6560	2500	4021	26139	2630
1934	9000	4850	2560	3964	13933	2460
1935	5280	4180	2166	2601	12631	4400

1	2	3	4	5	6	7
1936	3310	5050	3340	2577	9685	2380
1937	3220	4560	10100	3596	14528	2860
1938	3520	4290	2470	1345	12093	4950
1939	2340	4810	1220	3313	14726	2350
1940	2430	3570	2640	6145	11921	1725
1941	3130	5120	2410	8495	2860	2745
1942	6620	4740	6450	3596	15548	5000
1943	4400	3800	4250	8127	21552	5275
1944	4220	5400	2860	4191	14358	2300
1945	5100	6300	2750	3058	14613	9385
1946	4620	4560	4640	5437	20079	3820
1947	4340	6300	15970	1274	14358	7800
1948	4880	4480	3850	1812	12008	2355
1949	3610	4010	4220	1954	34097	3040
1950	6190	4433	3540	3794	21042	4860
1951	4760	8400	8780	2350		9200
1952	3810	4800	3710	2379		2365
1953	5470	3640	4360	3058		4810
1954	6375	6260	7140	1642		5660
1955	4610	2970	13200	4955		3670
1956	6365	2400	12750	4021		
1957	5420	4670	9000	1756		
1958	4050	5860	4190			
1959	5020	3280	5100			
1960	3880	4480	2520			
1961	4850	3790	4900			
1962	5750	3740	7830			
1963	6350	4105	8140			
1964	4010	11280	8330			
1965	2430	9050	2220			
1966	4750	7300	4150			
1967	5920	8430	7180			
1968	3320	8960	3900			
1969	5360	7000	3400			
1970	6310	7460	2400			
1971	5700	11380	7140			
1972	4920	8240	5750			
1973	7400		5220			
1974	5160		8200			
1975	3810		5100			
1976	6790		8180			
1977	5710		7100			

* After Ranjan et al (1985).

TABLE II

Statement showing the percentage deviations of the various frequency floods from long-term series.

Gauging Stations	Series	Percentage deviations from long-term series					
		By Method of Least Squares			By Method of Maximum Likelihood		
		100 yrs	500 yrs	1000 yrs	100 yrs	500 yrs	1000 yrs
	1	2	3	4	5	6	7
GH	1901-1910	11.3	5.5	3.7	18.1	12.8	11.1
	1911-1920	-03.5	-5.8	-6.5	21.4	21.2	21.1
	1921-1930	69.0	74.2	75.9	59.0	62.0	62.9
	1931-1940	-14.6	-13.2	12.7	-14.9	-14.1	-13.8
	1941-1950	-38.6	-40.9	- 41.7	-32.2	-34.2	-34.8
	1951-1960	-14.1	-44.2	-45.2	-38.1	-41.0	-42.0
	1961-1970	-32.5	-34.6	-35.2	-11.8	-12.0	-12.0
	1968-1977	-28.9	-31.6	-32.5	-13.5	-14.9	-15.4
	1901-1925	28.8	27.2	26.7	41.0	40.4	40.2
	1911-1935	20.4	20.9	21.2	21.0	21.4	21.5
	1921-1945	18.2	21.1	22.0	09.7	11.2	11.7
	1931-1955	-31.7	-32.8	-33.3	-25.2	-25.9	-26.1
	1941-1965	-41.5	-44.3	-45.2	-28.3	-29.8	-30.3
	1953-1977	-36.7	-39.7	-40.7	-77.1	-80.2	-80.1
	1901-1950	14.7	15.4	15.7	18.7	19.8	20.2
	1911-1960	-02.5	-02.0	-01.9	-05.7	-05.7	-05.7
	1921-1970	-09.7	-09.4	-09.3	-14.0	-14.4	-14.5
	1928-1977	-35.4	-37.5	-38.2	-26.1	-27.3	-27.7
	1901-1977	Long term series			Long term series		
AZ	1901-1910	13.6	12.7	12.4	30.4	30.4	30.4
	1911-1920	64.8	68.9	70.2	60.6	63.0	63.9
	1921-1930	-06.4	-05.8	-05.6	03.7	04.7	05.0
	1931-1940	-40.6	-42.4	-43.0	-28.8	-29.8	-30.0
	1941-1950	-44.0	-47.4	-48.6	-40.7	-44.0	-45.0
	1951-1960	-20.1	-20.1	-20.0	-14.7	-14.7	-14.7
	1961-1970	13.9	12.9	12.6	38.8	39.6	39.9
	1901-1925	23.4	23.9	24.0	25.2	25.5	25.6
	1911-1936	18.4	20.6	21.2	15.5	16.9	17.4
	1921-1945	-28.8	-30.1	-30.6	-21.2	-21.9	-22.1
	1931-1955	-37.1	-39.3	-40.0	-29.3	-30.8	-31.3
	1941-1965	-18.6	-19.3	-19.5	-18.8	-19.9	-19.9
	1901-1950	- 0.1	00.2	00.2	-02.8	-03.0	-03.0
	1911-1960	- 8.1	-07.9	-07.9	-12.1	-12.4	-12.6
	1921-1970	-19.1	-20.0	-20.0	-15.1	-15.7	-15.8
	1901-1972	Long Term Series			Long Term Series		
YT	1924-1933	30.0	32.0	33.0	32.2	33.8	34.3
	1934-1943	-19.3	-18.1	-17.8	-25.5	-25.3	-25.0
	1944-1953	17.2	19.0	19.5	00.6	00.4	00.5
	1954-1963	26.5	25.5	25.0	43.2	42.7	43.0
	1968-1977	-20.8	-23.2	-23.9	-03.5	-04.6	-05.0
	1924-1948	09.0	11.0	11.5	-00.4	00.5	00.8
	1934-1958	13.5	14.9	15.5	08.0	08.7	09.0
	1944-1968	11.7	11.8	11.8	10.9	10.6	10.5
	1953-1977	-03.4	-05.1	-05.6	-07.2	06.1	05.8

	1	2	3	4	5	6	7
	1924–1977	Long Term Series			Long Term Series		
	1905–1914	-24.7	26.3	26.8	23.7	24.6	24.8
	1915–1924	-07.2	-06.3	-06.0	-13.0	-13.1	-13.1
	1925–1934	04.0	01.7	01.0	-02.9	-06.2	-07.2
	1935–1944	35.3	36.8	37.3	44.8	46.4	46.9
BA	1948–1957	-28.9	-29.4	-29.6	-27.0	-27.8	-28.0
	1905–1929	11.0	11.7	11.9	09.4	09.7	09.8
	1915–1939	-06.5	-06.9	-07.0	-04.0	-04.4	-04.6
	1925–1949	11.6	11.7	11.7	19.1	19.5	19.6
	1933–1957	00.6	01.0	01.1	03.5	03.9	04.0
	1905–1957	Long Term Series			Long Term Series		
	1901–1910	23.9	25.7	26.3	04.1	03.3	03.1
	1911–1920	39.8	43.9	45.3	36.1	38.9	39.8
	1921–1930	-29.4	-33.0	-34.1	-21.3	-24.0	-24.8
	1931–1940	-20.2	-21.6	-22.1	-25.8	-27.8	-28.5
KV	1941–1950	17.3	19.2	19.7	38.7	41.7	42.7
	1901–1925	15.6	16.6	17.0	09.5	09.7	09.7
	1911–1935	01.6	01.6	01.6	-00.0	-00.4	-00.5
	1926–1950	-11.5	-12.2	-12.4	-00.1	-00.0	-00.1
	1901–1950	Long Term Series			Long Term Series		
	1912–1921	05.0	04.4	04.2	23.9	24.6	24.9
	1922–1931–	01.4	01.9	02.0	03.1	03.2	03.2
	1932–1941	26.8	-27.5	-27.7	-26.0	-26.9	-27.2
SB	1946–1955	43.4	-45.9	46.8	43.0	44.7	45.2
	1912–1936	-08.2	-09.0	-09.2	-02.4	-02.7	-02.8
	1922–1946	-16.1	-16.7	-16.9	-14.8	-15.4	-15.5
	1931–1955	06.4	07.6	08.0	01.2	01.7	01.86
	1912–1955	Long Term Series			Long Term Series		

GH = Ganga at Hardwar; AZ = A at Z; YT = Yamuna at Tajewala
BA = Baiterni at Akhupada; KV = Krishna at Vijaywada;
SB = Sutlej at Bhakra.

TABLE III

Statement showing variation due to sampling errors(Considering results for 1000 year return period only).
(Abstract from Table II)

Name of Gauging Station	C.A. (Sq.km)	SS*	SS** ±10%	SS** ±10–20%	SS** ±20–25%	SS*** ±25%
1	2	3	4	5	6	7
A-Gumbel's extreme value distribution by least square method						
Ganga at Hardwar	23388	18	4	2	2	10
A at Z	2530	15	3	5	2	5
Yamuna at Tajewala		9	1	5	2	1
Baiterni at Akhupada	10750	9	4	2	0	3
Krishna at Vijaywada	257340	8	1	3	1	3

1	2	3	4	5	6	7
Sutlej at Bhakra	57000	7	4	1	0	2
Total		66	17	18	7	24

B-Gumbel's extreme value distribution by method of maximum likelihood

Ganga at Hardwar	23388	18	1	7	2	8
A at Z	2530	15	2	5	1	7
Yamuna at Tajewala		9	5	1	1	2
Baiterni at Akhupada	10750	9	4	2	1	2
Krishna at Vijaywada	257340	8	4	0	1	3
Sutlej at Bhakra	57000	7	3	1	1	2
Total		66	19	16	7	24

*SS = No. of sample series
**SS = Sample series showing error within
***SS = Sample series showing error exceeding

TABLE IV

Average Standard error in estimates(for 1000 years only)

Gauge Stations	Method of Least Squares		Method of Maximum Likelihood	
	10 yr. Series	25 yr Series	10 yr. Series	25 yr. Series
Ganga at Hardwar	38.80	32.79	31.77	41.31
A at Z	37.57	28.07	37.35	23.74
Yamuna at Tajewala	24.46	11.61	26.98	07.50
Baiterni at Akhupada	24.60	09.06	27.67	11.40
Krishna at Vijaywada	30.95	12.18	31.16	05.64
Sutlej at Bhakra	27.28	13.22	29.20	06.36
Average	30.61	17.82	30.68	16.00

6. REFERENCES

Subramanya, K., 1984. Engineering Hydrology, Tata McGraw-Hill Publishing Co. Ltd.New Delhi PP.203-229

Varshney,R.S., 1979. Engineering Hydrology, New Chand & Bros., Roorkee (U.P.), pp.553-625.

Chow, V.T., 1964. Hand Book of Applied Hydrology, McGraw Hill, New York.

U.S.G.S. Flood Frequency Analysis-Manual of Hydrology, Part III, Water Supply Paper 1543-A, PP.21.

Central Water Commission, 1973. Estimation of Design flood peak, flood Estimation Directorate, Report No. 1/73, New Delhi, India.

Ranjan, B.H., Rao, K.N.S., 1985. Spillways and Hydraulic Energy Dissipators, CBIP, Tech. Report No. 37, New Delhi.

Gole, C.V.,Krishnamurthy, K., Bhatija, T.S. 1978. 'Extrapolation of peak Floods by Frequency Analysis,' Forty-Seventh Research Session of the CBIP Vol.II, Hubli-Dharwar, Karnataka (India).

Transactions of the Twelth International Congress on Large Dams, Vol.V,46 Preliminary Planning of Dam Developments.

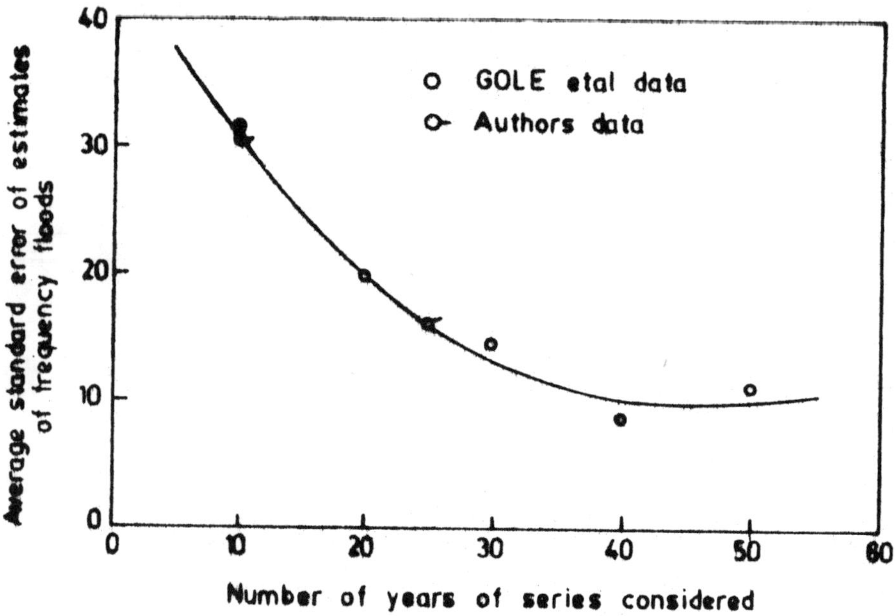

FIGURE 1. Standard error versus number of years of record considered

EFFECT OF DISCHARGE MEASUREMENT ERROR ON THE RESULTS OF FLOOD FREQUENCY ANALYSIS

Shuzheng Cong & Yinbo Xu
Department of Hydrology
East China Technical University of Water Resources
Nanjing
China

ABSTRACT. A mathematical model of the discharge measurement error was proposed, verified and calibrated. On the basis of the model, theoretical effect of the error on the population parameters was analysed by analytic method and the sample effect of error on the results of frequency analysis was studied by Monte Carlo simulation. Results obtained show that in the absence of historical outlier, error effects are small and may be neglected if the magnitude of the error is of ordinary order. When a historical outlier is considered, the results of flood flow frequency analysis can be improved if the error parameter of the outlier is less than 0.3.

1. INTRODUCTION

Observed hydrological data are usually thought of as being accurate when performing flood frequency analysis (FFA). Under this assumption, a great deal of researches has been conducted and a lot of literatures has been published. In contrast with this, researches on the effect of the discharge measurement error in the FFA are significantly inadequate.

The paper of Potter & Walker [9] is one of the major studies on this subject. They have studied the effects of measurement error on the population distribution function of the flood. They have shown that "the coefficient of variation, skewness and kurtosis of the distribution of the measured flood discharges are significantly higher than the corresponding coefficients of the parent flood distribution. This bias has important implications with regard to flood frequency analysis." In contrast with [9], it has been shown in [6] that "measurement errors are usually random and the variance introduced is usually small in comparison to the year-to-year variance in flood flows. The effects of the measurement errors may normally be neglected in flood flow frequency analysis." Thus, we have met two inconsistent conclusions, so a further study on this subject is necessary to make the problem more clear.

V. P. Singh (ed.), Application of Frequency and Risk in Water Resources, 175–190.

2. A MATHEMATICAL MODEL OF THE MEASUREMENT ERROR

2.1. Discussion of Potter and Walker's Model

If x denotes the true flood discharge under study, z denotes the measured value of x, according to [9] we have z=xs, where s is a dimensionless random variable reflecting the accuracy of the measurement. It is assumed in [9] that:

(1) both x and s are log normal distributed with two parameters; (2) s and x are independent of each other and (3) Es=1, Var s= σ^2. There are some problems which may be presented for discussion.

2.1.1. Due to the fact that a particular property of the log normal distribution with two parameters has been used in the derivation of some conclusions in [9], it is apparent that the conclusions may not be valid for other distributions. Hence there is a restriction on the application of the model, especially, when considering the fact that log Pearson type 3 distribution is recommended in the United States and Pearson type 3 distribution is commonly used in China.

2.1.2 According to the above assumptions, it is easy to show that the distribution of measurement error of flood flow v is not normal. This departs from the generally accepted principle that measurement error is normally distributed. This principle is supported by theoretical consideration [1] and wide experiences as well [5].

2.1.3. It has been shown in [9] that "It is usefull to consider first the case in which the standard deviation of the measurement error is a constant proportion of the actual discharge." It will be seen later that this point of view is basically in accordence with the practical situation of measurement error and is therefore reasonable. However, this argument is not really used in [9].

2.1.4. Only theoretical effects of measurement error, i.e. the effects of measurement error on the population distribution was studied in [9]. It is easy to understand that more important and essential is to investigate the effects of error on the results of FFA, or for short, the sample effect. It should be noted that improvement of the P-W model have been made in [10] by the same authors.

2.2. The Proposed Model

Let x denote the true value of discharge, z the measured value of x, v the random error of discharge measurement. Assuming that there is no systematic error in the measurement, we have the following measurement equation:

$$z=x+v \tag{1}$$

On the basis of practical situation of the discharge measurement, it is assumed that the conditional probability density function (pdf) of the random measurement error v satisfies the following three conditions:

2.2.1. $E(v|x)=0$ (2)
i.e. the conditional expectation of v given x is always zero and independent of x. Hence, we have

$$E(z|x)=x \quad (3)$$

2.2.2. $\sigma(v|x)=\sqrt{(E(v^2|x))}=\alpha(x)\cdot x$ (4)
in which $\sigma(v|x)$ denotes the conditional standard deviation of v given x. Eq.(4) means that $\sigma(v|x)$ is proportional to x, where $\alpha(x)$ is a proportional coefficient whose structure should be determined on the basis of measured data. The simplest case is

$$\alpha(x)=\alpha \quad (5)$$

Considering both Eq.(1) and (4), we have

$$\sigma(z|x)=\sigma(v|x)=\alpha(x)\cdot x \quad (6)$$

2.2.3. the conditional pdf of v given x is normal, i.e.

$$g(v|x)=1/(\alpha(x)\cdot x\sqrt{(2\pi)})\cdot\exp(-v^2/(2\alpha^2(x)\cdot x^2)) \quad (7)$$

If f(x) denotes the pdf of x, h(x,v) denotes the joint pdf of (x,v), we have

$$h(x,v)=f(x)g(v|x)=1/(\alpha(x)x\sqrt{(2\pi)}\exp(-v^2/(2\alpha^2(x)x^2))f(x) \quad (8)$$

2.3. Determination of the Model Parameter $\alpha(x)$

There are two ways to determine the model parameter $\alpha(x)$.

2.3.1. On the basis of analysing the Standards of discharge measurement. The accuracy of discharge measurement in China is stipulated in the STANDARDS OF HYDROMETRY [7] as follows: for the current meter measurement there should be at least 75% of the measurement of which the relative error lies in the range of ±5% as shown in Fig. 1; for measurement using float method the relative error is ±8%.

According to these stipulations, the model parameter $\alpha(x)$ can easily be determined. It is well known that the quantile of order p of the standard normal distribution is 1.15, if p=(1-75%)/2=12.5%. Therefore, we have $((x+5\%)-x)/\sigma(z|x)=1.15$, this means $\sigma(z|x)=5\%x/1.15=0.0435x$, i.e. $\alpha(x)=0.0435$, independent of x. Through a similar simple calculation, the following results may be obtained for confidence level 75%.

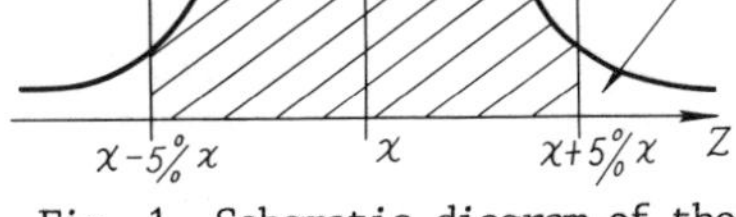

Fig. 1 Schematic diagram of the distribution of measurement error

Relative error	± 5%	± 8%	± 10%	±11.5%	±23.0%	±34.5%
α	0.0435	0.0696	0.0870	0.1000	0.2000	0.3000

Measurement carried out according to the relevant ISO Standard may be expected to have the following upper limits of uncertainties at the

95% confidence level [5]:

(1) Single determination of discharge ±7%,
(2) Standard error of the estimate of the rating curve ±10%,
(3) Standard error of the mean of the rating curve ±5%,
(4) Recorded discharge (15-min. value, ect.) ±5%,
(5) Daily mean discharge, monthly mean discharge or annual discharge ± 5%.

The values quoted above are equivalent to 2α in this paper. Thus, according to the Chinese and ISO Standards the value of the model parameter $\alpha(x)$ is independent of x and is generally less than 0.05. In order to aquire more information about the effect of measurement error, higher values of α, such as 0.1, 0.2 and 0.3 are also considered.

3.2.2. On the basis of analysing the actually measured data of flood flow. The accuracy stipulated in all Standards is essentially of subjective nature. Whether the stipulated accuracy can be reached in practice has to be tested through analysing the actually measured data of discharge. For this purpose, measured data at five stations in China with steady and single rating curve have been analyzed. The relevant information of the data used and the results of calculation are shown in Table 1.

Table 1 Analysis of discharge measurement error at five stations located in Fujian Province of China

Name of stations	Sample size	Aver. value of $\alpha(x)$	calculated χ^2	Max.obser. discharge (m/s)	River
Konxia	92	0.03666	4.7825	4370	Nanpuxi
Hongwei	100	0.02784	8.1585	3956	Jianxi
Tangban	76	0.03248	5.4858	2989	Aojian
Wuliting	84	0.02217	3.9938	4422	Dajian
Yongkou	83	0.01478	2.4858	7208	Futunxi

In Eq. (1) x denotes the true value of discharge and is represented here by the discharge obtained from the fitted rating curve. From Eq.(2) we have equivalently

$$E((z/x-1)|x)=0 \qquad (9)$$

Due to the limited amount of data, it is impossible to test the above equation with respect to any single value of x. But if we consider a short series of $(z/x)_i$ according to a consecutive series of x_i in descending order and calculate the mean of the series of $(z/x)_i$, it will be seen that the mean values are always very close to 1 and the variation of the mean values is very small. This justifies the first assuption.

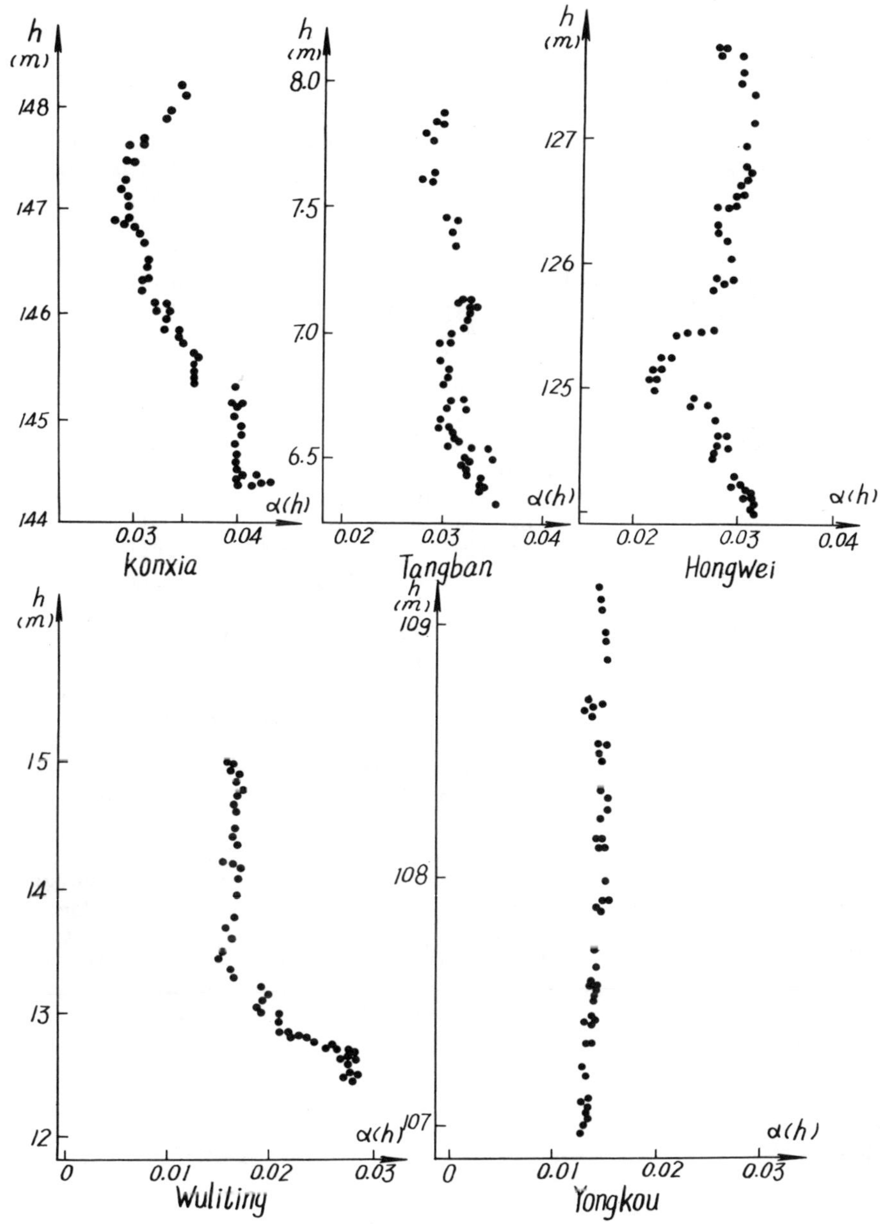

Fig. 2 Relation of α(h) versus h for five stations

As for the second assumption, it is clear that Eq. (6) is equivalent to the following one:

$$\sqrt{E((z/x-1)^2 \mid x)} = \alpha(x) \tag{11}$$

since x is treated here as a parameter rather than random variable. Using a similar reasoning procedure as used before, relations of $\alpha(x)$ versus h, water stage, for the five stations are obtained as shown in Fig. 2. Though the shapes of the five "curves" are not completely alike, they appear to have two common characteristics: (1) all values of $\alpha(x)$ at five stations do not exceed the limit of 0.0435 as stipulated by the Standard; and (2) the variation of $\alpha(x)$ against h (or equivalently x) is not essential, $\alpha(x)$ is generally greater to same extent in the lower part of the curve. On account of the two characteristics and considering the fact that only measurement error of flood discharge is analysed in this paper, it would be a reasonable simplification to take $\alpha(x)$ as a constant. In order to prevent from underestimating the effects of error, $\alpha(x)=0.05$ is accepted as a major index of the measurement error.

A difficult problem in determining the model parameter is to determine its value for historical outlier. The magnitude of historical outlier is usually determined by the slope area method. Using the Taylor expansion we can calculate the error of historical flood on the basis of errors of affecting factors, such as slope, cross area, roughness of the river bed, etc. Detailed determination of the value of α for historical flood needs special analysis and is beyond the scope of this paper. It is assumed here that the vlaue of α for historical outlier lies between 0.2-0.4.

There is an important problem which should be emphasized at the end of this section. The value of $\alpha(x)$ given in Eq.(6) denotes the relative conditional standard deviation, or the conditional coefficient of variation, of the individual measured discharge z with respect to its conditional expectation x. But the values of the annual flood discharges used in FFA are determined from the fitted rating curve y versus h (cf. Fig.3) according to the corresponding stage h of the flood discharge x. This means

$$y=x+u \tag{11}$$

in which x and y represent respectively the true value of discharge and its estimate obtained from the fitted rating curve, u is a random error. Theoretically speaking, instead of the conditional coefficient of variation $\sigma(z|x)/x=\alpha(x)$ of z, the conditional coefficient of variation $\sigma(y|x)/x$ of y should be used in the analysis of the effects of measurement error in the FFA. On the basis of the theory of regression analysis, it can be shown that

$$\alpha(x)/\sqrt{n} \le \sigma(y|x)/x \le \alpha(x) \tag{12}$$

where n is the number of items involved in the data used for determining the fitted rating curve $y \sim h$. Since no accurate formula for calculating $\sigma(y|x)/x$ is now available and, in addition, the error of determining the stage is neglected in the paper, it is assumed that $\sigma(y|x)/x=\alpha(x)$ in order not to underestimate

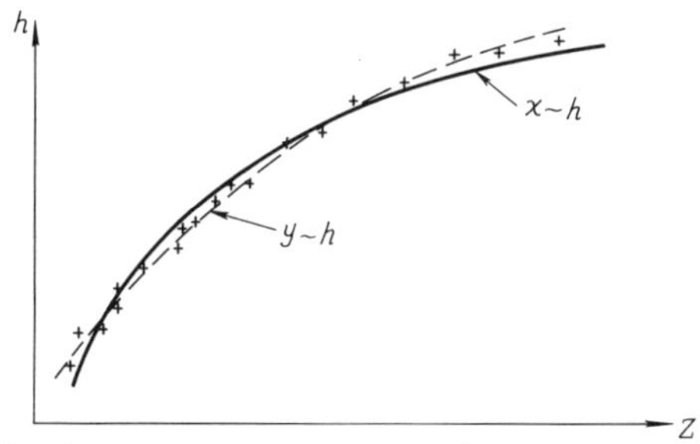

Fig. 3 Schematic diagram of the fitted rating curve

the effects of error. This means that the error u of y is the same in value as the error v of z. Consequently, Eqs. (2) through (4) are also valid if u is subsituted for v. For the sake of simplicity, z and v are still used later to represent y and u respectively.

3. STATISTICAL TEST OF THE DISTRIBUTION FUNCTION OF THE MODEL

Under the condition that only the mean and the standard deviation of some measurement error v can be determined on the basis of measurement data, the pdf of v can be derived to be normal according to the maximum entropy principle. The reasonableness of this distribution lies in the fact that its information content which we assign subjectively to the variable v is the least [1]. In order to be more prudent, the normal distribution has been tested using the observed data of the above mentioned five stations.

Eq. (7) can be simply denoted as $(z-x)|x \sim N(0, \alpha^2(x)x^2)$. As only z is a random variable while x is used as a parameter here, the above relation is equivalent to the following one:

$$(z/x)|x \sim N(1, \alpha^2(x)) \tag{13}$$

Due to the limited amount of data the test of goodness of fit can not be made for any single value of x. However, if the variation of $\alpha(x)$ versus x is not large, so that $\alpha(x)$ may be considered approximately as a constant, (it has been shown above this condition is basically satisfied), the test of goodness of fit can be carried out using the total data corresponding to various values of x. The test statistic used is the Chisquare χ^2. Grouping number is 5 for all five stations, the degree of freedom of χ^2 is $\nu=4$. The critical values of χ^2 corresponding to various significance level β are given below, while the actual values of χ^2 for the five stations are given in Table 1.

Significance level β	0.25	0.10	0.05	0.025	0.01
χ^2	5.384	7.779	9.448	11.140	13.28

If $\beta=0.05$ is accepted, the null hypothesis of normal distribution can be accepted for all 5 stations.

From the above discussion, it can be seen that three assumptions on the model of measurement error are not only theoretically justifiable but also in accordance with the practical situation of discharge measurement. Therefore, they may be used as a basis for further studies.

4. EFFECTS OF MEASUREMENT ERROR ON THE DISTRIBUTION OF POPULATION (THEORETICAL EFFECTS)

Given $\alpha(x)$ we can derive the effects of measurement error on the distribution of population using the error model given above. However, if $\alpha(x)$ is in a very general form, the derivation may be rather complicated and it is difficult to express the results in a simple form. It has been shown that for flood discharge $\alpha(x)$ may be approximately considered as a constant. Thus, if we take $\alpha=\max \alpha(x)=0.05$, the derivation is very simple and the effect of error will not be underestimated.

According to Eqs. (1) through (9), it is easy to obtain the following results.

(1) $Ez=E(x+v)=Ex+\iint vg(v|x)f(x)dxdv=Ex+\int E(v|x)f(x)dx$

From Eq. (2), we have

$$Ez=Ex \tag{14}$$

(2) $E(z-Ez)^2=E(x-Ex+v)^2=E(x-Ex)^2+Ev^2+2E(v(x-Ex))$

owing to $E(v(x-Ex))=\iint v(x-Ex)g(v|x)f(x)dxdv=\int E(v|x)(x-Ex)\cdot$ $\cdot f(x)dx=0$

and $\sigma^2(v)=Ev^2=\iint v^2g(v|x)f(x)dxdv=\int(\int v^2g(v|x)dv)f(x)dx$

$$=\int \alpha^2x^2f(x)dx=\alpha^2(\sigma^2(x)+(Ex)^2) \tag{15}$$

so that $\sigma^2(z)=(1+\alpha^2)\,\sigma^2(x)+\alpha^2(Ex)^2$ (16)

or equivalently, $cvz=\beta\, cvx,\ \beta=\sqrt{(1+\alpha^2(1+1/(cvx)^2))}$ (17)

in which cvz and cvx denote the coefficient of variation of z and x respectively. This result is identical to that given in [9].

(3) Reasoning in a similar manner we can obtain

$$\mu 3z=(1+3\alpha^2)\,\mu 3x+6\alpha^2 Ex\cdot\sigma^2(x)=\mu 3x\cdot(1+3\alpha^2+6\alpha^2/(cvx\cdot csx)) \tag{18}$$

in which csx is the coefficient of skewness of x, $\mu 3x$ and $\mu 3z$ denote the third central moment of x and z respetively. The above equation may also be written in the following form

$csz=\gamma\, csx$

where $\gamma=(1+3\alpha^2+6\alpha^2/(cvxcsc))/(1+\alpha^2(1+1/(cvx)^2))^{3/2}$ (19)

where csz denotes the skew coefficient of z.

According to the special property of the log normal distribution with two parameters, $csz=3cvz+(cvz)^3$, it is derived in [9] that $csz=3cvz+(cvz)^3=\gamma'\cdot csx$

where $\gamma'=\sqrt{(1+\alpha^2\cdot(1+1/(cvx)^2))}\cdot(1+\alpha^2(1+(cvx)^2)/(3+(cvx)^2))$ (20)

According to the above equations, the values of β, γ and γ' for various values of α, cvx and csx can be calculated, as shown in Table 2.

Table 2 Relation of β, γ and γ' against α, cvx, and csx

α	cvx	csx	β	γ	γ'
0.05	0.5	1.625	1.006	1.007	1.011
0.05	1.0	4.0	1.003	1.004	1.004
0.10	0.5	1.625	1.025	1.026	1.029
0.10	1.0	4.0	1.010	1.014	1.015
0.20	0.5	1.625	1.095	1.077	1.112
0.20	1.0	4.0	1.039	1.054	1.060
0.30	0.5	1.625	1.204	1.108	1.246
0.30	1.0	4.0	1.086	1.096	1.135

Table 2 shows that γ' is generally larger than γ and the larger α, the larger the difference between γ' and γ. In addition, if zp and xp denote the p-th quantiles of z and x respectively, i.e. p(z zp)=P(x$\geq$xp)=P, results of calculation show that, for log normal distribution with two parameters, the values of zp1/xp1 and zp2/xp2 are respectively smaller than (or close to) the corresponding values of β and γ'. Therefore, Table 2 shows that the effects of measurement error on the distribution of population is not significant.

5. EFFECTS OF THE MEASUREMENT ERROR ON THE RESULTS OF FFA (SAMPLE EFFECTS)

5.1. General description

Though the discussion about theoretical effects of the measurement error given above has some theoretical meanings, it is easy to understand that study on the effects of measurement error on the results of FFA (sample effects) is more important.

Besides measurement error, there are still many other factors which influence the results of FFA, e.g. the sample size, the method of parameter estimation, the values of parameters, and the distribution function of the population. Moreover, the criteria for evaluating the goodness of estimation results play also an important role. Consequently, the study of sample effects is much complicated than that of theoretical effect discussed before. Hence, it is difficult to use analytical method again and we have to make use of Monte Carlo technique.

5.1.1 Probability distribution of the population. Due to the fact that Pearson type 3 distribution is the principle one used in China, it is selected as the major distribution of population in this study. Considering log Pearson type 3 has been recommended in the United States, some calculations are also carried out for this distribution.

5.1.2 Values of parameters of the population. According to the practical situation of the FFA in China, two sets of population parameters are selected: (1) cvx=0.5, csx=2.0; (2) cvx=1.0, csx=3.5. The expectation may be assumed arbitrarily, say, Ex=1000.

5.1.3. Methods of parameter estimation. A curve fitting method is stipulated as a principal method of parameter estimation in the Chinese Standard of Design Flood [8]. For Pearson type 3 distribution, Cong et al. [3] has shown that the curve-fitting method can generally give better results than that given by the moment method and maximum likelihood method, if Weibull plotting position formula is used and the sum of absolute deviations of the fitted distribution function from the emperical points is accepted as the criterion of goodness of fit. Unfortunately, this method is very time consuming on the computer, so it is used only for a few cases, while moment method is applied for most cases. For log Pearson type 3 distribution, the moment method is used to the logarithms of the original data according to the GUIDELINES [6].

5.1.4. Criteria for evaluating the results of FFA. The unbiasedness and the efficiency of the estimators of the 100-year and 1,000-year design events, $\hat{x}p1$ and $\hat{x}p2$, are accepted as fundamental criteria for evaluating the results of FFA with different values of error parameter. This means that the expetation of $\hat{x}pi$, $E\hat{x}pi$, should be equal (or very close) to the corresponding quantile of population, xpi, and the root mean square error of $\hat{x}pi$, $\sigma(\hat{x}pi)= \sqrt{E(\hat{x}pi-xpi)^2}$, should be as small as possible. In the Monte Carlo simulation, $E\hat{x}pi$ and $\sigma(\hat{x}pi)$ are approximated by $m\hat{x}pi=1/k\cdot\Sigma\,\hat{x}pij$ and $s\hat{x}pi=\sqrt{1/k\cdot\Sigma(\hat{x}pij-xpi)^2}$ respectively, in which $\hat{x}pij$ denotes the estimated value $\hat{x}pi$ in the j-th sample.

5.1.5 Historical flood. For some computational cases, one historical flood (i.e. outlier) is considered. The calculation method considering historical flood is given in [3]. N' and α' in Table 3 and 4 represent respectively the return period and the error parameter of the historical flood.

5.2. Results Obtained by Using the Moment Method

5.2.1. In the absence of historical flood. According to the practical situation of discharge measurement in China, the largest peak flow in the systematic record can generally be practically measured on middle and large rivers; only in some particular cases it is determined through extrapolation of the rating curve. Therefore, the error parameter for the largest item in a sample may be considered as the same for other items, though the value of the first item may be much bigger than the values of the others.

(1) For the Pearson type 3 distribution. The results obtained by using moment method is given in Table 3. Analyzing the results the following conclusions may be drawn.

(a) In the range of usual precision of the measurement, i.e. $\alpha \leq 0.05$, both biases of $m\hat{z}p1$ and $m\hat{z}p2$ from the population values xp1 and xp2 are smaller than those of $m\hat{x}p1$ and $m\hat{x}p2$ respectively, at the same time the root mean square errors $s\hat{z}p1$ and $s\hat{z}p2$ are larger than $s\hat{x}p1$ and $s\hat{x}p2$ only in an amount of 1%. This is true for two sets of parameters and for all sample size n. Therefore, the effects of measurement error on the results of FFA are neglegible.

(b) The effect of the measurement error is increasing with the increase of α. However, even if α=0.1, for the case of cvx=0.5, csx=2.0 $s\hat{z}p1/s\hat{x}p1-1$ and $s\hat{z}p2/s\hat{x}p2-1$ are smaller than 5%; for the case of cvx=1.0 csx=3.5, they are less than 3%. More over, the biases of $m\hat{z}p1$, $m\hat{z}p2$ for both sets of parameters are still less than those of $m\hat{x}p1$ and $m\hat{x}p2$. Consequently, the effect of the measurement error is still allowable.

(c) For a given value of α , the increment of the sampling error $\Delta spi=s\hat{z}pi-s\hat{x}pi$ of $\hat{z}pi$ is related to the parameters of the population and sample size n. However, the effects of all of them are not significant.

It has been shown before that the error parameter α is generally smaller than 0.05 under a normal condition of discharge measurement; therefore, there is very little chance for α to exceed 0.1. Summarizing the above illustration, it can be concluded that the effects of measurement error of discharge on the results of FFA are completely permissible in the case without historical outliers.

(2) For log Pearson type 3 distribution. The results are shown in Table 4. It can be seen from the results that the effects are not only permissible but also positive if α =0.1: the biases of $m\hat{z}p1$ and $m\hat{z}p2$ are smaller than those of $m\hat{x}p1$ and $m\hat{x}p2$ respectively, and $s\hat{z}p1$ is basiclly the same as $s\hat{x}p1$, but $s\hat{z}p2/s\hat{x}p2$ is only 96% for n=50 and 87% for n=20. The reason of this lies (to be continued)

Table 3 Results of calculation with various values of error parameter in case applying Pearson type 3 distribution (Historical length N'=100)

No. of alternatives	α	α'	Variables	$m\hat{x}p1$ or $m\hat{z}p1$ or $m\hat{z}'p1$	$m\hat{x}p2$ or $m\hat{z}p2$ or $m\hat{z}'p2$	$s\hat{x}p1$ or $s\hat{z}p1$ or $s\hat{z}'p1$	$s\hat{x}p2$ or $s\hat{z}p2$ or $s\hat{z}'p2$	
1	0		x	2595	3527	719	1254	20
2	0.05		z	2601	3535	722	1260	20
3	0.10		z	2625	3573	733	1280	20
4	0.20		z	2724	3727	779	1361	20
5	0.30		z	2874	3956	864	1503	20
6	0.50		z	3265	4525	1159	1953	20
7	0	0	x	2691	3743	449	773	20
8	0.05	0.30	z'	2762	3918	518	1006	20
9	0.05	0.40	z'	2803	4018	569	1167	20
10	0		x	2643	3620	611	1093	30
11	0.05		z	2650	3631	617	1102	30
12	0.10		z	2677	3673	630	1126	30
13	0.20		z	2783	3842	684	1221	30
14	0.30		z	2944	4091	783	1384	30
15	0.50		z	3353	4693	1113	1890	30
16	0	0	x	2717	3785	414	734	30
17	0.05	0.30	z'	2775	3926	484	950	30
18	0.05	0.40	z'	2808	4007	531	1094	30
19	0		x	2696	3730	484	883	50
20	0.05		z	2702	3740	490	893	50
21	0.10		z	2730	3784	503	919	50
22	0.20		z	2840	3963	559	1021	50
23	0.30		z	3007	4222	665	1198	50
24	0.50		z	3425	4842	1014	1733	50
25	0	0	x	2726	3797	388	705	50
26	0.05	0.30	z'	2766	3893	446	877	50
27	0.05	0.40	z'	2789	3950	483	988	50
28	0		x	4412	6774	2114	3959	20
29	0.05		z	4414	6778	2120	3970	20
30	0.10		z	4430	6807	2139	4002	20
31	0.20		z	4504	6941	2210	4112	20
32	0.30		z	4632	7165	2321	4312	20
33	0.50		z	5007	7790	2647	4872	20
34	0	0	x	4806	7838	1289	2449	20
35	0.05	0.30	z'	4883	8091	1408	2894	20
36	0.05	0.40	z'	4942	8264	1484	3170	20
37	0		x	4581	7139	1826	3515	30
38	0.05		z	4584	7146	1836	3531	30
39	0.10		z	4603	7182	1859	3572	30
40	0.20		z	4690	7345	1943	3716	30
41	0.30		z	4837	7613	2068	3942	30
42	0.50		z	5264	8350	2423	4594	30
43	0	0	x	4876	7931	1208	2375	30
44	0.05	0.30	z'	4929	8119	1330	2787	30
45	0.05	0.40	z'	4978	8260	1402	3032	30

Table 3 (continued)

46	0		x	4769	7584	1462	2912	50
47	0.05		z	4773	7595	1473	2932	50
48	0.10		z	4795	7638	1495	2975	50
49	0.20		z	4893	7830	1574	3120	50
50	0.30		z	5056	8140	1695	3367	50
51	0.50		z	5521	8971	2062	4079	50
52	0	0	x	4904	7960	1155	2331	50
53	0.05	0.30	z'	4931	8068	1256	2651	50
54	0.05	0.40	z'	4962	8163	1313	2839	50

Note: In the above, cvx=0.5, csx=2.0, xp1=2805, xp2=3955,(No.1-27)
In the above, cvx=1.0, csx=3.5. (No.28-54)

perhaps in the fact that the moment method is applied to the logarithms rather than to discharges themselves. Bobee et al.[2] has demonstrated that the performance of the above method is poorer than that of the moment method applied to discharges.

5.2.2. In the presence of historical flood.
(1) For the Pearson type 3 distribution. Let α' denote the error parameter of the historical flood and all symbols below with a prime will refer to the case with a historical flood. The results of Monte Carlo calculations using moment method are given in Table 3. Comparing the results in cases with and without historical flood, the following conclusion may be drawn. (a) For $\alpha'=0.3$, the presence of historical flood can improve the results for both sets of parameters and all three values of n. (b) For $\alpha'=0.4$, the presence of a historical flood causes only 10% increase of $s\hat{z}'p2$ against $s\hat{z}p2$ for the alternative No. 27 in Table 3, while the remaining performances of $\hat{z}'p$ are still better than those of $\hat{z}p$. This means that even if α' is as large as 0.4, the presence of a historical flood is still effective for improving the results. (c) The effeciency (to be continued)

Table 4 Results of calculation with various values of error parameter in case of log Pearson type 3 distribution

No. of alter- natives	α	α'	Vari- ables	$m\hat{x}p1$ or $m\hat{z}p1$ or $m\hat{z}'p1$	$m\hat{x}p2$ or $m\hat{z}p2$ or $m\hat{z}'p2$	$s\hat{x}p1$ or $s\hat{z}p1$ or $s\hat{z}'p1$	$s\hat{x}p2$ or $s\hat{z}p2$ or $s\hat{z}'p2$	n
55	0		x	2832	6475	1597	13163	20
56	0.05		z	2832	6410	1594	12360	20
57	0.10		z	2844	6309	1598	11492	20
58	0.20		z	2913	6042	1602	9653	20
59	0	0	x	2834	4475	895	2490	20
60	0.05	0.30	z'	2885	4625	986	2690	20
61	0		x	2828	6158	1245	7960	30
62	0.05		z	2829	6112	1249	7730	30
63	0.10		z	2842	6019	1261	7434	30
64	0.20		z	2896	5715	1282	6637	30
65	0	0	x	2863	4512	878	2461	30
66	0.05	0.30	z'	2897	4618	960	2646	30

Table 4 (continued)

67	0		x	2832	5891	900	4460	50
68	0.05		z	2834	5844	907	4391	50
69	0.10		z	2843	5739	917	4280	50
70	0.20		z	2873	5362	933	3905	50
71	0	0	x	2875	4522	871	2449	50
72	0.05	0.30	z'	2895	4584	934	2592	50

Note: In the above, cvx=0.5, csx=2.0, N'=100.

Table 5 Results of calculation with various values of error parameter for Pearson type 3 distribution by using the curve-fitting method

No. of alter-natives	α	α'	Vari-ables or or	$m\hat{x}p1$ or $m\hat{z}p1$ or $m\hat{z}'p1$	$m\hat{x}p2$ or $m\hat{z}p2$ or $m\hat{z}'p2$	$s\hat{x}p1$ or $s\hat{z}p1$ or $s\hat{z}'p1$	$s\hat{x}p2$ or $s\hat{z}p2$ or $s\hat{z}'p2$	n
73	0		x	2868	4068	774	1377	20
74	0.05		z	2872	4093	786	1408	20
75	0	0	x	2889	4109	534	948	20
76	0.05	0.30	z'	2927	4198	630	1229	20
77	0		x	2854	4059	643	1148	30
78	0.05		z	2854	4056	656	1173	30
79	0	0	x	2812	3966	466	800	30
80	0.05	0.30	z'	2824	3989	479	828	30

Note: In the above, cvx=0.5, csx=2.0, N'=100.

of the historical flood decreases with the increase of sample size n. (2) For log Pearson type 3 distribution. Table 4 shows the presence of a historical flood may reduce $s\hat{z}p1$ and $s\hat{z}p2$ to different extent in accordance with n. Let $\Delta sri=(s\hat{z}pi-s\hat{z}'pi)/s\hat{z}pi$ represent the relative decrease of $s\hat{z}'pi$ with respect to $s\hat{z}pi$, then we have sr1=40% for n=20, sr1=23% for n=30 and sr1=-3% for n=50; sr2=78% for n=20, sr2=68% for n=30 and sr2=31% for n=50. In addition, the bias of $m\hat{z}'p1$ is also smaller than that of $m\hat{z}p1$. However, in contrast to the above benefit of historical flood, the bias of $m\hat{z}'p2$ has a substantial increase in comparison with the bias of $m\hat{z}p2$. Therefore, in case log Pearson type 3 distribution is used as population distribution, effects of historical flood appear to be more complicated and attention must be paid to this situation.

5.3. Results Obtained by Using a Curve-fitting Method

Since the curve-fitting method is very time consuming on the computer , this method has been used only for a few alternatives as shown in Table 5.

5.3.1. In the absence of historical flood. The results are similar to those obtained by moment method. The effect of measurement error is very slight as α =0.05. For instance, comparing the results of alternative No.78 to that of No.77, it may be seen that $m\hat{z}p1$ and $m\hat{z}p2$ are basically the same as $m\hat{x}p1$ and $m\hat{x}p2$ respectively; and $s\hat{z}p1/s\hat{x}p1=1.020$, $s\hat{z}p2/s\hat{x}p2=1.022$ (the corresponding results by moment method are 1.018

and 1.008 respectively), i.e. the increase of the root mean square error of $\hat{z}p1$ and $\hat{z}p2$ is less than 2%. Obviously, this should be considered as permissible.

5.3.2. In the presence of a historical flood. Comparing the results in the presence and absence of historical flood, it can be seen that the effect of historical flood is quite significant. As shown in Table 5, not only the biases of $m\hat{z}'p$ become less than those of $m\hat{z}p$, but also the root mean square error $s\hat{z}'p$ of $\hat{z}'p$ become less than $s\hat{z}p$ of $\hat{z}p$. For in-instance, comparing the results of alternative No.80 with those of No.78, $s\hat{z}'p1$ and $s\hat{z}'p2$ are respectively reduced in an amount of 27% and 29% with respect to $s\hat{z}p1$ and $s\hat{z}p2$. The corresponding results of moment method are 22% and 14% respectively. This means the positive role of historical flood with $\alpha' \leq 0.3$ in the curve-fitting method is greater than that in the moment method. Comparing the results in the presence of a historical flood with those in the absence of historical flood and the systematic record is accurate, it can be seen that the former results are also better than the latter one's.

6. CONCLUSION

6.1. The mathematical model of measurement error is the basis of the study. Different models will generally produce different results. Therefore, not only the structure of the model should be reasonable, theoretically justifiable and tesified with practical data, but also its parameter should be derived on the basis of practical data and able to reflect the practical situation of discharge measurement. Taking the above principle into full account, a mathematical model is proposed.

6.2. Based upon the proposed model and applying the Monte Carlo simulation, the following results are obtained.

6.2.1 For the Pearson type 3 distribution. If only systematic record is used in the FFA, the effect of the measurement error is slight and neglegible whatever parameter estimation method is used, if the error parameter $\alpha \leq 0.05$, i.e. if the error magnitude is of ordinary order. If, besides the systematic record one historical flood is also used in the FFA, and the error parameter α' of the historical flood is less than 0.3, consideration of the historical flood in the FFA can improve the results, especially if the curve-fitting method is used.

6.2.2. For the log Pearson type 3 distribution. In the absence of historical outlier and $\alpha \leq 0.10$, the effect of measurement error are not only permissible but also positive. If along with the systematic record one historical flood is also used in the FFA, the effects of measurement error appear to be complicated and attention must be paid to them.

6.3. The above statements are refered to the effects of measurement error of peak discharges. If instead of peak discharges, the flow volume in a given period is studied in the FFA, the effects of measurement error will decrease, and the longer the period, the smaller the effect.

6.4. The above conclusions are derived under the condition of a stable and single rating curve. This imposes certain restrictions on the applications of the conclusions. However, considering the fact that a stable and single rating curve is quite representative in the upper and middle reaches of rivers and also the fact that FFA are usually conducted just for stations located in these reaches, it may be expected that conclusions obtained in this paper will have a broad range of applications. As for stations with loop rating curve, effect of measurement error may be larger than what we described here. However, when we calculate discharge on the basis of stage in that case, different sections of the loop rating curve should be used according to some relevant parameters, such as time order, slope of water surface, etc. Therefore, the error is not so large as that of applying an average rating curve. Of course, further study on the error effect for loop rating curve is needed.

6.6. The determination of the error parameter α' for historical flood is beyond the scope of this study. However, it is worth while to make a special investigation in order to obtain a reliable information about α' and the effect of historical flood.

6.5. Besides the mathematical model, a more important factor which plays a fundamental role in this study is the practical situation of discharge measurement. For instance, the largest annual peak discharge in the systematic record on large and middle rivers can generally be practically measured in China, hence it is not necessary to extrapolate the rating curve (or extrapolate only to a short range) in order to determine the largest discharge. If this is not the case in another country,for example in U.S.A.[11], the effect of measurement error may be larger than what is demonstrated here.

REFERENCES

[1] Bard, Yonathan, 1974, Nonlinear parameter estimation, Academic Press.

[2] Bobee, B. and R.Robitanlle, 1977, The use of Pearson type 3 and log Pearson type 3 distribution revisited, Water Resources Research, 12(2)

[3] Cong Shuzheng & Tan Weiyan, 1979, Statistical testing research on the method of parameter estimation in hydrological computation, Proc. of the International Symposium on Specific Aspects of Hydrological Computations for Water Projects, UNESCO-IHP, Leningrad, USSR,3-7,Sept.

[4] Draft International Standards ISO/DIS 1070, 1982, Liquid flow measurement in open channels-slope-area method.

[5] Herschy, R.W., 1978, Chapter 10: Accuracy, in Hydrometry principles and practice, Edited by R.W. Herschy, A Wiley-Interscience Publication.

[6] Hydrology Subcommittee, 1982, Guidelines for determining floods flow frequency, Editorial Corrections, U.S. Department of the Interior,

Geological Survey, Office of Water Data Coordination
[7] Ministry of Water Resources and Electric Power, 1980, Handbook of hydrometry, **Vol. 3**, Beijing.
[8] Ministry of Water Resources and Electric Power, 1980, Standards of Design Flood Computation for Hydro and Hydroelectric Projects, Beijing.
[9] Potter, K.W. & J. F. Walker, 1981, A model of discontinuous measurement error and its effects on the probability distribution of flood discharge measurement, Water Resources Research, **17(5)**.
[10] Potter, K.W. & J.F. Walker, 1982, Modeling the error in flood discharge measurements, in Time Series Methods in Hydrosciences, by A.H.El-Shaarawi and S.R. Esterby (Editors).

SOME RESEARCHES ON DETERMINATION OF RETURN PERIOD FOR HISTORICAL FLOOD IN ANALYSIS OF FLOOD FREQUENCY

Minggany Li . Senior Hydrological Eng.
Qiusheng Yang Hydrological Eng.
Mid-South Design Institute for Hydroelectric Proj., MWREP, PRC.
Guitang, Changsha, Hunan, China

ABSTRACT. This paper emphasizes the important effect of historical flood on flood frequency analysis and points out that the considerable swayed range of the exceedance frequency of high extraordinary floods may bring about some difficulties for the upper fitting. The authors suggest that the representativity of computing series including observed and historical data be demonstrated through analysing the magnitude and frequency of high extraordinary flood which has occurred from further years to the present and the plotting position for the first terms in computing series calculated by use of the average return period so as to strengthen the stability of sample.

INTRODUCTION

The method of frequency analysis is mainly used to determine the criteria of flood control for hydro projects on the river in China. However, due to new extraordinary flood to be introduced as time goes on, the results of design flood for hydro project are usually changed, thus bringing certain difficulty to design and operation of projects. Therefore how to improve the technique of frequency analysis and keep the relative stability of the result of design flood is an important problem for analysis of flood frequency at present.

Practice has proved that to make the results of design flood reasonable and relative stable mainly lies in basic data, i.e. the observed data should meet the requirement of proper accuracy and the computing series possesses suitable representativity.

The problem of accuracy of observed data won't be specially discussed. This paper only presents some opinions about the representativity of computing series.

THE MEANING OF REPRESENTATIVITY AND HOW TO STRENGTHEN IT

The representativity of computing series means the representative degree of available series for the population distribution, namely the stability of exceedance frequency (plotting position) of available every

V. P. Singh (ed.), Application of Frequency and Risk in Water Resources, 191–197.

term flood. If the plotting positions of every flood don't vary obviously as the increase of series length, then the results of analysis of flood frequency will be kept relative stability, thus they might be considered rational and reliable. Generally speaking, the shorter the flood series is, the more poor the representativity would be, also the bigger the sampling error of statistic parameters and results of frequency analysis. In order to reduce the sampling error and increase the series representativity, it is most important to consider the effect of historical flood as far as possible. That historical high outlier of long period is put in frequency analysis could sufficiently extrapolate the data and increase the effect of representativity for computing series. Especially, if a number of historical floods could be investigated in a longer period, then it would not only reflect the distribution characteristics of high outlier and second outlier in a basin and contribute to choice of the type of frequency curve but also reduce greatly the subjectivity in fitting curve.

Above mention emphasizes the effect of historical flood on frequency computation when a number of historical flood in a longer period put in computing. On the other hand, no matter how long the investigated and verified period would be, for a computing series of limeted sample, the exceedance frequency for every flood always has a considerable swayed range because of sampling error in it, especially for the first (or last) terms, the swayed ranges are more evident and projecting. In the light of introduction in concerned paper* if the condition is adopted that the average of sampling distribution for exceedance frequency in confidence interval is equal to the expected value m/(n+1) and the adopted confudence probability equals 50%, the confidence interval of exceedance frequency P_a---Pb for the first three terms (m=1,2,3) could be calculated as follows:

Table 1. The confidence interval of exceedance frequency for the first three terms in various length of series

Ordered sequence	Series length 100	200	300	500
1	0.43-1.89%	0.22-0.95%	0.15-0.64%	0.087-0.38%
2	1.16-3.12%	0.58-1.57%	0.39-1.05%	0.23-0.63%
3	1.94-4.28%	0.97-2.16%	0.65-1.44%	0.39-0.87%

SOURCE: Adapted from (1)

It is well known that the representativity of computing series strengthen with increase of term number of series (year). However, it may be seen from above table that the swayed range of exceedance frequency of the first three terms in the series obviously widens with increase of term number of the series. When series length is 100 terms for confidence probability of 50%, the confidence interval of exceedance frequency of the first term is 0.43-1.89%, i.e. the return period may be from 53 to 233 year with a variation range of 180 years. While series length is 500 for the same confidence probability, the confidence interval is 0.087-0.38% and corresponding return period will be from 263 to 1149 year with a variation range up to 886 years. The above results well indicate instability of position of the first term when term number of the series increases. This feature is rather unfavourable for the upper fitting of frequency curve. Moreover, it may be seen that although variation range of position of the first term become larger after increase of the series length, probable variation range of positions thereafter each term obviously decrease than that of the first term, thus providing a new possibility to decrease sampling error of extraordinary flood after extension of the series. That is to say, if it may be adjusted through investigation and verification of the historical flood that exceedance frequency of the first extraordinary flood doesn't rank first in several years from someone year (including that year when it occurred), but rank 2nd, 3rd, etc in a longer period, the sampling error of exceedance frequency determined in this way would be less (than the first term), hence its position may be considered to be relatively stable. This method to deal with exceedance frequency of extraordinary flood is of practical significance for resolving the upper fitting and extraextension of the frequency curve.

METHOD OF DETERMINING EXCEEDANCE FREQUENCY FOR HISTORICAL EXTRAORDINARY FLOOD

This paper takes analysis and determination of the historical flood at Wuqianxi Hydro Project as an example to describe the application of above mentioned viewpoint and method.

The Available Data and Past Computation of Exceedance Frequency for Historical Flood

The Wuqianxi Hydro Project is located in the lower reach of the Yuanshui River with a catchment area of 83800km². There is available recorded discharge data of 55 years (1925-1944, 1947-1981). Additionally, information of the five historical floods of 1766, 1834, 1878, 1911, 1912 in near past period have been obtained through in-situ investigation and visit upstream and downstream, and the stone-inscription trace for the flood events of 1766, 1834, and 1911, the ink mark for that of 1878 and the finger-pointed mark for that of 1912 may be found respectively. The flood elevation of every event is rather reliable and may be accurately transfered into the section at the hydrologic station. In consideration of following characteristics: stable control conditions at hydrologic station; no wash out and deposit variation on the section; and a stripped observed-point distribution on the stage-discharge curve in past years, when high stage is extended to the historical extraordinary flood, the maximum percentage only account for 40% of variation range of the recorded stage, with comparatively high accuracy of extrapolated discharge curve. Therefore the historical flood of every year was all calculated from stage-discharg curve and the peak discharge were 41700, 37300, 34000, 33500, 33000 m³/s respectively.

In the past, the return period of historical flood of every year was only based on investigation and calculated from the time when the farthest historical flood event occurred to date (1981), i.e, the exceedance frequency of the flood of 1766, P=1/(216+1), those of 1834, 1878,1911 and 1912 were then calculated respectively in their sequences during this period.

Recent Researches and Handling Method

Because of a long history of the Yuanshui drainage, most of the counties along the Yuanshui River have written records of historical floods and dry calamities. According to historical documents and investigation data, large floods occurred in 25 years since the Song dynasty, i.e. 1189, 1369, 1469, 1512, 1532, 1571, 1608, 1611, 1612, 1613, 1618, 1684, 1725, 1742, 1748, 1765, 1766, 1831, 1833, 1834, 1844, 1848, 1878, 1911, 1912, etc. The in-situ investigation indicated that among above-mentioned floods, only five of 1189, 1369, 1571, 1618 and 1766 belong to extraordinary ones. With the exception of Q_{1766} to be accurately determined, all of other peak discharges were Larger than Q_{1766}. In consideration of instability of position of the first term in above analysis, the exceedance frequency of extraordinary flood of 1766 may be calculated as the 5th since 1189, the 4th since 1369 or the 3rd since 1571 in order to decrease the sampling error. Seeing that there were only simple records of flood/dry calamities between 1369 and 1571 and extraordinary flood might be missed; while those since 1571 were complete and less missed, the flood of 1766 was considered as 3rd extraordinary flood since 1571 with the frequency P_{1766} =3/(411+1). In this way, not only the verified duration of historical flood was adequately increased, but also instability of exceedance frequency of the first term flood in the calculation series decreased.

As for determination of the exceedance frequency of the Secondary extraordinary floods such as those of 1834, 1878, 1911, 1912, etc, because their magnitudes are less and easy to be missed in the historical records, the adopted verified historical duration is not traced back to too far, and it may be considered to be adequate to analyse the flood events occurred since 1766.

The thorough investigation in situ and detailed verification of the related historical records shown that since 1766, with the exception of above mentioned four events, other secondary extraordinary floods occurred in 1831, 1833, 1844, and 1848 respectively with peak discharges being 38200, 37900, 34500 and 38100 m^3/s. Consequently, the exceedance frequency of the flood of 1834 should rank the 5th, not the 2nd in the 216 years since 1766. And in the same way, those of 1878, 1911 and 1912 should rank the 7th, 8th and 9th, but not the 3rd, 4th and 5th as identified before.

Considering that the flood/dry calamities records in this drainage area since 1571 are more complete, the determined frequency of the secondary extraordinary flood were further verified by use of the flood records since 1571.

Based on degree of the flood calamities described in the historical records and their magnitudes to be determined accurately, the secondary extraordinary floods may be divided into 4 grades as follows: the first is refered to those equal to or larger than the flood of 1834; the 2nd is larger than the flood that "boats could sail on the streets"; the 3rd is equal to or larger than the flood of 1878; and the 4th is equal to or larger than the flood that "large volume of water entered the town" with the responding discharges being 37300, 36600, 34000 and 30000m^3/s respectively. Then the occurrence number and return period of the flood of each grade in long(since 1571) and short (since 1766) durations were classified respectively. The results are given in Table 2.

Analyses of the frequencies (return periods) of each grade floods in long and short durations shown that most grades have a comparatively close frequency. Therefore it may be judged that the exceedance frequencies of the secondary extraordinary floods, which were determined according to occurrence number of each grade floods since 1766, are basically stable in the long duration since 1571.

Table 2. The Contrast between return periods of each grade flood in long and short durations

Grade	Corresponding discharge(m^3/s)	Duration	
		216 years(1766-1981)	411 years(1571-1981)
		Return period	Return period
I	37300	43	46
II	36600	43	37
III	34000	31	29
IV	30000	17	16

Concerning the general grade floods, although there are recorded data of 55 years, they are only a small sample with larger sampling error. Therefore, the attention should be paid to the influence of the continuous flood and dry periods on the small sample.

If we regard the Q=34000 m^3/s of 1878 as a standard, it may be seen that the flood (this magnitude) happened six times in the 19th century, but no one in the 20th century; otherwise, taking Q=30000 m^3/s of the flood that "large volume of water entered the town" as the standard, the flood (this magnitude) happened six times before 1950s, but no one from the 1950s up to now. Above analyses indicated that the sample of available floods of 55 years lie in obvious medium/ small water section, and the distribution parameter calculated with the recorded data of 55 years might be lower. Considering the poor representativity of 55 years series, the positions of the first several terms in the continuous series may be adjusted in accordence with the magnitude sequences of the floods in long duration, in order to increase representativity of the computation series. And it is unnecessary to modify term by term the rank sequences of the others besides the first several terms due to lack of evidence of the occurrence frequencies for the medium/small floods and less influence on the caculation results.

Based on above analyses and treatment, the swayed range of return period for the first extraordinary flood Q_{1766} has been reduced from 380 y. to 120 y. and the others were also reduced when the confidence probability equals 50% as compared with before. So we have reason to consider that the present computation series possesses better representativity and the results of frequency calculation is rational and reliable. The comparison between the calculation results obtained at present and in the past are given Figures 1 and 2.

SUMMARY

Considering the fact that the large and small floods occurred in 18th-19th and 20th century respectively, the historical floods must be put in analysing flood frequency as many as possible when analysing flood frequency so as to reduce the uncertainty of return period for flood. However, in view of the larger sampling error for the exceedance frequency of first terms in a series, the average return period should be adopted instead of some specific interval in order to avoid the fortuity in determining the plotting position of first floods. The design result on this basis will certainly have better precision and stability.

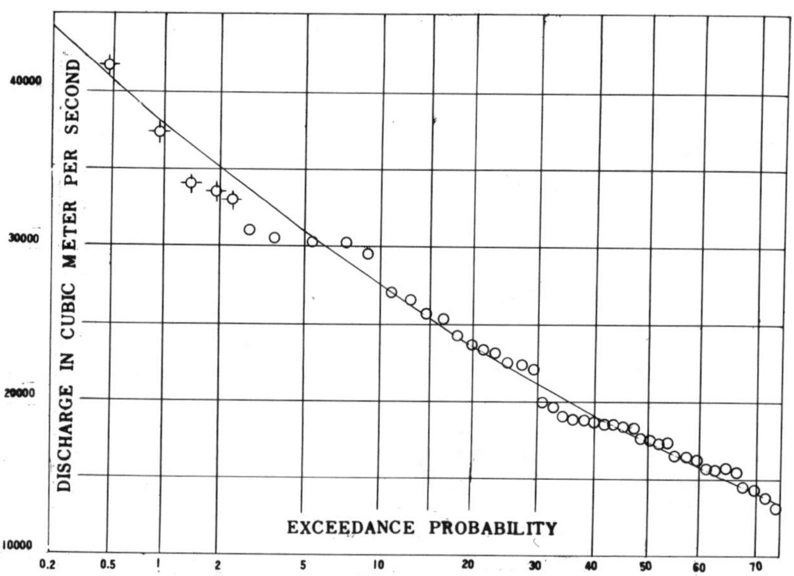

Fig 1. Preliminary Frequency Curve For Yuan River at Wuqianxi Damsite

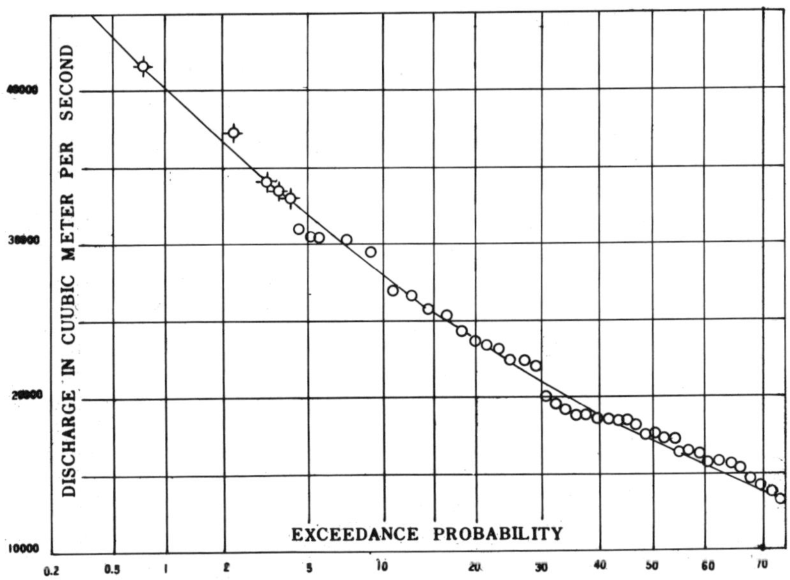

Fig. 2 Final Frequency Curve For Yuan River at Wuqianxi Damsite

ACKNOWLEDGEMENTS

The authors would like to express their thanks to many colleages for flood computing of Wuqiangxi Hydroproject, specially to those persons who took part in a large number of work on investigation and verification of historical flood in situ. It is because of their hard work that such rich information about historical flood may be obtained at present.

REFERENCES

(1) Hydrological group of Design Institute of Fourth and Eleventh Engineering Bureaus for Hydroloectric Projects, MWREP,1975. Some Problems on Analysis and Computation for Flood Frequency PP. 26-34

(2) Jin Guang Yan 1963. Statistic Principle and Method for Hydrology, Publishing House of Chinese Industry, Beijing. PP. 183-186

(3) Linsely Ray K, Kohler Max A, Paulhus Jaseph L, H. 1975. Hydrology for Engineerings, McGraw-Hill Book Co. New York, PP. 340-342

(4) Ministry of Water Resources and Electric Power 1979. A Computation Standard of Design Flood for Hydro Project, Water Conservancy Publishing House, Beijing, PP. 67-70

POTENTIAL AND APPLICATION OF FLOOD FREQUENCY DETERMINATIONS

Leo R. Beard
Espey, Huston & Associates, Inc.
P.O. Box 519
Austin, Texas 78767
USA

ABSTRACT. Limitations of uncertainty in estimating flood frequencies and means of minimizing the uncertainty are discussed. The value of reducing uncertainty by improved use of data is discussed in relation to risk of future flooding that cannot be eliminated or reduced except by structural measures. The value and uses of reliability criteria are much misunderstood. An arbitrary safety factor is infeasible, whereas the concept of expected probability is a logical result of considering uncertainties. This concept has been used to assess impacts of uncertainties in mean and variance of flood values. When it can be extended to include uncertainties in higher order moments, such as the skew coefficient, light might be shed on the reason why extreme events occur that are far beyond any reasonable expectation. Applications of flood frequency estimates in the flood insurance program and in general flood mitigation projects are discussed. Most important, stress is placed on the most fundamental use of flood frequency information -- assessment of expected damages and benefits of all kinds for various alternative programs of flood management under all possible flood conditions.

UNCERTAINTY

Estimates of flood probabilities or frequencies are necessarily based on flood experience, and consequently their reliability is dependent on the extent of that experience and how effectively the information is used, and their accuracy depends on the degree to which past experience is representative of future potential.

There are several techniques for maximizing reliability through maximizing the effective use of available information. Primarily, a characteristic function is developed that is intended to transform an array of individual flood observations at a specific location into a continuous function of magnitude versus exceedance frequency. The function is intended to be most representative of the true natural relationship and is based on studies of functions that best agree with observations at a great many locations. The function currently most used for flood flow frequencies is the logarithmic Pearson-3 function, which has the undesirable characteristic that logarithms of

V. P. Singh (ed.), Application of Frequency and Risk in Water Resources, 199–205.

flood flows are bounded for non-zero skew and that the estimate of such bound is highly suspect, since it is sensitive to minor data variations.

Secondly, a method of fitting the adopted function to data is developed so as to extract maximum information from data at a single location. Currently, the first three moments of the logarithms of annual maximum streamflows at a specific location are used to define the best-fit function. This is not the maximum-likelihood method of fitting the Pearson-3 function; but is considered superior (more robust) because the Pearson-3 is probably not accurately representative of the population.

Thirdly, there is great potential for increasing reliability of frequency estimates through transfer of information from many locations where data are available to one location of particular interest. This is usually accomplished by correlating frequency statistics with drainage-basin characteristics, but most of such studies to date have not greatly increased reliability.

The net result of studies of uncertainties of flood frequency estimates is that standard errors of estimating flood magnitudes corresponding to specified frequencies are very high -- in the order of 10 to 50 percent, depending on the stream characteristics and amount of data.

RISK

The chance that a 1000-year flood will occur in the next 50 years is essentially the same as the chance that the past 50 years of record contains a 1000-year flood. Consequently, reliable and accurate estimates of flood frequencies are no guarantee that an area is safe from flooding. This poses the question as to whether great effort should be expended to increase reliability of flood estimates, since the actual risk cannot be changed.

If 100-year protection is intended but only 50-year protection provided because of uncertainty, investment is saved if the 50-year flood does not happen to be exceeded. Also, if a 200-year flood happens to occur, perhaps not much is lost because of the uncertainty. On the other hand, if 200-year protection is actually provided when 100-year protection is intended, one will have over-invested if the 50-year flood is not exceeded but will benefit if the 100-year or 200-year flood occurs and the 200-year flood is not exceeded.

Essentially, then, benefits from protection are a function of investment and are otherwise independent of uncertainties in estimating flood frequencies. Reducing uncertainties will permit one to determine more accurately the degree of protection or zoning that will yield maximum benefits, but one finds that variations in net benefits are small in the region of optimum. In other words, moderate changes in protection or zoning do not change net benefits greatly -- the benefit function has a broad or relatively flat peak.

It is not the intention here to discourage efforts to improve reliability of flood frequency estimates, but simply to put into perspective the importance of such changes in reliability. If we are failing in flood management today, it is through lack of recognition by the general public that there is virtually no way to eliminate flood dangers, while our programs often are interpreted to do so.

EFFECTIVE ASSESSMENT OF FLOOD FREQUENCY

While the great uncertainty in flood frequency estimation cannot be denied, it is also recognized that small differences in flood estimates can be crucial in the implementation of flood regulations and in the design of remedial measures. Consequently, we have emphasized uniformity in flood frequency techniques, hoping to maximize accuracy and reliability while assuring that different investigations will arrive at essentially the same answers, thus minimizing controversy. However, it is important not to gain uniformity through substantial loss of accuracy.

The United States Water Resources Council, in Bulletin 17B recommends fitting flood flow frequency data to the log Pearson 3 function using the first 3 moments of the logarithms of annual maximum stream flows. Guidance is given on the selection of the third moment (skew) and on treatment of zero flows, outliers and historical flows. These procedures are, in my opinion, the best currently available, but leave much to be desired.

The most important draw-backs are due to uncertainty or lack of knowledge about certain aspects of the problem. This will be discussed in the following sections.

In considering alternative approaches, it is important to face the fact that we will probably never know the true nature of a universal function of flood flow magnitude versus frequency. Uncertainties in the functional form are extremely difficult to evaluate and to apply in the manner discussed below. Furthermore, fitting by method of maximum likelihood can be hazardous when it is not known that observations actually derive from the functional form being fitted. Least-squares fit may be satisfactory, but the method of moments applied to a data transform where that transform has very low skewness should be fairly robust and consequently fairly dependable. The transform selection must not be based on a single sample, however, because this would make the process less robust.

USES OF RELIABILITY CRITERIA

While it is always desirable to know how reliable engineering determinations are, it is difficult to make effective use of such information. In flood frequency evaluations, it is not feasible simply to add safety factors. To illustrate, if one adds a safety factor to the 100-year flood estimate in order to make it more safe for design purposes, the estimated benefits associated with that design are being exaggerated, or, in the case of flood insurance, the premiums would be too high.

The most effective use of reliability information is in the application of mathematical expectation theory. After making a so-called best probability estimate, all possible true probabilities associated with a specified magnitude are weighted in proportion to their likelihood of being true, as indicated by reliability criteria, and the weighted value is used as the average that will in fact occur as the number of applications approaches infinity. Thus, although in each application, we do not know the true probability, the differences between our estimates and the true probabilities will average toward zero as the number of applications increases. This, of course, is excellent for applications such as a flood insurance program or a widespread flood-protection program. In the case of a single application, however, the logic is not so obvious. However, since any management scheme is a gamble, it appears only logical to follow the usual gambler's rule that each gamble is best managed as though it is one of a large number. There is no absolutely clear decision, and the best procedure is to consider the odds of all possible consequences and the consequences associated with each.

It is essential to work with expected probability corresponding to a specified magnitude rather than expected magnitude corresponding to a specified probability, since consequences (damages or benefits) are a linear function of frequency and not of magnitude. Thus, we must solve for an unbiased frequency or probability corresponding to each magnitude rather than an unbiased magnitude corresponding to each frequency.

This has been done for the normal distribution as given in WRC Bulletin 17B, considering only uncertainties in the mean and standard deviation of the logarithms of flow values. These are applied to Pearson-3 functions having small skew (near-normal). However, this ignores uncertainty in the skew and uncertainty in the form of the function (or higher-order moments). At present, this application represents the lower limit of uncertainties and thus the lower limit in removing bias of frequency estimation. The net result is that large floods will probably occur more frequently, on the average at many locations, than is estimated using expected probability procedures in Bulletin 17B.

DISTORTION OF THE FUNCTIONAL FORM

It can be noted that the expected probability estimates of frequencies based on samples drawn from a normal distribution do not, in themselves, follow the normal distribution. Adjustment for bias depends on the sample size and distorts the resulting function.

Likewise, considerations of uncertainty will result in distortion of the Pearson-3 function. In particular, since there is uncertainty in the skew, solutions showing negative skew are accompanied by some likelihood that the true skew is positive, and vice versa. Thus, while the Pearson-3 function is bounded for non-zero skew, the expected probability solution is not. Bulletin 17B does not take this into consideration, nor does any alternative procedure known to the writer.

Recognition of this distortion indicates that it is extremely dangerous to extrapolate Pearson-3 frequency functions having substantial negative skew, since probabilities decrease very rapidly as the upper bound of magnitude is approached. Thus, we may be subject to major flooding that is indicated to be virtually impossible in the frequency analysis.

THE CRITICAL-MASS PHENOMENON

History is replete with disasters that are far more severe than was ever anticipated. While some statisticians use the term "outlier" to characterize these phenomena, most consider that they are events that simply are very rare and happen to have occurred at this time and place. This may be hard to prove or disprove, but it is recognized that in a thousand independent cases, a 1000-year event is expected once every year, a 10,000-year event once every 10 years someplace, etc. On the other hand, events such as the Great Depression or World War occur that are far beyond any condition that could be anticipated by simply extrapolating observations of every-day or every-year phenomena.

Some extreme events in nature, as well as in society, require a critical mass to initiate an extreme event far beyond extrapolated values. Consider starting a log fire with a single piece of kindling or a single sheet of newspaper. A match is applied, and the fire lasts only a few seconds. As kindling or paper is added in successive trials, the fire burns brighter and lasts longer. At some trial, the logs will ignite and continue to burn, creating an event that is far greater in duration and intensity than could have been forecasted by extrapolation.

This phenomenon is understandable in cases like the Great Depression where final loss of faith caused a cascading economic crash, and in war where a final act changes negotiation into hostility. A natural log fire or forest-fire phenomenon is easy to understand. Perhaps similar phenomena occur in meteorology and hydrology, where interaction of severe conditions are mutually sustaining and mutually augmenting after a critical stage is reached.

We cannot discount this possibility completely and should recognize that somewhere along the extrapolation of frequency functions, magnitude may increase very rapidly or abruptly. Thus, while we may extrapolate frequency functions with some degree of confidence for most applications, estimating a probable maximum event corresponding to a very rare frequency is hazardous.

LOGICAL USES OF FREQUENCY ESTIMATES

The relationship of flood magnitude versus frequency of future events is used fundamentally to integrate the consequences of floods that can occur in the future under specified management plans. Ideally, a selected management plan should maximize net annual benefits -- the difference between benefits and costs.

Expediency and political and legal considerations often require specific frequency determinations, such as a 5-year flood for storm-drain inlet design or a 100-year flood for flood-plain zoning. In these cases, it is usually tacitly assumed that the frequency of larger and smaller floods is not pertinent. However, it must be remembered that the consequences of all flood magnitudes and their associated frequencies should be considered in the selection of any management plan where it is feasible to do so.

Usually such integrations are made using a best-estimate frequency curve -- not the true frequency curve. Ideally one might consider all possible frequency curves and weight the integration results of each in accordance with the likelihood that each represents the true frequency curve, but this is not practicable. Fortunately, a single expected-probability frequency curve, which averages all possible frequencies for each magnitude in proportion to their respective likelihoods, can be used. This is true only because average annual damages are linearly related to frequency.

FLOOD INSURANCE

Although the flood-control program of the Corps of Engineers since the last century and particularly since the 1936 Flood Control Act has greatly enhanced the use and productivity of flood plains, population pressures and increased per-capita gross national product have resulted in ever-increasing flood damages. Not until the National Flood Insurance Program has action been taken at the national level to manage uses within flood-prone areas. In general, the program works to shift the responsbility for disaster relief from governmental agencies to insurance companies. In doing so, it has been necessary to make actuarial studies involving flood frequency estimates for tens of thousands of locations.

By congressional edict, the areas to be relieved of flood damage are those below the 100-year floodplain. As the program proceeds through the years, these damages should be virtually eliminated.

In the areas above the 100-year flood level, costs of flooding will greatly increase, since facilities will now be concentrated at and above that level, since few flood precautions are now taken because the area is considered safe, since insurance premiums exceed physical damages by very substantial overhead costs, and since rehabilitation costs are much higher when paid by insurance. It is hoped that this increased cost can be more than offset by decreased losses in the 100-year flood plain, but this is highly questionable.

Flood frequency evaluations are used to set the 100-year flood levels and to establish insurance rates at levels above the 100-year. In so doing, the principle of expected probability is not applied, and consequently the 100-year levels established will be exceeded more frequently than once in 100 years on the average. For the same reason, insurance rates will be set too low, at least until considerable experience is gained. Most importantly, there is no overall mechanism or objective to minimize total average annual damages or maximize overall average annual benefits of the program. It appears that new concepts, new and more comprehensive engineering analyses, and new legislation are needed.

FLOOD DAMAGE MITIGATION

Floodplains are our most valuable and most productive land. In some regions, they are our only lands useful for occupation and for commercial, industrial and agricultural purposes. We need a comprehensive program of floodplain management that includes restrictive zoning as well as flood protection, but which is flexible enough to allow maximum freedom to use and develop one's own property consistent with wise and effective use of the floodplain by all concerned. The principles and application guides for flood frequency determinations, as discussed here, are fundamental essentials of such a program.

FLOOD ANALYSES PERTAINING TO DESIGN OF NUCLEAR POWER PLANTS

Yung-Chi Chang
Supervisor of Hydrological Analysis Group
Stone & Webster Engineering Corporation
Boston, MA 02107

ABSTRACT. This paper discusses the regulatory requirement of flood analysis as part of the licensing procedure of nuclear power plants. Flood analyses for three nuclear power stations which are representative of the coastal, estuarine, and inland sites are presented. The flood analyses include combinations of events, methodology, and results. Finally, the paper discusses the need for further research to balance the safety and economic goals of nuclear power plants.

INTRODUCTION. A nuclear power station utilizes nuclear reaction to generate electricity. The public has to be protected from potential releases of radioactive material from the power station. A nuclear power station shall be protected from natural hazards including flooding. The safety-related structures, systems, and components shall be designed to withstand the probable maximum flood (PMF) without loss of capacities to perform their safety functions. This paper will provide a brief review of the regulatory requirements of flood analyses, present flood analyses of three nuclear power stations, and discuss the need for further research.

REGULATORY REQUIREMENTS OF FLOOD ANALYSIS.

The Atomic Energy Act of 1954, as amended, is the law regulating the industry. The AEC had regulatory responsibility until 1974, when the regulatory arm of the AEC was reorganized and became the new Nuclear Regulatory Commission (NRC).

Following Section 50.34 of 10CFR50, the licensing procedure of a nuclear power station takes the form of submittal to the NRC and review by that agency of a Safety Analysis Report (SAR). During its review, the NRC may address interrogatories to the applicant, requiring additional information. The NRC then issues a Safety Evaluation Report (SER) summarizing its review. The SER may concur with the applicant's analysis or identify the outstanding issues. A

V. P. Singh (ed.), Application of Frequency and Risk in Water Resources, 207–217.

Preliminary SAR (PSAR) and a Final SAR (FSAR) are required and safety issues must be resolved before the construction permit and operating license are issued. Cooperation of applicants and the NRC staff is important to shorten the licensing process and to ensure plant safety.

Section 2.4, Hydrologic Engineering, of an SAR contains the flood analysis required for a nuclear power station. In general, the regional flood that affects the site grade and major flood protection requirements is addressed in a PSAR. The flood analysis in the PSAR is upgraded in the FSAR, if necessary. A study of local flooding at the site, reflecting the final site grade, is included in the FSAR.

To assist the industry in the preparation of SARs, guides (AEC, 1972; NRC, 1975a; NRC, 1975b) were issued. The standard Format and Content identifies the analysis required. The Standard Review Plan shows the NRC's review procedure and acceptance criteria. Specifically pertaining to flood analysis, Regulatory Guide 1.59, Design Basis Floods for Nuclear Power Plants (AEC, 1973; NRC, 1976) was issued.

In addition to the NRC guides, Working Group ANS 2.8 of the American Nuclear Society (ANS) was formed in 1973 to develop guidelines to establish design basis flooding at Power Reactor Sites. The Group issued the ANS-2.8 ANSI N170-1976 American National Standards for Determining Design Basis Flooding at Power Reactor Sites (ANS, 1976). This standard covers material that meets the requirements of Section 2.4, Hydrologic Engineering, of the Standard Format and Content of Safety Analysis Reports for Nuclear Power Plants, Regulatory Guide 1.70, Rev. 2, 1977.

As can be seen from the dates of the NRC Regulatory Guides, the Guides were constantly revised and updated in the early seventies. As it finally settled down, Regulatory Guide 1.59 and ANSI N170-1976 provide the most important guidelines in performing flood analysis for nuclear power plants.

In general, flood analysis shall be performed for the following events, depending on the water body where the plant site is located.

- Probable Maximum Flood From Precipitation
- Nonhydrologic Dam Failures
- Probable Maximum Surge and Seiche Flooding
- Ice Flooding
- Tsunami Flooding
- A Reasonable Combination of Lesser Floods from Above Events

Methodologies for performing flood analyses of flood-causing events except tsunami are documented in great detail in Regulatory Guide 1.59 and ANSI N170-1976. For detailed information, the reader should refer to the above-mentioned documents.

FLOOD ANALYSES OF SAMPLE NUCLEAR POWER PLANT SITES

Three sample nuclear power plant sites were selected from the safety analyses reports available to the pulbic to illustrate flood analyses for coastal estuarine, and inland sites, respectively. The flood analyses for the Jamesport (LILCO, 1975) and Greene County (New York

State Power Authority, 1968) Stations were performed in the mid-seventies at the PSAR stage. Analyses for the River Bend Station (Gulf State Utilities, 1985) were performed as part of an amendment to the FSAR in 1985.

In preparing an SAR, the applicant would use the applicable Regulatory Guides at the time of preparation. However, the NRC could ask the applicant to use more recent technology. For example, the NRC requested the use of HMR Nos. 51 and 52 (NOAA, 1978; NOAA, 1982) for the River Bend PMF analyses in the FSAR interrogatories, even though HMR No. 33 (U.S. Weather Bureau, 1956), (which was superseded by HMR Nos. 51 and 52) was used in the PSAR stage.

Jamesport Nuclear Power Station

Probable maximum hurricane (PMH) analysis. The Jamesport Nuclear Power Station is located on the north slope of Long Island, approximately 26 miles east (41.9 km) of Port Jefferson (LILCO, 1975). Because the site is located on Long Island Sound, it was analyzed for the Probable Maximum Hurricane (PMH). The hurricane characteristics in HUR 7-97 (U.S. Weather Bureau, 1968) were used. A two-dimensional finite difference model (SWEC, 1975) was employed to predict the maximum storm surge at the site. The hourly surge level of the whole of Long Island Sound was simulated by moving the PMH along a storm track, beginning at the Continental Shelf. Several storm tracks and PMHs of different sizes and forward speeds were used to simulate the surge level. The grid size used was 16,667 ft (5,083 m).

In the application of this model, a sensitivity analysis of the grid size and verification of the model were performed. A finer grid of 10,000 ft (3,050 m) in Long Island Sound was used to simulate both astronomical tide and wind surge conditions. For verification of the model, the simulated surge elevations for several gage stations for the 1938 and 1944 hurricanes were compared with the recorded data. The sensitivity and verification indicated the model has the required accuracy and could reproduce the recorded hurricane surges in Long Island Sound.

Different combinations of hurricane storm parameters in HUR 7-97 were used with the storm tracks shown in Figure 1. The calculations indicate that the large-radius, slow-speed storm will produce the highest still water level. This storm has a radius to maximum wind of 48 nautical miles (87.8 km), a forward speed of 15 knots (27.4 km/hr), and a maximum gradient wind of 131 mph (210.9 km/hr).

In addition to the still water level, a sea level anomaly and a ten-percent exceedance of spring high tide were also added to the still water level. The maximum water level is predicted to reach El. 25.6 ft (7.8 m) (MLW).

The site grade for all structures except the circulating and service water pump house is El. 61 ft (18.6 m) (MLW). All safety-related equipment, except the service water pumps, is protected from the PMH flooding by the high plant grade.

Of the safety-related equipment, only the service water pump (located between two jetties at sea level) could be potentially affected by the PMH. An additional analysis was performed for the PMH waves. Potential waves along different fetches across the Sound were generated for various PMHs. In the wave analysis, the storms were conservatively assumed to make a 90-degree turn after crossing the sound, in order to generate on-shore winds at Jamesport. Wave refraction was also considered in the analysis. Maximum waves entering the intake canal are controlled either by the maximum generated wave or the maximum wave allowed by bottom topography. An envelope of the maximum wave height was constructed with the same time base as the PMH surge. Because of the configuration of the water body and the storm track direction, the maximum waves arrive within a certain period of time after the maximum surge.

The intake structure is further analyzed for the wave runup. For the runup analysis, it was found that the combination of the surge and wave for the medium speed storm, at approximately 1.0 hours past the peak, was most critical. Hence, the walls of the pump house were designed to be watertight to El. 48.5 ft (14.8 m) (MLW) to protect the pump motors and associated electrical equipment from surge and wave action.

Greene County Nuclear Power Plant

The Greene County Nuclear Power Plant is located on the Hudson River estuary at River Mile 108 (173.9 km). The Hudson River estuary extends 154 miles (247.9 km) from The Battery in New York City to the dam at Troy, New York, so the site is subject to potential flooding by both coastal and riverine sources (PASNY, 1977).

Combined flood events. In determining the maximum possible water level at the site, the following combined events were considered:

1. Combination of PMF at Green Island, 25-year hurricane surges at The Battery, and wind-wave activity at the site.

2. Combination of standard project flood coincident with seismic failure of dams, routed to Green Island, 25-year hurricane surge at The Battery, and the corresponding wind-wave activity.

3. Combinations of a 25-year flood at Green Island, PMH at The Battery, and the wind-wave activity at the site.

4. Combination of one-half PMF at Green Island, standard project hurricane surges at The Battery, and the corresponding wind-wave activity at the site.

To develop the flood level at the site for each combination, the following steps were performed:

- The Hudson River basin above the site was divided into subbasins or hydrologic units on the basis of tributary area and location of hydraulic controls. Unit hydrographs were then developed for each subbasin.

- The Probable Maximum Precipitation (PMP) (U.S. Weather Bureau, 1956) or Standard Project Precipitation (SPP) (COE, 1965) was applied to the unit hydrograph with the appropriate infiltration losses to develop the estimated flood hydrograph for each subbasin.
- The hydrographs for dam breaching by overtopping were computed, wherever applicable.
- Starting at the uppermost reaches in each subbasin, the subbasin flood hydrographs, including dam breaching, were combined in their proper time sequence and routed downstream to Green Island.
- The combined flow at Green Island was then routed with a one-dimensional tidal model to the site (Harleman and Lee, 1969), in conjunction with a corresponding hurricane surge at The Battery and local inflow between the site and Green Island.
- The wave height and wave runup were calculated using the wind speed of the corresponding hurricane. The wave runup was superimposed on the maximum water level at the site.

Inland flood. In determining the runoff at the river basins, HEC-1 Flood Hydrograph Package (COE, 1973) was used. For selected gaged subbasins of the Upper Hudson and Mohawk River basins, reconstitutions of the flood hydrograph for the storms of December 1948 and October 1955 were performed using recording and nonrecording precipitation data and recorded flow rates. The loss rate and the unit hydrograph characteristics were determined as part of the optimization processes. For ungauged subbasins, the lag time (t) for the Snyder unit hydrograph was computed from the shape and the slope of the basin. Values for the coefficient, c, were selected from the values where loss functions were optimized using HEC-1 in similar subbasins. For flow- through reservoirs and river channels, the modified Puls and Muskingum methods were used, respectively.

For generating the runoff, the PMP and SPP were distributed using the spatial distribution of the maximum historical storm of 1948. An antecedent storm, half of the main storm, was also included. For each basin, the hydrograph at the uppermost subbasin was first generated, then routed through the channel downstream and combined with local runoff or dam-breaching hydrographs. The generating and combining processes were repeated through each successive subbasin, as required, toward Green Island. The 25-year flood was calculated by frequency analysis of the flood flow data at Green Island.

There are 24 dams in the Upper Hudson River Basin and 15 dams on the Mohawk River. Except for the Conklingville Dam, the storages of these dams were considered insignificant comparatively and were disregarded.

To account for the failure of Conklingville Dam by earthquake, a dam break hydrograph was computed using the formula for embankment dams derived by the U.S. Army Corps of Engineers (COE, 1960). To produce the greatest possible discharge value, at the site, the Conklingville Dam was assumed to fail, such that when the peak flow of the dam failure hydrograph arrives at Green Island, it coincides with the peak flow of the SPF from all these subbasins. The Sacandaga Reservoir was assumed to start full and to fail completely. The SPF into the reservoir was routed through the dam using the spill way and reservoir characteristics. The dam failure hydrograph was then calculated, using the water surface elevation and the storage behind the dam at the time of failure.

The dam breach hydrograph was assumed to be represented by a triangular shape. The peak discharge was calculated, assuming that the whole dam was instantaneously removed. The duration of the flow was computed as the time to empty the entire stored volume of water above the breach.

Coastal Flood. Calculation of the tidal surge height was accomplished using a computerized bathystrophic storm surge model (Bodine, 1971). The surge is computed along the path of the locus of maximum wind as the storm moves on-shore. In determining the maximum surge at The Battery, the locus of maximum winds is brought on-shore along a track producing the maximum average height at the mouth of New York Bay, and consequently at The Battery. The hurricane characteristics in HUR 7-97 were used to evaluate the storm surge at The Battery. Because HUR 7-97 gives three different values for both radius-to-maximum wind and speed of translation, it was necessary to compute nine different surge levels, using all of the possible combinations of meteorological parameters. These calculations indicate that a 45-nautical-mile radius (82.3 km), 51-knots (93.3 km/hr) speed or translation storm yields the highest surge level at The Battery. Calculations were made for the PMH, SPH, and a 25-year hurricane.

Water level determination. The water level at the site was determined by a one-dimensional tidal model (Harleman and Lee, 1969), modified to suit the particular situation for this study. The model predicts the time and longitudinal variation of the water level and flow rate. Input to the model includes channel characteristics, the hydrograph at the inland end, the water elevation at the ocean end, and the lateral inflow between the site and Green Island.

This tidal model was first calibrated for normal flow conditions with tidal table prediction, and the result verified by observed data. For flood conditions, the same procedures were repeated by utilizing historical floods in 1936, 1948, 1950, and 1955.

Using the calibrated and verified mathematical tidal model, the maximum possible water level at the site was determined by inputting the combined flow of both Upper Hudson and Mohawk Rivers at Green Island and water elevation at The Battery due to hurricanes. The wind shear is also considered in the model.

Wave runup. The coincident wind-wave activities were determined in accordance with the Shore Protection Manual (U.S. Army Coastal Engineering Research Center, 1973). The hurricanes were assumed to move inland from The Battery northward over the Hudson River. The wind speed was also reduced by land friction. The effective fetch along the critical fetch was calculated. The wave run up was calculated on the 1:4 permeable rubble mound structure, using the one-percent wave. This wave run up was then superposed on the maximum water elevation.

Probable maximum flood elevation. The flood levels for the four combined events were evaluated. The results, summarized below, show that the combination of PMF at Green Island, 25-year hurricane surge at The Battery, and wind-wave activity at the site is the most critical event. The site grade was set at 30 ft (9 m) MSL which is above the probable maximum flood elevation.

Combined Events	Max. Flow at Green Island (10^6 cfs(cms))	Max. Surge Level at Battery (ft (m)MSL)	Max. Level at Site (ft (m)MSL)	Wave Runup (ft(m))	Total Flood Elev. (ft (m)MSL)
1	0.926(0.086)	16.1(4.9)	25.0(7.6)	3.5(1.1)	28.5(8.7)
2	1.190(0.110)	16.1(4.9)	20.6(6.3)	1.9(0.6)	22.5(6.9)
3	0.162(0.015)	24.7(7.5)	17.2(5.2)	4.6(1.4)	21.8(6.6)
4	0.463(0.043)	18.4(5.6)	20.3(6.2)	4.0(1.2)	24.3(7.4)

River Bend Nuclear Power Station

The River Bend Station Site is located adjacent to the Mississippi at about River Mile 262 (421.8 km). The original ground grade at the site is about El. 110 ft (33.6 m) MSL. The finished ground grade is at El. 95 ft (29 m) MSL.

The site is drained by Grant's Bayou on the east and Alligator Bayou on the west. Numerous unnamed intermittent streams cross the site and drain to either Grant's or Alligator Bayou. Grant's Bayou enters Alligator Bayou to the south of the site. The latter then flows south into Thompson Creek, which enters the Mississippi River approximately 7 miles (11.3 km) downstream of the River Bend Station embayment.

Mississippi river flood. The U.S. Army Corps of Engineers has made extensive studies of Mississippi River Flood hydrology and has determined a project Design Flood (PDF). The PDF is based upon floods predicted by the U.S. Weather Bureau as the "maximum possible" and by the Mississippi River Commission as "maximum probable."

The total unregulated PDF flow at the latitude of Red River Landing is 3,330,000 cfs (309,773 cms). Upstream reservoir storage would provide a reduction in peak flow to about 3,030,000 cfs (281,865 cms). About 1,530,000 cfs (142,328 cms) would be diverted to the Morganza and West Atchafalaya Floodways and the Atchfalaya River upstream of the site, and about 1,500,000 cfs (139,537 cms) is estimated to pass the site. The estimated flood level at the site for this flow is 54.5 ft (16.6 m) MSL. The levee elevation along the west bank opposite the site is about 57.5 ft (17.5 m) MSL, 3 feet (0.9 m) above the PDF crest level. Usually, a Standard Project Flood (SPF) is considered to be 40 to 60 percent of the PMF. In this case, the PDF is considered more severe than the SPF. A PMF for the river at the site was conservatively determined by considering that the PDF is 60 percent of the PMF. The estimated unregulated flow PMF at the latitude of Red River Landing is 5,500,000 cfs (511.637 cms) . With this flow, the reduction in PMF peak flow due to upstream reservoir storage is assumed to increase to 400,000 cfs (37.210 cms). Similarly, the diversion of flow through the floodway will also increase. However, at the site, the flow would overtop the levees and enter the vast low-lying storage area between the main stem of the river near the site and the western extent of the flood way system. The ground level in the overbank area is 20-35 ft (6.1-10.7 m) MSL. Considering the volume of this storage area together with the fact that it would not become flooded until just prior to peak flood flow, it is extremely unlikely that the peak water level near the site would ever exceed 60 ft 18.3 MSL, about 25 feet (7.6 m) below plant grade. No plant safety-related equipment would be jeopardized in the event of a Mississippi River PMF.

Local stream floods. The potential flooding due to the local streams was analyzed for PMF, ½ PMF plus Operational Bases Earthquake (OBE), and 25-year flood plus Safe Shutdown Earthquake (SSE). Because the OBE was determined not to fail the side slopes, the ½ PMF plus OBE was less severe than the PMF.

PMP values for the Grant's Bayou basin and subbasins were based on the data contained in HMR Nos. 51 and 52. One-half PMP was assumed to occur one day prior to PMF. However, the antecedent storm would have impact only on loss rate. The flow would have left the site before PMF arrived.

Surface runoffs were calculated by the Soil Conservation Service (SCE) method. The unit hydrographs were determined by applying drainage area runoff characteristics to three independent methods of unit graph preparation, obtaining an average unit hydrograph for each basin. Basin lag times were also estimated from local basis data and several independent empirical methods.

The PMF values were then obtained by applying the PMP runoff to the unit hydrographs. The 25-year peak flows from the local basins were determined by methods developed from regression relationship based on a large amount of Louisiana Data. The peak flows are:

	Grant's Bayou Basin (cfs (cms))	Grant's Bayou Above West Creek Confluence (cfs (cms)	West Creek (cfs (cms))
PMF	42,690 (3.971)	35,346 (3.288)	6,699 (523)
25-yr	6,760 (628)	4,364 (406)	842 (78)

Water elevations were determined by the HEC-2 Water Surface Profiles Computer Program (COE, 1979). The Grant's Bayou PMF was conservatively assumed to coincide with the PDF on the Mississippi River. The peak river stage is El. 54.5 ft (16.6 m) MSL. This was used as the starting elevation for Grant's Bayou backwater profile calculations.

Flood flows for West Creek and Grant's Bayou below West Creek were reduced in the upstream direction at each cross-section to account for the reduction in contributing drainage area. The flow for Grant's Bayou above West Creek was conservatively assumed to exist undiminished upstream.

The local streams are spanned by railroad and road bridges with piers located adjacent to and in the stream bed. These streams are subject to debris accumulation. For these reasons, each bridge crossing downstream of the station was assumed to be 50-percent clogged at the occurrence of the PMF and ½ PMF and OBE, and 100-percent clogged for the 25-year flood and SSE. For the SSE, the site slope at the site is also assumed to fail with a 20H:IV maximum inclination.

The maximum water elevation on Grant's Bayou, adjacent to the cooling water yard, varies from 95.3 (29.1 m) to 101.8 ft (31.1 m) MSL. The yard is at about 104 ft (31.7 m) MSL. The maximum water level on West Creek near the plant is about 94.3 ft (28.8 m) MSL, which is below the top of the adjacent railroad spur at 95.0 ft (29.0 m) MSL, and plant area flooding would not occur.

The PMF at the plant site was also analyzed using the rational method and back water analysis. It was concluded that the flood levels at entrances to safety-related buildings are all below the door sills. The plant is adequately protected from potential flooding.

NEED FOR FURTHER RESEARCH

From the above three examples, it can be seen that many combinations of site-related floods were very conservatively analyzed and the worst case was selected for the design of nuclear power stations. The NRC guidelines at the time of SAR preparation were closely followed. Hence, the safety of the plants against potential flooding was assured.

The design basis events for nuclear power stations have been defined clearly in the NRC guidelines. In general, there is a basic agreement on the events to be considered. However, when a specific flood analysis is carried out, different assumptions, such as the antecedent storm and time sequence of combined events, could make a significant difference in the final flood level. It has been pointed out that the probability of a PMP plus an antecedent storm, expressed as a percentage of the PMP, for five watersheds in the Tennessee Valley region was estimated to be 15^{-12} to 10^{-21} (Newton, 1983). This probability is way below the probability of 10^{-6} set by the update to ANSI N170. In addition, there has never been a systematic analysis to bring all design flood events or a combination thereof to a same probability level.

The probabilities of the combined events are likely to be significantly below the 10^{-6} level. It has been stated policy that NRC will review and accept a flood analysis if the analysis meets the NRC's acceptance criteria. However, a deviation from the NRC guidelines (which might accompany a less conservative approach) will assure the NRC's detailed and lengthy review. In reality, an applicant is likely to choose a conservative approach to expedite the licensing of a plant. A conservative guideline will thus increase the cost of a plant.

The NRC incorporated ANSI N170 into Regulatory Guide 1.59. It has also requested the use of HMR Nos. 51 and 52 in the most recent PMF analyses. New meteorological criteria for SPH and PMH (NOAA, 1979) have been developed in which the direction of the storm and wind speed are correlated. This is a significant improvement over HUR 7-97. Undoubtedly, the NOAA report will be incorporated into future updates of the NRC Guides. Studies of the probability of sequential storm events and combined events will continue with the goal of incorporating the resulting data into the NRC Regulatory Guides, in order to establish resonable design basis flood events, to achieve safety at reasonable cost.

CONCLUSIONS

1. The Nuclear Regulatory Agency (NRC) and the nuclear industry have worked together on the safety issues and stringent requirements on flood protection have been set for nuclear power plants.
2. The NRC Regulatory Guides and flood analysis technology are evolving constantly. Updating the flood analysis for a nuclear power station may be required by the NRC.
3. Cooperation of applicants and the NRC staff is important to shorten the licensing process and to ensure plant safety.
4. The NRC Regulatory Guides were closely followed by the flood analyses of the three sample nuclear power station sites. The safety of the plants against potential flooding was assured.
5. Further research is needed to establish reasonable probabilities for the occurrence of flood-causing events, either singly or acting together, at nuclear plant sites. Incorporating reasonable design basis flood events into the NRC Regulatory Guide will assure safety of nuclear plants at reasonable costs.

REFERENCES

AEC (Atomic Energy Commission) 1972. Standard Format and Content of Safety Analysis Report of Nuclear Power Plants (Revised October 1972).

AEC (Atomic Energy Commission) 1973. Design Basis Floods for Nuclear Power Plant, Reg. Guide 1.59.

ANS (American Nuclear Society) 1976. American National Standard for Determining Design Basis Flooding of Power Reactor Site, ANS 2.8, ANSI N170.

Bodine, B.R. 1971. Storm Surge on the Open Coast: Fundamentals and Simplified Prediction, U.S. Army Corps of Engineers, Coastal Engineering Research Center, TM No. 35.

COE (U.S. Army Corps of Engineers) Waterways Experiment Station 1960. Floods Resulting from Sudden Breached Dams, Conditions of Minimum Resistance, Miscellaneous Paper No. 2 -374.

COE (U.S. Army Corps of Engineers) 1965. Standard Project Flood Determinations, EM 1110-2-1411.

COE (U.S. Army Corps of Engineers) 1973. Flood Hydrograph Package, HEC-1.

COE (U.S. Army Corps of Engineers) 1979. Water Surface Profiles, HEC-2.

Gulf States Utilities 1985. River Bend Station Final Safety Analysis Report, Amendment 17.

Harleman, D. R. F. and Lee, C. H. 1969. The Computation of Tides and Currents in Estuaries and Canals, Technical Bulletin No. 16, Hydrodynamic Laboratory, MIT.

LILCO (Long Island Lighting Company) 1975. Jamesport Nuclear Power Station, Units 1 and 2, Preliminary Safety Analysis Report, Amendment 5.

Newton, D. W. 1983. Realistic Assessment of Maximum Flood Potentials, ASCE Proceedings, Journal of Hydrology, Vol. 109, No. 6.

NOAA (National Oceanic and Atmospheric Administration) 1978. Probable Maximum Precipitation Estimates, United States East of the 105th Meridian, HMR No. 51.

NOAA (National Oceanic and Atmospheric Administration) 1979. Meteorological Criteria for Standard Project Hurricane and Probable Maximum Hurricane Windfields, Gulf and East Coasts of the United States, NOAA Technical Report NWS 23.

NOAA (National Oceanic and Atmospheric Administration) 1982. Application of Probable Maximum Precipitation Estimates, United States East of the 105th Meridian, HMR No. 52.

NRC (Nuclear Regulatory Commission) 1975a. Standard Format and Content of Safety Analysis Report of Nuclear Power Plants. Revision 2, September 1975; Revision 3, 1978.

NRC (Nuclear Regulatory Commission) 1975b. Standard Review Plan for the Review of the Safety Analysis Reports for Nuclear Power Plants - LW Edition, NUREG-75/087.

NRC (Nuclear Regulatory Commission) 1976. Design Basis Floods for Nuclear Power Plants, Regulatory Guide 1.59. Revision 1, April 1976; Revision 2, August 1977.

PASNY (Power Authority of the State of New York) 1977. Greene County Nuclear Power Plant, Preliminary Safety Analysis Report, Supplement 16.

SWEC (Stone & Webster Engineering Corporation) 1975. Two-Dimensional Coastal Storm Surge Model, SWECO 7501-P.

U.S. Weather Bureau 1956. Seasonal Variation of the Probable Maximum Precipitation East of the 105th Meridian, HMR No. 33.

U.S. Weather Bureau 1968. Meteorological Characteristics of the Probable Maximum Hurricane, Atlantic and Gulf Coast of the United States, Interim Report, HUR 7-97.

SOME APPLICATIONS OF FLOOD FREQUENCY AND RISK INFORMATION IN FOREST MANAGEMENT

William F. Hansen
Francis Marion & Sumter National Forests
P. O. Box 2227
Columbia, SC 29202

ABSTRACT. Floods have been responsible for the damage or destruction of numerous forest road, channel, and water diversion structures. Some of the losses are to be expected because it is too costly to design for all circumstances. Other losses occur because of little or no flood design, debris accumulations, slope instability, and poor awareness and understanding of the problem. Presented are some applications of flood frequency and risk information in forest land management. Some of the problems of application are discussed. A simple approach to refine water resource records to meet local needs is included as a tool to check other flood analyses or provide design assistance.

1. INTRODUCTION

The damage caused by flooding includes untold losses to property and lives. As in Noah's time, many of these losses could be avoidable, but the decision to prevent loss falls on skeptical or unbelieving ears. Sometimes those affected by floods even rebuild in the floodplain with the assumption that such a flood will not happen again. Those involved with the study and application of information relative to floods have long been concerned with the disregard to flood impacts. Symposiums such as is, are important to inform others that something can be done about changing the damage that floods do.

The President of the United States (Carter, 1977) showed his concern for the needless loss to property and lives by signing an Executive Order (EO 11988) dealing with floodplain management. The direction found in EO 11988 includes the prevention of losses through avoidance of floodplains, design to minimize flood damage and notification of potential victims. Since this time, more emphasis has been placed on flood frequency and risk analysis, as well as floodplain delineation.

Flood frequency and risk analysis is needed in land management. However, the funding to collect streamflow information in forested areas may be in jeopardy due to budget cuts and/or greater needs elsewhere. One way to compensate is to be more efficient at applying the present

V. P. Singh (ed.), Application of Frequency and Risk in Water Resources, 219–226.

data to fit local forest conditions. Some managers may not be using any flood frequency or risk information in the design of road or other structures. Some may not care that the structure is flood damaged or gone in a short time. Possible reasons for this lack of concern include poor awareness about the need, savings in design and material costs, and insurance coverage or tax write-off in case of loss. Whatever the reasons, the lack of flood frequency and risk analyses costs our nation heavily. This would surely be true of most other nations as was evidenced in numerous examples presented at the symposium.

This paper will center on some of the applications that can be made of flood frequency and risk information in forest land management. Presented will be examples of simple approaches in the use and refinement of water resource data to meet specific local needs. These methods are not to be used instead of detailed analysis on critical projects. They can be used where analysis is not done or to check more rigorous approaches. Other land managers may also be able to use this information. In most cases, flood frequency and risk will be separated as two steps in the process of making an informed decision.

Flood frequency analysis combines information related to how often and to what magnitude of flooding is likely to occur. This information is largely based on historical records of hydrometeorological events and/or certain watershed or channel characteristics. The flood frequency information is useful in designing structure or project needs and estimating costs.

Risk analysis is a calculated decision to weigh a project's chances of success or failure through time. The analysis is a probabilistic evaluation. The risk analysis will help determine the design flood frequency for a project. Experience is not a good substitute for risk analysis. Severe floods are infrequent and easily forgotten.

2. EXISTING INFORMATION

For the most part, forest managers must use existing data to formulate and apply flood frequency and risk information. The U. S. Geological Survey, State Water Resource Agencies, Universities and others collect a wealth of streamflow records and analyses which can be applied to decision making and design of structures within streams or floodplains. Also, the Department of Housing and Urban Development publishes the Flood Hazard Boundary Maps for many areas in the United States. The Soil Conservation Service prepares county soil maps which include delineation of flood-prone soils. Related hydrometeorological information is published by the National Weather Service and/or State Climatology Offices. The key is to be aware and familiar with the types of data that are available.

Other information which may have some application include empirical formulas or models which have been developed to define or predict flood frequency relationships or processes. Visual observations or measurements of stream channel or watershed conditions may also be important to the application of flood frequency and risk information.

3. INFORMATION SELECTION

The care and detail with which flood information will be selected and analyzed will depend on the cost and the potential risk of the project to life, property or the environment. Essentially all in-channel structures should receive some analysis for flood risk. In most cases, there exists one or more formulas which can easily be applied to get an estimate of peak flows. Formulas are developed from local or areawide flood records or are rational type formulas which require some estimates, assumptions or measurements to apply them.

However are the published formulas for flood prediction related to or representative of local forest conditions? Forested areas in mountainous terrain may need adjusted equations to better fit local conditions. Local rainfall and streamflow conditions may vary dramatically from other areas where the flood data were collected. Flood frequency data developed from selected stations within the vicinity of the land management need to be analyzed. Comparisons could then be made to other flood predictive tools. Some considerations in selecting flood stations include data proximity to the management area, length of data record (best if 10 years or more), and watershed conditions as geology, topography, soils, land stability, and past management practices (especially system and logging road drainage and logging debris in streams).

4. INFORMATION APPLICATION

The land manager may rely on a specialized person (hydrologist, engineer, etc.) to determine any necessary flood frequency criteria. Where a specialist is not available, assistance from state and federal agencies as well as the private sector can be sought. General steps and some examples in applying the information follow.

4.1 Determine Flood Frequency

Flood frequencies for different return periods are estimated for the selected stations. This is typically done using a log-Pearson Type III analysis (uses a log transformation and probability assessment of annual peak flow data). Computer programs have been developed to do the calculations, but it can be calculated with a scientific calculator and the proper Pearson Type III distribution table.

Graphical methods can also be used to determine flow frequency using log- probability paper and ordering the annual peak flows or all flood flows (partial duration series) in a given period. Procedures may be found in texts dealing with analytic hydrology.

4.2 Analysis of Flood Frequency Data

Figures can be made for flood data from the local selected stream gaging stations by plotting watershed area versus peak discharge for each return period using several cycle log-log paper. The result is a scatter of points with a linear trend (assuming data from numerous watersheds of various sizes was used). Furthur stratification of the data points by rainfall regimen, geology, etc. may reduce some of the variability. When the data points are numbered or otherwise identified, stations nearest an individual project area can be visually separated and help identify unusual conditions for further investigation.

The plot data are further analyzed by doing a linear regression on the log transformed area and peak discharge data pairs. The resultant line is in the following form and can be plotted on log-log paper:

$$\log Q = a + b \log A \tag{1}$$

where Q is flood discharge in cubic feet or metres per second
A is watershed area in square miles or kilometres

The equation or the plotted line can be used to predict peak flows from ungaged watersheds in the management area. It should be noted that the correlation coefficiant value may be high, but the accuracy of predicted values is limited because log transformed data are used. This procedure provides a visual and mathematic expression of actual streamflow data near the management area. The spread of data points can be visualized and aid in project analysis as well as give a validity check of more rigorous methods.

4.3 Flood Frequency Analysis with Insufficient or Limited Records

When long-term streamflow data are not available, individual flood data for numerous stations can be helpful in estimating project design flow. These data are sometimes collected after a flood by the U. S. Geologic Survey or others. Plotting and regression would be similar to the long-term data, but the return period would not be known and may vary somewhat from watershed to watershed.

As an example, this was done with 32 stations in or adjacent to the Siskiyou National Forest in southwest Oregon using the 1964 flood data (Waananen et al, 1971). Comparison of the individual flood data from the 1964 flood was made for local stations with the long term flood frequency data (Harris et al, 1979).

A problem became evident in using one unusual flood event to determine design criteria. Almost always the flood frequency on small watersheds was lower than expected. Large regional storms that produce widespread flood damage are generally not noted for the local intense conditions that may occur on small watersheds. Extreme care must be exercised if limited data as this is to be applied as design criteria, especially on small watersheds.

If streamflow records from a short-term station are to be used, the period of record should be compared with records from nearby long-term stations to assure abnormal flooding did not occur within the period. If unusual flooding occurred, an adjustment could be made in the flood frequency analysis. If close correlation exists over a wide range of flood peaks, some data generation may be considered. Interpreting the data from the short-term station may be useful to the flood frequency analysis for the local conditions.

Another way to use short-term station data is to compare individual storm or flood flows with records from nearby long-term stations. When comparing the flows, it is often helpful to use a "normalized" flow by dividing the floodpeaks (cubic feet or metres per second) by the area (square miles or kilometres) to get cubic feet per second per square mile (CSM) or cubic metres per second per square kilometre. This takes out the watershed size differences when comparing values and makes the relationship easier to visualize. If the comparison is consistent, a correction factor can be applied to the long-term station data for application on the area with short-term records.

4.4 Selection of Acceptable Risk and the Design Return Period

The risk assessment is calculated by using the equations found in most hydrology texts:

$$\text{probability of failure (F)} = 1 - (1 - 1/T)^n \qquad (2)$$

$$\text{probability of success (S)} = (1 - 1/T)^n \qquad (3)$$

Where T is the design flood return period in years and n is the design or expected project life. Probability values are between 0.00 and 1.00.

A precalculated risk table can be used in the analysis to determine the individual project design criteria. This is especially helpful in estimating what is realistic. For instance, if you are only willing to take a 1% risk of failure for a project with a life of 20 years, you would design for the 1990 year flood. However, if acceptable failure rate was 20% (success 80%), the 90 year flood would be used. An example calculation follows:

$$T - \frac{1}{1 - \text{antilog}(\log S/n)} \qquad (4)$$

$$T = \frac{1}{1 - \text{antilog}(\log .80/20)} = 90 \text{ years}$$

Risk information can help in real problems in land management. The analysis process is a logical step-by-step approach to fill in unknown or uncertain values which are reasonable and defendable. Assumptions can be easily changed to examine other alternatives.

4.5 Examples of Flood Frequency and Risk Application

An analysis was done to determine whether a culvert or bridge was needed at a stream crossing. Logging debris and channel instability were noted above the stream crossing. Anadromous fish passage was also a concern at the site. The present culvert had been installed after the

1964 flood in southwest Oregon had washed out a slightly smaller culvert. The present culvert invert was worn after only 20 years due to bedload movement in high flows. Should a culvert or bridge be installed? Additional data include the following:

	Culvert	Bridge
Initial Cost	$50,000	$100,000
Expected Life	25 years	50 years
Ability To Pass Flood	50 year	100 year
Average Maintenance Costs	high ($1000/year)	low ($100/year)
Expected Success Probability		
25 years	60%	78%
50 years	36%	61%
Total Present Costs	$77,412 *	$101,576
(discounted at 6%)	$115,760 **	

* Culvert added at year 25 at cost of $50,000.
** Culvert added at year 25, but cost increased at 6%/year to $214,593.

Before an analysis of the data was done, the culvert was being favored over a bridge. The final decision to install the bridge was made for several reasons:

1. Culvert installation costs were keeping up with inflation, so what appeared to be an initial savings was likely a long-term cost.
2. The upstream logging debris and channel instability were likely to effect the culvert installation.
3. The cost of structure replacement times the expected failure rate was similar for both structures.
4. The bridge was more consistent with fishery needs.

On another occassion, numerous stream crossings were analyzed for culvert size needs in southwest Oregon. Watershed sizes also varied over a range of hundredths to tens of square miles. Comparison of estimated culvert sizes was done using the U. S. Geologic Survey (USGS) 30 year flood methods and by using the selected local station data (Harris et al., 1979) reanalyzed as in equation 1. Under most conditions, both methods predicted the need for a culvert of similar size (usually the same or within one size class). The agreement between the two methods provided a desirable validity check on each other.

Flood frequency information was also used to determine culvert spacing needs for a road. Above the road were few defined channels. The road was located in the late 1950s on steep side slopes. During the 1964 flood in southwest Oregon, several sections of the road failed because the culverts could not handle the flow. The culvert spacing at the time of the analysis ranged from about 500 feet to 1500 feet. The area of the watershed above the road was determined. A few areas with defined channels had to be analyzed separately. The flood volume for the road section was estimated by the extrapolation of the area above the road versus flood frequency curves. Standard culverts which were used as road cross drains could handle 7 cfs. A design flow of 4 cfs per culvert was used to allow for possible sediment or debris problems and minimize environmental effects downslope. The total road segment flow was divided by the design flow for each culvert (4 cfs). The average spacing need was determined to be about 500 feet between culverts. Additional culverts were added during a road reconstruction project for sections having insufficient culverts.

Flood frequency information can be used in the design of other stream channel or water diversion structures. Manning's formula for open channel velocity

$$V = 1.486/n\ R^{2/3}\ s^{1/2} \tag{5}$$

(where V = the average velocity in cfs, n = the roughness coefficient, R = the hydraulic radius defined as the cross section area/wetted perimeter, and s = the channel slope)

is often used in conjunction with the design flow information to estimate unknown components in the design of the structure. For channel structures, the velocity is often the quantity of concern. For water diversion structures, the size, slope, and spacing are all necessary components which needs to be determined. <u>Design is an important part of stream and water structures</u>. Many alternatives or designs can be considered by adjusting the risk or the assumed values. The flood frequency information has also been useful in the assessment and design of activities as floodplain management, gully rehabilitation, fishery enhancement structures, and riprap sizing.

4.6 Factors that Influence Floods and Structure Design in Forests

Numerous papers have been written and are available on the effect of forest practices on storm peaks. Conclusions from the research are not altogether consistent but need to be weighed in the final project analysis. Several factors which can be listed as being <u>potentially important</u> to flood magnitudes or the damage which they can do include:

1. Over approximately ten percent of a watershed in roads.
2. Inadequate surface drainage on roads and logging trails.
3. Abnormal accumulations of woody debris in streams (especially from logging or trees blown down).
4. Stream channel and watershed slope instability.
5. Watersheds where a substantial percentage of the area has been harvested in a short time.
6. Locations with an abrupt drop in stream channel slope.

4.7 Practices to Reduce Flood Damage

Infrequent failure will occur even when structures are properly designed. Practices need to be designed to allow for possible failure. In some cases this can be done for little or no additional cost. For example, dip road sections down across streams. If the culvert fails, the road will fail at that point instead of washing down the road.

Activities should be designed to reduce impacts to the soil, provide for surface water drainage, and stream protection. Adequate surface water drainage on system and logging roads are of primary importance to prevent concentrating water and the effects of accelerated erosion. Directional falling of timber or stream cleanout are practices to prevent or mitigate excessive logging debris in streams.

5. CONCLUSIONS

Hopefully, this paper will encourage more interest in using flood frequency and risk analysis techniques in international forest management. Storm and flood damage can be avoidable in many instances when projects are adequately assessed and designed.

Flood frequency and risk analyses use a combination of related data and numerous assumptions to predict design criteria for a project. Other site or watershed conditions should also be considered in the project design. Expectations from the analysis should be realistic. A conservative approach is recommended in interpreting data and in adjusting for potential problems encountered in forest management. Structures should be designed with the understanding that failure could occur someday. Also, forest management practices including road building, timber harvest, and regeneration need to be consistent with soil and water conservation needs. Several examples were presented at the symposium where improved roading and timber harvest practices would have helped to prevent downstream flood damage.

6. REFERENCES

Carter, J. 1977. `Floodplain Management. Executive Order 11988.' The White House, Washington, D. C.

Harris, D. D., Hubbard, L. L., and Hubbard, L. E. 1979. `Magnitude and Frequency of Floods in Western Oregon. U. S. Geological Survey. Open-File Report 79-553.' Portland, Oregon.

Waananen, A. O., Harris, D. D., and Williams, R. C. 1971. `Floods of December 1964 and January 1965 in the Far Western States: Part 2 Streamflow and Sediment Data. U. S. Geological Survey Water Supply Paper 1866-B.'

ACKNOWLEDGEMENTS

Special thanks to David Muir, Vicki Glover and Terri Taylor in the typing and editing of this paper.

REVIEW OF ADVANCES IN RISK AND RELIABILITY ANALYSIS FOR HYDRAULIC STRUCTURES

Larry W. Mays
Professor of Civil Engineering
Department of Civil Engineering
The University of Texas at Austin
Austin, Texas 78712

ABSTRACT. Recently, considerable emphasis has been given to the problem of the state of decay in the nation's infrastructure because of its importance to the needs of society and industrial growth. Large capital expenditures will be needed to bring the concerned systems to higher levels of serviceability. These systems include many types of hydraulic structures found in urban areas such as storm sewer systems, flood levees, water distribution systems, culverts, bridges, flood bypass tunnels, detention systems, etc. Conventional design methods for hydraulic structures basically fail to recognize and to systematically analyze and account for the effect of the various uncertainties that prevail in the design. Because of the existence of uncertainties, the hydraulic structure has an associated risk or probability of failure and an associated reliability or probability of not failing. There is much confusion among engineers in using the concepts and terms of uncertainties, risk, and reliability as applied to the design of hydraulic structures. For hydraulic structures subjected to the conveyance of flood waters, many engineers only think of the uncertainty of hydrologic events as being attributable to risk. For hydraulic structures such as water distribution systems, there is no universally accepted definition for risk or reliability. A great deal of work related to risk and reliability has been reported in the electronics, mechanical engineering, chemical engineering, and operations research literature; however, only a limited amount of work has been reported in the water resources literature. Several works have been reported dealing with hydrologic uncertainties and only a few works have been reported on the risk and reliability evaluation of hydraulic structures. The purpose of this paper is to review the work that has been done in the risk and reliability evaluation of hydraulic structures and to identify research areas that need to be explored.

1. BASIC CONCEPTS OF RISK, RELIABILITY, AND UNCERTAINTY ANALYSIS

1.1. Uncertainties In Hydraulic Structure Design

Various kinds of uncertainties exist in common engineering projects during the planning and design, the construction, and the operation and

V. P. Singh (ed.), Application of Frequency and Risk in Water Resources, 227–262.

maintenance stages. The existence of uncertainties for flow conveyance structures such as storm sewers, culverts, and flood levees is mainly ascribed to the lack of knowledge of the natural hydrologic process and the uncertainties generally involved related to the hydrologic, hydraulic, structural, and economic aspects (Tung and Mays 1980a). Hydrologic uncertainties originate from the inherent randomness of natural hydrologic processes and the uncertainties in selecting the probability model and in evaluating the corresponding parameters used in the probability model. Hydraulic uncertainties are attributed to the simplification of mathematical models in describing natural physical phenomena, imprecise dimensions of hydraulic structures, nonuniformity of construction materials, and various operational conditions. Structural uncertainties result from potential weakness of hydraulic structures, such as the slope failure of a levee, the slide of a foundation, etc., and underestimation of the resistance capacity of hydraulic structures. Economic uncertainties can result from false estimation of construction costs, damage costs, and operation and maintenance costs, from improper financial budgeting, and from unpredictable economic crises. There have been many attempts at defining hydrologic risk reported in the literature (e.g., Davis et al. 1972; Vicens et al. 1975; Wood and Rodriquez-Iturbe 1975a,b; Bodo and Unny 1976; Wood 1977; Castano et al. 1978; Duckstein and Bogardi 1981). Only a few attempts have incorporated other uncertainties simultaneously (Tang et al. 1975; Szidarovszky et al. 1976; Duckstein and Borgardi 1981; Tung and Mays 1980a, 1981a).

In view of the uncertainties, each quantified element related to a system is considered to be a random variable having an associated probability distribution. The term "loading" is used for those external influences upon a system, whereas "resistance" is usually considered as the strength or capacity to withstand these loadings. For a water distribution system, the daily peak demands or emergency loading for fire demand could be looked upon as the loading, whereas the capacity of supplying this demand is the resistance. Another example is for flood levee systems along natural river channels or manmade channels where the levees are designed to prevent flooding. In this case, the design flood is the loading and the levee capacity is the resistance. Occasionally, a single resistance can respond to several loadings, such as the water supply to municipal, industrial, and agricultural demands. Conversely, a single loading can also induce distinct resistances such as various hydraulic structures (dams, levees, pumping stations, etc.) used in a flood mitigation project. More complex interactions between the loadings and the resistance are still possible.

Yen and Tang (1971) defined two types of uncertainties, objective and subjective uncertainties, that are involved in defining the risk of failure of a hydraulic structure. Objective uncertainties are inherent in the population characteristic of a random phenomenon. The objective risks are measurable and/or quantifiable such as sampled statistical variables. Subjective uncertainties have no quantitative factual information and must be handled subjectively using judgment, experience, and intuition. These uncertainties include unmeasurable inaccuracies of engineering analysis, uncertainties arising from transforming rainfall data into runoff information, and the errors associated with forecasting future conditions of a hydraulic system. The paper emphasized the idea of extended reliability theory to incorporate the subjective uncertainties in order to define a total system risk.

Yevjevich (1977) defined basic risk as "the probability of exceedence or nonexceedence of any critical design value, which is related to random phenomena of the physical world to which a hydraulic structure belongs." A total risk, or hydraulic risk, was defined as "a composition of all the random variables involved in the definition of the basic risk and all the unavoidable uncertainties, whether they are random variables, deterministic functions, or both. The total risk considered the load and resistance as two independent processes. The loads were defined in very general terms as the pressures or subpressures on the structure that would jeopardize its safety or performance. Resistance was defined as the ability of the structure to withstand the design load and/or perform as designed. These concepts were basically borrowed from the many ideas developed in risk and reliability analysis for engineering design (Kapur and Lamberson 1977).

The concept of total risk discussed by Yevjevich (1977) has had many previous engineering design applications in the mechanical, geotechnical, and structural engineering literature. However, only a few models have been reported in the water resources literature. Tang and Yen (1972), Tang, Mays, and Yen (1975), and Yen and Tang (1976) applied these concepts to storm sewer design, combining the hydrologic and hydraulic uncertainty analysis to develop a total risk or overall risk. Since these earlier works, several new methods and applications have been developed and reported in the literature.

1.2. Reliability

Reliability can be defined in general terms as the probability that a system (hydraulic structure) performs its mission within specified limits for a given period of time in a specified environment. Depending upon the type of hydraulic structure, different reliability concepts and reliability definitions are applicable. The type of reliability measures are dependent upon the various types of uncertainties that exist in common hydraulic designs. For example, in the design of a storm sewer system or a flood levee system, hydrologic and hydraulic uncertainties are of concern to define the reliability based upon the hydrologic input and the ability of the structure to handle the flows. Because of the different types of hydrologic models used for the storm sewer (rational formula) and the flood levee (flood flow frequency analysis) different methods are used to evaluate the hydrologic uncertainties, but similar methods can be used to evaluate the hydraulic uncertainties. As a result, stress-strength type analysis, defined later, is used for these types of structures, in which flow conveyance reliability is important. In contrast, a water distribution system has uncertainties as to when component failures will occur causing demands not to be met. The component failures include pipe breaks, pump failures, valve failures, storage malfunction, etc. The reliability of each of these components has an effect upon the overall system reliability. Time-to-failure analysis, defined later, can be used for reliability assessment of water distribution systems.

Determination of system reliability is also important. In the case of a storm sewer system, the component reliability refers to the reliability of a sewer pipe, and in a levee system the component reliability refers to a section of levee, whereas in a water distribution system the components refer to pipes, pumps, valves, storage facilities, etc. The reliability of the various components has an effect upon the overall system reliability. Network reliability analysis methods have been successfully developed and applied to electrical, chemical, and mechanical engineering processes but are nonexistent in water distribution systems analysis. For a large system, with many interactive subsystems (such as a water distribution system), it is extremely difficult to analytically compute the mathematical reliability. Calculation of a mathematical reliability requires knowledge of the reliability of the basic subsystems or components and the impact on mission accomplishment caused by the set of all possible subsystem (component) failures.

Work has been reported on developing component reliability models for hydraulic structures but very little work has been performed in developing system reliability models for hydraulic structures.

2. STATIC RELIABILITY MODELS – STRESS-STRENGTH ANALYSIS

2.1. Risk-Reliability Definitions

The strength of a component is defined as the ability of the component to accomplish its required mission satisfactorily without a failure when subjected to an external loading and the environment. Stress is the loading of the component and tends to produce failure of the component. When the strength of the component is less than the stress imposed on it the failure occurs. This type of analysis can be applied to the reliability analysis of hydraulic structures. The reliability of a hydraulic component, subsystem, or system is defined as the probability of the strength or resistance, r, exceeding the stress or loading, ℓ, i.e., the probability of survival. The mathematical representation of the reliability, R, is expressed as

$$R = P_r(r > \ell) = P_r(r - \ell > o) \tag{1}$$

in which $P_r(\ )$ refers to probability. Since both loading and resistance can be random variables, a knowledge of the probability distributions of $\underline{r}$ and ℓ are required to develop risk and reliability models. The risk, $\bar{R}$, of a hydraulic component is the probability that the loading exceeds the resistance, so that the relationship between risk and reliability is

$$R = 1 - \bar{R} \tag{2}$$

Static reliability computations consider the worse single stress, or load applied. The loading applied to many hydraulic structures is random, and in addition, the number of times of loading is unknown and random. Following the definitions given above, the reliability of a hydraulic structure is expressed as

$$R = \int_0^\infty f_r(r) \left[\int_r^\infty f_\ell(\ell)d\ell\right]dr \tag{3}$$

where $f_\ell(\ )$ and $f_r(\ )$ are the probability density functions of the loading and the resistance, respectively.

2.2. Safety Factor

The success of an engineering design is accomplished by adopting safety factors to compensate for the presence of uncertainties. Tang et al. (1975) used the estimated mean value of resistance, $\bar{r}$, and the design loading ℓ_T to define the safety factor SF for storm sewer design as

$$SF = \bar{r}/\ell_T \tag{4}$$

introduced by Yen (1979) for hydraulic design. The safety factor determines the location of the resistance probability distribution once the design loading is chosen.

Figure 1 illustrates the relationships among SF, $\bar{r}$, and ℓ_T. The loading distribution is fixed and the resistance distribution is allowed to shift. The variances of both distributions are assumed constant. Two different safety factor values, 1.5 and 2.0, arbitrarily selected, are used to demonstrate the effects of incorporation of the safety factor in risk computations. As seen from the graphs, the larger the safety factor, the smaller the interference between the loading and resistance distributions, which means less overall risk involved for the design.

2.3. Treatment of Hydraulic Uncertainties

The reliability of the engineering design of a hydraulic structure is a function of several variables and parameters which are often random variables for which the performance of a system is dependent. In general, a hydraulic model to define structure capacity, Q_c, or resistance, r, is a function of several random variables, $w_1, w_2, \ldots, w_p$, whose properties can be estimated, and is expressed as

$$r = Q_c = G(w_1, w_2, \ldots, w_p) \tag{5}$$

As an example of a hydraulic model, Manning's equation can be expressed as a function of several random variables (Manning's roughness, cross-sectional area, wetted perimeter, and slope). The problem is to find certain properties of Q_c, which is also a random variable because it is a function of random variables $w_1, w_2, \ldots, w_p$. In practice, it is usually difficult to determine the uncertain feature of the system as a whole. Alternatively, it is easier and more economical to analyze the uncertainty feature of the parameters or components and then derive the properties of uncertainty of the entire system. First-order analysis is a method of providing an estimate of the uncertainty for a deterministic model involving variables with uncertainty. The first-order analysis

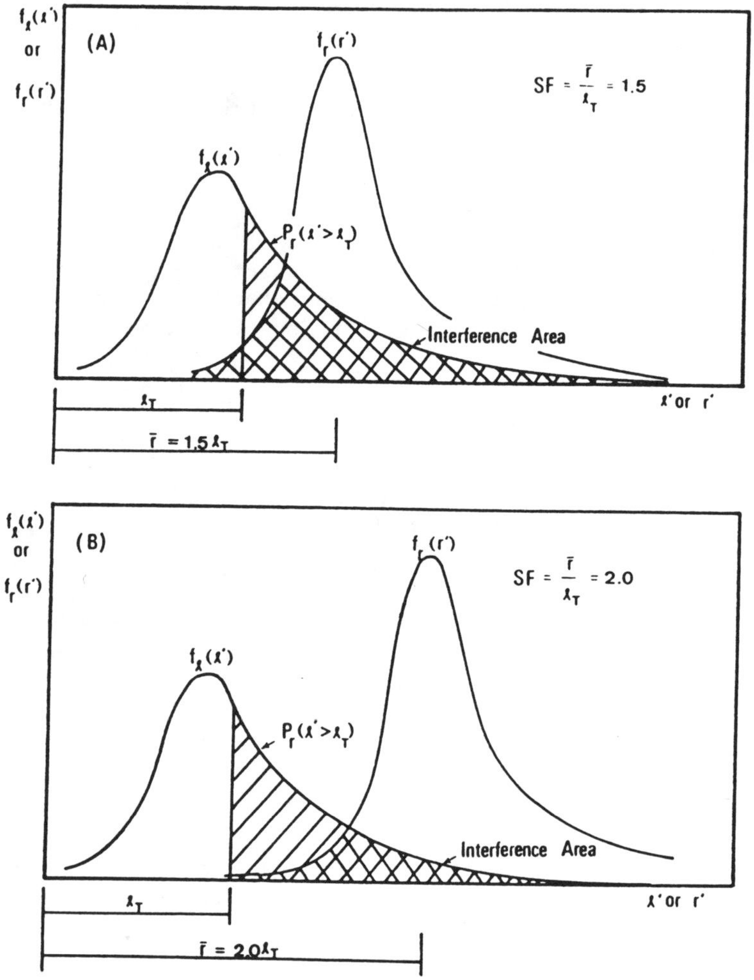

Figure 1. Relationships Among the Safety Factor SF, the Design Loading ℓ_T, and the Mean Resistance, $\bar{r}$.

provides an estimate of the mean $\bar{r}$, and the variance (or standard deviation) Ω_r to describe the probability model $f_r(r)$. Through the first-order analysis, the combined effect of uncertainty resulting from use of uncertain variables in a predictive hydraulic model can be estimated.

The mean is expressed as

$$\bar{r} = \lambda\, G(\bar{w}_1, \bar{w}_2, \ldots, \bar{w}_p) \tag{6}$$

where λ is a model error term that can be introduced to consider the error introduced by using a simplified hydrologic and/or hydraulic equation (Yen 1975). The coefficient of variation is expressed as

$$\Omega_r^2 = \Omega_\lambda^2 + \frac{1}{\bar{G}^2} \sum_p \left(\frac{\partial G}{\partial w_p}\right)^2 \bar{w}_p^2\, \Omega_{wp}^2 \tag{7}$$

2.4. Applications of Static Reliability Models

Several applications of the above concepts have been reported in the literature to develop static reliability models. Some of the first applications of these concepts were to storm sewers (Tang, Mays, and Yen 1975; Yen and Tang 1976). These applications were based upon using the rational formula for the loading and Manning's equation or the Darcy-Weisbach equation for the resistance. These risk models were based upon assuming the resistance and loading as log normal variates, to express the reliability as

$$R = 1 - P(\ell > r) = 1 - \phi\left[\frac{\ln\frac{\bar{\ell}}{\bar{r}}}{(\Omega_\ell^2 + \Omega_r^2)^{1/2}}\right] \tag{8}$$

in which ϕ is the cumulative standard normal distribution function. First order analysis of uncertainties was used to determine the statistical parameters for the loading and resistance. For the loading, the parameters are determined using

$$\bar{r} = \bar{\lambda}_r\, \bar{C}\, \bar{i}\, \bar{A} \tag{9}$$

$$\Omega_r = (\Omega_{\lambda r}^2 + \Omega_C^2 + \Omega_i^2 + \Omega_A^2)^{1/2} \tag{10}$$

in which C is the runoff coefficient, i is the design rainfall intensity, and A is the area drained by the sewer. The resistance parameters are determined from Manning's equation,

$$Q_c = \frac{1.49}{n} (4\pi)\, R^{8/3}\, S^{1/2} \tag{11}$$

$$\bar{r} = 18.72\bar{\lambda}\ \bar{S}^{0.5}\ \bar{n}^{-1}\ \bar{R}_h^{\ 2.67} \tag{12}$$

$$\Omega_r = (\Omega_\lambda^2 + 0.25\ \Omega_S^2 + \Omega_n^2 + 7.1\ \Omega_{Rh}^2)^{1/2} \tag{13}$$

where n is Manning's roughness coefficient, S is the friction slope, and R_h is the hydraulic radius. Several methods and applications of determining the various parameter (C, i, A, S, n, R_h) uncertainties have been reported in the literature (e.g., Yen 1975; Tang, Mays, and Yen 1976; Yen and Tang 1976). The analysis can be used to derive risk-safety factor relationships as presented in Figure 2.

Applications of static reliability models have also been applied to other hydraulic structures such as culverts. Yen, Cheng, and Tang (1980) used a similar approach as presented above for reliability computations in the hydraulic design of culverts. Tung and Mays (1980) developed static reliability models for culvert design based upon equation (3) in which the loading model was a Type I extremal distribution as opposed to the deterministic type model (rational formula) used by Yen, Cheng, and Tang (1980). The resistance was modeled using the Manning's formula and performing a first order analysis of the uncertainties. Typical risk-safety factor curves for culvert design developed by Tung and Mays (1980) are shown in Figure 3.

Tung and Mays (1981) developed static reliability models for flood levee design considering the loading model using different probability distributions such as the log-normal, Type I extremal, Pearson III, and log-Pearson III distributions. The resistance model was based upon a first-order analysis of Manning's equation,

$$Q_c = 1.49\ \lambda_m \left[\frac{1}{N_c} A_c^{\ 5/3} P_c^{\ -2/3} + \frac{2}{N_b} A_{bh}^{\ 5/3} P_{bh}^{\ -2/3} \right] S^{1/2} \tag{14}$$

where λ_m is the parameter accounting for the error in using Manning's equation to describe the unsteady, nonuniform flows of natural rivers; N_c and N_b are roughness coefficients for the channel and flood plain, respectively; S is the friction slope; A_c and P_c are the cross-sectional area and wetted perimeter of the channel; and A_{bh} and P_{bh} are the cross-sectional area and wetted perimeter of the overbank flow.

A rather detailed analysis of the hydraulic uncertainty was performed by Lee and Mays (1983a). To perform the sensitivity analysis for the hydraulic uncertainty resulting from the change of individual uncertainty, each Ω_i is increased by a certain quantity while the rest of the Ω_i's are kept unchanged. The purpose of this analysis is to monitor the percentage change of Ω_{Qc} (the overall hydraulic uncertainty) caused by the change of each Ω_i. Figure 4 shows the results of the study by Lee and Mays (1983a) using typical hydraulic data for flood levee design. For this set of data, S, N_b, N_b, N_c and λ_m accounted for 95 percent of the total hydraulic uncertainty resulting in the simplified expression for the coefficient of variation of the resistance (levee capacity)

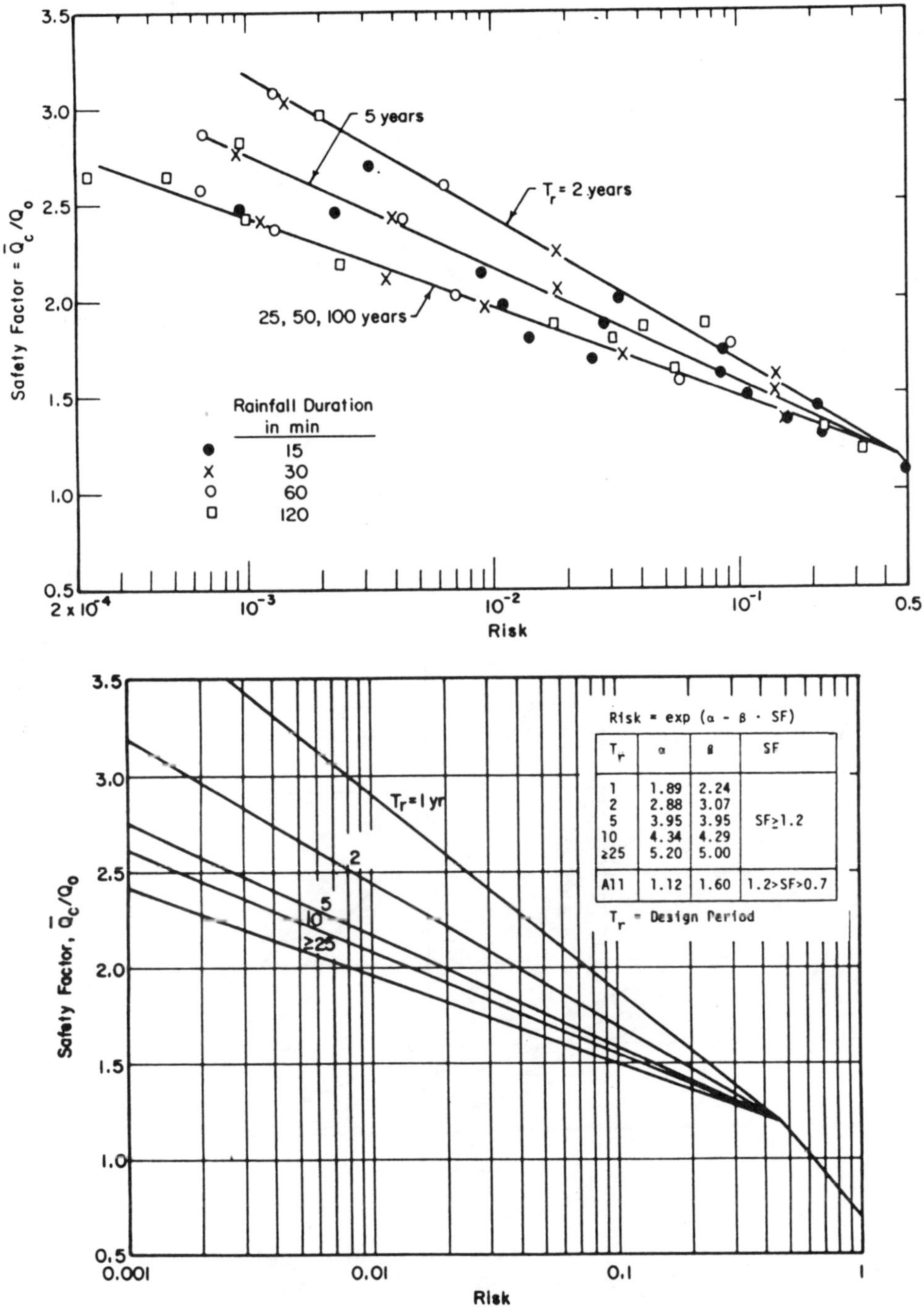
Safety Factor = $\bar{Q}_c/Q_o$
5 years
T_r = 2 years
25, 50, 100 years
Rainfall Duration in min
15
30
60
120
Risk
Safety Factor, $\bar{Q}_c/Q_o$
T_r=1 yr
2
5
10
≥25
Risk = exp (α - β · SF)
T_r | α | β | SF
1 | 1.89 | 2.24 |
2 | 2.88 | 3.07 |
5 | 3.95 | 3.95 | SF≥1.2
10 | 4.34 | 4.29 |
≥25 | 5.20 | 5.00 |
All | 1.12 | 1.60 | 1.2>SF>0.7
T_r = Design Period

Fig. 2

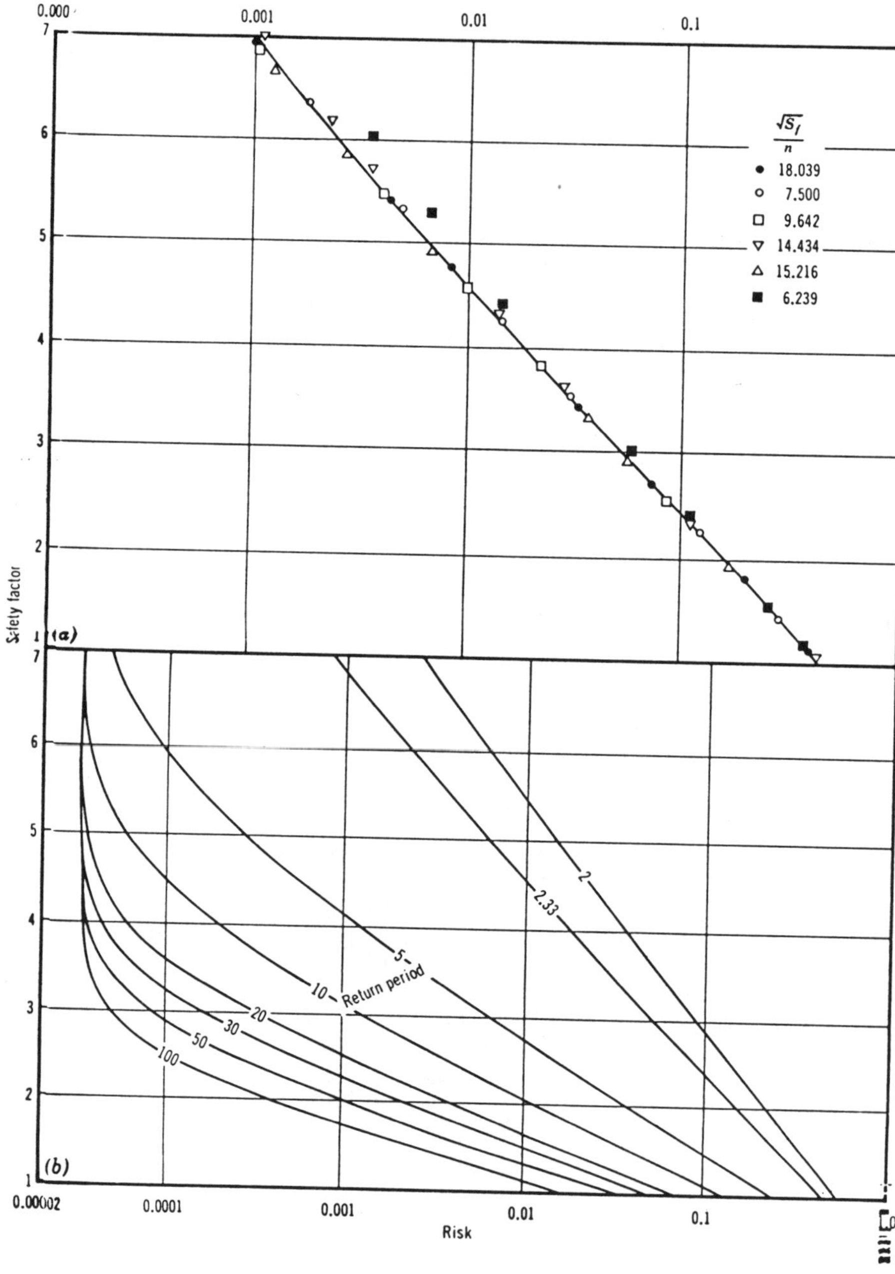

Figure 3. (a) Sensitivity of Risk-Safety Factor Relationship; (b) Risk-Safety Factor Curves for Culvert Design (Static Model).

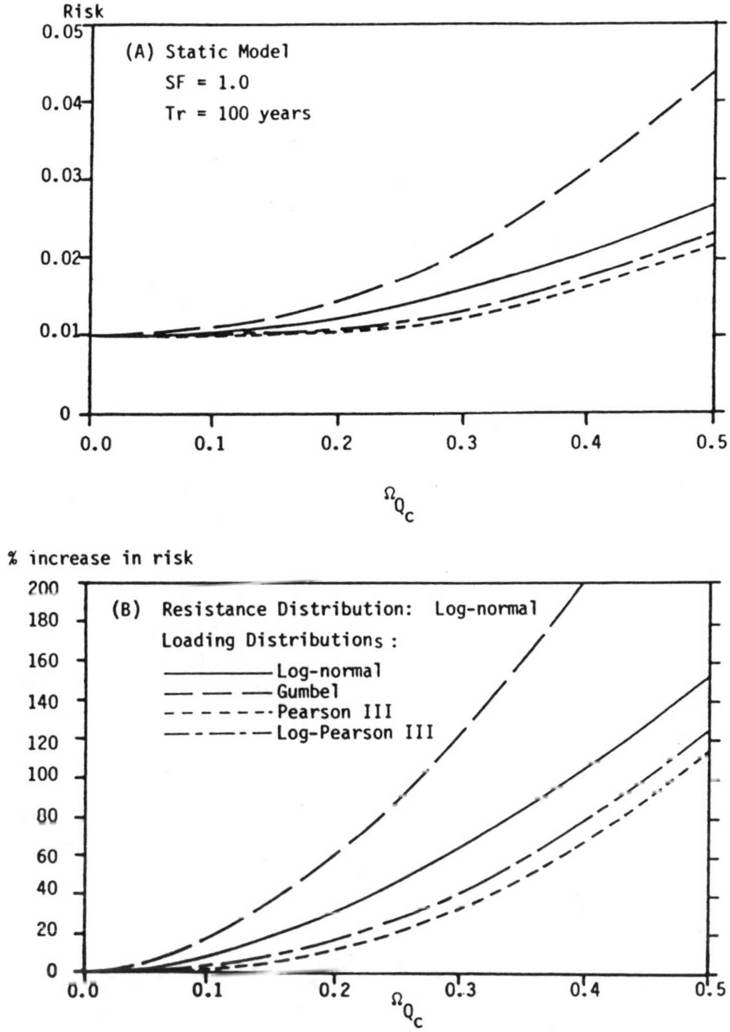

Figure 4. Effect of Hydraulic Uncertainty on Static Risk Computation, Tr = 100 years.

$$\Omega_{r_c}^2 = \Omega_{\lambda_m}^2 + \frac{1}{4}\Omega_S^2 + \frac{1}{\Psi^2}\Omega_{N_c}^2 + \frac{4}{\phi^2}\Omega_{N_b}^2 \tag{15}$$

where

$$\Psi = 1 + 2\ \frac{N_c}{\bar{N}_b}\left[\frac{A_{bh}}{\bar{A}_c}\right]^{5/3}\left[\frac{\bar{P}_c}{\bar{P}_{bh}}\right]^{2/3} \quad \text{and} \quad \phi = \frac{2\Psi}{\Psi - 1} \tag{16 a,b}$$

Examples of the static risk model developed by Lee and Mays (1984) for levee design are shown in Figure 5 along with a figure showing the effect of hydraulic uncertainty on the static risk computation.

2.5. Treatment of Hydrologic Uncertainties

Hydrologic uncertainties can be classified into three types (Wood 1975): (1) the inherent uncertainty due to the stochastic nature of the hydrologic process, (2) the model uncertainty resulting from a limited amount of data available for assessing the true random mechanism of the hydrologic process, and (3) the parameter uncertainty due to an insufficient amount of data from which the parameters in an assumed model are estimated. Bayesian techniques have been proposed by other researchers for treating the hydrologic uncertainties. The procedure by Mays and Tung (1981) does not minimize parameter uncertainty but instead develops generalized values of the parameters (mean, standard deviation, and skew coefficient) through the use of a weighting between sample and regional parameter estimates. The weights are defined as a function of the variance of both the sample parameter estimates and the regionalized parameter estimates such that the variance of the generalized parameter estimate is minimized.

Given two parameter estimates, θ_s (a sample estimate) and θ_u (a regional estimate), a weighted parameter estimate, θ, is defined by combining θ_s and θ_u as

$$\hat{\theta} = W\theta_s + (1 - W)\theta_u \tag{17}$$

where W is the weight. The commonly used expression for measuring the uncertainty feature of an estimator is the variance. Assuming that θ_s and θ_u are independent and unbiased, the optimal weight, w, which minimizes the variance of $\hat{\theta}$ is

$$W = \frac{\mathrm{Var}(\theta_u)}{\mathrm{Var}(\theta_s) + \mathrm{Var}(\theta_u)} \tag{18}$$

The variances of the parameter estimates from the sample data, $\mathrm{Var}(\theta_s)$ can be determined using either of two nonparametric statistical methods called the bootstrap or the jack-knife (Tung and Mays 1980b). While the variance of the regional parameter estimate, $\mathrm{Var}(\theta_u)$, can be derived using regression analysis relating θ_u to physiographical and meterological characteristics of the watershed or river basin considered. This

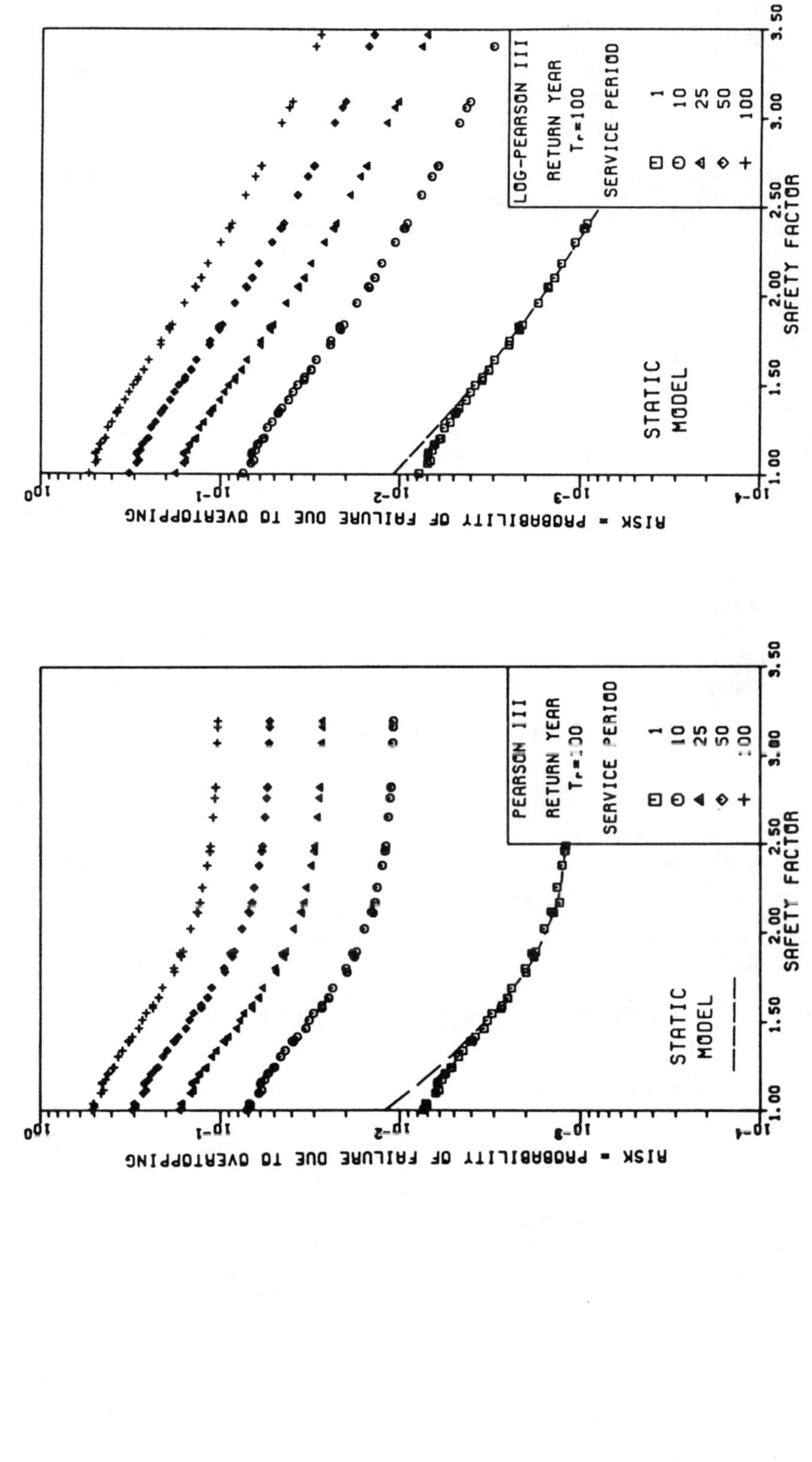

Figure 5. Comparison Between Static and Dynamic Risk-Safety Factor Curves Using Pearson III and Log Pearson III Loading Distributions, T = 100 years.

treatment of hydrologic parameter uncertainty can be very effective in improving flood frequency estimates (Tung and Mays 1980b, 1981).

The hydrologic model uncertainty results from the assessment of a representative model to describe the random mechanism of the hydrologic process based upon a limited amount of data. In other words, the uncertainty is associated with the choice of probability model based on limited data. Several nonparametric statistical goodness-of-fit techniques could be used. Another approach is to apply Bayes theorem to compute the posterior probability of a candidate model using

$$P''(m_i) = \frac{L(\underline{X} \quad \hat{\underline{\theta}}, m_i) \cdot P(m_i)}{\sum_{m_i \varepsilon M} L(\underline{X} \quad \hat{\theta}, m_i) \cdot P(m_i)} \tag{19}$$

in which $P''(m_i)$ is the posterior probability for model m_i, $P'(m_i)$ is the prior probability for model m_i. $L(\underline{X}\ \underline{\theta},\ m_i)$ is the likelihood function for model m_i with weighted parameter estimates $\underline{\theta}$ and observations $\underline{X}$ such that the treatment of hydrologic parameter uncertainty is inherently built in. M is the set of candidate models (normal, log-normal, Gumbel, Pearson III, log-Pearson III, etc.).

The value of the posterior probability is a measure of the likelihood for a model being the "true" one among all the models considered based on the available data and prior knowledge. This approach enables the derivation of a composite probability distribution (Bogardi et al. 1977; Wood and Rodriquez 1975) for the loading which is obtained by weighting the associated posterior probability for each candidate model

$$f_{\ell,m_c}(Q_d) = \sum_{m_i \varepsilon M} P''(m_i) \cdot f_{\ell,m_i}(\ell) \tag{20}$$

where $f_{\ell,m_c}(\ell)$ is the composite probability distribution of the loading and $f_{\ell,m_i}(\ell)$ is the probability distribution of the loading for candidate model m_i.

Composite static reliability models (consider two or more hydrologic loading models) can be expressed in general form as

$$R = \sum_{m_i \varepsilon M} P''(m_i) \int_0^\infty f_r(r) \left[\int_r^\infty f_\ell(\ell)\, d\ell\right] dr \tag{21}$$

3. DYNAMIC RELIABILITY MODELS – STRESS-STRENGTH ANALYSIS

Dynamic or time-dependent reliability models consider repeated application of the loading over the expected service life of the hydraulic structure. This type of model can consider uncertainty about the loading and resistance random variables with respect to time and also the uncertainty about the loading cycles. The objective of the dynamic reliability computations is to determine the reliability after n cycles or occurrences of the loading, R_n, i.e., the probability of not having a failure on any one of the n cycles or loadings. The loading on a hydraulic structure could be deterministic under normal loading conditions and random under emergency loading conditions. The time-dependent reliability model for deterministic cycles can be expressed as

$$R_n = \int_o^\infty f_r(r)\ [\int_o^r f_\ell(\ell) d_\ell]^n\ dr = \int_o^\infty f_r(r)\ [F_r(\ell)]^n\ dr \qquad (22)$$

The time-dependent reliability model for random loading times is expressed in terms of time as

$$R(t) = \pi_n(t)\ R_n \qquad (23)$$

where $\pi_n(t)$ is the probability of n loadings occurring in the time interval [o,t]. Using a stationary Poisson process to describe loadings for natural hydrologic events, the reliability model is expressed as

$$R(t) = \sum_{n=o}^{\infty} \frac{e^{-at}(at)^n}{n!} \int_o^\infty f_r(r)\ [\int_o^r f_\ell(\ell) d\ell]^n\ dr \qquad (24)$$

The simplest of the dynamic reliability models ignores the hydraulic uncertainties and only considers the hydrologic uncertainties. Probably the most common model is based upon the binomial distribution.

$$R_1(n,\ T,\ SF) = [\ 1 - \frac{1}{T(SF)}\ ]^n \qquad (25)$$

where $R_1(n,T,SF)$ is the flow conveyance reliability of a hydraulic structure over an expected service life of n years using a design return period of T years associated with a safety factor, SF. Yen (1979) used equation (25) to develop a series of curves relating risk to the structure return period and the expected service life with a safety factor of one for hydraulic structures.

A second model that ignores the hydraulic uncertainties utilizes the Poisson distribution for the reliability evaluation and can be expressed as

$$R_2(t,\ T,\ SF) = \exp[-t/T(SF)] \qquad (26)$$

in which t is the time period of interest which could be the expected service life of the hydraulic structure. On the basis of (26), Hall and Howell (1963) presented a risk evaluation procedure. It is known that for a fixed service life (T or n) and safety factor (SF), the flow conveyance reliablity calculated by (26) will asymptotically converge to that calculated by (25) as T gets larger. Chow and Takase (1977) developed a model using the nonparametric approach which also can be applied to estimate the risk of hydraulic structures without considering hydraulic uncertainties.

The development and application of dynamic reliability models that integrate both the hydrologic and hydraulic uncertainties have only been done recently (Tung and Mays, 1980b; 1981a). The original dynamic reliability model developed by Tung and Mays (1980a) considered the annual maximum flood series as the loading. This model was expressed as

$$R(t) = \int_0^\infty f_r(r) \cdot (1 - \exp\{-\alpha t[1 - F_\ell(r)]\})dr \tag{27}$$

Because the annual maximum series was considered as the loading, this particular model probably has a drawback in that it is unable to clearly define the rate of occurrence of the design event.

Lee and Mays (1983a) more recently addressed the problem and attempted to improve the model formulation using conditional probabilities. The probability of n successive independent, identically distributed loadings not exceeding the resistance r was expressed as

$$R_n = [P_r(\ell < r \; \ell \geq \ell_T]^n \tag{28}$$

where ℓ_T is the magnitude of the loading corresponding to the T-year flood with a mean occurrence rate α. The reliability, R_n, considering only the hydrologic uncertainties is

$$R_n = [\frac{P_r(\ell_T \leq \ell < r)}{P_r(\ell_T \leq \ell)}]^n = [\int_{\ell_T}^r f_\ell(\ell)\, d\ell\,]^n \tag{29}$$

Considering the hydraulic resistance then the reliability is expressed as

$$R_n = \int_{\ell_T}^\infty f_r(r)\, [\int_{Y_T}^r f_\ell(\ell)d\ell]^n \tag{30}$$

where $f_r(r)$ and $f_\ell(\ell)$ refer to the conditional distribution of resistance and loading.

The final form of the model is derived by substituting (30) into (23) and using the Poisson distribution to solve for (Lee and Mays 1983a)

$$R(t) = \frac{\int_\ell^\infty f_r(r) \exp\{-t[1 - F_\ell(r)]\})dr}{1 - F_r(\ell_T)} \tag{31}$$

Tung (1985) claims that the above model is still incomplete in that it does not lead to a formulation for the total probability of survival or failure. Defining ℓ_T as the magnitude of the design hydrologic event, Tung (1985) partitioned the magnitude of the future hydrologic event into two complementary subsets, i.e., $\ell \leq \ell_T$ and $\ell > \ell_T$, with each representing different recurrence interval hydrologic processes. Then, the structural flow conveyance reliability subjected to the ith hydrologic loading occurring in the future was expressed using the law of total probability as

$$R_i = Pr(\ell_i \leq r) = Pr(\ell_i \leq r \; \ell_i > \ell_T)Pr(\ell_i > \ell_T)$$
$$+ Pr(\ell_i \leq r \; \ell_i \leq \ell_T)Pr(\ell_i \leq \ell_T)$$
$$= Pr(\ell_T \leq \ell_i \leq r) + Pr(\ell_i \leq r, \; \ell_i \leq \ell_T) = P_1 + P_2 \quad (32)$$

More explicitly, the terms P_1 and P_2 can be expressed as

$$P_1 = \int_{\ell_T}^{\infty} \int_{\ell_T}^{r} f(r, \ell)d\ell \; dr \quad (33)$$

$$P_2 = \int_{o}^{\ell_T} \int_{o}^{r} f(r, \ell)d\ell \; dr + \int_{\ell_T}^{\infty} \int_{o}^{\ell_T} f(r, \ell)d\ell \; dr \quad (34)$$

where $f(r,\ell)$ is the joint probability density function of the resistance and loading. If r and ℓ are considered independent, then $f(r,\ell) = f(r)f(\ell)$, in which case $f(r)$ and $f(\ell)$ are the marginal probability density functions for the flow capacity of the hydraulic structure and the magnitude of a hydrologic event based on the annual maximum series, respectively.

Using the above partitioning of the future flood events into two complementary subsets, the justification of the rate of occurrence of the design hydraulic event becomes very clear, which was not the case in the original dynamic reliability formulations by Tung and Mays (1980a); and the later model by Lee and Mays (1983a) does not satisfy the law of total probability.

Tung (1985) used the binomial law to derive the flow conveyance reliability of a hydraulic structure. This model considers the flow capacity determined on the basis of a T-year design hydrologic event and a specified safety factor, SF, under n loadings or over an n year service period, expressed as

$$R_3(n, T, SF) = R_n = \sum_{x=0}^{n} \binom{n}{x} P_1^{\,x} P_2^{\,n-x} \quad (35)$$

When only the inherent hydrologic uncertainty is considered, equation (35) reduces to (26).

Tung (1985) developed a second generalized dynamic formulation using the Poisson distribution. By using the Poisson distribution the probability of n occurrences of a hydrologic event over a period of [0,t], the dynamic reliability model is

$$R_4(t, T, SF) = \sum_{n=0}^{\infty} \Phi_n(t)R_n$$
$$= \sum_{n=0}^{\infty} \frac{e^{-t}t^n}{n!} \sum_{x=0}^{n} \binom{n}{x} P_1^{\,x} P_2^{\,n-x} \quad (36)$$

Again, (36) reduces to (26) when hydraulic uncertainties are ignored.

Tung (1985) presented a comparison of the above models (Figure 6) which shows the relationship between risk and service life for a levee using a design return period of 50 years and a safety factor of one. Figure 7 shows the resulting risk-safety factor curves for a levee with a design return period of 100 years derived from the two generalized reliability models, equations (35) and (36).

4. TIME TO FAILURE ANALYSIS OF COMPONENT RELIABILITY

Time-to-failure analysis considers the problem of characterizing the reliability parameters for repair-to-failure process or installation-to-failure process. The reliability, R(t), is defined as the probability that the component experiences no failure during the time interval (0,t) given that it is new or repaired at time zero. The reliability is defined as

$$R(t) = \int_0^\infty f(t)dt \tag{37}$$

where f(t) is the probability density function. The mean time to failure, MTTF is the expected value of the time to failure, given as

$$MTTF = \int_0^\infty t\ f(t)dt \tag{38}$$

Now consider a process starting with a failure and ending at the first repair. The time axis is shifted appropriately so that t - 0 is the time when the component failed. The repair probability, g(t), is the probability that the repair is completed before time t, given that the component failed at time zero. The time to repair is a random variable B that the expected value of the time to repair is the mean time to repair defined as

$$MTTR = \int_0^\infty tg(t)dt \tag{39}$$

The mean time between failures is the expected length of time between two consecutive failures defined as

$$MTBF = MTTF + MTTR \tag{40}$$

The mean time between repairs is the expected length of time between two consecuive repairs and equals the MTBF. Probably the most widely used distribution is the exponential distribution because of its simplicity for performing reliability analysis.

The time-to-failure analysis considers a process consisting of repetitions of the repair-to-failure process and the failure-to-repair processes. The probabilistic concepts for the whole process includes the availability, A(t) which is the probability that a component is normal at time t, given that it was as good as new at time zero. Reliability generally differs from availability because reliability requires the continuation of the normal state over the whole interval (0,t).

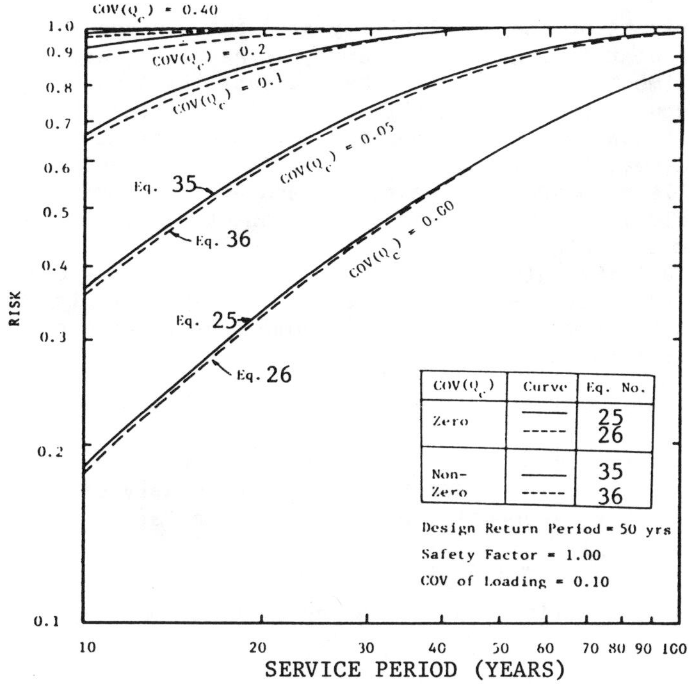

Figure 6. Comparison of Reliability Models (Tung 1985).

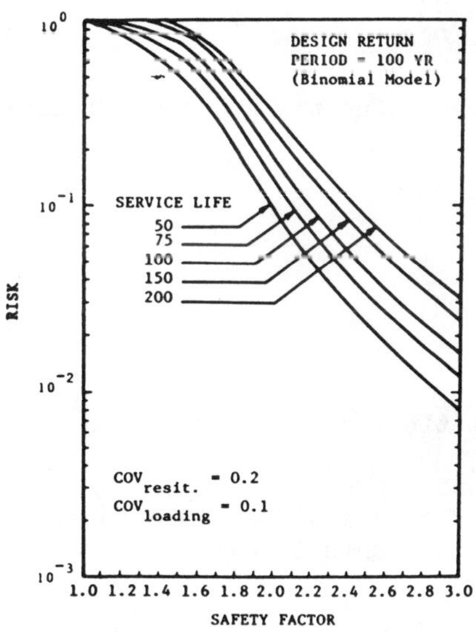

(a) Generalized Dynamic Reliability Model Using Binomial Distribution.

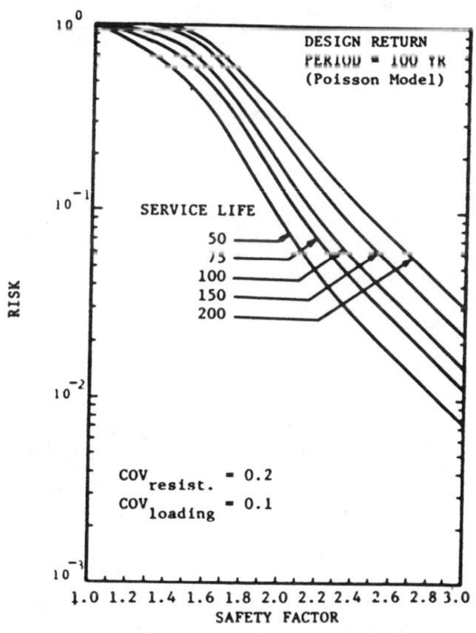

(b) Generalized Dynamic Reliability Model Using Poisson Distribution.

Figure 7. Risk-Safety Factor Curves Derived from the two Generalized Dynamic Reliability Models Using Lognormal Loading and Lognormal Resistance.

Components contribute to the availability A(t) but not the reliability R(t) if the component failed before time t, is repaired, and is then operational at time t. As a result, the availability A(t) is always larger than or equal to the reliability, $A(t) \geq R(t)$, for a repairable component. For a nonrepairable component, it is normal at time t only if it has been normal to time t, so that A(t) - R(t). The availability of a nonrepairable component decreases to zero as t becomes larger, whereas the availability of a repairable component converges to a non-zero positive number (Figure 8).

The unavailability, Q(t), at time t is the probability that a component is in the failed state at time t, given that it started into the normal state at time zero. Because a component is either in the normal state or failed state at time t, then

$$A(t) + Q(t) = 1 \tag{41}$$

Because of its relative simplicity for performing reliability analyses, the exponential distribution is probably the most widely used failure density function. The exponential density function is:

$$f(t) = \lambda e^{-\lambda t} \qquad t \geq 0, \ \lambda > 0 \tag{42}$$

where λ is a constant failure rate. The reliability is simply

$$R(t) = e^{-\lambda t} \tag{43}$$

The MTTF is

$$\text{MTTF} = \int_0^\infty t\lambda e^{-\lambda t}\, dt = \frac{1}{\lambda} \tag{44}$$

The mean time to repair can also be defined using the exponential density function

$$g(t) = ue^{-ut} \tag{45}$$

so that

$$\text{MTTR} = \int_0^\infty t\ ue^{-ut}\, dt = \frac{1}{u} \tag{46}$$

The MTTR can be estimated using an arithmetic mean of the time to repair data for various types of components.

For a constant failure rate and a constant repair rate, the analysis of the whole process can be simplified to analytical solutions. Henley and Kumamoto (1981) used Laplace transforms to derive the unavailability as

$$Q(t) = \frac{\lambda}{\lambda + u} \left[1 - e^{-(\lambda+u)t}\right] \tag{47}$$

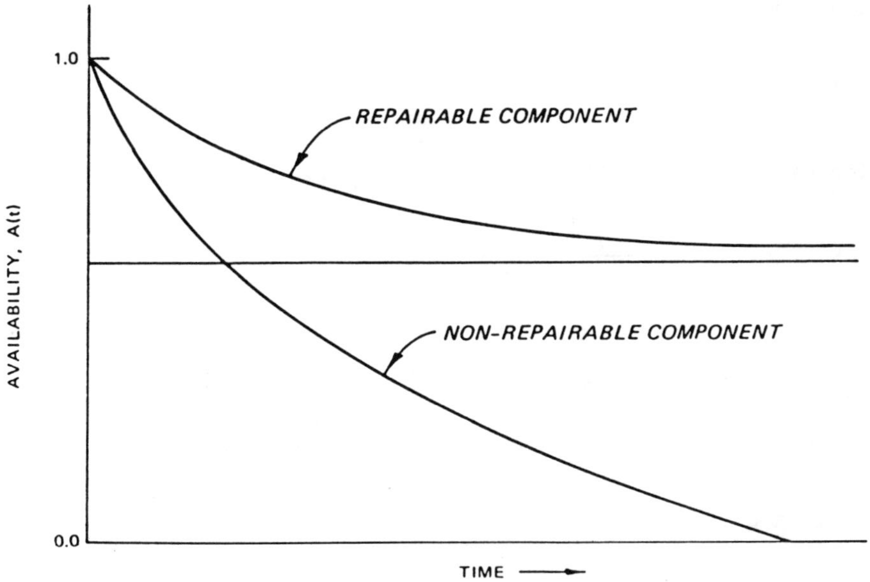

Figure 8a. Availability for repairable and nonrepairable components

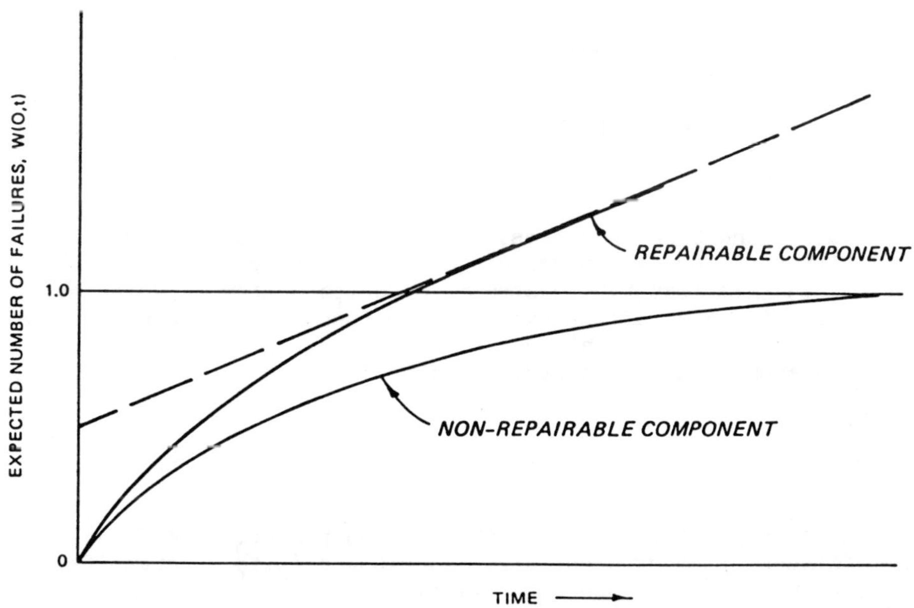

Figure 8b. Expected number of failures for repairable and nonrepairable components

and the availability

$$A(t) = 1 - Q(t) = \frac{u}{\lambda + u} + \frac{\lambda}{\lambda + u} e^{-(\lambda+u)t} \tag{48}$$

The stationary unavailability $Q(\infty)$ and the stationary availability $A(\infty)$ are

$$Q(\infty) = \frac{\lambda}{\lambda + u} = \frac{MTTR}{MTTF + MTTR} \tag{49}$$

$$A(\infty) = \frac{u}{\lambda + u} = \frac{MTTF}{MTTF + MTTR} \tag{50}$$

The following relation is also true

$$\frac{Q(t)}{Q(\infty)} = 1 - e^{-(\lambda+u)t} \tag{51}$$

Figure 9 is the reliability curve for a pump system with $\lambda = 0.0008$/hr and Figure 10 presents the availability curves for $\lambda = 0.0008$/hr and u = 0.02, 0.01, and 0.005/hr. For repairable systems the availability is always greater than or equal to the reliability as illustrated in Figure 11 for $\lambda = 0.0008$/hr and u = 0.01/hr. The reliability curve for a 5-mile water main of sandspun cast iron pipe is shown in Figure 12. The break rate per year for the water main was determined to be 0.3185 exp(0.0137(t-k)) (Mays and Cullinane 1985).

5. SYSTEM (NETWORK) RELIABILITY EVALUATION

5.1. Simple Systems

Most hydraulic structures (systems) are composed of several subsystems. The reliability of the system depends on how the components are interconnected. The simplest type system is a series system in which every component must function if the system is to function. Considering the random variable of the time of failure as t_i for the i-th component, then for a system of n components, the system reliability is

$$R_s(t) = \prod_{i=1}^{n} P(t_i > t) = \prod_{i=1}^{n} R(t) = (R_1)(R_2)\ldots(R_n) \tag{52}$$

where $R_i(t)$ is the reliability for the i-th component. For a system that has failure times exponentially distributed (with constant failure rates) so that the i-th component reliability is $e^{-\lambda_i t}$ then the system reliability is

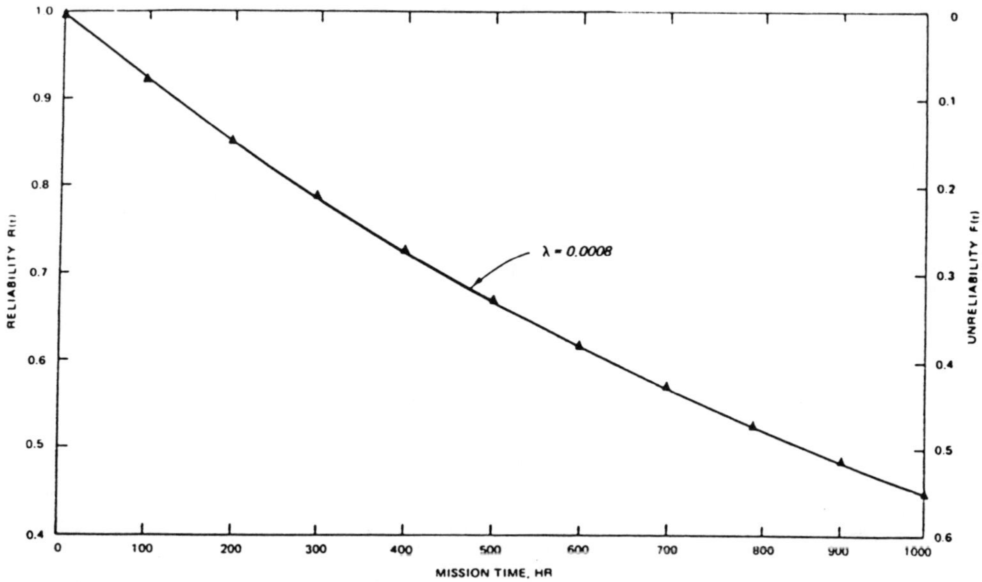

Figure 9. Reliability curve for a pump system with $\lambda = 0.0008/hr$

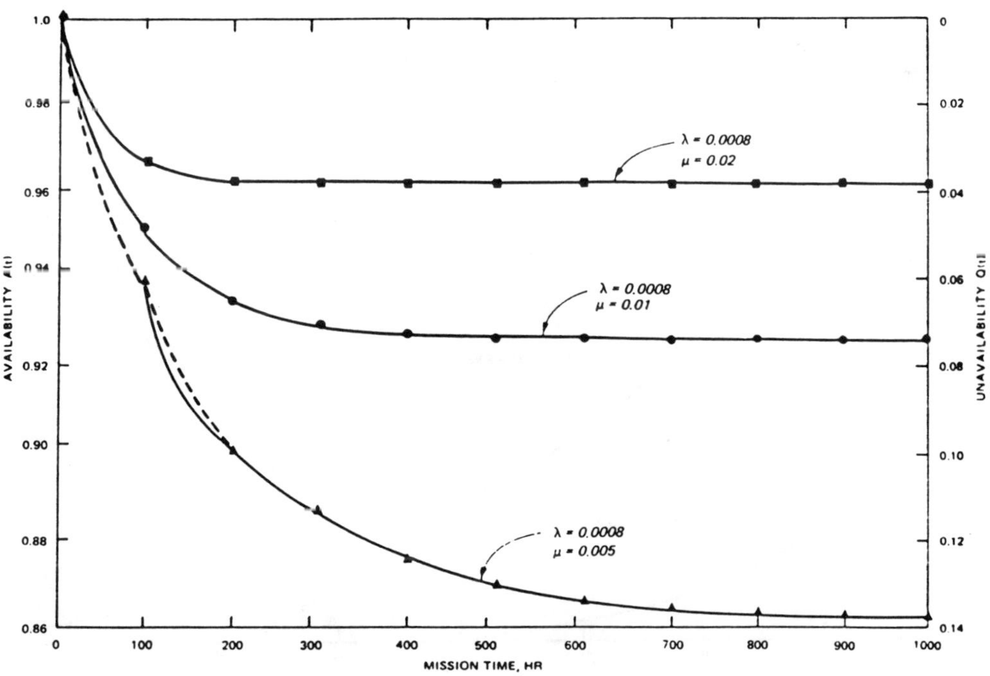

Figure 10. Availability curves for a pump system with $\lambda = 0.0008/hr$ and $\mu = 0.02$, 0.01, and 0.005/hr

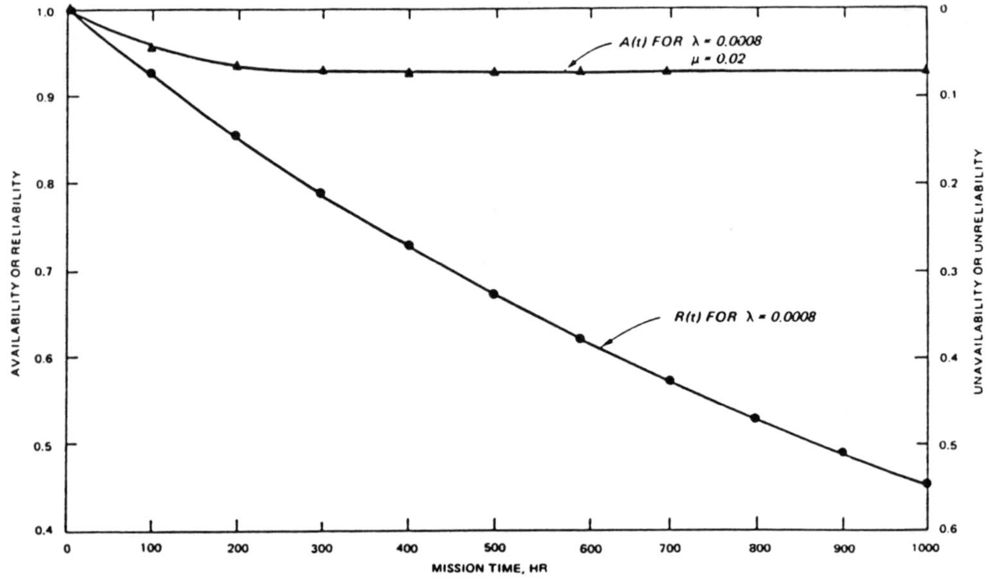

Figure 11. Comparison of reliability and availability for $\lambda = 0.0008/hr$ and $\mu = 0.02/hr$

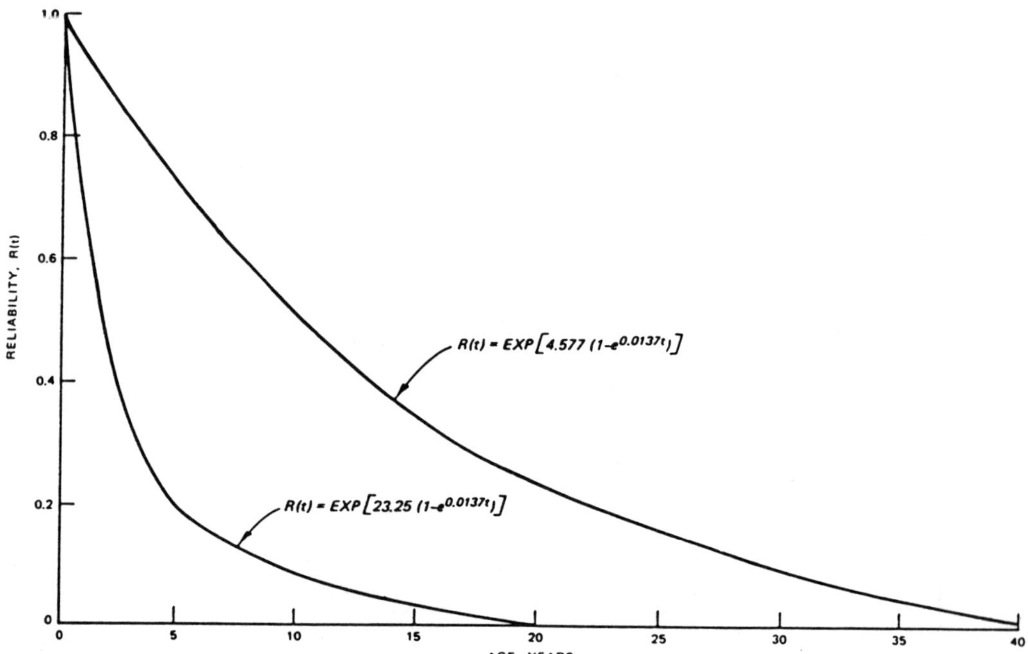

Figure 12. Reliability curves for pipe evaluation example

$$R_s(t) = \exp\left(- \sum_{i=1}^{n} \lambda_i t\right) \tag{53}$$

The MTTF is

$$\text{MTTF} = \int_0^\infty \exp\left(-\Sigma\lambda t \; dt\right) = \frac{1}{\sum_{i=1}^{n} \lambda_i} \tag{54}$$

A series chain model is a series system such that if any one component fails, the system will fail. This model is based on the idea of a chain composed of n links where the chain will break if the applied stress x exceeds the strength y of any one link. This model is also referred to as a weakest link model. The system reliability is then

$$R_s = \min_i R_i \tag{55}$$

A parallel system is defined as one which will fail if and only if all units in the system fail or malfunction. The pure parallel system is one in which all components are initially activated, and any component can maintain the system operation. The system reliability is then expressed as

$$R_s(t) = 1 - \prod_{i=1}^{n} [1 - R_i(t)] \tag{56}$$

For a system with exponentially distributed time to failure and a constant failure rate for each component of the system, the system reliability is

$$R_s(t) = 1 - \prod_{i=1}^{n} (1 - e^{-\lambda_i t}) \tag{57}$$

and the MTTF for a system with identical components is

$$\text{MTTF} = \frac{1}{\lambda} \sum_{i=1}^{n} \frac{1}{i} \tag{58}$$

5.2. Complex Systems

In most practical situations, such as water distribution systems, the system (networks) have nonseries-parallel configurations. The reliability evaluation for series-parallel systems is straight forward; however, for nonseries-parallel configurations the evaluation is much more difficult. There have been many techniques developed for system reliability evaluation, including state enumeration methods, network reduction methods, and path enumeration methods, which probably are the most promising.

Path enumeration methods could be valuable tools for water distribution system reliability evaluation. A path is a set of elements (components) which form a connection between input and output when traversed in a stated direction. A minimal path is one in which no node is traversed more than once in going along the path. The i-th minimal path will be denoted as P_i, i=1,. . . , M. Assuming any path is operable, the system performs adequately, then the system reliability is

$$R = P_r \left[\bigcup_{i=1}^{m} P_i \right] \tag{59}$$

where $P_r[\ ]$ represents probability and U denotes the union.

A cut set is defined as a set of elements, which if they fail, cause the system to fail regardless of the condition of the other elements in the system. A minimal cut is one in which there is no proper subset of elements whose failure alone will cause the system to fail. In other words, a minimal cut is such that if any component is removed from the set, the remaining elements collectively are no longer a cut set. The minimal cut sets are denoted as C_i, i.e., the failure of all elements of the cut C_i. The system reliability is

$$R = 1 - P_r\left[\bigcup_{j=1}^{m} \bar{C}_j \right] \tag{60}$$

Hwang, Tillman, and Lee (1981) presented a review of literature related to system reliability evaluation techniques for small to large complex systems. A large system was defined as one which has more than 10 components and a moderate system as one which has more than 6 components and less than 10. Complex systems were defined as ones which could not be reduced to a series-parallel system. The authors concluded that for a large complex system, computer programs should be used that provide the set of minimal cuts and calculate the minimal cut approximation to system reliability. Minimal paths can be generated from minimum cuts. Based on minimal path (tie) sets, reliability approximation can then be obtained for large-complex networks.

5.3. System Reliability: Fault Tree Analysis

Any failure of any system, whether it be a lawn mower or a water distribution system, can be depicted by a fault tree. A fault tree is a logical diagram of the consequences of basic (primary)failures on the system failure (top failure). Basic steps to develop fault trees include first to define the system failure (top event) of the system. The analyst must thoroughly understand the system and its intended use. One must determine the higher order functional events and continue the fault event analysis to determine the logical relationship with lower level events. Once the above has been accomplished, then the fault tree can be constructed. The major objective of fault tree construction is to represent the system condition, which may cause system failure, in a symbolic manner.

The fault tree method requires determination of the cut-sets, so this method would probably be applicable only as long as the system is not too complex. Identifying all the cut-sets can be very complicated and computationally expensive.

6. OPTIMAL-RISK OR RELIABILITY BASED DESIGN PROCEDURES

6.1. Models for Flow Conveyance Structures

Conventional practice in risk-based water resource engineering design considers only the inherent hydrologic process. In addition to the annual installation cost, AIC, of the hydraulic structure, the annual expected damage, E(D), is taken into account. This could be expressed in the form of an objective function, by minimizing annual total expected cost, ATEC, for a structure with n = 1,. . . , N subsystems or components.

$$\min \text{ATEC} = \sum_{n=1}^{N} [\text{AIC}_n + E(D)_n] \tag{61}$$

The annual expected damage, considering only the hydrologic uncertainties, can be expressed as

$$E(D) = \int_0^\infty D(\bar{r}, \ell)\, f_\ell(\ell)d\ell \tag{62}$$

where $D(\bar{r}, \ell)$ is the damage-discharge relationship for a particular design having a mean resistance, $\bar{r}$.

In the above objective function, the hydraulic structure capacity is deterministic, ignoring hydraulic uncertainties. The neglect of hydraulic uncertainty can lead to an underestimate of the overall risk (Tung and Mays 1981a; Lee and Mays 1983b), and consequently the annual expected damage. The annual expected damage that considers both the hydrologic and hydraulic uncertainties can be expressed as

$$E(D) = \int_0^\infty [\int_0^\ell D(r, \ell)\, f_r(r)dr]\, f_\ell(\ell)d\ell \tag{63}$$

where the damage-discharge relationship $D(r, \ell)$ is a function of r instead of $\bar{r}$ as in (62).

Tang, Mays, and Yen (1975) developed an optimal risk-based design model for the design of storm sewer systems using dynamic programming (DP). Tung and Mays (1981b) developed an optimal risk-based design models for flood levee systems using DP and discrete differential dynamic programming (DDDP). These models considered both hydrologic and hydraulic uncertainties.

The U.S. Department of Transportation Federal Highway Administration (Corry, Jones, and Thompson 1980) developed procedures for the design of encroachments on flood plains using risk analysis. This procedure actually computes economic risks or expected damage costs along with the annual capital costs to determine total expected cost for alternative designs. The objective is to determine the design which has the least total expected cost (LTEC). Basically, this procedure considers the inherent hydrologic uncertainty and ignores the hydraulic uncertainties, as illustrated by equation (62).

Tung and Mays (1982) developed an optimal risk-based design procedure for the design of bridges. This procedure did not consider the hydraulic uncertainty but included the hydrologic (inherent, parameter, and model) uncertainties. The optimization was based upon the Hook and Jeeves scheme.

6.2. Reliability Optimization

Reliability allocation models assign reliability requirements to individual components to attain a specified system reliability. These types of design models are basically nonexistent in the area of hydraulic structure design.

Two basic procedures for optimal reliability optimization would be: (1) maximize system reliability subject to cost constraints, and (2) minimize cost subject to reliability constraints. Most of the earlier literature in the area of system reliability optimization considered only series-parallel systems. For an N-stage series model the problem is stated as

$$\text{Maximize } R_s = \prod_{j=1}^{N} R_j \tag{64}$$

subject to

$$\sum_{j=1}^{N} g_{ij}(R_j) \leq b_i \qquad i = 1, \ldots, m \tag{65}$$

here R_s is the system reliability, R_j is the component reliability of the j-th stage, $g_{ij}(R_j)$ is the resource i at stage j, and b_i is the total amount of resource i available.

The above problem could also be stated as finding the optimal number of redundancies which maximize system reliability,

$$\text{Maximize } R_s = \prod_{j=1}^{N} R_j(X_i) \tag{66}$$

subject to

$$\sum_{j=1}^{N} g_{ij}(X_j) \leq b_i \qquad i = 1, \ldots, m \tag{67}$$

The above problem could be restated as

$$\text{Minimize } C_s = \sum_{j=1}^{N} R_j(x_j) \geq R_r \tag{68}$$

where C_s is the total system cost, C_j is the cost of the j-th stage which is a function of the number of components in each stage, x_j. The constraint says that the system reliability must be greater than or equal to the required systems reliability, R_s.

System reliability for complex systems must be obtained using the approaches previously mentioned. The optimization problem would be stated as

$$\text{Maximize } R_s = f(R_1, R_2, \ldots, R_N) \tag{69}$$

subject to

$$\sum_j g_{ij} (R_j) \leq b_i \qquad i = 1, \ldots, m \tag{70}$$

6.3. New Methodologies for Optimal Network Design

The development of optimal-reliability based design models for water distributions has only been very recent. A great deal of optimization model development has been reported in the literature. Coals and Goulter (1985) presented three approaches by which the probability of failure of individual pipes can be related to a measure of the overall system reliability in a least cost analysis. Linear programming is used to minimize cost in each of the approaches.

Lansey and Mays (1985) are developing a methodology based upon optimal control theory to optimize the design of water distribution networks using reduced gradient techniques with a network simulator. The problem is reduced in complexity by incorporating a network simulator into the optimization model. This allows a more general problem to be solved including pipe sizing for new and expanding systems, pump sizing, valves, and storage facilities. In addition, the reduced problem can consider large networks under a number of loading conditions.

The overall optimization problem for water distribution system design and analysis can be stated as functions of the heads, H, and flow rates, Q, as:

Objective:

Minimize Cost	Min $f_c(H,Q)$	(71a)
or		
Maximize Reliability	Min $f_R(H,Q)$	(71b)

Subject to:

a. Conservation of flow constraints

b. Energy equation $\quad g(H,Q) = 0 \quad$ (72)

c. Component reliability constraints $h(H,Q) = 0$ (73)

d. System reliability constraints $r(H,Q) = 0$ (74)

e. Cost constraints $c(H,Q) = 0$ (75)

f. Head bounds (lower $\underline{H}$ and upper $\bar{H}$) $\underline{H} \leq H \leq \bar{H}$ (76)

g. Reliability bounds (lower $\underline{R}$) $R \geq \underline{R}$ (77)

This new methodology for the design and analysis of new and existing water distribution systems that brings together unrelated technologies: water distribution simulation models, numerical optimization techniques, and network reliability analysis models. The technologies for each have been developed to differing levels of success: (a) the water distribution simulation models are highly developed with a history of successful use; (b) the numerical optimization techniques for water distribution systems are not as highly developed, with little successful use in water distribution design and analysis, and (c) the network reliability analysis models have been successfully developed and applied to electrical, chemical, and mechanical engineering processes but are nonexistent in water distribution system analysis. The overall methodology is based upon the concepts of optimal control theory in which the generalized reduced gradient method makes up the overall optimization framework along with a simulation model and a network reliability model used to perform function evaluation at each iteration of the optimization. Such a methodology allows very detailed analysis of the system components, various emergency loading conditions, and reduces the constraint size so that very large water distribution systems with thousands of components can be designed for new systems and/or analyzed and operated for existing systems.

7. WHERE DO WE GO FROM HERE?

Risk and reliability analysis can involve complicated probability theories and laborious mathematical analysis. The risk and reliability values generated from those theories are not always easily appreciated. This paper has presented a summary of the various risk and reliability models that have been reported in the literature. Basically, this summary identifies the fact that a great deal of research work lies ahead before the concepts of uncertainty analysis, risk analysis, and reliability analysis become a part of conventional design practice in hydraulics and hydrology.

The following is a partial list of ideas that the author has compiled concerning the research direction that needs to be taken in the future.

1. Develop accepted definitions and measures of reliability for various types of hydraulic structures that could be universally accepted.

2. Develop conceptual frameworks or methodologies to apply reliability concepts to the design and evaluation of various types of hydraulic structures that can be utilized in conventional design practice.

3. Design methods should be developed that explicitly generate and evaluate various hydraulic structure designs in terms of their ultimate importance to both the total system cost and the total system risk.

4. A great deal of research work concerning reliability has been published in the electronics literature, power literature, chemical engineering literature, mechanical engineering literature, and operations research literature. The results of these works could be very important so that these ideas could be extended to the design and analysis of hydraulic structures.

5. For hydraulic structures subjected to loadings as a result of hydrologic events, considerable work has been performed in analyzing the hydrologic uncertainties but little work has been performed in analyzing the hydraulic uncertainties. A great deal of work needs to be done in this area including the following.

a. Determine the design condition under which both the hydraulic and the hydrologic uncertainties need to be considered.

b. Determine the relative importance of each uncertainty in the design.

c. Analyze in detail the hydraulic model error term. This is one area which has not been investigated to any great degree.

d. Determine the reliability variations from using simplified hydraulic models such as Manning's equation to using more sophisticated models such as unsteady flow models.

6. Develop methods for determining system or network reliabilities for hydraulic structures with many components such as storm sewer systems and water distribution systems.

7. Develop optimal-risk-reliability based models for the design of hydraulic structures. One such direction that should be given consideration is the method of the optimal control framework that can incorporate nonlinear optimization models with hydraulic simulation models.

8. Many hydraulic structure systems should possibly be designed and analyzed using not only the concepts of risk and reliability, but also the concepts of availability and unavailability. An example of this is a water distribution system.

9. Time-to-failure type of analysis should also be investigated for possible adoption into design procedures, particularly systems such as water distribution systems.

10. The idea of fault tree analysis should also be investigated. This method has proven to be useful in the field of chemical engineering for the reliability analysis of various types of fluid flow process plants. Of particular interest is the ability to handle time delays such as storage tanks and standby redundancy.

11. Of course, to accomplish the above, it is mandatory that detailed data bases on hydraulic structures be developed and maintained.

REFERENCES

Bodo, B. and T.E. Unny, 1976. 'Model Uncertainty in Flood Frequency Analysis and Frequency-Based Design,' Water Resources Research, **12(6)**, pp. 1109-1117.

Castano, E., L. Duckstein, and I. Bogardi, 1978. 'Choice of Distribution Functions for Hydrologic Design,' Water Resources Research, **14(4)**, pp. 643-652.

Chow, V.T. and N. Takase, 1977. 'Design Criteria for Hydrologic Extremes,' Journal of the Hydraulics Division, American Society of Civil Engineers, **103(HY4)**, pp. 425-436.

Coals, A. and I.C. Goulter, 1985. 'Approaches to the Consideration of Reliability in Water Distribution Networks,' Proceedings of the 1985 International Symposium on Urban Hydrology, Hydraulic Infrastructures, and Water Quality Control, University of Kentucky, Lexington, Kentucky.

Corry, M.L., J.S. Jones, and P.L. Thompson, 1980. 'The Design of Encroachments on Flood Plains Using Risk Analysis,' Hydraulic Engineering Circular No. 17, Federal Highway Administration, U.S. Department of Transportation.

Davis, D.R., C.C. Kisiel, and L. Duckstein, 1972. 'Bayesian Decision Theory Applied to Design in Hydrology,' Water Resources Research, **8(1)**, pp. 33-41.

Duckstein, L. and I. Bogardi, 1981. 'Application of Reliability Theory to Hydraulic Engineering Design,' Journal of the Hydraulics Division, American Society of Civil Engineers, **107(HY7)**, pp. 799-815.

Hall, W.A. and D.T. Howell, 1963. 'Estimating Flood Probabilities Within Specific Time Intervals,' Journal of Hydrology, **1(1)**, pp. 265-271.

Henley, E.J. and H. Kumamoto, 1981. Reliability Engineering and Risk Assessment, Prentice Hall, Englewood Cliffs, N.J.

Hwang, C.L., F.A. Tillman, and M.H. Lee, 1981. 'System Reliability Evaluation Techniques for Complex/Large Systems—A Review,' Institute of Electrical and Electronics Engineers, Transactions on Reliability, Vol. **R-30**, No. **5**.

Kapur, K.C. and L.R. Lamberson, 1977. Reliability in Engineering Design, John Wiley, New York.

Lansey, K. and L.W. Mays, 1985. 'A Methodology for Optimal Network Design, Proceedings of Specialty Conference, Computer Applications in Water Resources,' Sponsored by the Water Resources Planning and Management Division, ASCE, Buffalo, New York.

Lee, H.L. and L.W. Mays, 1983a. 'Improved Risk and Reliability Model for Hydraulic Structures,' Water Resources Research, **19(6)**, pp. 1415-1422.

Lee, H.L. and L.W. Mays, 1983b. 'Advances in Risk and Reliability Analysis for Hydraulic Structures,' 1983 ASCE Hydraulics Division Specialty Conference, MIT, Cambridge, Massachusetts.

Mays, L.W., 1979. 'Optimal Design of Culverts Under Uncertainties,' Journal of the Hydraulics Division, ASCE, **105(HY5)**, pp. 443-460.

Mays, L.W. and M.J. Cullinane, 1985. 'A Review and Evaluation of Reliability Concepts for Design of Water Distribution Systems,' Miscellaneous Paper EL-85, U.S. Army Engineering Waterways Experiment Station, Vicksburg, Mississippi.

Szidarovszky, F., I. Bogardi, L. Duckstein, and D. Davis, 1976. 'Economic Uncertainties in Water Resources Project Design,' Water Resources Research, **12(4)**, pp. 573-580.

Tang, W.H., L.W. Mays, and B.C. Yen, 1975. 'Optimal Risk-Based Design of Storm Sewer Networks,' Journal of the Environmental Engineering Division, ASCE, **101(EE3)**, pp. 381-398.

Tang, W.H. and B.C. Yen, 1972. 'Hydrologic and Hydraulic Design Under Uncertainties,' Proceedings International Symposium on Uncertainties in Hydrologic and Water Resources Systems, Vol. 2, pp. 868-882, Tucson, Arizona.

Tung, Y.K., 1985. 'Evaluation of Water Distribution Network Reliability,' Proceedings 1985 Hydraulics Division Specialty Conference, Orlando, Florida.

Tung, Y.K., 1985. 'Models for Evaluating Flow Conveyance Reliability of Hydraulic Structures,' Water Resources Research, **21(10)**, pp. 1463-1468.

Tung, Y.K. and L.W. Mays, 1980a. 'Optimal Risk-Based Design of Water Resource Engineering Projects,' Tech. Report CRWR-171, Center for Research in Water Resources, The University of Texas at Austin, Austin, Texas.

Tung, Y.K. and L.W. Mays, 1980b. 'Risk Analysis for Hydraulic Design,' Journal of the Hydraulics Division, ASCE, **106(HY5)**, pp. 893-913.

Tung, Y.K. and L.W. Mays, 1981a. 'Risk Models for Flood Levee Design,' Water Resources-Research, **17(4)**, pp. 833-841.

Tung, Y.K. and L.W. Mays, 1981b. 'Optimal Risk Based Design of Flood Levee Systems,' Water Resources Research, **17(4)**, pp. 842-852.

Tung, Y.K. and L.W. Mays, 1982. 'Optimal Risk-Based Hydraulic Design of Bridges,' Journal of the Water Resources Planning and Management Division, ASCE, **108(WR2)**, pp. 191-203.

Yen, B.C., 1970. 'Risks in Hydrologic Design of Engineering Projects,' Journal of the Hydraulics Division, ASCE, **96(HY4)**, pp. 959-966.

Yen, B.C., 1979. 'Safety Factor in Hydrologic and Hydraulic Engineering Design,' In: Reliability in Water Resources Management, edited by McBean et al., Water Resources Publications, Fort Collins, Colorado.

Yen, B.C. and A.H.S. Ang, 1971. 'Risk Analysis In Design of Hydraulic Projects,' Paper presented at 1st International Symposium on Stochastic Hydraulics, University of Pittsburgh, Pittsburgh, Pennsylvania.

Yen, B.C., S-T. Cheng, and W.H. Tang, 1980. 'Reliability of Hydraulic Design of Culverts,' Proceedings, IAHR Conference on Water Resources Development, Taipei, Taiwan, Republic of China, pp. 991-1001, May 12-14.

Yen, B.C., H.G. Wenzel, Jr., L.W. Mays, and W.H. Tang, 1976. 'Advanced Methodologies for Design of Storm Sewer Systems,' Research Report No. 112, Water Resources Center, University of Illinois at Urbana-Champaign, Illinois.

Yen, B.C. and W.H. Tang, 1976. 'Risk-Safety Factor Relation for Storm Sewer Design,' Journal of the Environmental Engineering Division, ASCE, **102(EE2)**, pp. 509-516.

Yevjevich, V., 1977. 'Risk and Uncertainty in Design of Hydraulic Structures,' In: Stochastic Processes in Water Resources Engineering, Water Resources Publications, Fort Collins, Colorado.

Vicens, G.J., I. Rodriquez-Iturbe, and J.C. Starcke,Jr., 1975. 'A Bayesian Framework for the Use of Regional Information in Hydrology,' Water Resources Research, **8(1)**, pp. 33-41.

Wood, E.F., 1977. 'An Analysis of Flood Levee Reliability,' Water Resources Research, **13(3)**, pp. 665-671.

Wood, E.F. and I. Rodriquez-Iturbe, 1975a. 'Bayesian Inference and Decision Making for Extreme Hydrologic Events,' Water Resources Research, **11(4)**, pp. 533-542.

Wood, E.F. and I. Rodriquez-Iturbe, 1975b. 'A Bayesian Approach to Analyzing Uncertainty Among Flood Frequency Models,' Water Resources Research, **11(6)**, pp. 839-843.

FLOOD RISK EVALUATION OF URBAN RIVERS WITH STANDARD EQUI-RISK LINES

Takeharu Etoh
Dr. Eng., Prof.
Department of Civil Engineering, Kinki University
Kowakae, Higashi-Osaka, Osaka 577, JAPAN.

Akira Murota
Dr. Eng., Prof.
Department of Civil Engineering, Osaka University

Masanori Nakanishi
Department of Civil Engineering, Kinki University

ABSTRACT. The authors have presented the equi-risk line theory for storage facilities with the constant release rule. It is extended to be applicable to those with the following general storage-release relation: $q = az'^p$, where q is the release discharge, and z' the volume of stored water. The release rule is characterized by the value of p. The shape parameter of the equi-risk line is expressed in terms of p. A standard equi-risk line for urban flood control systems is proposed, which can be applicable in common within a basin to calculate the necessary and sufficient capacities of any kind of flood control facilities, regardless of their location.

1. INTRODUCTION

A flood control system is designed on the basis of risk, i.e., the chance of flooding in the basin, estimated for some assumed capacities of the facilities. If floods are controlled by only a drainage facility, such as a channel or drainage pump, the drainage capacity is determined to be equal to the peak discharge whose frequency corresponds to a tolerable risk level. In addition, when a storage facility, such as a dam, detention, or underground reservoir, is introduced to the design, the relation among three essential indices, i.e., the risk level, the drainage, and the storage capacities, should be determined and presented to the decision maker. In this case, since the chance of flooding is dependent on the shape of hydrographs as well as the peak, the design is based on a more complicated analysis. In view of the variety of the shape of hydrograph, the expression of the relation was analytically derived and named the equal-risk line [Murota and Etoh, 1984.]

V. P. Singh (ed.), Application of Frequency and Risk in Water Resources, 263–282.

A standard procedure is presented to apply the theory of the equi-risk line to the design of flood control facilities in urban rivers. Some secondary devices are always necessary to apply any theory in practice, even a simple and explicit theory. In the present paper, devices used in the standard procedure for the modifications are listed and summarized; we, however, handle the simplest flood control system, which consists of one drainage and one storage facilities for a simplicity in explanation. Examples of standard equi-risk line applications are also given to show their practical utility.

The conventional technique of evaluating the risk in flood control is shown in Figure 1. Assuming a specific rainfall pattern and various combinations of the capacities of the facilities, the relation between the capacities and the efficiency of the system is examined through simulations of runoff and flooding. This technique, however, has following deficiencies.

(1) There are innumerable patterns of hyetgraphs; flat, sharp, forward- or backward-leaning, etc. Thus, as long as the design hyetgraph is fixed at one pattern, the rationality of the assumption is always suspect.

(2) Simulations for many cases have to be done individually for each design for each basin to determine the relation. No proposal has been presented to integrate the vast experience and information necessary to express the relation in a general and explicit form.

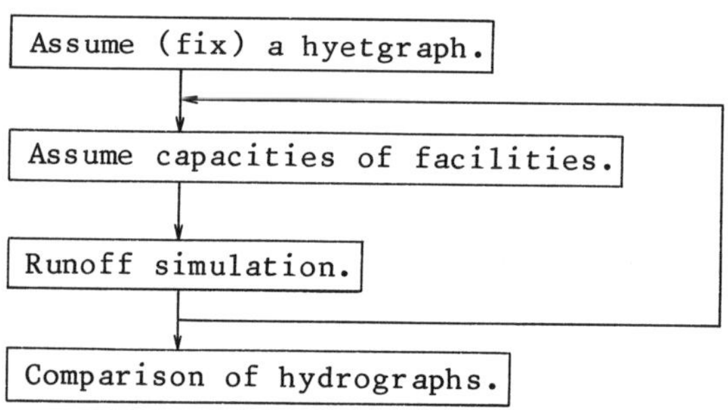

Figure 1. Conventional technique of risk evaluation for flood control systems with storage facilities

The synthesized probability hyetgraph shown in Figure 2(a) is widely employed as a fixed design hyetgraph. It is made so that the total amount or average intensity around the peak for any duration is equal to the depth or average intensity estimated from historical records of a rainfall sequence for a certain frequency or an average recurrence interval. Namely, the hatched area and the dashed line represent the T-year probability rainfall depth and average intensity for duration t_p, respectively.

It is reasonable to evaluate peak discharges based on the synthesized hyetgraph, since

(1) According to the rational formula:

$$q_p = 1/3.6\, f_p \bar{r} A,$$

the peak discharge q_p is proportional to the average rainfall intensity $\bar{r}$ during the concentration time t_p of the basin, where f_p and A are peak runoff ratio and the basin area,

(2) The recurrence interval of the average intensity $\bar{r}$ of the synthesized hyetgraph for any concentration time is equal to the given recurrence interval which was used to make the hyetgraph, and

(3) Thus, the recurrence interval of the peak discharge q_p calculated by runoff simulation with the hyetgraph is always expected to correspond to the assumed risk level, if there is no storage facility.

The authors have pointed out that it is not reasonable to use the hyetgraph to determine the capacity of a storage facility. To simplify the demonstration, we shall assume the similarity of the shape of a hyetgraph to that of the resulting hydrograph, and that the flood control system has one drainage and one storage facility. If the drainage capacity is y_0, the discharge exceeding y_0 has to be stored in the storage facility. The volume which is stored is denoted by z'. If $z' > z_0$, where z_0 is the storage capacity, the storage facility becomes full and fails to control the flood. Thus, to keep the frequency of flooding within the T-year probability level, the storage capacity z_0 should be larger than the T-year probability level of z'. There is, however, no theoretical background to assure that z' for the synthesized hyetgraph shown in Figure 2(b) is equal to the T-year probability volume of excess discharge for the drainage capacity y_0. Consequently, it is concluded that a new design procedure or concept should be developed for the design of flood control systems with storage facilities. For greater credibility we shall develop a new

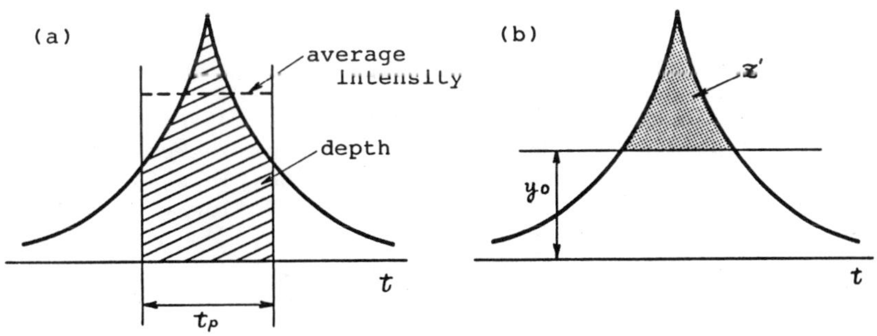

t_p : duration (concentration time)
y_0 : drainage capacity
z' : volume of stored water

Figure 2. T-year probability synthesized hyetgraph

procedure that admits a variety of hyetgraph patterns. This may make the process of mathematical manipulation much more complicated initially, but we shall finally express the relation between the risk level, the storage, and the drainage capacities in a simple and general form in view of the probability of the occurrence of various hyetgraph patterns. Along this line, the authors mathematically derived the expression of the relation, the equation of the equi-risk line, and presented its practical approximate expression [Murota and Etoh, 1984].

Some difficulties were found in the application of the equi-risk line theory to the practical design of the flood control system of an urban river in Osaka. Modifications were devised to reduce the difficulties. The revised procedure to apply the equi-risk line theory is presented here.

The major difficulties with the devices for them are briefly explained in the following articles, a - e.

a. Standard Equi-Risk Line

It may be troublesome to draw an equi-risk line for each flood control facility or for each target point to be protected from floods. Especially in urban river networks, there are generally a lot of facilities and a lot of protective targets in a relatively small basin. Thus, it is expected to present a standardized or regionalized equi-risk line for each basin, which can be converted to apply to the evaluation of the risk level of any protective target, and to determine the capacity of any flood control facilities.

The equi-risk line based on hourly rainfall sequence may be one of the most promising alternatives to the standard equi-risk line. Some devices are presented to draw the standard equi-risk lines using hourly rainfall sequences. Examples of practical applications are also presented, which prove the high utility of the concept of the standard equi-risk line. Hereafter in this paper, "equi-risk line" is used in reference to hourly rainfalls.

b. Threshold Rainfall Intensity

As sketched in Figure 3, for smaller drainage capacity y_0, an equi-risk line depicted directly from historical records increases much more steeply than its theoretically predicted line. This is because relatively weak rainfalls which precede and follow a cluster of severe rainfalls must be stored in a storage facility for a smaller drainage capacity, while, in theory, a triangular hyetgraph is assumed, neglecting preceding and following weak rainfalls.

To eliminate the deviation shown in Figure 3, it is practical to neglect weak rainfalls below a threshold, which should be assumed small enough so that the effect of this threshold is negligible on the design of the flood control system.

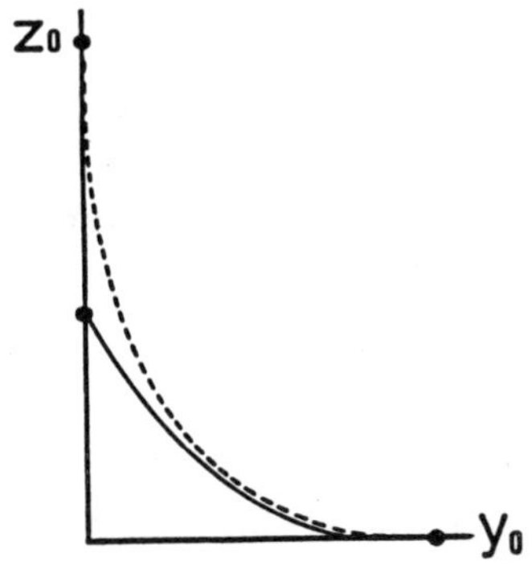

Figure 3. Equi-risk lines depicted on the basis of historical records (broken line) and theoretical prediction (solid line)

c. The Rainy Spell

The intersection of an equi-risk line and the vertical axis, i.e., the z_0-axis, represents the T-year probability total amount (or depth) of rainfall of a single storm [Murota and Etoh, 1984]. However, since it is difficult to strictly define a single rainstorm, the rainy spell is substituted for it in practice. The rainy spell is defined as a rainy period with the preceding and the following n consecutive unit intervals without rain, and may contain relatively short periods of $1 \sim (n-1)$ consecutive unit intervals without rain. The arbitrary number should be decided in view of the concentration time, the level of the threshold defined in b, above, and so forth.

d. Frequency Distribution of Depth

The frequency distribution of the annual maximum of the total depth of a single storm is reasonably represented by the SQRT-ET-max distribution [Etoh, Murota, and Nakanishi, 1986], which implies the occurrence of much larger total rainfalls of single storms than predicted by conventional distribution functions. Thus, the SQRT-ET-max distribution is introduced to evaluate the frequency distributions of the volume z' of rainfall more than y_0, as well as the total volume z .

e. Hourly vs 60-minute Rainfall Maxima

Generally, the 60-minute rainfall maxima are larger than the hourly rainfall maxima. However, much longer historical records of hourly precipitation are available than those of 10-minute precipitation, by which the 60-minute rainfall maxima are usually calculated. Thus, the T-year probability hourly rainfall is estimated from historical hourly records, and converted to the T-year probability 60-minute rainfall, which is used for the probability peak rainfall intensity.

The presented procedure is applied mainly to hourly precipitation records in Osaka during June and October from 1900 to 1983.

2. BRIEF DESCRIPTIONS OF THEORY AND PROCEDURE

2.1. Expression of an Equi-Risk Line

The equi-risk line is defined as the curve which represents the relation between the drainage and the storage facilities, y_0 and z_0, when the risk is fixed at a certain level. It is approximated by the following expression,

$$\frac{z_0}{z_0^u} = \left(\frac{y_0^u - y_0}{y_0^u}\right)^S \qquad (1)$$

in terms of three parameters, y_0^u, z_0^u, and S, and two variables, y_0 and z_0. z_0^u and y_0^u respectively denote the probability total volume and the probability peak discharge of floods for a given risk level, and can be individually and thus easily estimated from the frequency distributions of the total values and the peaks, without any information about their joint distribution.

The shape of an equi-risk line is represented by S. The value of S is dependent on the following two characteristics:

(1) The hydraulic characteristics of the storage facility, which are integratively represented by the exponent p in the storage-release relation,

$$q = a z'^p \qquad (2)$$

where q and z' are the release discharge and the volume of stored water, and a is the storage constant, and,

(2) The probabilistic characteristics of the shape of the hydrographs, which are represented by the joint distribution of the characteristic variables as the peak, the duration, the volume, etc.

The joint distribution of the characteristic variables of discharge is characterized by that of rainfall, and the runoff characteristics of the basin. Thus, when the point where the risk is evaluated is specified in a basin, the joint distribution is specified and the value of S is expressed only in terms of p as follows:

$$S = S_\infty + (S_0 - S_\infty)\, e^{-\sqrt{p}} \qquad (3)$$

where S_0 and S_∞ are the values of S for $p = 0$ and $p = \infty$, i.e., the constant release and no spill-out rules, respectively.

The following values have been proposed so far by the authors [Etoh and Murota, 1985]:

(1) When the peak and the duration of the hydrographs are independent,

$$S_0 = 3, \quad S_\infty = 0.7 \qquad (4).$$

(2) When they are proportional, i.e., linearly and completely dependent,

$$S_0 = 2, \quad S_\infty = 0.4 \qquad (5).$$

Generally, S_0 and S_∞ take intermediate values between those of Eqs. (4) and (5).

2.2. Standard Equi-Risk Lines

The concentration time of floods in urban rivers ranges mostly from 20 minutes to three hours, and averages around one hour. The runoff ratio is roughly constant and approximately equal to the ratio of impervious area throughout the duration of a flood. Therefore, the shape of a hyetgraph may be similar to that of the resulting hydrograph, when they are observed hourly. Thus, the shape parameter S may take the same value, regardless of the points within a basin, as estimated using the probabilistic characteristics of the shape of the hyetgraph. This is the basis of the proposal of the standard equi-risk line.

Furthermore, the value of S may be specified, since the correlation coefficient between the peak and the duration of hyetgraphs has been proved to be low, i.e., 0 - 0.5 [Etoh and Murota, 1984]. Namely, if we neglect the correlation, we can employ Eq. (4) to determine the value of S of the standard equi-risk lines. Hereafter, we shall assume $S_0 = 3$ and $S_\infty = 0.7$.

In the expression of the standard equi-risk line, the drainage and the storage capacities, y_0 and z_0, and their upper bounds, y_0^u and z_0^u, are all expressed in units of rainfall. Let the subscript r designate the capacities expressed in the units of rainfall as in y_{or} (mm/hr) and z_{or} (mm). They are converted to the capacities expressed in the original discharge units, y_0 (m^3/sec) and z_0 (m^3) by the following simple equations,

$$y_0 = C_y y_{or}, \quad C_y = \frac{1}{3.6} f_p A i_s \tag{6}$$

$$z_0 = C_z z_{or}, \quad C_z = 1000 f A \tag{7}$$

where Eq. (6) is the rational formula, normally used for estimating the peak discharge, and Eq. (7) converts depths into volumes by multiplying the depths by the runoff ratio f, the basin area A, and the unit-coversion coefficient 1000. Both the peak runoff coefficient f_p and the runoff ratio f are approximately equal to the ratio of the impervious area. The coefficient i_s is for the compensation of hourly precipitation based on the DD (depth-duration) relation. Namely, since the peak discharge is proportional to the average rainfall intensity during the concentration time t_p and the average intensity is larger for smaller duration (the DD relation), if t_p is smaller than one hour, the discharge should be larger than that calculated from hourly rainfall. Then, i_s takes a value larger than unity. Similarly, $i_s < 1$ for $t_p > 1$. As shown later, i_s is equivalent to the standardized hourly rainfall intensity.

y_{or}^u and z_{or}^u are also convertible to y_0 and z_0 by means of Eqs. (6) and (7).

All we have to do beforehand to draw the standard equi-risk lines (the equi-risk lines for hourly rainfall) is to determine the individual frequency distributions of the peak intensity and the total depth of rainfall of single storms. Then, y^u_{or} and z^u_{or} for a given average recurrence interval are estimated separately from the distributions, where the recurrence interval represents the risk level. The expression of the standard equi-risk line is the same as Eq. (1), i.e.,

$$\frac{z_{or}}{z^u_{or}} = \left(\frac{y^u_{or} - y_{or}}{y^u_{or}}\right)^S \qquad (8).$$

When the value of p is estimated from the hydraulic characteristics of the storage facility, the value of S is determined by substituting the value of S_0 and S_∞ in Eq. (4) and p to (3). Then, y_{or} and z_{or} are determined from Eq. (8), and are converted to the drainage and the storage capacities, y_0 and z_0, by means of Eqs. (6) and (7).

The concept of the standard equi-risk line is summarized in Figure 4. It is assumed that among the various characteristics of a rainfall sequence, the peak intensity, the total amount, and the DD relation integrate those which are essential to the designs of flood control systems. Similarly, it is recognized that the runoff characteristics are integrated in the basin area, the runoff ratio, the peak runoff coefficient and the concentration time, and the hydraulic characteristics of the storage facility in exponent p of the SR (Storage-Release) relation. Consequently, the standard equi-risk line can be regarded as an integrated expression of the relation among the risk level and the storage and drainage capacities, considering all conditions vital to flood control planning.

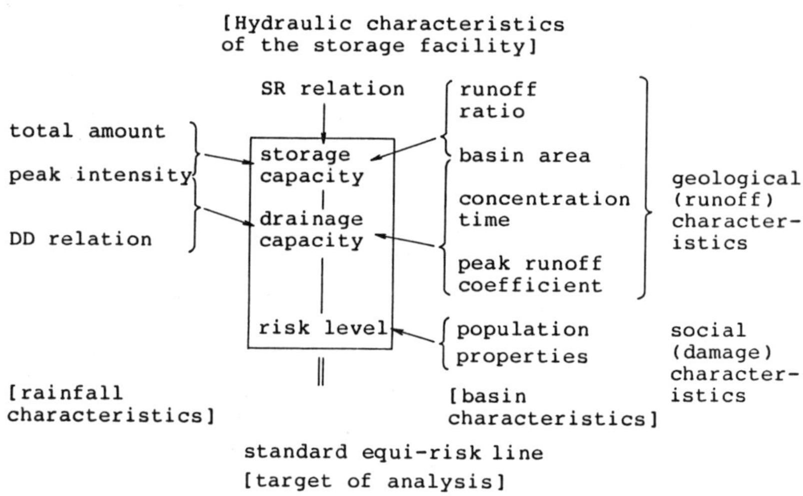

Figure 4. Conceptual explanation of standard equi-risk line

2.3. Frequency Analyses

The authors derived the SQRT-ET-max distribution for the distribution of the annual maxima of the total amounts of single rainstorms. It is as simple in expression as the Gumbel distribution, but tails much longer than the latter, which implies the occurrence of single storms with much larger total rainfall depth. Actually, the outlier problem in annual maximum daily rainfalls in Japan has been reasonably explained by the SQRT-ET-max distribution. On the other hand, distribution of the annual maximum peak intensities is well approximated by the Gumbel distribution. Thus, we shall employ the SQRT-ET-max and Gumbel distributions for the respective frequency analyses of the total depths, and the hourly peak intensities of single rainstorms.

2.4. A Practical Definition of Threshold Value

Weak rainfalls accompanied by a cluster of severe rainfalls are disregarded by subtracting a threshold value y_B from an original rainfall sequence and equalizing the resulting negative rainfalls to zero. y_B should be set at a small value, not to compromise the reliability of the designs in practice. To compensate for the effect, y_B is added to drainage capacity y_0 of the obtained standard equi-risk lines, or, schematically, the equi-risk lines are moved by y_B parallel to the y_{0r}-axis.

The authors recommend the following criterion to determine y_B: "y_B is determined to be the 5% level of the average intensity during the concentration time, in which the average intensity corresponds to the design risk level."

In urban rivers, the concentration time is around one hour, and the design risk level is about the 10-year recurrence interval. Thus, it is recommended to set y_B at the 5% value of the 10-year probability hourly rainfall. In Osaka, the value was estimated as 2.47mm/hr.

2.5. Definition of a Single Rainstorm

As previously described, the rainy spell is substituted for the single rainstorm. The chief problem is now to determine the number n, which represents the length of the dry periods preceding and following a rainy spell. The authors recommend the following guidelines:

(1) The hourly rainfalls greater than y_B are employed as effective ones.

(2) Then, rainy spells are collected for $n = 12$.

The value, $n = 12$, was determined by the following information.

(i) We shall denote the rainfall depth of a rainy spell for $n = 3 \sim 48$ by z_3, ..., z_n, ..., z_{48}. Then, it is shown by a calculation using historical records that the annual maxima of z_6 and z_{24} are approximately equal to those of z_{12}. Namely, the depths of rainy spells are not sensitive to the value of n, when n is about 12.

(ii) The average duration of the rainy spell is 10 to 20 hours in Osaka [Etoh and Nishimura, 1984]. Thus, it is reasonable to regard rainfalls after a $10 \sim 20$-hour no-precipitation period as a separate rainy spell. Twelve is within 10 ~ 20.

(iii) The annual maxima of z_{12} are approximately equal to those of 24-hour rainfall. In Japan, the total design rainfalls of most rivers are based on 24-hour rainfalls. Thus, z_{12} approximately corresponds to the basic design criterion of river planning in Japan.

(iv) In the designs of sewage systems in Japan, rainy spells for n = 3 are used for single rainstorms, but, in this definition, the threshold value y_B is not considered. The rainy spells are collected on this criteria, i.e. n = 3 and y_B = 0mm/hr from the hourly rainfall records of Osaka, and, then, those whose peaks are more than y_B (=2.47mm/hr) are selected. The number of selected rainy spells is 2142. On the other hand, the number of rainy spells collected by the definition, n = 12 and y_B = 2.47mm/hr, is 2144. The numbers agree closely. Thus, it may be concluded that rainy spells for n = 12 correspond to those used in sewage planning in Japan.

Generally, the locality should be taken into account when deciding the value of n.

2.6. Hourly vs 60-min Rainfall Maxima

The probability peak intensities estimated using hourly rainfall records are compensated by being multiplied by a coefficient C_1. This means the standard equi-risk lines obtained from hourly records are expanded at the rate of C_1 in the direction of y_{or}.

The average ratio of the largest to the fifth largest of the annual maxima of the hourly rainfall sequence to those of the 60-minute annual maxima in Osaka is shown in Table I, while the data available for this analysis is limited (1933 - 1967). As C_1 = 1.138, the 60-minute maxima are about 14% larger than the hourly maxima. This value should be re-examined considering the locality.

TABLE I Ratio of 60- to 360-min rainfall to 1- to 6-hour rainfall

	Ratio
60 min/1 hr	1.138
120 min/2 hr	1.031
180 min/3 hr	1.011
360 min/6 hr	1.000

In the table, the ratio of 2 ~ 6-hour rainfall maxima to 120 ~ 360-minute maxima are also listed. For longer duration, it approaches unity.

For the compensations to the introduction of a threshold y_B and the utilization of hourly rainfall records, the following expressions can be used.

$$y_{or}^{u'} = c_1 (y_{or}^{u} + y_B) \tag{9}$$

$$z_{or}^{u'} = z_{or}^{u} \left(\frac{y_{or}^{u} + y_B}{y_{or}^{u}} \right)^3 \tag{10}$$

where y_{or}^{u} and z_{or}^{u} are the probability peak rainfall intensity and the probability total depth for a given risk level, estimated from hourly rainfall records from which y_B is subtracted. $y_{or}^{u'}$ and $z_{or}^{u'}$ are the compensated pair, and, hereafter, are expressed by y_{or}^{u} and z_{or}^{u} for simplicity in expressions, unless confusion is anticipated.

2.7. Summarized Procedure

The modifications explained above are summarized in the following step-by-step procedure.

(1) Make an annual maximum sequence of hourly rainfall.
(2) Estimate the 10-year probability hourly rainfall by applying the Gumbel distribution to the annual maximum hourly rainfall records, and then, set y_B at the 5% level of the 10-year rainfall.
(3) Subtract y_B from all hourly rainfall records and set the resultant negatives at zero.
(4) From the records made by subtracting y_B, pick up the annual maxima and apply the Gumbel distribution to them to estimate the probability peak rainfall intensities for some recurrence intervals. The probability peak intensities are equal to y_{or}^{u} for the given risk levels.
(5) Rainy spells are set apart by n consecutive hours of no rainfall, where n = 12 for Japan; and pick up the annual maxima of the depths of the rainy spells to estimate z_{or}^{u} for the given recurrence intervals by the application of the SQRT-ET-max distribution.
(6) Compensate y_{or}^{u} and z_{or}^{u} with Eqs. (9) and (10).
(7) Estimate p based on the hydraulic characteristics of the storage facility, and determine the value of S with Eqs. (3) and (4).
(8) Draw the standard equi-risk lines with Eq. (8).

It is recommended to draw standard equi-risk lines for each basin beforehand for specific values of p as 0, 0.5, 1, ...

3. EXAMINATION OF THE ASSUMPTIONS

3.1. Aims

The validity of the standard equi-risk lines should be verified by examining all assumptions involved in the concept and the procedure. In addition, it should be reconfirmed by other engineers and researchers. In this section, the assumptions which should be examined are listed, and the guidelines for verification are presented. Thus, this section is irrelevant to the utilization of standard equi-risk lines in practice, and those who are interested only in the applications can skip this section. The expression of the standard equi-risk line, i.e., Eq. (8), has three parameters. Among them, y_{or}^u and z_{or}^u are based on the statistical evaluations of the peak and the depth of a single rainstorm. Thus, the validity of the methods of the statistical estimations, and the substitution of the rainy spell for rainfall of a single storm should be examined at first. Both of them have been verified, i.e., the former in a related paper [Etoh, Murota, and Nakanishi, 1986] and the latter in the previous section. Then, the form of the expression of Eq. (8), itself, and the value of S which characterizes the shape should be checked.

Some other assumptions should be verified relating to the conversion of y_{or} and z_{or} to y_o and z_o. Most of them, including the rational formula, the approximately constant runoff ratio and peak runoff coefficient, have been confirmed through world-wide hydrologic experiences. In this section, the assumption that the standardized rainfall intensivity i_S is approximately constant regardless of the recurrence interval is examined.

Consequently, the shape of the present expression of the standard equi-risk line is examined, and, additionally, the constant i_S is checked as below:

(1) Verification of the form of Eq. (8).
(2) Comparison of the value of S calculated from historical hourly rainfall records with the theoretical prediction from Eqs. (3) and (4).
(3) Check of the similarity between the standard equi-risk line for hourly rainfall and the equi-risk line of discharge in urban rivers, i.e., the comparison of their values of S.
(4) Check of the assumption of the constant i_S.

The first three concern the shape of the standard equi-risk line. The first two are examined at the same time. The method is explained briefly below.

To draw a standard equi-risk line, we have only to determine both ends, (y_{or}^u, 0) and (0, z_{or}^u) from historical records. The curve between the ends is simply drawn as a parabolic curve with the power of $0.7 \sim 3$. For the verification of the curve, we have to obtain from the records some points which should be plotted on the equi-risk line. The procedure exemplified here is for the constant-release rule, i.e., $p = 0$.

(1) Fix the risk level at a certain level.
(2) Fix y_{or} at a certain level.
(3) Calculate the total amount z' of rainfalls over y_{or} for all rainy spells.
(4) Estimate z_{or} for the assumed risk level by the statistical analysis of the annual maxima of z'. Then, one point is obtained, (y_{or}, z_{or}), which should be on the equi-risk line for the given risk level.
(5) Change the value of y_{or} and repeat processes (2) to (4) to obtain more points.
(6) Similarly, by changing the risk level, points on the equi-risk lines for several risk levels are obtained.

In the above procedure, it is unknown how z' distributes. Thus, the problem in the following section should also be investigated. The procedures to obtain the points for $p \neq 0$ are described elsewhere [Nakanishi, Etoh, and Murota, 1985].

Hereafter, we shall suppose that the basic records are made by subtracting y_B.

3.2. Some Notes on Statistical Analyses

Both the Gumbel and the SQRT-ET-max distributions were applied to the annual maxima of z' {(see Figure 2(b)} calculated from the records of Osaka. The SQRT-ET-max distribution showed a higher likelihood than the Gumbel, and thus is employed for the statistical analysis of z'.

When the coordinates of the points are calculated for a general SR relation, $p \neq 0$, rather than the constant-release rule, $p = 0$, the volume of stored water z' maximizes at the time of peak discharge q_p, as suggested by Eq. (2). This maximum volume is the necessary storage capacity. Thus, a peak discharge uniquely corresponds to the necessary storage capacity. Hitherto, it had been recommended that peak discharges and storage capacities be evaluated respectively by the Gumbel and the SQRT-ET-max distributions. However, when the peak corresponds uniquely to the storage, a criterion is necessary to choose one of these distributions. Intuitively, when storage constant a is small, the shape of the hyetgraph is not much transformed, and thus the distribution applied to the peak, i.e., the Gumbel distribution, may be appropriate. On the other hand, for larger storage coefficients, the SQRT-ET-max distribution may be suitably employed. The intuitive prediction was confirmed for the case of the linear reservoir, i.e., the case where $p = 1$ and the peak discharge was proportional to the necessary storage. The Gumbel distribution showed a higher likelihood for $a \leq 1$ (hour), and the SQRT-ET-max was superior for $a > 1$ [Nakanishi, Etoh, and Murota, 1985]. In Figure 5(b), explained later, the change of the applied distribution from the Gumbel to the SQRT-ET-max according to the value of a is due to this reasoning. Therefore, for general cases apply both of the distributions at first to either the peak discharge or the storage volume to find the critical value of a at which the likelihoods of the distributions intersect, and choose the distribution based on whether a is smaller or larger than the critical value. Values estimated with the Gumbel distribution should be multiplied by C_l.

When the assumed value of y_{or} in process (2) described above is large, no peaks exceed y_{or} and no sample of z' is obtained for some years. Namely, the sample size of the annual maxima of z' reduces, and modified statistical analyses considering the reduction of sample size have to be applied [Nakanishi, Etoh, and Murota, 1985].

When the obtained points (y_{or}, z_{or}) are plotted, the coordinates y_{or} and z_{or} should be compensated with Eqs. (9) and (10).

3.3. Shape of the Standard Equi-Risk Line

The results are shown in Figure 5, in which the compensation y_B is not applied to either the theoretical predictions or the points calculated from historical records, i.e., y_B is set to zero in Eqs. (9) and (10).

The actual curves and points should be moved parallel to the y_{or}-axis by y_B.

It is concluded from observation of these figures that the proposed standard equi-risk lines show good agreement, from a practical point of view, with their counterparts based on data.

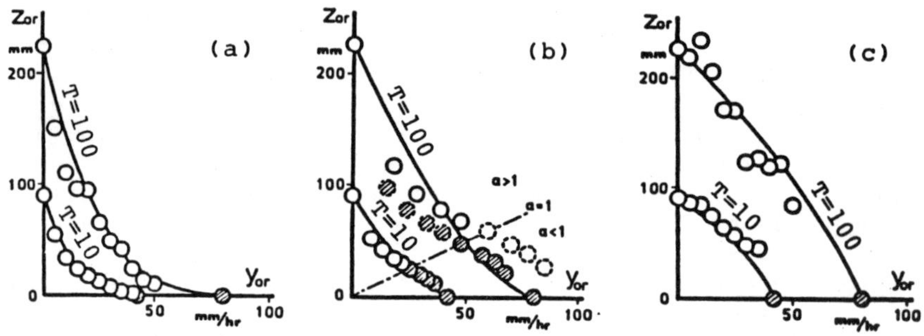

Rainfall less than y_B is set to 0.
T: average recurrence interval (years)
a: p = 0, S = 3 b: p = 1, S = 1.55 c: p = ∞, S = 0.7
o : Estimation by SQRT-ET-max distribution
● : Estimation by Gumbel distribution

Figure 5. Agreement of expression of standard equi-risk line {Eq. (8)} with the points calculated with historical records

3.4. Similarity between Hyetgraph- and Hydrograph-Based Equi-Risk Lines

To check the similarity, the value of S estimated from the points calculated from hyetgraphs is compared with that from hydrographs which are made by the runoff analysis, with the hyetgraphs as the input.

For the case where equi-risk lines based on discharge data were made for the design of a detention reservoir in Osaka, the value of S was estimated as 3.5 for the constant release rule, while the theoretical prediction was slightly larger than 3.0. As shown in Figure 5(a), points based on observed hyetgraphs are plotted slightly below the theoretically predicted curves, and curves for S_0 = 3.5 actually fit better to the points. Although only the one example above has been presented so far, the similarity seems to be assumable. Theoretically, S_0 is slightly larger than 3, but is fixed at 3 for safety.

3.5. Standardized Rainfall Intensity

The probability rainfall intensities in Osaka for some recurrence intervals are calculated for various durations. They are divided by the probability one-hour rainfall. The results are shown in Table II and Figure 6. The ratios are confined in relatively narrow ranges for the same duration, regardless of the recurrence interval, as shown in the figure and the table. A standardized rainfall intensity curve, the

TABLE II Relationship between probability rainfall intensity and duration

T / Duration	5	10	20	50	100	200
10 min	101.05 (2.2556)	119.84 (2.2918)	137.86 (2.3181)	161.20 (2.3437)	178.68 (2.3588)	196.10 (2.3715)
1 hr	44.80 (1.0000)	52.29 (1.0000)	59.47 (1.0000)	68.78 (1.0000)	75.75 (1.0000)	82.69 (1.0000)
2 hr	27.08 (0.6045)	32.30 (0.6177)	37.70 (0.6339)	45.25 (0.6579)	51.32 (0.6775)	57.71 (0.6979)
3 hr	21.38 (0.4772)	25.52 (0.4882)	29.80 (0.5011)	35.78 (0.5202)	40.59 (0.5358)	45.66 (0.5522)
6 hr	13.64 (0.3045)	16.07 (0.3073)	18.57 (0.3122)	22.05 (0.3206)	24.84 (0.3279)	27.77 (0.3358)
12 hr	8.60 (0.1920)	10.24 (0.1958)	11.94 (0.2008)	14.30 (0.2079)	16.20 0.2139)	18.21 (0.2202)
24 hr	5.41 (0.1208)	6.48 (0.1239)	7.58 (0.1275)	9.13 (0.1327)	10.38 (0.1370)	11.69 (0.1414)
48 hr	3.23 (0.0721)	3.90 (0.0746)	4.60 0.0773	5.58 (0.0811)	6.37 (0.0841)	7.21 (0.0872)

T: average recurrence interval

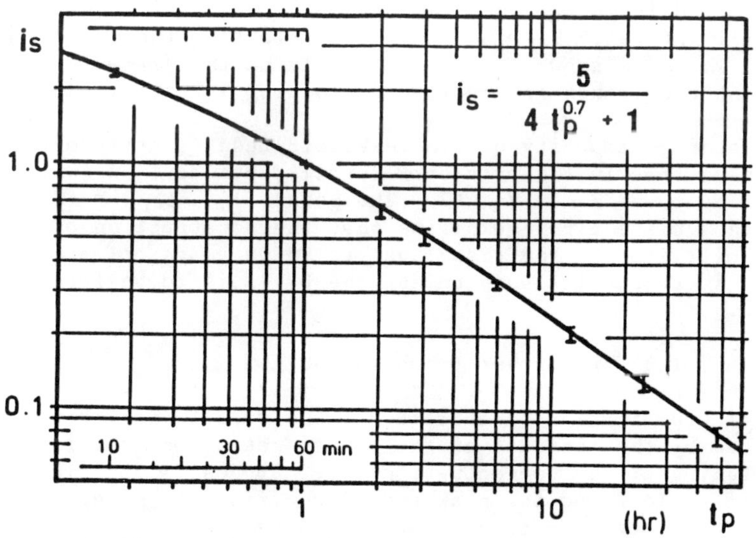

Figure 6. Standardized rainfall intensity

solid line in Figure 6, is estimated to pass through the ranges for all recurrence intervals. The curve for Osaka is expressed as follows:

$$i_S = 5 / (4t_p^{0.7} + 1) \quad (11)$$

where t_p is the duration. The approximate probability rainfall intensity for arbitrary duration is calculated by multiplying i_S by the probability one-hour rainfall.

It is noted that 10-minute and one-hour rainfall intensities are estimated by the Gumbel distribution, and 2- to 48-hour rainfall intensities by the SQRT-ET-max distribution.

4. APPLICATIONS

Examples of application of the standard equi-risk line in practical designs are given here.

Suppose a low-lying basin with the area of $A=10\text{km}^2$, and that a drainage pump with the capacity $y_0 = 50\text{m}^3/\text{sec}$ has been installed at the downstream end of the basin. A constant-release detention reservoir with the capacity $z_0 = 200{,}000\text{m}^3$ is planned for the same place for further urbanization of the basin.

The runoff characteristics are supposedly represented by the concentration time $t_p = 1.25$ hours, and both the runoff ratio and the peak runoff coefficient, $f = f_p = 0.5$.

4.1. Current Risk Level

Substituting $t_p = 1.25$ in Figure 6 or Eq. (11), $i_s = 0.88$. Substituting y_0, A, f_p, and i_s in Eq. (6), the drainage (pump) capacity in the unit of rainfall intensity $y_{0r} = 40.9$mm/hr. By plotting the point (y_{0r}, z_{0r})= (40.9, 0) which is shown by a solid circle on the y_{0r}-axis in Figure 7, the current recurrence interval T of flooding is estimated to be about 5 years.

The storage capacity of the reservoir under consideration is converted to that in the unit of rainfall depth as $z_{0r} = 40$mm by substituting f and z_0 in Eq. (7). The point $(y_{0r}, z_{0r}) = (40.9, 40)$ shown with a white circle explains that the construction of the storage facility will improve the risk level to T=200 years or more, and make the basin impractically safe unless the basin is to be further urbanized.

4.2. Counterplan for Further Urbanization

Effects of urbanization of a basin, or the runoff characteristics, are represented by the increase of the runoff ratio, or coefficient, and the shortening of the concentration time. Suppose a case in which only the runoff ratio f and the peak runoff coefficient f_p increase to 0.8, while the concentration time t_p stays constant.

Then the capacities in the units of rainfall decrease, i.e.,

y_{or} = 25.6mm/hr and z_{or} = 25.0mm from Eqs. (6) and (7). So does the recurrence interval of flooding, shown by the solid arrow in Figure 7, by T = 200 to 15 years. On the other hand, if only t_p shortens from 1.25 to 0.75 hours, i_s = 1.17, y_{or} = 30.8mm/hr and thus, T ≒ 50 years as shown by the dashed arrow. Actually, the shortening of t_p and the increase of f and f_p take place simultaneously. Then, T reduces to 7 years (the double circle), which seems quite low and thus supports the urgent construction of the storage facility.

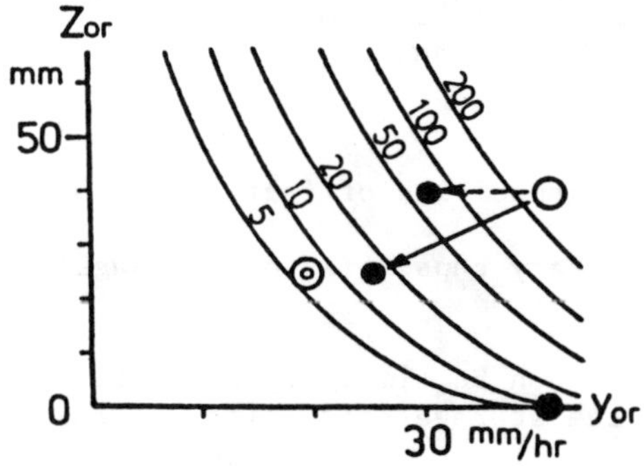

● present state
○ after underground reservoir has been constructed
Solid line: increase of runoff ratio f and peak runoff coefficient f_p
Broken line: reduction of concentration time
◎ both increase of f and f_p and reduction of t_p

Figure 7. Example application: Increase in flood risk caused by urbanization

4.3. Estimation of Necessary Capacities

The capacities of the existing facilities are plotted by the white circle in Figure 8, i.e., the present risk level is expressed by T=20. We shall raise the risk level to T=50 by the augmentation of the drainage and/or storage capacities. As described in Figure 8, if we increase only the drainage capacity, the capacity must be raised by 7mm/hr in the rainfall unit, and for only the storage capacity, it must be augmented by 22mm/hr.

When the runoff characteristics are represented by the same figures as those used in the above example, the additional capacities are estimated by y_o = 8.6m^3/sec and z_o = 110,000m^3 using Eqs. (6) and (7). For example, either a new drainage pump with the capacity 8.6m^3/sec or a new detention pond with the capacity 110,000m^3, should be installed to raise the risk level to T = 50.

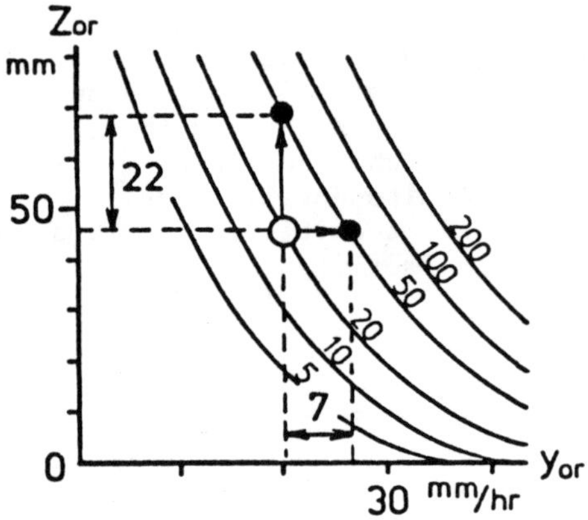

o present state
● drainage/storage capacities of facilities to achieve T=50

Figure 8. Evaluation of necessary capacities of drainage/storage facilities

The additional capacities can be similarly estimated for cases where both the drainage and the storage capacities are augmented.

4.4. Evaluation of the Benefit of Flood Controls

Another explanation is possible for Figure 8. Namely, for the present facilities, floods with the peak flooding discharge 7mm/hr, or the volume of inundating water 22mm/hr, both in the units of rainfall, are expected by the 50-year recurrence interval. Since the amount of flood damage is usually expressed in terms of the peak flooding discharge and/or the volume of inundating water, the projected damage can be evaluated by the summation of the product of the damage and the frequency.

4.5. A Detention Pond with $p = 0.5$

One example is shown to evaluate the risk level of a system with a storage facility whose release rule is not the constant release rule, i.e., $p \neq 0$. When a detention pond is surrounded by vertical walls and has a hole at the bottom of the front wall as the outlet, the release discharge from the hole is proportional to the 0.5th power of the volume of stored water, i.e., $p = 0.5$. From Eqs. (3) and (4), $S = 1.83$ for $p = 0.5$. The standard equi-risk lines are made for the value of using Eq. (8). When the drainage capacity y_0 of the downstream channel of the detention pond is given, the necessary capacity in the rainfall unit z_{or} can be estimated by converting y_0 to y_{or} and substituting it in Eq. (8) or a diagram similar to Figure 7 or 8 for $S = 1.83$. Then, z_{or} is converted to z_0. Substituting y_0, z_0 and p in Eq. (2), the value of

a is obtained, and the necessary size of the hole can be determined from the value of a.

4.6. Extended Applications

Some expansions of the theory and application of the equi-risk line have been presented elsewhere mainly in Japanese, including the application to the stage construction of facilities, the proposition of the hyetgraph equivalent to the equi-risk line, the expansion of the theory applicable to the system with confluence from another basin downstream of the storage facility, the application to the determination of economically optimum capacities with equi-cost lines, etc.

Further progress is expected in theory and applications.

5. CONCLUDING REMARKS

A standard procedure is presented to apply the theory of the equi-risk line to the design of flood control facilities in urban rivers. Some secondary devices are always necessary to apply a theory in practice, even a simple and explicit theory. The present paper lists and summarizes the devices used in the procedure. Examples of standard equi-risk line applications are also given to demonstrate their practical utility.

6. REFERENCES

6.1. Journals and Periodicals

Etoh, T., and Murota, A., May 1984, 'A Probabilistic Model of Rainfall of a Single Storm', Proc. of JSCE (Japanese Society of Civil Engineers), No. 345, Vol. II-1 pp.101-109 (in Japanese).

Etoh, T., and Murota, A., Feb. 1985, 'Equi-Risk Lines Applicable to a General Storage-Release Relation', Proc. Annual Conference of Hydraulic Committee, JSCE (in Japanese).

Etoh, T., Murota, A. and Nakanishi, M., 1986, 'SQRT-Exponentital Type Distribution of Maximum', Proc., Int. Sympo. on Flood Frequency and Risk Analyses

Etoh, T., and Nishimura, K., Feb. 1984, 'An MPP Model of Daily Precipitation Series', Proc. JSCE, No. 342, pp. 171-178 (in Japanese).

Murota, A., and Etoh, T., 1984, 'Application of the Equi-Risk Line Theory to the Design of a Detention Reservoir', Natural Disaster Science, Vol. 6, No. 1, pp. 17-30.

Nakanishi, M., Etoh, T., and Murota, A., Sept. 1985, 'Standard Equi-Risk Lines in Osaka', Research Reports of Science and Engineering of Kinki Univ., Vol. 21 (in Japanese).

THE EVALUATION OF FINANCIAL RISK

Max Beran
Institute of Hydrology
Wallingford, Oxon, UK.

ABSTRACT. Risk in hydrological design is often based on the return period of the design flood. This measures the average interval between events and hence presents only a partial picture. A method of assessing financial risk is presented which recognises the potentially greater impact of flooding which occurs near the start of a project's life or in proximity to a preceding flood. The fundamental concept is that of a flood-fighting fund. The fund starts with an initial capital which in time accrues interest. In years when flooding occurs the fund is depleted to pay for agricultural losses caused by the inundation. Attention focuses on the size of the capital sum which will reduce the risk of extinction of the fund to an acceptable level within some specified design horizon. The practical utility of the method will be illustrated using a recent UK case where floodplain storage was to be provided to alleviate downstream urban flooding. The financial risk concept was used to arrive at a fair level of compensation for occupants of the area to be flooded. Parallels are drawn with insurance risk and the further applicability of the technique to conventional flood protection schemes is discussed.

1. INTRODUCTION AND SUMMARY

Hydrologists normally measure flood risk as the probability of experiencing a given flood magnitude within some specified period. The dependence of risk (r) on design life (L) and return period (T) has been well researched and at its simplest is given by the formula:

$$r = 1 - (1-1/T)^L \tag{1}$$

By expanding the right hand side of equation 1 it can be shown that for large T there is a near proportionality between risk and the annual flood probability.

Simply to equate risk with the probability of a flood event understates the generality of the concept of risk; part of the skill of successful design practise is to identify relevant risk measures which accord with the perception of those who experience the flood hazard. This paper illustrates the construction of one such alternative risk measure which is tied to the idea of the financial loss due to flooding and which is tailored to the problem of flood plain occupancy.

It differs from the equation 1 risk in that it is sensitive to the placement of flood events in time, in particular to inundations near the start of project life. Another

V. P. Singh (ed.), Application of Frequency and Risk in Water Resources, 283–291.

difference is that it recognises the impact of flood events which occur close together. Also, as will appear later, it does not have a simple linear relationship with return period. For convenience the concept is termed 'financial flood risk' although it does not exhaust the possibilities of money-based measures of flood risk.

1.1. Washland Storage Schemes

Financial flood risk arises naturally in the context of a washland storage scheme although it is capable of rather wider application in quantifying the economic benefits of flood mitigation.

Purpose built flood detention reservoirs in headwater areas have not proved popular in Britain due to the shortage of suitable sites and the environmental impact of reservoir projects. Instead it is common to provide temporary flood storage on the floodplain. Inlet and control structures can be designed to increase the available storage beyond that naturally available and so attenuate flood peaks.

This solution has a number of advantages:

1. land is taken which is in any case subject to intermittent flooding;
2. the storage can be sited close to the problem reach thus removing many sources of operational uncertainty;
3. except when flooded the storage area is dry so can be used for agricultural or recreational purposes.

Disadvantages are that:

1. the reservoir is shallow so the land take can be high;
2. special inlet facilities are needed in order to deliver the water at a sufficient level to fill the reservoir;
3. likewise it may be necessary to pump water out of the reservoir after the flood;
4. for an offstream storage a considerable length of bund is normally required to surround the reservoir;
5. there are legal difficulties with a scheme which entails deliberate inundation.

Despite these disadvantages the controlled washland approach to flood mitigation has proved very popular in Britain, especially in the flatter eastern half of the country. A recent example is Lincoln on the River Witham. There had been a history of river improvement and bank building upstream of Lincoln which, while providing high security upstream, had aggravated the flood problem within the city. The floodplain is nowadays used for high grade agriculture; cereals, vegetables, seeds, and soft fruit. An economic evaluation of alternative solutions had proved that an 11 km^2 washland storage would be the optimal solution for reducing flood flows through Lincoln. The storage area was designed so that it would come into use with a 10 year return period and would be filled by a 100 year flood.

2. ECONOMIC CONSIDERATIONS

2.1. Basis for Compensation

Occupants of land within the planned washland storage were to be compensated by means of an 'easement to flood'; a sum paid at the outset of the scheme which gives the managing flood authority the right to flood. The question remains how was such an easement to be valued. Strictly the basis for compensation is the diminution of freehold value based either on market experience (not available in this case) or on what is theoretically equivalent, the value of future lost production.

The pre-feasibility studies had adopted a deliberately conservative value based on the purchase price of the land. The first serious attempt at arriving at an accommodation with floodplain occupants centred on the present value of future flood damage (PV). PV is approximately inversely proportional to the frequency of flooding, and hence increases proportionately with return period. On this basis there would be a ten to one ratio between the most and the least compensated land. This proved unacceptable to occupiers who rightly considered all parts of the storage to be to some extent blighted by the scheme, although accepting that compensation must bear some relationship to the frequency of inundation.

2.2 Flood Fund Concept

A fresh approach to the compensation problem was suggested by the nature of the problem and considered the financial risk accepted by the occupier in receiving any given level of compensation. In this approach the compensation can be regarded as the initial capital of a flood fighting fund. This fund is considered to be separate from the working capital of the farm and its sole function is to pay for losses due to inundation. Each year the fund increases with interest earned on the capital (calculated at a real rate) but is diminished in years when floods occur by the value of losses.

The performance of this fund through time is the basis of the financial risk concept described in this paper. It is related to an insurance company's fund into which premiums are paid and from which claims are withdrawn (Beard et al., 1983). It has been discussed in the water resources literature by Schaake and Fiering (1967) in connection with the US flood plain insurance programme. However assumptions and simplifications which apply in those cases and render the problem in part algebraically tractable do not apply here and a simulation solution had to be adopted.

3. DATA REQUIREMENTS

Basic information is needed on the dimensions of the scheme, including the area inundated by floods of any given return period, and the area under different crops and soil type. In addition hydrological and agricultural data are needed to operate the financial risk procedure. The hydrological data concern flood frequency and the agricultural data relate to the losses caused by inundation. The provision of these data is rendered more difficult by the need to consider their seasonal variation, both of the occurrence of flooding and of the consequences to agriculture.

3.1 Seasonal Flooding

Data were available at four gauging stations in the vicinity of the study area with a total record length of 58 years. Peak over threshold data were collected from these sites and the monthly probability was obtained from a simple frequency analysis. Table I shows the relative frequency of exceedences. More detailed analysis of contingent distributions revealed no tendency for the flood magnitude distribution to vary with season so the same marginal flood distribution was applied to each month.

TABLE I Probability of monthly inundation

Month	Monthly inundation probability
January	0.16
February	0.14
March	0.11
April	0.06
May	0.03
June	0.02
July	0.03
August	0.03
September	0.04
October	0.08
November	0.14
December	0.16

The washland area is divided into return periods zones. For example one zone spanned the interval 30 to 50 years and was considered to be affected by 40 year return period floods. The probability of inundation of this zone in any year is 0.025 and the monthly probabilities are obtained by scaling this by the values in Table I. Discretisation is not essential to the procedure but it provided convenient units for purposes of calculating the compensation in this particular application.

Probabilitics arc of thc annual maximum type but it is nevertheless necessary to allow for multiple floods in a year. The probability was obtained using the Langbein equivalence which has been shown empirically to provide an excellent approximation to UK flood exceedences (Beran and Nozdryn-Plotnicki, 1977). As will be described in the next section discrimination into different patterns of multiple flooding was not necessary. Such years were allowed for by treating them within the analysis as equivalent to a thirteenth month with low probability and high damage.

3.2 Agricultural Losses

The estimation of the agricultural losses due to flooding was undertaken by Lawrence Gould Consultants Ltd. In computing the impact month by month the following factors were considered:

a) effect of flooding on the topsoil, nutrients and subsoil including correction for long term damage due to waterlogging;
b) estimation of yield reduction by crop, soil unit and duration;
c) financial implications in the light of typical farm operations and crop treatment.

Examples follow of particular assumptions that were made. Any inundation causes total crop loss and a requirement for reseeding. This is recognised to be conservative. Flooding in June and July reduces the yield in the following winter by 10% and 20%. Yields were set at current rates but sensitivity tests were carried out to evaluate the impact of higher yields. Multiple flooding is always allowed to occur in such a way as to maximise losses; again a conservative assumption but one which simplifies the hydrological probabilities.

The losses were calculated as the following sum: potential value of crop (expressed as gross margin) + variable and overhead costs + difference in income between crop and substitute crop - income from substitute crop (if appropriate) + yield loss on subsequent crop + management time. Losses are heaviest following summer flooding and drop sharply after harvest and prior to land preparation. Table II shows the average losses weighted by crop and soil type for the thirteen cases considered.

TABLE II Summary of agricultural losses at 1984 prices

Month of flood	Average agricultural losses £/ha
January	698
February	698
March	760
April	790
May	811
June	804
July	931
August	415
September	394
October	686
November	692
December	697
Multiple flood	1810

4. FINANCIAL RISK ANALYSIS

As stated above the financial risk faced by flood plain occupants is viewed through the medium of a flood fund. The analysis of the fund through time makes it possible to assess the viability of any initial capital level. The study focussed on the following specific properties: the performance of the fund capitalised by the present value of future flooding; the risk of extinction during the early years; and the long term performance of the fund.

4.1 Computer Program

In most years the balance contained in the fund grows according to a compound interest rule

$$B(i+1) = B(i)\ (1+r/100)$$

where B is the balance after each year and r is the real rate of interest.

In years during which the washland storage is flooded the balance is reduced by the value of the agricultural loss. If a number of floods recur in near successive years then the fund may be extinguished. Conversely the fund may grow to a point when the annual interest exceeds the maximum loss even under worst conditions. After this state has been reached there is no further possibility that the fund can be extinguished.

Although the transition probabilities are fully determined there is no known derivation for the marginal distribution of the time series or the distribution of time to extinction. The departure from classical time series lies in the multiplicative element in the disturbance term and its dependence on the current state. Schaake and Fiering (1967) quote analytical results for the distribution of the balance contained in the fund under restrictive assumptions; viz Normal disturbances, and the fund contents large compared with the annual change. Some results can be obtained by enumeration of possibilities for short periods when the seasonal variation in losses can be ignored. However in the present case this approach is not practicable (except as a basis for non-random simulation) because of the prodigious variety of temporal patterns.

For this reason the time series was modelled through computer simulation. The program was written in BASIC on a personal microcomputer. Its inputs are the seasonal probabilities and losses of Table I and II, and the real interest rate and initial capitalisation. It was found that at least 1000 random simulations were necessary in order to obtain stable values for extinction probabilities. It is possible that a non-random procedure that mimics the enumeration approach would reduce the sampling requirement.

Use was made of the graphical capabilities of the microcomputer to produce summaries of the simulations such as shown on Figure 1. The top graph shows the current fund time series. The other graphs show histograms and cumulative distributions of the state of the fund after 10, 20, 30, 50, 100 and 200 years. In all cases the money units are multiples of the initial capitalisation. Fund expectation and standard deviation are printed out with the probabilities of remaining solvent and of a capital gain.

With the initial capitalisation of £894 (1.33 times the present value of future damages) adopted in Figure 1 and 5% real rate of interest the histograms reveal the tendency for the fund to migrate either towards extinction or to unlimited growth. The balance after 10 years shows some 7% of extinctions and an expected capital growth of 20%. The extinction probability increases with time to 22% after 200 years.

4.2 Program Results

For most studies the annual real rate of interest was taken to be 5%, the trial discount rate for public sector schemes. Particular attention was paid to the state of the fund after 10 years, this being regarded as the time horizon for an individual farmer's financial planning. Figure 2 shows the extinction probabilities for various levels of capitalisation for each of the return period zones. As expected the more frequent the flooding the larger the capitalisation to retain fund viability. However the ratio between the lowest and highest zones for a given level of risk is much less than the ratio of their return periods. This reflects the considerable risk of a flood, even a 100 year flood, within a ten year period. Therefore for fund viability all zones require capitalisation at least to cover the average loss.

After inspection of these results and sensitivity tests on yield and discount rate a 5% risk was chosen as the permitted level. Table III shows the necessary capitalisation as well as the extinction probability and expected profit after 30 years. This is a significant date as it is intended to review the performance of the scheme at that point.

Supplementary experiments were made to investigate the fund performance starting from the pre-feasibility and the PV initial capitalisations. The PV case displayed a variable pattern of extinction and profit probability for each zone and an extinction probability from less than 10% to over 20%. Such risk levels were rightly regarded as too high and also the variation was judged to be undesirable. Table III confirms that the capitalisation that

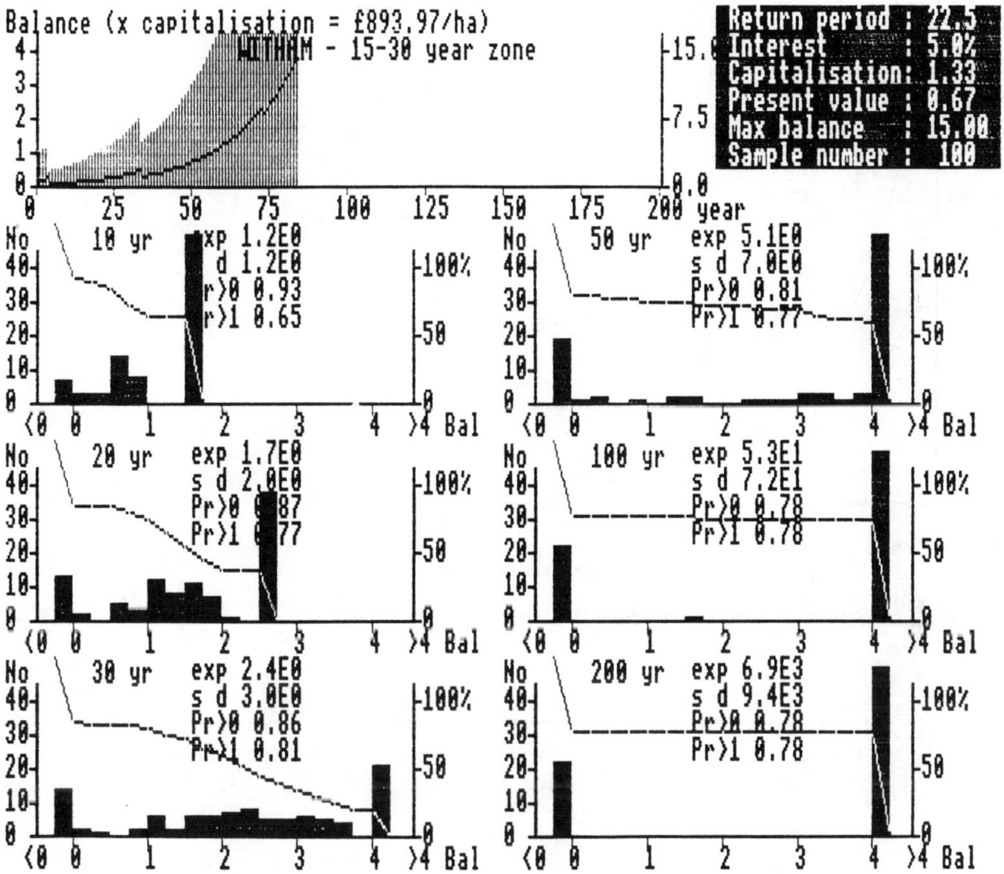

Figure 1 Graphical output from simulation program

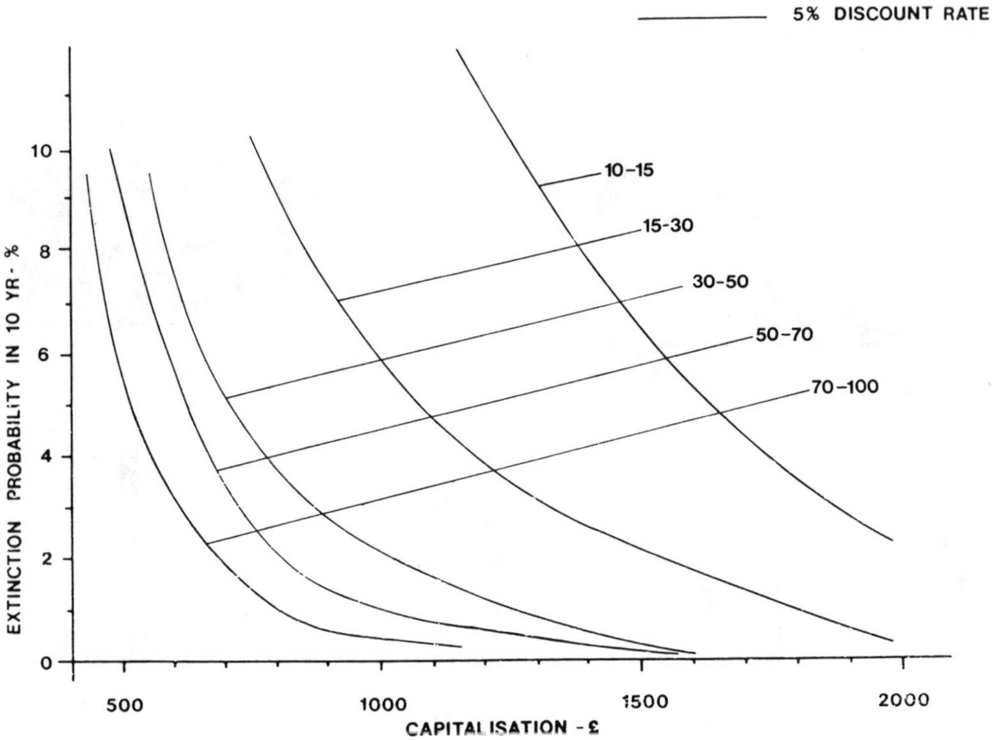

Figure 2 Extinction probability for given level of capitalisation

TABLE III Fund behaviour of 5% risk in ten years

Return period zone years	Present value £/ha	Capitalislation for 5% risk £/ha	Profit after 30 years Probability	Expectation £/ha
10 - 15	1255	1630	0.71	1600
15 - 30	664	1090	0.79	1600
30 - 50	363	700	0.76	1200
50 - 70	239	620	0.83	1300
70 - 100	168	510	0.85	1200

gives 5% risk exceeds the PV in all zones. The equality of risk and approximate 3:1 ratio in the compensation rate between the lowest and highest zones was regarded as much more acceptable by flood plain occupants as, unsurprisingly, were the higher levels of compensation.

As explained the performance of the fund after 30 years is also of interest. The capital gain, equivalent to an internal rate of return of around 3%, is nearly constant from zone to zone and gives an acceptable level of assurance that the 30 year review will not be adverse from the Water Authority's point of view.

5. CONCLUSIONS

The simulation analysis of the flood fund has allowed a number of useful conclusions to be reached. The most important was that risk could be related very directly to terms which mirrored the main concerns of those experiencing the hazard. The basis for compensation which had been proposed prior to this analysis had proved unacceptable despite the fact that it varied linearly with flood risk, the conventional measure used in hydrological design.

Sensitivity analyses were used to demonstrate the acceptability of the 5% risk criterion in the face of changed assumptions about discount rate and agricultural yield. The long term performance of the fund could be used to quantify the risk to the Authority of adverse results from the 30 year review.

Although the analysis has been presented in the rather special circumstances of a deliberate enhancement of the flood risk there is no reason why the financial risk criterion should not be applied in the more conventional circumstances of a river improvement scheme for flood mitigation. In such a case the flood risk is reduced but at a price. This price could be used as an initial capitalisation for a flood fund and the implied financial risk evaluated for the proposed project. This is equivalent to setting up the fund as a shadow alternative and using it to quantify the true value of flood protection to affected individuals.

6. ACKNOWLEDGEMENTS

Thanks are due to Mr David Brown of Lawrence Gould Ltd whose staff was responsible for the evaluations of seasonal agricultural losses. The work described was performed for the Lincoln River Division of Anglian Water Authority and part funded by the Ministry of Agriculture Fisheries and Food.

7. REFERENCES

Beard, R.E., Pentikainen, T. and Pesonen, E. 1983. Risk Theory. Chapman and Hall, Monographs in Statistics and Applied Probability. London.

Beran, M.A. and Nozdryn-Plotnicki, M.J. 1977. 'The estimation of low return period floods.' Hydrological Sciences Bulletin, **22**, (3), 275-282.

Schaake, J.C. and Fiering, M.B. 1967. 'Simulation of a national flood insurance fund.' Water Resources Research, **3**, (4), 913-929.

UNCERTAINTIES IN FLOOD PLAIN MAPPING

B.W. Oegema and E.A. McBean
University of Waterloo
Department of Civil Engineering
Waterloo, Ontario, N2L 3G1, Canada

ABSTRACT. Two of the sources of uncertainty associated with the delineation of floodplains, those of estimation of design discharge and channel capacity, are examined via the joint probability distribution. The uncertainty of the channel capacity is determined using first-order uncertainty analysis of Manning's equation. The variations of the flow area, wetted perimeter, and the friction slope, are expressed in terms of the variances of vertical, lateral, and longitudinal measurements. The variation of Manning's n is characterized by a triangular distribution. The uncertainty information is used to derive confidence limits for the water surface elevation of a flood of a given return period for a channel. In a case study application to the Grand River, Ontario, the standard deviation due to the two sources of error was estimated as 0.50 meter.

INTRODUCTION

With hydraulic models such as HEC II, the generation of water surface or backwater curves have become commonplace and straightforward for the practicing engineer. However, the magnitude of the uncertainty of predicted flood levels found using these hydraulic models has received little attention. Several sources of uncertainty associated with backwater calculations exist. In determining the design discharge, uncertainty arises from the fact that the hydrologic model chosen to fit the data record may not be the true distribution for the hydrologic series. Also, since the record is only a sample of the hydrologic series, the parameters in the distribution model have uncertainty associated with them, the degree of uncertainty depending on the record length. The uncertainties of the hydrologic estimates of the design discharge have been recognized for some time (e.g. Gumbel (1958) Chow (1964)).

In more recent years, research has focussed on sources of error in the backwater computations. McBean and Perkins (1975a) investigated the numerical errors of solving the gradually-varied flow equation. Studies by Burges (1979), Tung and Mays (1980) and McBean et al (1984)

V. P. Singh (ed.), Application of Frequency and Risk in Water Resources, 293–303.

considered the effect of the uncertainties associated with the parameters in Manning's equation. Other concerns in using the backwater equations include the modelling errors which may be evoked in representing the channel flow with a one-dimensional flow equation and with using Manning's equation to calculate channel velocities, and the assumption of steady flow conditions where flood events may be better characterized as unsteady.

This paper addresses two of the sources of uncertainty associated with the delineation of floodplains, these being the uncertainty in both the estimate of the design discharge and the channel capacity. A case study application to the Grand River is included.

Hydrologic Uncertainty

The true probability distribution of any hydrologic series can never be known. However, several distributions seem to fit most series reasonably well; these include the type I extremal, the log-Pearson type III, and a subset of the log-Pearson, the lognormal distribution.

Following the notation of Chow (1964), most distributions are of the form:

$$\hat{x}_T = \bar{x} + ks \tag{1}$$

where $\hat{x}_T$ is the estimate of the event of recurrence interval T, $\bar{x}$ is the sample mean, k is the frequency factor, and s is the standard deviation. Since both $\bar{x}$ and s are sample estimates only, the value of $\hat{x}_T$ is also somewhat uncertain.

A general formula for the variance of the estimate of the T-year event, s_T^2, for the commonly-used distributions is

$$s_T^2 = \frac{S^2}{N} [1 + ak + b\ k^2] \tag{2}$$

where N is the sample size and 'a' and 'b' are fitting coefficients.

For the type I extremal distribution fitted by the method of moments a, b and k have the following values:

$$a = 1.14 \tag{3}$$

$$b = .6 + \frac{N}{2(N-1)} \tag{4}$$

$$k = .7797y - .45 \tag{5}$$

where y is the transformed variate given by

$$y = - \ell n\ (-\ell n\ (1 - \frac{1}{T})) \tag{6}$$

For the type 1 extremal distribution fitted by Gumbel's fitting method a, b and k are

$$a = 1.14\ (\frac{N-1}{N})^{1/2} \tag{7}$$

$$b = 1.1 - \frac{.6}{N} \tag{8}$$

$$k = \frac{y(T) - \bar{y}N}{S_y} \tag{9}$$

For the two-parameter lognormal distribution the values for a, b and k are

$$a = 0 \tag{10}$$

$$b = .5 \tag{11}$$

$$k = Z(T) \tag{12}$$

where Z(T) is the variate on the normal distribution when the area under the curve equals 1 - 1/T.

Further, Moran (1957) and Condie (1977) develop techniques for estimating the variance of estimates using the Pearson type III and log-Pearson type III distributions respectively, for 50 and 100 year return intervals.

Case Study Area

To demonstrate the methodology, the Grand River at Galt (Cambridge), Ontario was used as a test reach. Figure 1 shows the Grand River basin and the location of the study area. At Galt, the drainage area is 3520 km^2. The maximum daily flows from 1945 to 1983 were used in the analyses to follow.

Included in Table 1 for comparison, are discharge estimates obtained by alternative formulae. The two-parameter lognormal distribution is seen to be conservative in both its estimate of the 50 and 100 year events and their corresponding variances. The type 1 extremal distribution provides estimates of the 50 and 100 year events which are close to the estimates provided by the three-parameter models. Of the two fitting-methods for the type 1 extremal distribution, Lowery and Nash (1970) concluded that the method of moments is more efficient.

The type 1 extremal, moments method with the 100-year estimate of 1271 m^3/s with a standard deviation of 156 m^3/s will be used to calculate the floodline confidence intervals.

Hydraulic Uncertainty

Theoretical Considerations. Burges (1979) outlines a procedure for quantifying the uncertainty in placement of the floodplain fringe. The procedure involves applying first-order uncertainty analysis to Manning's equation. The resulting uncertainty equation is then solved iteratively to find the variance of the flow depth and, subsequently, the variance of the floodplain width (see Oegema (1985) for further details).

A major difficulty in Burges' approach is that he equates the uncertainty of the design discharge to the uncertainty of the channel capacity. While the design discharge determined from the hydrologic record is numerically equal to the channel design capacity, their variances are independent. The variance of the design discharge is deduced by examination of the hydrologic record whereas first-order uncertainty analysis of Manning's equation results in an expression for the variance of the channel capacity. Therefore, an alternative procedure is necessary.

To assess the hydraulic uncertainty of a delineated flood channel, first-order analysis will be applied to Manning's equation for a composite channel. The discharge for a composite channel with three subsections is given by

$$Q = \left[\sum_{i=1}^{3} k_i \right] s_f^{1/2} \qquad (13)$$

where s_f is the friction slope, and k_i is the conveyance of the ith flow subsection and m is the number of subsections employed (m=3 for ease of subsequent presentation). The values of k_i are given by

$$k_i = \frac{A_i^{5/3}}{n_i \; P_i^{2/3}} \qquad (14)$$

where A_i is the flow area, n_i the Manning's coefficient and P_i the wetted perimeter, all for the ith subsection.

The variance of the discharge, Q, from first-order analysis principles (e.g. see Benjamin and Cornell (1970)), becomes

$$\mathrm{Var}(Q) = \left[\frac{\partial g}{\partial s_f}\right]^2 \mathrm{Var}(s_f) + \sum_{i=1}^{3} \left[\left[\frac{\partial g}{\partial A_i}\right]^2 \mathrm{Var}(A_i) + \right.$$

$$\left. + \left[\frac{\partial g}{\partial n_i}\right]^2 \mathrm{Var}(n_i) + \left[\frac{\partial g}{\partial P_i}\right]^2 \mathrm{Var}(P_i)\right] \qquad (15)$$

assuming the correlations between s_f, A_i, n_i and P_i are negligible.

Using the coefficients of variation, V, to simplify the presentation of variance expressions (V = standard deviation/mean), Eqn. (15) may be written as (after considerable manipulating):

$$\mathrm{Var}(Q) = \frac{1}{4} Q^2 V_{s_f}^2 + s_f \sum_{i=1}^{3} k_i \left[\frac{25}{9} V_{A_i}^2 + V_{n_i}^2 + \frac{4}{9} V_{P_i}^2 \right] \qquad (16)$$

Further, using Eqn. (14) for the conveyance of a flow subsection, Eqn. (5) simplifies to

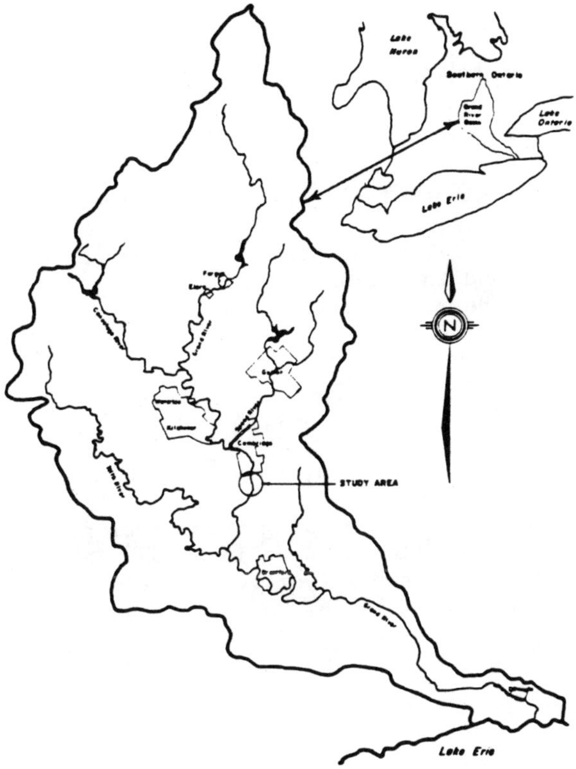

Figure 1 Grand River Basin, Ontario

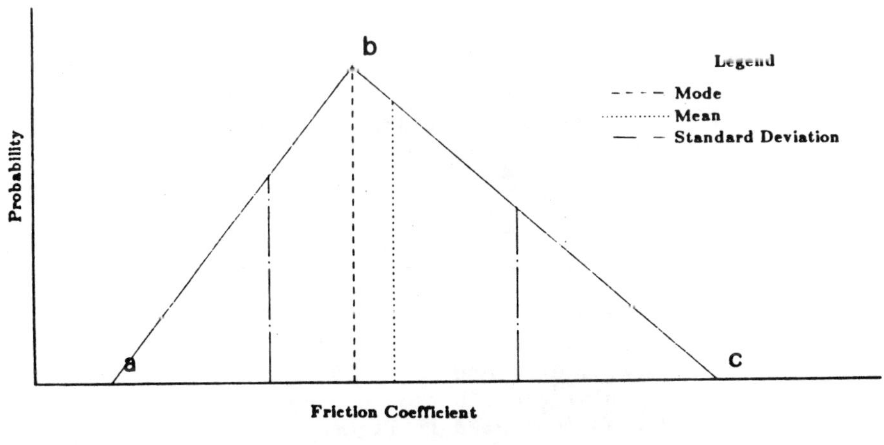

Figure 2 Triangular Distribution for Friction Coefficient

$$V_Q^2 = \frac{1}{4} V_{s_f}^2 + \sum_{i=1}^{3} \left[\left[\frac{k_i}{k_t} \right]^2 \left[\frac{25}{9} V_{A_i}^2 + V_{n_i}^2 + \frac{4}{9} V_{P_i}^2 \right] \right] \tag{17}$$

where k_t represents the total conveyance.

(i) friction slope coefficient of variation, V_{s_f} – the variation of the friction slope depends on the variation of the reach length, the water surface elevation, and the velocity head. If the variation in the velocity head is considered negligible, then V_{s_f} can be estimated as

$$V_{s_f}^2 \simeq V_{\Delta x}^2 + V_y^2 \tag{18}$$

(ii) flow area coefficient of variation, V_{A_i} – the value can be considered as having two components arising from vertical uncertainty and lateral uncertainty in the flow area delineation. Assuming that the coefficient of variation in the y (vertical dimension), V_y, is independent of the similar entity in the horizontal offset direction z, V_z, then V_{A_i} may be written as

$$V_{A_i}^2 = V_y^2 + V_z^2 \tag{19}$$

(iii) wetted perimeter coefficient of variation, (V_{P_i} – assuming the transverse slope of each line segment making up the perimeter is mild, allows the approximation

$$P_i = \sum_{j=1}^{J} \Delta z_j \tag{20}$$

where there are J line segments in the ith subsection. Thus,

$$V_{P_i}^2 \simeq V_z^2 \tag{21}$$

(iv) coefficient of variation of Manning's n, V_{n_i} – the friction coefficient is not measured directly; rather, it is calculated from known discharges, channel geometry and friction slope. To quantify uncertainty in Manning's n, Tung and Mays (1980) assumed the friction coefficient had a triangular distribution defined by the mode and the upper and lower bounds, as indicated in Figure 2.

The modal value is the value which is used to characterize the section. The bounds represent the reasonable limits that the friction coefficient may assume. The mean of the distribution is given by

$$\bar{n} = \frac{1}{3}(a + b + c) \tag{22}$$

The coefficient of variation of the distribution is given by

$$V_n = \left[\frac{1}{2} - \frac{1}{6\bar{n}^2}(ab + bc + ca)\right]^{1/2} \tag{23}$$

TABLE I MAXIMUM DAILY FLOW DISTRIBUTION

Distribution	Variance Method	X_{50} (cms)	S_{50} (cms)	X_{100} (cms)	S_{100} (cms)
type 1 extremal	Moments	1136	134	1271	156
type 1 extremal	Gumbel	1215	145	1365	170
2-P[1] lognormal	2-P lognormal	1322	203	1532	257
3-P lognormal		1117		1238	
Pearson III		1124		2343	
log Pearson III		1096		1170	

Application Considerations. Eight cross-sections were available for this research. The sections have natural banks and an undeveloped floodplain. To calculate the hydraulic uncertainty of a river reach requires estimates of the coefficients of variation of Manning's n, the reach length, vertical distances, and lateral distances.

The reach length variation of 5% translates to plus or minus 50 m. in a reach length of one thousand metres. This section of the Grand River is generally straight so that uncertainty of flow path length does not include the distance around meanders. The lateral and vertical distance variations used are as indicated in Table 2.

TABLE 2 COEFFICIENTS OF VARIATION FOR LENGTHS

Variable	Coefficient of Variation
Reach Length	.05
Overbank Vertical	.10
In-bank Vertical	.025
Overbank Lateral	.01
In-bank Lateral	.02

The coefficients for Manning's n values were calculated assuming the triangular distribution discussed above. The associated mean and coefficient of variation are as indicated in Table 3.

TABLE 3 MANNING'S n VALUES

Flow Section	Lower Bound (a)	Median (b)	Upper Bound (c)	Mean (m)	V
Overbanks	.055	.080	.110	.082	.138
In-Bank	.045	.055	.065	.055	.074

The values listed in Table 4 list the section-by-section breakdown of the discharge variations (note that the contributions are percentages of V_Q^2 and not percentages of V_Q). The values show that the discharge variation is most strongly dependent on the variation of Manning's n.

TABLE 4 BREAKDOWN OF DISCHARGE COEFFICIENT OF VARIATION

Section	Discharge V %	Discharge V	Contributions to V: Reach %	Vertical %	Lateral %	Manning's n %
1	6.98	.00487	12.8	8.0	0.7	78.5
2	6.87	.00472	13.2	9.8	2.2	74.8
3	7.58	.00575	10.9	7.9	0.8	80.5
4	7.04	.00496	12.6	9.9	1.1	76.3
5	7.15	.00511	12.2	6.1	1.4	80.2
6	7.11	.00506	12.4	7.7	1.6	78.4
7	7.19	.00517	12.2	8.9	3.1	75.9
8	7.58	.00575	10.9	6.7	2.6	79.8
Average	7.19		12.1	8.1	1.7	78.0

Combined Effect of Hydrologic and Hydraulic Uncertainties

Theoretical Considerations. To derive a methodology for placing confidence limits on the flood elevations, consider first the hypothetical case where the hydraulic uncertainty is zero. The discharges that would define the confidence elevations would then be functions of the design flow and the hydrologic uncertainty only. The discharge that defines the floodline for the lower 95% confidence limit is then

$$Q_L = \hat{Q}_T - 1.96\, S_{Q_T} \tag{24}$$

and the upper limit is

$$Q_R = \hat{Q}_T + 1.96\, S_{Q_T}$$

(assuming a normal distribution).

The analysis is extended to cases in which the hydraulic capacity uncertainty is not zero, as

$$\frac{\alpha}{2} = \int_{o}^{\infty} f_T(q) \left[1 - F_{Q_L}(q) \right] dq \tag{26}$$

$$1 - \frac{\alpha}{2} = \int_{o}^{\infty} f_T(q) \left[1 - F_{Q_u}(q) \right] dq \tag{27}$$

(Note that Equations (24) and (25) are specific realizations of (26) and (27)). The term f_T is the density function of the hydrologic event. The right-hand side of each equation is a joint probability of the occurrence of a specific hydrologic event and the probability of the flood channel being able to carry that discharge.

The lower limits of integration in Equations (26) and (27) should be negative infinity to be mathematically correct. The limits are shown as zero because discharge cannot be negative. This presents no problem in that the lower tail of the distribution is practically zero at a discharge of zero.

In essence, the problem becomes one of finding the discharges Q_L and Q_u which will satisfy the above equations. Q_L will be that discharge whose floodline incurs a 97.5% change of being exceeded by the floodline of the T-year event. Q_u will be that discharge whose floodline incurs a 2.5% change of being exceeded by the floodline of the T year event.

Application Considerations. To determine either Q_L or Q_u requires that four parameters be known, specifically, $\hat{Q}_T$, $S_{\hat{Q}}$, α, and the hydraulic uncertainty at Q_L or Q_u. Since the hydraulic uncertainty at a discharge is not known until the discharge itself is known, an iterative procedure must be followed.

Starting with a given α level, $\hat{Q}_T$, $S_{\hat{Q}}$ and the hydraulic variance at $\hat{Q}_T$, initial estimates of Q_L and Q_u are made. A bisection search pattern (see McBean and Perkins, 1975(b)) was used to find the discharge which satisfies the probability parameter.

Case Study - Combined

The limits for a 70% interval were examined; a Q_L of 1091 for the lower 70% limit and a Q_u of 1464 for the upper limit, with the hydraulic COV of 7%. The upper and lower confidence limit elevations for all eight sections are included in Table 5.

TABLE 5 CONFIDENCE LINE ELEVATIONS

Section	Elevation (m) Lower	Water Surface	Upper
1	263.94	264.45	264.96
2	263.95	264.46	264.97
3	264.05	264.55	265.08
4	264.15	264.65	265.15
5	264.30	264.77	265.25
6	264.56	265.02	265.49
7	264.75	265.18	265.64
8	264.73	265.18	265.65

Top widths for the confidence lines and water surface for the eight sections are tabulated in Table 6.

TABLE 6 CONFIDENCE LINE FOR WATER SURFACE TOP WIDTHS

Section	Top Width (m) Lower	Water Surface	Upper
1	530.35	533.96	537.56
2	521.90	532.09	543.68
3	536.45	544.47	552.99
4	530.43	536.45	542.45
5	327.78	344.33	431.37
6	427.77	457.98	463.52
7	387.71	439.01	450.05
8	227.84	277.82	364.66

CONCLUSIONS

The variance of the estimate of the T-year discharge has been used in the development of a methodology for quantifying the uncertainty of a flood level.

By itself, this technique can be used to analyze the relative contribution of each term in the variance expression. With this analysis, rational decisions can be made to most effectively reduce the hydraulic uncertainty with additional data collection.

The variance of the capacity estimate can also be combined with the variance of the design discharge estimate to develop confidence limits for the predicted water surface elevation.

The third use for this variance of the capacity estimate is in calculating the probability of flows exceeding the channel capacity. This risk may then be used for insurance purposes or for evaluating the flood risk of a site within the floodplain.

REFERENCES

Benjamin, J.R., and Cornell, C.A., Probability, Statistics and Decision for Civil Engineers, McGraw-Hill Book Co., New York, New York, 1970.

Burges, S.J., 'Analysis of Uncertainty in Flood Plain Mapping', Water Resources Bulletin, Vol. 15, No. 1, pp 227-243, February 1979.

Chow, V.T., Handbook of Applied Hydrology, Section 8, McGraw-Hill Book Co., New York, N.Y., 1964.

Condie, R., 'The Log Pearson Type 3 Distribution: the T-year Event and its Asymptotic Standard Error by Maximum Likelihood Theory', Water Resources Research, Vol. 13, pp 987-991, 1977.

Gumbel, E.J., Statistics of Extremes, Columbia University Press, New York, 1958.

Lowery, M.D., and Nash, J.E., 'A Comparison of Methods of Fitting the Double Exponential Distribution', Journal of Hydrology, Vol. 10, pp 291-298, 1970.

McBean, E.A., Penel, J., Siu, K-L, 'Uncertainty Analysis of a Delineated Floodplain', Canadian Journal of Civil Engineering, 11, pp 387-395, 1984.

McBean, E.A. and Perkins, F.E., 'Numerical Concerns in Water Profile Computations', ASCE - Journal of Hydraulics Division, Vol. 101, No. HY11, November, pp 1389-1403, 1975(a).

McBean, E.A., and Perkins, F.E., 'Convergence Schemes in Water Profile Computations', ASCE - Journal of Hydraulics Division, Vol. 101, No. HY10, October, pp 1380-1385, 1975(b).

Moran, P.A.P., 'The Statistical Treatment of Flood Flows', Transactions of the American Geophysical Union, Vol. 38, pp 519-523, 1957.

Oegema, B.W., 'Uncertainty Analysis for Delineated Floodplains', M.A.Sc. thesis, Dept. of Civil Engineering, University of Waterloo, Waterloo, Ontario, 1985.

Tung, Y., and Mays, L.W., 'Risk Analysis for Hydraulic Design', ASCE - Journal of the Hydraulics Division, Vol. 106, No. HY5, pp 893-913, May 1980.

PROJECT RISK CONSIDERING SAMPLNG UNCERTAINTIES AND A FINITE PROJECT OPERATION PERIOD

Upmanu Lall
Assistant Professor
Department of Civil Engineering
University of Utah
Salt Lake City UT 84112

ABSTRACT. The risk or probability of failure associated with a flood control project is a random variable, even if perfect information, rather than sample estimates of flood frequency are used for the design process. This paper presents procedures for the estimation of the expected value of this risk and its distribution where the flood control project has a finite operation period. Of interest are both the variability of the flood process during project operation and uncertainties inherent in the flood frequency estimation process. The traditional design procedure for flood control projects considers a finite project operation period, as far as the amortization of economic values is concerned. No analysis on the likely exceedance of the design magnitude during an actual run of events in a finite duration, is however considered. As far as the likelihoods of project failure are concerned, the project operation period is thus implicitly considered to be infinite. Design choices are likely to be quite different where the likelihood of project failure during a specified period (over a certain random set of future events) is considered. Where a finite operation period is considered and flood frequency relationships are estimated from annual maximum flow records, the probability of one or more exceedances of the design level is a function of the sample size for estimation and the length of the project operation period. General procedures in a Bayesian framework for the estimation of this probability are presented. Possible design implications are discussed.

1. INTRODUCTION

Uncertainties as to hydrologic, hydraulic, structural and economic aspects are associated with the design and analysis of flood control projects. These uncertainties stem from the inherent randomness of the processes studied, as well as from inadequate information on these processes. Risk due to hydrologic uncertainties (inherent and informational) related to (1) the inherent variability of the flood process during project operation, and (2) the estimation of flood frequency relationships, is primarily focused upon. For the purposes of this paper, the hydrologic risk (or risk) associated with a flood control project is defined as the probability that the design flow magnitude will be exceeded at least once in the design life (operation period) of the project. The objectives of this presentation are to (1) determine the hydrologic risk when the flood control project is designed on the basis of flood frequency estimates from historical data, and (2) examine its design implications. It is assumed that annual maximum flows (floods) are independent and are identically distributed. Expressions for the expected value of the hydrologic risk are developed in terms of the length of the operation period, the level of design (return period), and the moments of the probability distribution of the estimate of the design probability of exceedance. The probability distribution of this risk is also inferred for some cases. It is observed that the expected value of risk is lower when a flood frequency relationship is estimated from sample data, than when there is perfect information on the probability distribution governing flood flows. Some possible explanations for this phenomena are advanced. The possible use of expected risk in the design of flood control projects is then reviewed.

V. P. Singh (ed.), Application of Frequency and Risk in Water Resources, 305–318.

1.1 Flood Frequency Analysis

A T-year flood X_T, with a return period T, is defined (Viessman et al (1977)) as the annual maximum flow level that is equalled or exceeded, on the average, once in T years, in the long run. This definition implies that the annual probability of exceedance p, associated with a flow of magnitude X_T is given as :

$$p \quad = \quad 1/T \quad = \quad \int_{X_T}^{\infty} f(x)\, dx \qquad (1)$$

where x is a random variable representing annual maximum flow
f(x) is the probability distribution of x

Flood frequency analysis focuses on the estimation of a relationship between p (or T) and X_T, using a sample of historical flood flows. Traditionally, a parametric probability distribution is used to represent the annual maximum flow (flood) process, and its parameters are estimated from historical data using either the method of moments or the method of maximum likelihood. Recently, nonparametric procedures (Yakowitz (1985), Schuster and Yakowitz (1986), Adamowski (1985)) for flood frequency estimation have also been presented. In this paper, the flood frequency estimation problem is defined as the estimation of p', the probability of exceedance of a given flow X_T, from a sample of annual maximum flows of size m. The probability p' is estimated for each X_T of interest from equation (1), where f(x) is replaced by f'(x), the probability distribution function (p.d.f.) of x, as estimated from the sample. A desirable property of the estimate p' is that it equal p in expectation and have minimum variance. Lall (1981) reviews some unbiased estimators of p' for parametric models (Normal, Gamma, Exponential) of f(x). It is beyond the scope of this paper to review the merits of various flood frequency estimation schemes.

1.2 Hydrologic Uncertainty and Risk

Several investigators (Davis et al (1972); Vicens et al (1975); Wood and Rodriguez-Iturbe (1975a,b); Bodo and Unny (1976); Sidarovzsky et al (1976); Wood (1977); Castano et al (1978); Tung and Mays (1980); Duckstein and Bogardi (1981)) have focused on the analysis of hydrologic uncertainties. Most of their work has been in a Bayesian framework. The analyses reported on in the literature focus on inherent, parameter and model uncertainties. Inherent uncertainties relate to the random nature of the flood process in the design life. Model uncertainties relate to the lack of knowledge as to the form of the proper probability distribution function (e.g. Log Normal, Pearson III) for the flood process. The estimation of the parameters of the the p.d.f. of floods x, from finite samples of historical flood information leads to parameter uncertainties. The effects of sampling uncertainties with respect to the parameters of the probability model, have been incorporated into a risk assessment framework for selected probability models (Normal, Log Normal) by several of the above investigators. The reduction of parameter uncertainty by using regional information has also been the focus of some. In this paper model and parameter uncertainties are not explicitly considered. Instead, the work presented considers just one measure of informational uncertainty - the uncertainty associated with the estimate p' of the design probability of exceedance. It is presumed that the estimation of the flood frequency relationship either explicitly accounts for model (and parameter) uncertainties through a Bayesian estimation framework (Wood and Rodriguez-Iturbe (1975a,b); Vicens et al (1977), Lall (1981)), or that nonparametric estimators are employed for flood frequency estimation. The distribution of the statistic p' (corresponding to a design level X_T) then reflects the total informational uncertainty associated with that design level. The objective of this paper is to develop a framework that is independent of the procedures used to estimate p'.

In general, the reliability of a flood control structure may be defined (Tung and Mays (1980)), as the probability that the resistance y of the structure (i.e.the flood it is subjected to) exceeds the loading x on the structure. For a levee the resistance is its flow capacity, while for a reservoir it is the available storage at the time of the flood. The resistance y is thus also a random variable. Tung and Mays (1980), Lee and Mays (1983), and Tung (1985) discuss static and dynamic risk models. Risk is the complement of reliability, and may hence be defined as the probability that the loading x exceeds a design level $y = x^*$ at least once in the n years of design life. Static risk models consider a single loading over a unit time period, while dynamic risk models consider repeated applications of loading with resistance varying as a function of time over the project operation period. In the context of these definitions, the results presented in this paper consider a dynamic risk model with the assumption that resistance y is fixed.

2. RISK ESTIMATION

A flood control structure is usually designed to avert flooding from a particular flood or annual maximum flow x^*. Since the flow x^* corresponds to a probability of exceedance of flows, in the context of flood control design, it may be thought of as a design level. An estimate of the probability of exceedance p' of this flow is first made using historical data. A sampling distribution of the statistic p' may be developed corresponding to this estimate. Further, there is uncertainty as to the probability p" with which this flow will be exceeded during the design life of the project. If a number of similar projects (at the same design level) are considered, it is meaningful to think of p" as a random variable. This section develops estimators for the probability of one or more exceedances of the design flow in the design life of the project, considering the uncertainties in the estimation of p' as well as the uncertainty represented by p".

2.1. Perfect Knowledge

Even if flood control projects are designed on the basis of precisely known probabilities, they may be under or over designed relative to the actual run of events in their design life. The probability of exceedance of the design flow during project operation is thus a random variable, as is the risk that at least one exceedance of the design level will occur during the operation period. Consider that the probability distribution f(x) of floods is precisely known (i.e. not estimated from sample information), and that the interest lies in estimating the risk due to the inherent uncertainty in the flood process. Let a probability p_o correspond to a design level x^* through equation (1). Then the probability distribution of r exceedances of x^* in a n year design life, with an average probability of exceedance p_o is the Binomial $Bi(p_o,n)$.

$$\Pr(r \mid p_o,n) = \Gamma(n+1)/(\Gamma(r+1)\Gamma(n-r+1))\, p_o^{\,r}\, (1-p_o)^{n-r} \qquad (2)$$

where $\Gamma(.)$ is a Gamma function with argument (.)

The risk R_n (probability of 1 or more exceedances of x^*, in n years) is then

$$R_n = \sum_{r=1}^{n} \Gamma(n+1)/(\Gamma(r+1)\Gamma(n-r+1))\, p_o^{\,r}\, (1-p_o)^{n-r}$$
$$= 1-(1-p_o)^n \qquad (3)$$

The result above is well known.

The distribution of this risk is next evaluated. Define the random variable p" as the probability of exceedance of x^* in a number of sequences, each of length n. Then following Ang and Tang (1975), p" has the Beta distribution Be(a,b).

$$f(p'') = \{1/B(a,b)\}\, p''^{\,a-1}\, (1-p'')^{b-1} \qquad (4)$$

where B(a,b) is the Beta function with parameters a and b
Expected value of p" $= E[p''] = p_o = a/(a+b)$
Variance of p" $= V_n[p''] = p_o(1-p_o)/n = ab/\{(a+b)^2 (a+b+1)\}$

Then rewriting equation (3) as Risk R_n defined as a function g(p") of p", the distribution $f(R_n)$ can be derived using

$$f(R_n) \quad = \quad f(g^{-1}(R_n)) \;\; | \, dg^1(R_n)/dR_n \, | \tag{5}$$

where $g^{-1}(R_n)$ (the inverse function of g(p")) = $1 - (1-R_n)^{1/n}$

After the substitution of $g^{-1}(R_n)$ for p' in equation (4) and differentiating $g^{-1}(R_n)$ with respect to R_n, equation (5) yields the following result for the distribution of risk :

$$f(R_n) \quad = \quad \{1/(nB(a,b)\} \; (1-(1-R_n)^{1/n})^{a-1} \; (1-R_n)^{(b/n)-1} \tag{6}$$

The evaluation of the expected value of risk and its distribution is thus quite straightforward when only inherent uncertainties are considered, and perfect knowledge as to the flood distribution f(x) is presumed.

2.2. Risk Estimation - Sample Information

The deliberations of the previous subsection are now extended to include the consideration of informational uncertainties concerning p_o.Following the earlier discussion it is presumed that it is possible to construct an unbiased estimator p' of p_o from a historical sample of size m. The uncertainty associated with p' is then represented through a probability distribution function f(p'|f'(x),m). The probability of r exceedances of x* in a n year project operation period, where the probability of exceedance of x* is estimated from a sample of size m, using a fitted p.d.f. f'(x) for x, is then given as

$$Pr\,(r \mid p',n,f'(x),m) \quad = \quad \int_0^1 Pr\,(r \mid p',n)\; f(p'|f'(x),m)\; dp' \tag{7}$$

Procedures for the evaluation of the terms in equation (7) are developed first for a simple nonparametric estimator of p', and are then extended to the more general case.

(1) Nonparametric Estimation of f'(x)

Consider a nonparametric specification of f'(x) following the rule that a flow magnitude x*, that was equalled or exceeded i times in the m year sample, has an estimated probability of exceedance p', given as (i/m). This estimation scheme may be considered valid where m >> T, the return period corresponding to p'. Chow and Takase (1977) present derivations similar to the ones that follow, for the same situation. Their derivation considers that the ranked flow x*, with rank i, is exceeded i-1 times (rather than equalled or exceeded i times) in a sample of size m. By definition the probability of exceedance of a flow magnitude is defined as the probability of the flow being equalled or exceeded. This is reflected in the derivations that follow.

The kernel of the likelihood function of the sample can be written (Raiffa and Schlaifer (1968)) as $p'^{i}(1-p')^{m-i}$. The Beta distribution Be(i,m-i), the conjugate of the likelihood function, is then the appropriate choice for the distribution f(p'|f'(x),m).

$$f(p'|f'(x),m) \quad = \quad \{1/B(i,m-i)\}\; p'^{i-1}\; (1-p')^{m-i-1} \tag{8}$$

where B(i,m-i) is the Beta function with parameters i and m-i
Expected value of p' = E[p'] = p_o = i/m
Variance of p' = $V_m[p']$ = $i(m-i)/m^3$

All terms in equation (7) are now known and after appropriate substitutions Pr (r | p',n,f'(x),m) is evaluated as

$$Pr\,(r \mid p',n,f'(x),m) \quad = B(r+i,n+m-r-i)\Gamma(n+1)/(\Gamma(r+1)\Gamma(n-r+1)\,B(i,m-i)) \tag{9}$$

where B(.,.) is a Beta function with arguments (.,.)

The risk R is now evaluated as the probability of at least one exceedance (the complement of the probability of no exceedances, i.e. r=0), as

$$R_{mn} = 1 - B(i,n+m-i) / B(i,m-i) \tag{10}$$

The risk R_n is defined as a function g(p') of the probability p' by noting that n and m in equation (10) are fixed parameters for a particular situation and that the random variable i is expresses as mp'. Equation (10) may then be rewritten as

$$R_{mn} = g(p') = 1 - B(mp',n+m-mp') / B(mp',m-mp')$$

$$= 1 - \{\Gamma(m) / \Gamma(n+m)\}\{\Gamma(n+m-mp') / \Gamma(m-mp')\} \tag{11}$$

(2) General Case

In the general case consider that an estimate p' of the probability p_o associated with a design level x* is obtained from a sample of size m, using an appropriate estimation procedure. It is assumed that the estimation procedure provides p' as an unbiased estimator of p_o and also provides the variance of estimate $V_m(p')$. Such procedures are available for the Normal (Zacks and Milton (1971)), Gamma (Woo (1979)), Pearson III (Lall (1981), and the Exponential (Olkin (1978)), probability models. Where other procedures are used and analytical results ensuring $E[p'] = p_o$ (i.e. unbiasedness) and providing $V_m(p')$ are not available, the use of the Jacknife (Miller (1974)) in conjunction with the estimation procedure used could satisfy these requirements.

The conjugate of the sample likelihood function - the Beta distribution Be(i,m-i) - was selected as appropriate for the distribution f(p'|f'(x),m) in the previous subsection . The properties of the Beta distribution function make it an attractive choice (Bickel and Doksum (1976), Ang and Tang (1975)) for approximating the true distribution function f(p'|f'(x),m), where f'(x) is estimated on the basis of a procedure (model) θ. The distribution f(p'|f'(x),m) is then described as a Beta distribution Be(c,d), with the parameters c and d estimated by equating the first two moments of Be(c,d) to E[p'] and $V_m(p')$ (estimated from the sample of size m) respectively. Note that the expressions of the previous subsection are obtained by substituting i for c, and (m-i) for d, in the expresssions below.

$$f(p'|f'(x),m) = \{1/B(c,d)\}\, p'^{c-1} (1-p')^{d-1} \tag{12}$$

where B(c,d) is the Beta function with parameters c and d
$E[p'] = p_o = c/(c+d)$
$V(p') = V_m(p') = cd/\{(c+d)^2(c+d+1)\}$

The Beta distribution parameters c and d are expressed in terms of p_o and $V_m(p')$ as

$$c = (\alpha / V_m(p') - (1+\alpha)^2) / (1+\alpha)^3 \tag{13}$$

$$d = \alpha c \tag{14}$$

where $\alpha = (1-p_o)/p_o$

The probability Pr (r | p',n,f'(x),m) of r exceedances of x* in n years, where p' is estimated from a sample of size m and f'(x) is the estimated distribution f(x) is then derived using equations (2) , (7) and (12) as

$$Pr(r \mid p',n,f'(x),m) = B(r+c,n+d-r)\Gamma(n+1) / (\Gamma(r+1)\Gamma(n-r+1)\, B(c,d)) \tag{15}$$

The risk R_{mn} is then derived from equation (15) using r = 0 as

$$R_{mn} = 1 - B(c,n+d) / B(c,d)$$

$$= \quad 1 - (\Gamma(n+d)\,\Gamma(c+d)) / (\Gamma(n+c+d)\,\Gamma(d)) \qquad (16)$$

The evaluation of the distribution of risk $f(R_{mn})$ is more complex where p' is an estimate from a historical record, since as can be seen from equations (11) and (16) the inverse function $g^{-1}(R_{mn})$ is not easy to evaluate explicitly in terms of p'. If desired, this distribution could be evaluated numerically by applying equation (5), after the numerical computation of a relationship between p' and R_{mn}.

2.3. Examples

The risk R_n for a n year design life at a design level x*, with a probability of exceedance p_o is obtained using equation (3) when one has perfect knowledge of the distribution of flows f(x). Where the distribution f(x) is estimated from a sample of size m, the risk is given by equation (16) as R_{mn}. Note that the evaluation of risk R_{mn} using equation (10) and the non-parametric estimation of f(x) is a special case of equation (16). Where another procedure θ is used for the estimation of f(x) the risk R_{mn} computed using equation (16) should be closer to R_n than that computed using equation (10), if the estimation procedure θ is efficient and exploits the proper parametric family(ies) for f(x). This section offers a numerical comparison of the risk in three situations - (1) perfect knowledge of f(x), (2) f(x) is unknown, and is estimated by the non-parametric procedure of the previous section, and (3) x is known to be Normally distributed with unknown mean and variance, and an unbiased maximum likelihood estimator p' of p_o is used. The third situation was selected primarily for illustrative purposes, since analytical results for the unbiased estimation of p' and its variance $V_m(p')$ are readily available (Zacks and Milton (1971)) for this specification. This selection does not reflect any bias of the author towards the appropriateness of the Normal distribution as a model for the annual maximum flow process.

The results of the evaluation of risk for the three situations (1 - Perfect Knowledge of f(x); 2 - Nonparametric (Binomial) estimation; 3 - Normal model with unknown mean and variance) delineated above are summarized in Figures 1 through 4. For the first two situations, these results were obtained using the relevant equations presented earlier. Zacks and Milton's (1971) estimators for E[p'] and $V_m(p')$ were used (these are not reproduced in this paper because of their complexity) in the third situation with equations (12) through (14) and (16) for the third situation. Figures 1 and 2 present the relationships between risk R (R_n or R_{mn} as appropriate) and the design life n of the project, for probabilities of exceedance p_o of 0.1 and 0.05, respectively. The results for the second and third situations are parameterized by the sample size m (10 and 50). Figures 3 and 4 present relationships between the risk R and the sample size m, for probabilities of exceedance p_o of 0.1 and 0.05, respectively. The results for the second and third situations are parameterized by the project design life n (10 and 50). The following general conclusions may be derived from an inspection of these figures:

(1) For a fixed probability of exceedance p_o, the risk R increases as the project design life n increases.

(2) For fixed p_o and n, the risk R is <u>lower</u> for the case where p_o is estimated from a sample, than for the case where there is perfect knowledge as to f(x). This is a very interesting observation, since one intuitively expects the risk R to be higher when one does not have perfect information. This observation is consistent with the observations of Chow and Takase (1977) and Lee and Mays (1983). Chow and Takase's (1977) results exhibit the same behavior, but they do not offer any explanations. Lee and Mays (1983) use Chow and Takase's (1977) derivations (among others), and do note this behavior. They indicate that this is perhaps due to biases in the estimation of p_o, using the non-parametric scheme defined here as situation (2). They suggest that perhaps this scheme yields overestimates of the design flood x* corresponding to p_o, and that the 'true' value of p_o is perhaps lower than estimated. The values of risk yielded by Chow and Takase's (1977) results is actually even lower for the same situation, than those computed by equation (10) in this paper. While Lee and Mays (1983) explanation seems plausible, the observation that even when the model is presumed to be Normal and an unbiased estimator of p_o is used (i.e. situation (3) above), the same behavior is

observed, calls for further insights. A possible explanation is discussed at the end of this section.

(3) Compared with the risk from the non-parametric (Binomial) procedure and equation (10) the risk associated with the Normal distribution case is closer to the risk with perfect knowledge for all cases where p_o is estimated. This was expected and is also consistent with the observation that the variance of p' estimated in situation (3) is consistently less than the variance of p' estimated in situation (2). Where the estimators p' of p_o from both situations are presumed to be unbiased, the one with lower variance should indeed provide a better approximation to the 'true' risk relationship.

(4) For fixed p_o and n, the risk curves for situations 2 and 3, asymptotically approach the risk for situation 1, as the sample size m, increases. Once again this is related to the fact that the variance of p' decreases as m increases.

An explanation of conclusion (2) above is now presented on the basis of two observations on the nature of the estimates of p' and their associated distribution f(p'|f'(x),m).

The first observation is due to Stedinger (1983) who compares two flood frequency estimators for the Normal distribution. The first is an estimation procedure that yields an unbiased estimator of x* for a given probability of exceedance p_o. The second procedure provides an unbiased estimate p' of the probability p_o for a specified design value x*. Stedinger (1983) notes that while both procedures are asymptotically unbiased they yield biased estimators of the 'true' flood frequency relationship, where the sample size m is small. The first procedure leads to the selection of design magnitudes that have expected probabilities of exceedance in excess of the desired values. The second procedure yields design values that are unbiased as far as the probability of exceedance is concerned but are upwardly biased as far as the design magnitude is concerned. For a fixed probability of exceedance p_o the flow magnitude x' that provides an unbiased estimate p' asymptotically approaches the true flow x* from above as the sample size m increases. The second procedure considered by Stedinger (1983) is equivalent to the Zacks and Milton (1971) procedure presented earlier. The implication here is that when unbiased estimators of p_o are used the project is designed for increasingly higher floods as the sample size m decreases. This statement is true not just for the Normal distribution assumption of f(x), but also for any estimation procedure that incorporates a consideration for sampling uncertainties in estimating the flood frequency relationship.

The second observation is that the distributions f(p'|f'(x),m) are increasingly, positively skewed (for p' <0.5), as the sample size m decreases. Using the Beta distribution Be(c,d) as the model for f(p'|f'(x),m), as in equation (12), the skewness γ_p of f(p'|f'(x),m) can be estimated as (Ang and Tang (1975)) (The reader may wish to note that Abramowitz and Stegun (1972), page 930, also present an expression for γ_p, that is incorrect) :

$$\gamma_p \quad = \quad 2\,(c-d) \,/\, \{\, (c+d)\,(c+d+2)\,\sqrt{V_m(p')}\,\} \qquad (17)$$

Figures 5 through 8 illustrate this skewness through plots of the density function f(p'|f'(x),m). Table 1 provides values of c, d and γ_p for selected values of p_o, m and $V_m(p')$ for both estimation procedures (Binomial and Normal) for p' considered earlier. One notes that as the sample size m increases (or rather the variance $V_m(p')$ decreases) for fixed p_o, the skew of the distribution f(p'|f'(x),m) approaches 0, and the mass of the distribution is concentrated around the mean value of p'. Also for the same sample size m, the distributions of p' exhibit a higher skew as p_o decreases. The distributions f(p'|f'(x),m) are less skewed for comparable situations where the Normal model is used instead of the Binomial estimation procedure. This follows from the fact that the variance $V_m(p')$ is lower for the Normal case. The conclusion then is that when the variance $V_m(p')$ is higher (because of either a smaller sample size m, or an inefficient estimation procedure), the distribution f(p'|f'(x),m) has a higher positive skew, and hence has higher likelihoods associated with values of p' smaller than p_o. As the perfect knowledge scenario is approached, the distribution f(p'|f'(x),m) has skew and variance approaching zero. The likelihood of p' taking on an extreme value less than p_o is thus considerably smaller in this situation.

The two observations on the behavior of p' are clearly related. In order to obtain p' unbiased, an increased upward bias results in the corresponding design flow x*, reflected by the increased positive skew of the distribution of p', as the sample size m decreases or the variance $V_m(p')$ increases. Unbiased estimation of p' from a sample thus results in a distribution f(p'|f'(x),m) that has higher likelihoods for more extreme (smaller than p_o) probabilities of exceedance p', as the sample size m decreases. The probability Pr (r | p',n,f'(x),m) of r exceedances in n future years given a sample estimate p' from a m year record, is obtained using equation (7), by convoluting the distribution f(p'|f'(x),m) with the probability Pr(r|p',n). The resulting probability of r (for r not equal to 0) exceedances is then lower because of the positively skewed nature of the distribution f(p'|f'(x),m) and the associated higher likelihoods for lower probabilities of exceedance p'. The corresponding risk R_{mn}, defined as the probability of one or more exceedances in n years is then lower as the sample size m decreases and the positive skew of f(p'|f'(x),m) increases.

3. DISCUSSION

The traditional design procedure for flood control entails the minimization of the total expected cost TEC of flood control defined as

$$TEC = \int_{x^*}^{\infty} D(x)\, f(x)\, dx + C(x^*) \tag{18}$$

where D(x) is the per event damage corresponding to a flow x
C(x*) is the annual cost of controlling a flood of magnitude x*

The value of x* that minimizes TEC is selected as the appropriate design level, and the estimated distribution f'(x) is used in place of f(x) in equation (18). The annual cost C(x*) is computed by amortizing over the n year project design life. The actual risk of flooding in the n year life is not considered. Only the annual estimated probability of exceedance p' for a design level x* is considered through the specification of f'(x). Thus, while the project costs are amortized over a n year period, the average annual damages as represented by the first term in equation (18) are implicitly considered over an infinite time period. Inequities in the comparison of flood control alternatives with unequal design life thus arise. This design procedure is likely to be conservative, since it does not consider the risk of the likely number of exceedances in the design life of the project. This statement and the earlier observation, that the risk is lower in the design life when probabilities of exceedance of design flows are estimated from a historical sample, are however offset to an extent, by another aspect of the traditional design procedure. The flood frequency estimation procedure commonly used corresponds to the first one identified by Stedinger (1983). That procedure seeks to determine an unbiased estimate x' of the design flow x* corresponding to a probability of exceedance p_o. As Stedinger (1983) and Lall (1981) point out, the procedures typically used for the purpose do not succeed in yielding x' as an unbiased estimate of x*, and also lead to a corresponding expected probability of exceedance p'' of x' that is larger than p_o, when the sample size m is small. Where these procedures are used, it is quite likely that the actual risk faced in the design life of the project is comparable to, or larger than, the risk corresponding to perfect knowledge. Using Stedinger's (1983) results, or those of Zacks and Milton (1971), with the results of this paper, it is possible to determine the risk in the design life corresponding to the traditional flood frequency estimation procedure. The traditional design procedure of equation (18) would, however, still need to be modified to account for the risk in the design life.

Ang and Tang (1984) discuss the use of the type of procedures developed earlier in this paper in engineering design. They reproduce the work of Chow and Takase (1977) and also employ asymptotic distributions of extremes to derive the risk that the largest value in a record of m years will be exceeded in a n year future period. The design problem is then the use of this risk to determine a design level that leads to a specified risk in the future period. The relationships developed in this paper are in the same spirit, and address a more general problem.

The expressions developed for risk and for the probability Pr (r | p',n,f'(x),m) of r exceedances in a n year period given an estimate p' of the probability of exceedances of a design level x*, are also potentially useful in the development of general, dynamic reliability models.

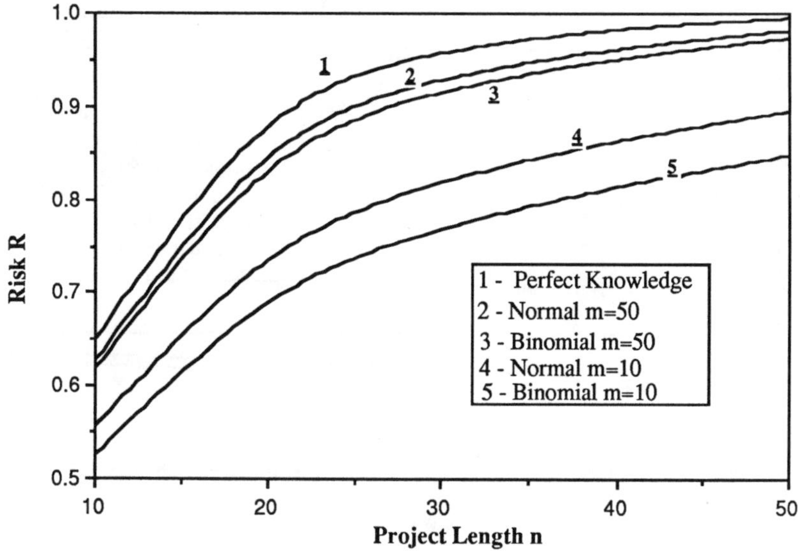

FIGURE 1
Risk R vs. Project Length n
Probability of Exceedance p_o = 0.1

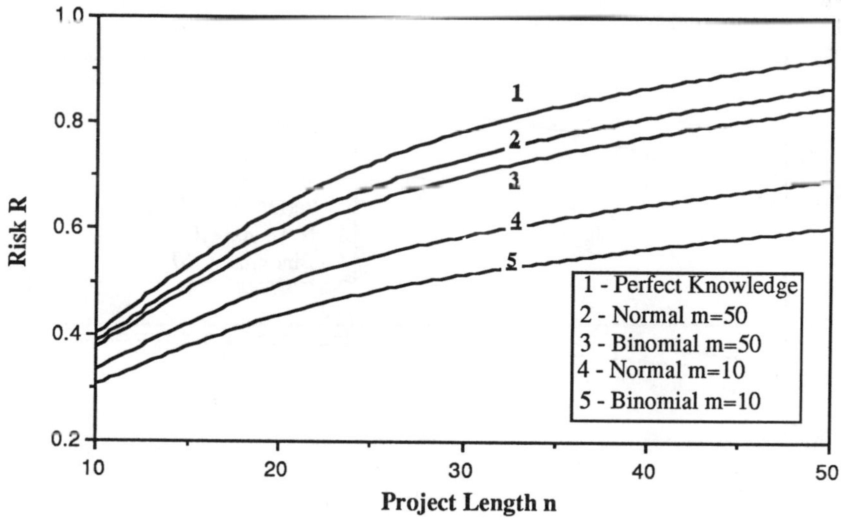

FIGURE 2
Risk R vs. Project Length n
Probability of Exceedance p_o = 0.05

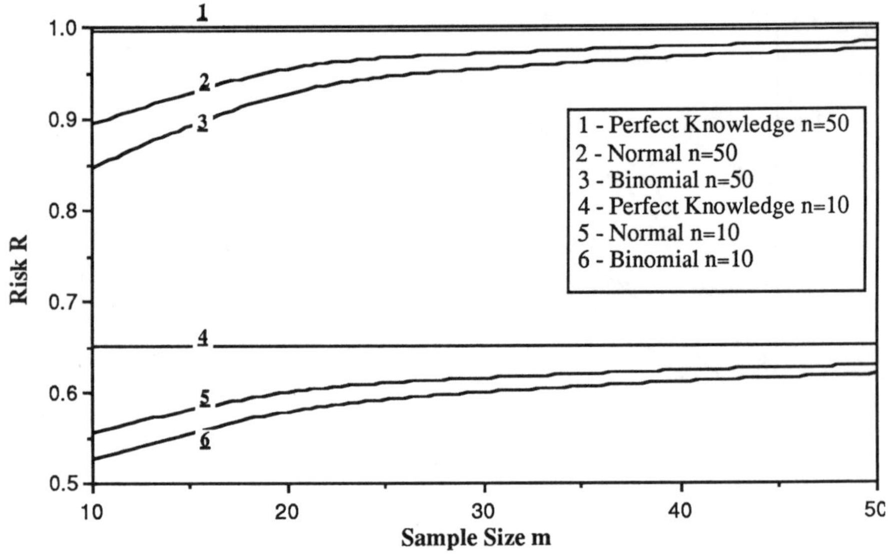

FIGURE 3

Risk R vs. Sample Size m

Probability of Exceedance p_o = 0.1

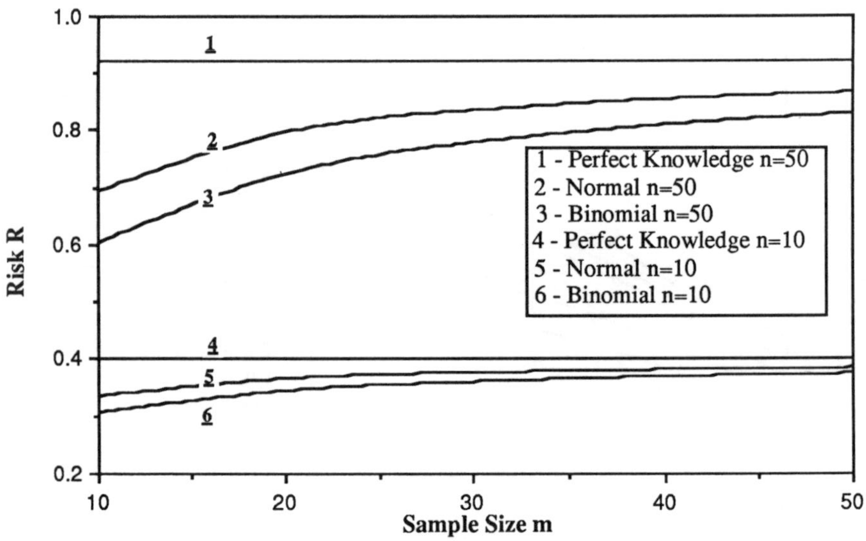

FIGURE 4

Risk R vs. Sample Size m

Probability of Exceedance p_o = 0.05

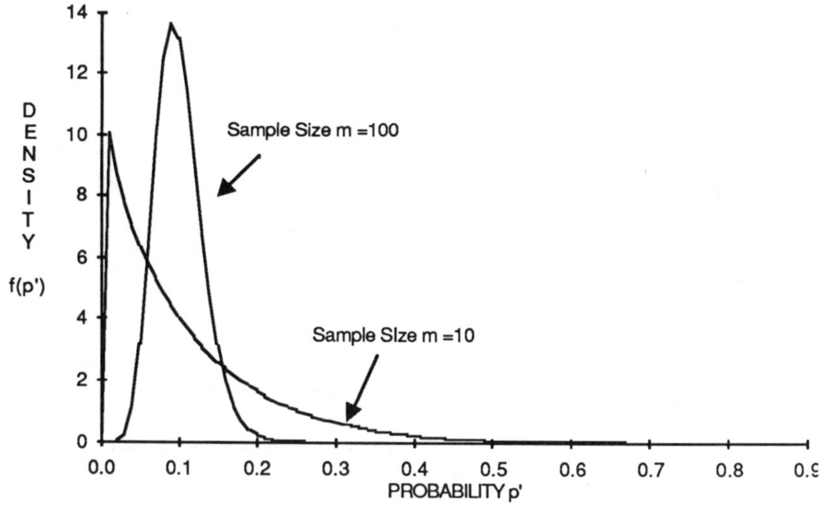

FIGURE 5

Density f(p'|f'(x),m) for $p_o = 0.1$
Nonparametric (Binomial) Estimation Procedure

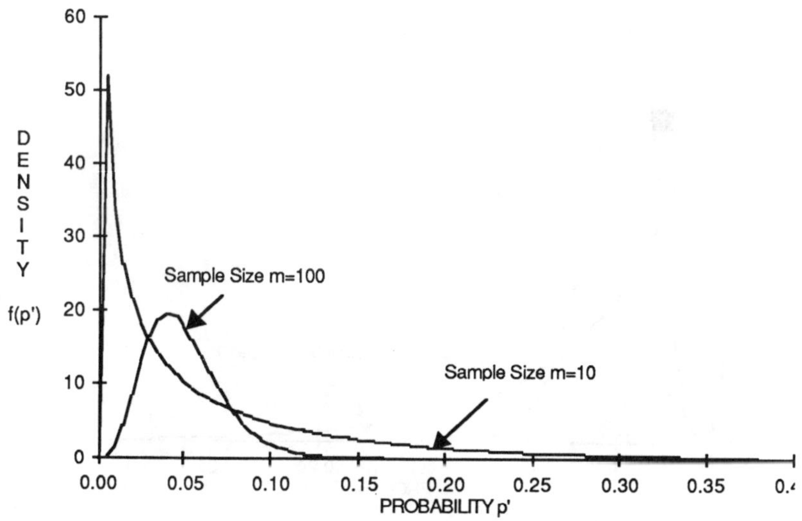

FIGURE 6

Density f(p'|f'(x),m) for $p_o = 0.05$
Nonparametric (Binomial) Estimation Procedure

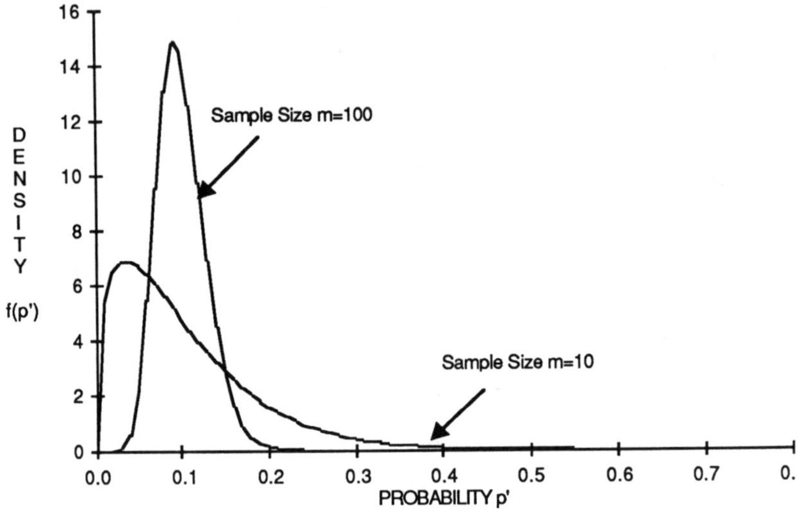

FIGURE 7

Density $f(p'|f'(x),m)$ for $p_o = 0.1$
Normal Distribution Estimation Procedure

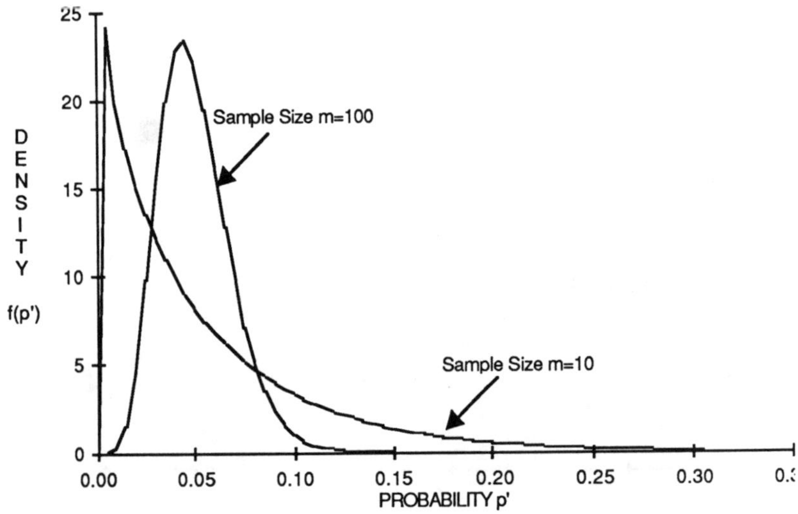

FIGURE 8

Density $f(p'|f'(x),m)$ for $p_o = 0.05$
Normal Distribution Estimation Procedure

TABLE 1

Beta Distribution Parameters Corresponding to Selected Probabilities p_o and Sample Sizes m

Probability p_o	Sample Size m	c	d	Variance $V_{nm}(p')$	Skew γ_p	Comments
.10	10	.90	8.10	9.0000E-03	1.533	Binomial
.10	100	9.90	89.10	9.0000E-04	0.528	Binomial
.05	10	0.45	8.55	4.7500E-03	2.374	Binomial
.05	100	4.95	94.05	4.7500E-04	0.8177	Binomial
.10	10	1.44	12.92	5.8604E-03	1.278	Normal
.10	100	12.07	108.60	7.3950E-04	0.480	Normal
.05	10	0.83	15.68	2.7133E-03	1.867	Normal
.05	100	7.41	140.70	3.1850E-04	0.672	Normal

Tung (1985) uses the Poisson distribution to evaluate the probabilities of r exceedances in a n year period, as part of a dynamic reliability model. The use of the relations developed in this paper is an attractive alternative for the same purpose, since they are conditioned on sample information and recognize the informational uncertainties associated with the estimates p'. The use of the Poisson distribution presumes that the mean rate of occurrence (i.e. p') of the event is 'known', and does not consider the uncertainty in estimating p'.

In conclusion, the observation that the estimated risk is lower than the risk with perfect knowledge was a most interesting phenomena, that possibly needs further explanation. The framework developed for the estimation of risk is potentially quite useful, since it is independent of the choice of the procedure used to estimate the probabilities of exceedance of floods. The incorporation of the developments of this paper into formal procedures for risk based design and assessment of reliability is feasible and is being pursued by the author. Of particular interest in that regard are problems relating to the random variability (e.g. available flood storage) of the resistance of the project to flooding during its life.

REFERENCES

1. Abramowitz, M. and Stegun, I.A. 1972. Handbook of Mathematical Functions, Dover Publications, New York, pp. 930.
2. Adamowski, K. 1985. Nonparametric Kernel Estimation of Flood Frequencies, Water Resources Research, Vol. **21**, No.11, pp. 1585-1590.
3. Ang, A. H-S., and Tang, W.H. 1975. Probability Concepts in Engineering Planning and Design, Vol. **I**, Basic Principles, John Wiley and Sons, New York, pp. 352-353.
4. Ang, A. H-S., and Tang, W.H. 1984. Probability Concepts in Engineering Planning and Design, Vol. **II**, Decision, Risk and Reliability, John Wiley and Sons, New York, pp.246-255.
5. Bickel, P.J. and Doksum, K.A. 1976. Mathematical Statistics, John Wiley, New York. pp.76.
6. Bodo, B. and Unny, T.E. 1976. Model Uncertainty in Flood Frequency Analysis and Frequency Based Design, Water Resources Research, Vol. **12**, No. 6, pp. 1109-1117.
7. Castano, E., Duckstein, L., and Bogardi, I. 1978. Choice of Distribution Functions for Hydrologic Design, Water Resources Research, Vol. **14**, No. 4, pp. 643-652.
8. Chow, V.T. and Takase, N. 1977. Design Criteria for Hydrologic Extremes, Journal of the Hydraulics Division, American Society of Civil Engineers, Vol. **103** (HY4), pp. 425-436.

9. Davis, D.R., Kisiel, C.C., and Duckstein, L. 1972. Bayesian Decision Theory Applied to Design in Hydrology, Water Resources Research, Vol. **8**, No.1, pp.33-41.
10. Duckstein, L. and Bogardi, I. 1981. Application of Reliability Theory to Hydraulic Engineering Design, Journal of the Hydraulics Division, American Society of Civil Engineers, Vol.**107** (HY7), pp. 799-815.
11. Lall, U. 1981. The Value of Data in Relation to Uncertainty and Risk. Phd. Dissertation, Department of Civil Engineering, University of Texas at Austin.
12. Lee, H.L., and Mays, L.W. 1983. Improved Risk and Reliability Model for Hydraulic Structures, Water Resources Research, Vol. **19**, No. 6, pp.1415-1422.
13. Miller, R.G. 1974. The Jacknife - A Review, Biometrika, Vol. **6**, No. 1, pp. 1-17.
14. Olkin, I. 1978. Estimating the Tails of a Distribution of Known Form, NSF Technical Report **138**, Department of Statistics, Stanford University.
15. Raiffa, H. and Schlaifer, R. 1968. Applied Statistical Decision Theory, The M.I.T. Press.
16. Schuster, E. and Yakowitz, S. 1986. Parametric/Nonparametric Mixture Density Estimation with Application to Flood Frequency Analysis, Water Resources Bulletin, to appear.
17. Stedinger, J.R. 1983. Design Events with Specified Flood Risk, Water Resources Research, Vol..**19**, No.2, pp.511.
18. Sidarovzsky, F., Bogardi, I., Duckstein, L. and Davis, D.R. 1976. Economic Uncertainties in Water Resources Project Design, Water Resources Research, Vol. **12**, No. 4, pp. 573-580.
19. Tung, Y.K. and Mays, L.W. 1980. Optimal Risk Based Design of Water Resource Engineering Projects, CRWR Report **-171**, University of Texas at Austin.
20. Tung, Y.K. 1985. Models for Evaluating Flow Conveyance Reliability of Hydraulic Structures, Water Resources Research, Vol. **21**, No. 10, pp. 1463-1468.
21. Vicens, G.J., Rodriguez-Iturbe, I., and Schaake, J.C., Jr. 1975. A Bayesian Framework for the use of Regional Information in Hydrology, Water Resources Research, Vol. **8**, No.1, pp. 33-41.
22. Viessman, W., Knapp, J.W., Lewis, G.L., and Harbaugh, T.E. 1977. Introduction to Hydrology, Harpers and Row, New York.
23. Woo, J. 1979. Jacknife Estimates of the Right Tail Probabilities in Small Samples from a Gamma Distribution, PhD. Dissertation, Department of Pure and Applied Mathematics, Washington State University.
24. Wood, E.F. and Rodriguez-Iturbe, I. 1975a. Bayesian Inference and Decision Making for Extreme Hydrologic Events, Water Resources Research, Vol. **11**, No. 4, pp. 533-542.
25. Wood, E.F. and Rodriguez-Iturbe, I. 1975b. A Bayesian Approach to Analyzing Uncertainty among Flood Frequency Models, Water Resources Research, Vol. **11**, No. 6, pp. 839-843.
26. Wood, E.F. 1977. An Analysis of Flood Levee Reliability, Water Resources Research, Vol. **13**, No. 3, pp. 665-671.
27. Yakowitz, S. 1985. Markov Flow Models and the Flood Warning Problem, Water Resources Research, Vol. **21**, No. 1, pp. 81-88.
28. Zacks, S. and Milton, R.C. 1971. Mean Square Errors of the Best Unbiased and Maximum Likelihood Estimators of Tail Probabilities in Normal Distributions, Journal of the American Statistical Association, Vol. **66**, No. 335, pp. 590-593.

OPTIMAL PLANNING OF FLOOD CONTROL SYSTEMS BASED ON RISK ANALYSES OF FLOOD INUNDATION

T. Kojiri	S. Ikebuchi	T. Hori
Civil Engineering	WRRC, DPRI	Civil Engineering
Univ. of Gifu	Univ. of Kyoto	Univ. of Kyoto
Yanagido	Gokasyo	Sakyoku
Gifu, 501-11	Uji, 611	Kyoto, 606
Japan	Japan	Japan

ABSTRACT. The aim of this study is to establish the optimal flood control system accounted for the comprehensive criteria on the whole river basin. Especially, for the system consisted of multi-sub basins and multi-defense or reference points, we will examine the effects of the flood control projects on the flood inundation probability in time and space. Project planning is divided into two parts, such as the site and scale problems and the scheduling problem. They are formulated through the screening model, the simulation model and the sequential model in order to be optimized by using Experimental Planning Method, and Dynamic Programming, Branch-bound Method.

1. INTRODUCTION

The objective of project planning for flood control is to decide the most effective system which defends human life and property against the flood. To attain this objective, it is important to know the facts concerning flood and to predict how much damage may be sustained in future. Both the optimal flood control system and the working policy should be planned in harmony with art and economy based on this hydrological information.

In the flood control project planning, the factor of design high water level is often calculated only at one terminal reference point for the design rainfall occuring at the river basin. This procedure, however, makes it impossible for some defense or reference points in that basin to have the proper safty rate against flooding because of the rapid transfiguration of the basin, such as i) increase of defense points, ii) construction of dam reservoirs and other projects, and iii) progress of urbanization. That is to say, it means that the safty rate of the basin must not be estimated only at the most important defense point, but the safety rates at all defense points should be made equal or proper to each other taking advantage of interaction of the operational projects and correlation between precipitations in sub-basins.

Therefore, computing the flood inundation probability in time and space for a system consisted of multi-dam reservoirs and multi-defense

V. P. Singh (ed.), Application of Frequency and Risk in Water Resources, 319–333.

points, we will establish the decision procedure for construction site, scale and time sequence or expansion of projects under a required safety rate based on the computed probability of flood inundation.

2. DECISION PROCEDURE FOR FLOOD CONTROL PROJECTS PLANNING

2.1. The Objective of Flood Control Planning

The objective of flood control is to prevent the flood inundation and reduce the flood damage. Concretely, there are i) reduction of the flood damage in the whole basin, ii) reduction of the occurence frequency of flood inundation, and iii) reduction of other factors, such as, the duration periods or intensity of flood inundation as follows.

flood damage;

$$OFD = \sum_{t=1}^{TF} \sum_{m=1}^{M} ODA_m(t) \longrightarrow \min \tag{1}$$

probability of flood inundation;

$$OIP = \max_{(m)} [\ PFA_m\ ] \longrightarrow \min \tag{2}$$

probability of flood duration;

$$ODP = \max_{(m)} [\ PFD_m\ ] \longrightarrow \min \tag{3}$$

$$PFD_m = \sum_{t=1}^{TF} (TFD_m(t)/(TF-t+1))/TF \tag{4}$$

flood intensity;

$$OFI = \max_{(m)} [\ (QP_m - QD_m)/QD_m\ ] \longrightarrow \min \tag{5}$$

Where, $ODA_m(t)$ is the flood damage at defense point m (m=1,2,...,M ; M is the total number of defense points), PFA_m, the probability of flood inundation, $TFD_m(t)$, the time periods of flood duration after control time t, QP_m, the peak flow, TF, the total number of control time periods, QD_m, the allowable flow discharge at defence or reference point m, respectively. Commonly, the first objective is represented as the expectation of damage. The control plicy is determined through economic evaluation such as benefit-cost analysis. The second one is getting more important because it clarifies the limit of control effect through the inundation probability estimated at each defense point. The third and fourth ones may be included in the first and/or second ones in some sense. Thus, we use the minimization of project cost as the planning objective and the second policy as the planning constraint to extract the most effective alternative under condition of keeping the probability less than the required rate or the exceedance probability at any defence point.

2.2. Decision Procedure for Planning

Flood control project planning is mainly made of two parts. One of them is the extraction of the optimal system satisfying the control objective from many alternatives. It is called the problem of project site and scale planning. The other is to order in time the projects contained in the extracted alternative more effectively without harming the safety rate under constraint of the limit budget. It is called the problem of schedule planning. For the site and scale planning, many methods have been proposed and the 0-1 mixed interger programming is one of the most powerful methods. But the more the projects under consideration increase, the more rapidly the execution feasibility decreases. As the counterplan for this problem, we apply the method proposed by Ikebuchi (1980), where optimal site, scale and schedule planning is gained through three optimization models. In the first step, some of the better solutions whose construction cost are cheaper and whose inundation probabilies satisfy the required rate, are extracted from the simplified basin model within limits which take congnizance of the reality of flood phenomena. In the second step, the final solution is determined by the simulation method based on the more exact basin model from the above better solutions. In the schedule planning, the objective is defined as the problem for minimizing the cumulative value of flood inundation

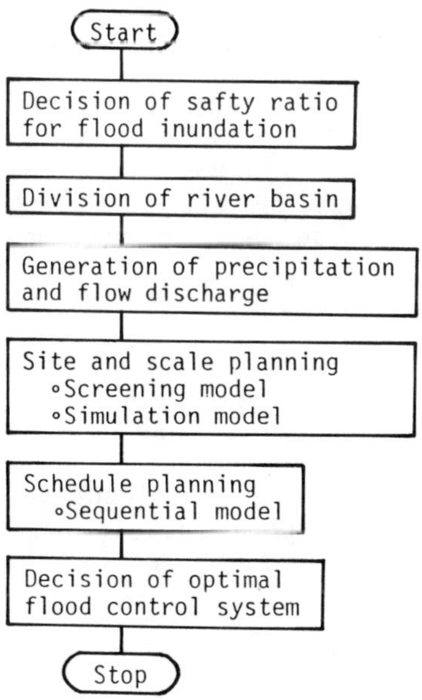

Figure 1. Flowchart of flood control planning

probability at each construction time stage. It may contain the same computation steps as the site and scale planning does. These models arecalled i) Screening model, ii) Simulation model and iii) Sequential model, and constitute the system of flood control planning as shown in Figure 1.

3. PRECIPITATION AND FLOW DISCHARGE

3.1. Generation of Precipitation

The flood rainfall varies with the complex phenomena in time and space. In this paper, the first-order nonhomogeneous Markov chain can be assumed to be the stochastic process of hourly precipitation in time on each sub-basin (Chow, 1971). And the regional dependence of hourly precipitation between sub-basins can be assumed to be the stochastic process with zero and first-order delayed correlation. The conditional probability of the continuous precipitation for TR hours is represented as follows (Ikebuchi, 1981).

$$f_R(r_{u+1}(t),r_{u+1}(t+1),\ldots,r_{u+1}(t+TR)\,|\,r_u(t),r_u(t+1),\ldots,r_u(t+TR))$$
$$= \frac{f_R(r_{u+1}(t),r_u(t))f_R(r_{u+1}(t+1),r_u(t+1)\,|\,r_{u+1}(t),r_u(t))\ldots \ldots f_R(r_{u+1}(t+TR),r_u(t+TR)\,|\,r_{u+1}(t+TR-1),r_u(t+TR-1)))}{f_R(r_u(t))f_R(r_u(t+1)\,|\,r_u(t))\ldots f_R(r_u(t+TR)\,|\,r_u(t+TR-1))} \quad (6)$$

Where, $r_u(t)$ is the precipitation at time t on sub-basin u, $f_R(\cdot|\cdot)$, the conditional probability function and TR, total number of time priods on flood rainfall, respectively. As the hourly precipitation obeys the log normal distribution, each term of eq. (6) is composed of the four dimensional and log normal distribution function (Kojiri and et al, 1985). A lot of hyetographs can be computed through the simulation method where the random numbers are generated obeying the above conditional distribution function.

3.2. Generation of Flow Discharge

The flow discharge is important to estimate the basin-wide safety rate against the flood inundation. Therefore, after simulating the hyetographs, the hydrographs are simulated through the runoff analysis. Since flood control system is ordinarily consisted of multi-sub basins model, we apply Nash's Instantaneous Unit-hydrograph Method in screeing model and Storage Function Method in simulation model as follows.

$$\begin{aligned} &SC_u(t)-SC_u(t-1)= dt\cdot(re_u(t)-QI_u(t-1)) \\ &SC_u(t+TL_u)= KW_u\cdot(QI_u(t+TL_u))^{**}PW_u \end{aligned} \quad (7)$$

Where, $SC_u(t)$ is the storage volume of water discharge in sub-basin u,

$QI_u(t)$, the runoff discharge from sub-basin u, $re_u(t)$, the effective rainfall at time t, dt, the time interval in computation, TL_u, lag time, KW_u and PW_u, inherent parameters to sub-basin u, respectively.

4. PROJECTS SITE AND SCALE PLANNING

4.1. Screening Model

In screening model, the planning problem is simplified. The problem objective is applied as the single objective optimization, because the preliminary superior alternatives are abstracted. We mean that the superior alternatives are the optimal and the approximatively optimal solutions satisfying the control objective. The flow discharge of sub-basin is expressed as the conditional probability between the neighboring sub-basins. The frequency of flow discharge is gained by discretizing the simulated hydrographs with the arbitary width. The probability of the most upstream sub-basin is represented as the occurence probability and the probabilities of other sub-basins are represented as the conditional probability given the flow discharge of neigboring sub-basin. Moreovor, the flow discharge passing through dam reservoirs, dikes where flood discharge overflows, river channel where flood discharge is stored, are expressed as matrixes (as shown in Figure 2, Takasao and et al, 1982). Where A(t) is the incremental storage volume

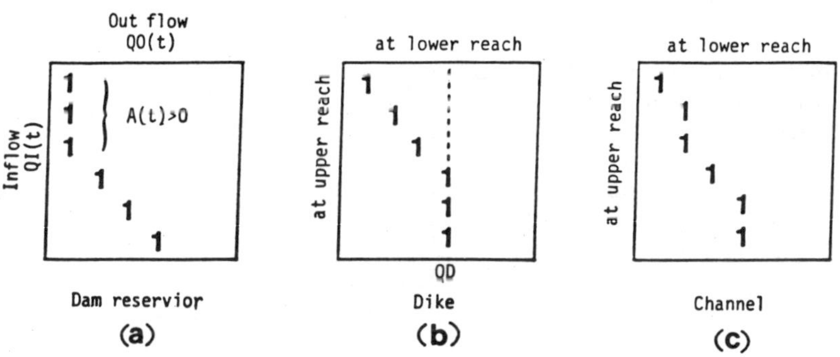

Figure 2. Transformation matrixes in river basin

of reservoir from time stage t-1 to t (= S(t)-S(t-1) ; S(t) is the storage volume) When the river basin is consisted of several sub-basins with no project, the conditional probability distribution of the first confluence discharge $\mathbf{QJ}_1(t)$ is expressed as follows.

$$\mathbf{QJ}_1(t) = \mathbf{QP}_1(t)*\mathbf{QP}_2(t) \qquad (8)$$

At the first sub-basin, $Q_{ij}(t)$ (the element of the conditional probability of flow discharge j given flow discharge i) is zero if $i \neq j$. Shift Operation Method consists of computing the probability of the confluence discharge and shifting the element at row i and column j to the position of row i+j-1 and column j for the product of their probability (Conover, 1965). The symbol of "*" indicates the excution of Shift Operation. If only one tributary river confluences the main river or another tributary river, the confluence point number is expressed as the number of sub-basin minus one. The conditional probability of the confluence discharge $\mathbf{QE}_u(t)$ is computed repeatedly as follows.

$$\mathbf{QJ}_{u+1}(t) = \mathbf{QJ}_u(t) * \mathbf{QP}_{u+2}(t) \tag{9}$$

$$\mathbf{QE}_u(t) = \mathbf{QJ}_u(t) \cdot \mathbf{E}_u \tag{10}$$

The probability of flood inundation is

$$PF_u(t) = 1 - \sum_{i=1}^{qfu} QE_u^i(t). \tag{11}$$

Where, $\mathbf{E}_u$ is the unit vector (Jx1), qfu, the discretized value of designed high water discharge, and QE_u^i, the element of $\mathbf{QE}_u(.)$. When the flood inundation is evaluated at each confluence point, the sub-script u is equal to the above mentioned defence point m. As the flood inundation probability is gained in time and space, the estimate value of flood control system is defined as follows.

$$PFA_m = \max_{(t)} [\ PF_m(t)\ ] \tag{12}$$

The computational procedure of $PF_u(t)$ must be changed according to the confluence types of the main river with the tributary river having some flood control projects (Kojiri and et al, 1985). If the allowable limit of flood inundation probability PFU_m from the viewpoint of the whole river basin is given, the objective function of project site and scale planning is written as the minimizing problem for summing up the construction cost as follows.

$$OB_{sqr} = \sum_{n=1}^{N} CO(n) \longrightarrow \min \tag{13}$$

subject to

$$PFA_m \le PFU_m \quad (m=1,2,\ldots,M) \tag{14}$$

Where, CO(n) is the contruction cost of the project implemented at site n, and N, the total number of projects, respectively. In order to reduce the execution amount of computation, Experimental Planning Method is introduced (Yoshikawa and et al, 1983). Concretely, regarding the sites of planning projects as the factor and their scales as the levels, the orthogonal layout is arranged. Then, after evaluating all of restricted alternatives according to the combinations in its table, the superior alternatives are extracted to satisfy the constraint under the condition of the given flood inundation probability (Kojiri and et al, 1984). Because of simplicity and rapidity on computation, Screening model has a lot of characteristic as follows;

(i) know the dangerous time and position with the high probability of flood inundation;
(ii) get the criteria denoting whether it is need to construct the additional projects or not;
(iii) get the information of flood control effect through the disparsion analysis.

4.2. Simulation Model

In simulation model, the optimal alternative is decided among the above superior alternatives. To represent the flow discharge and control effect with high accuracy, Storage Function Method is introduced as the runoff analysis. The dam reservoir is operated abidingly by real rule such as the releasing policy from gate with constant rate. The system is evaluated through eqs. (2) and (3). If there is some excess of budget, however, the probability curves of flood inundation and flood durability may be drawn as shown in Figure 3. Therefore, the problem becomes not

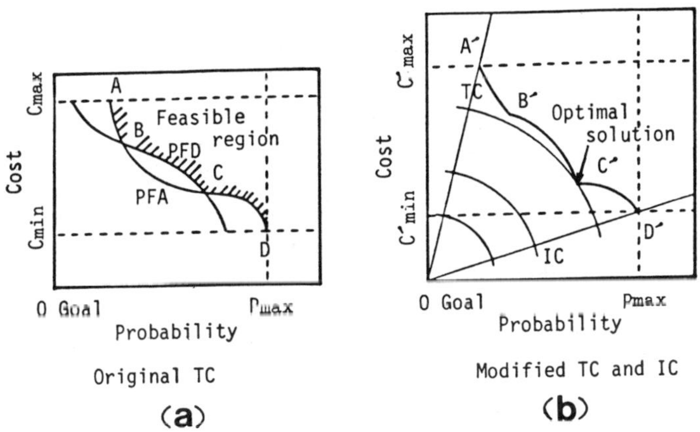

Figure 3. Optimal solution in simulation model

single but multi-optimization one. Using vector optimization, the synthesized curve which is generated by linking with their inferior parts means Transformation Curve (TC). In this case, the optimal solution is the point where Indifference Curve comes in contact with TC and the interval of indifference from Goal to TC denotes the level of non-modification. We proposed the transformation procedure that the original coordinates is changed into the new coordinates where the distance of OA' is equal to the distance of OB', because the worst point A of TC on the construction cost and B on the flood inundation probability are regarded as the same weight to Goal (Ikebuchi, 1979). After modified, concentric circles with center at the origin are drawn and the point where a circle having minimum radius comes in contact with transformed TC is determined to be the optimum point. In constraint method, TC is gained as follows. When N is the number of extracted projects, VA_n^h is the project scale ($h = 1,2,\ldots,H_n$; H_n is the total number of discretized scale on project n) and PFA^b and PFS^b are the additional constraints to the probabilities of flood inundation and durability, the final site and scale planning is

$$OB_{sim} = \sum_{n=1}^{N} CO(n, VA_n^h) \cdot X_n^h \longrightarrow \min \tag{15}$$

subject to

$$\sum_{h=1}^{Hn} X_n^h = 1 \tag{16}$$

$$X_n^h = 0 \text{ or } 1$$

$$PFS = \max[\, PFA_m,\ PFD_m] \leq e^b \tag{17}$$

Where, e^b denotes the satisfactory level of constraint. The above problem can be solved by Branch-bound Method. To reduce the computational time, the search tree to be solved can start from the nodes of superior alternatives in screening model and only project scale is treated as the variables. Moreover, the following items are checked.
i) New node where another project will be constructed satisfies all of constraints.
ii) New node where the downstream project will be constructed modifies the basin-wide evaluation value consisted of the values at the upstream points.
Ultimately, from the viewpoint of multi-objective and the exact evaluation, the optimal site and scale is decided.

5. CONSTRUCTION SCHEDULE PLANNING

5.1. Sequential Model

As there is a limit on the budget invested to flood control planning at every year, flood control projects may be constructed or expanted in sequential steps. Assuming that the flood control effects is generated as soon as the additional project is constructed, three objectives in sequential model are considered as follows.
(i) minimize the accumulative value of the expected damage sustained for the construction periods.
(ii) minimize the accumulative value of the flood inundation probability evaluated in the whole river basin for the construction time periods.
(iii) construct the additional project in the defence or reference point at every construction time stage where the flood inundation probability is the highest among all of periods.
We apply the second objective because of accordance with objective of project site and scale planning. At first step, the trial trajectory is extracted by using the simple basin model, that is to say, screening model. Then, at second step, the final schedule is determined through simulation model.

When PFT(s) denotes the flood inundation frobability at construction time stage s, the PFT(s) curve is lowering in sequential steps. Ignoring the discount rate in time, the objective function OB_{seq} is represented as follows.

$$OB_{seq} = \sum_{s=1}^{TS} PFT(s)\cdot ds \longrightarrow \min \tag{18}$$

Where, TS is the total number of construction time stage and ds is the unit time stage, respectively. On the other hand, the constraints is that the flood inundation probability $PFT_m(s)$ at any reference point must not be higher than the value of the previous time stage as follows.

$$PFT_m(s) \leq PFT_m(s-1) \qquad (m=1,2,\ldots,M;\ s=1,2,\ldots,TS) \tag{19}$$

5.2. Formulation by Dynamic Programming

In order to explain the formulation process easily, the following assumptions are given.
(i) one unit in scale of each project is constructed in one time stage.
(ii)the sum of the budget invested to the whole river basin is equal at every time stage.
The state of flood control system at any time stage depends on not the sequence of construction schedule till considered time stage but the combination of constructed projects (Ikebuchi and et al, 1985). Thus, the system is represented by following vector.

$$\mathbf{Y}(s) = [SY_1(s),\ SY_2(s),\ \ldots,\ SY_N(s)] \tag{20}$$

$$SY_n(s) = k \quad \text{if k-th scale of project n is constructed } (k=1,2,\ldots,K(n))$$
$$= 0 \quad \text{if project n is not constructed}$$

Where, $SY_n(s)$ denotes the constructed scale of project n and time stage s, K(n) is the total number of discretized project scale, respectively. As the flood inundation probabilities at the initial and the last state are independent on the construction schedule, the sequence of construction schedule is to optimize from the first time stage to the last time stage minus one. Therefore, the recursive equation of DP is formulated as follows.

$$f_s(\mathbf{Y}(s)) = \min_{X(s)} [\ PFT(s,\ \mathbf{Y}(s)) + f_{s-1}(\mathbf{Y}(s-1))\ ] \tag{21}$$

where, X(s) is the feasible scale of project to be constructed at time stage s and PFT(s, $\mathbf{Y}$(s)) is the flood inundation probability under the condition of the combination $\mathbf{Y}$(s). As eq. (21) means the principle of optimality, the optimal construction schedule can be gained. Of course, it is easy to apply the above formulation process to the case that several projects are constructed at the same time stage. Moreover, assuming that the state variable is the number of considered projects, the approximative solution will be gained in spite of a great reduction of state variables.

5.3. Formulation by Linear Programming

Schedule planning is also formulated by Linear Programming. Especially, Branch-bound Method is a powerful algorithm. The search tree in that method is composed of the nodes which represent the combination of feasible construction projects added to the previous state of construction projects (previous node). At node z, new node is decided by using the following enumeration criteria.
(i) it is impossible to enumerate the new node when the considered project is already constructed or the previous work to construct the project is not finished.
(ii) the new node is abandaned when the flood inundation probability will increase at some reference point.
(iii)the new node is also abandaned when the accumulative value of flood inundation probability till that time stage will not be modified.
Since all of feasible combinations of construction sequences are enumerated, this algorithm needs more computation time than DP. Using the constraint (i) - (iii) and decomposing the computation process into several blocks, however, it will be possible to solve this problem successively.

6. A CASE STUDY

We will apply the screening model on the project site, scale and schedule planning to the real river basin. The river basin is composed of 10 sub-basins (as shown in TABLE I), six sites of dam and seven defense points. The dike is constructed at the same place as defense point (as shown in later Figure 5). The higher flow discharge than the allowable discharge overflows from that defense point.

6.1. Conditional Probability of Flow Discharge

We analyzed the stochastic structure of rainfall from 25 observed hyetographs occurring flood. Arranging hyetographs according to the serial correlation to the basic hyetograph, parameters of four-dimensional and log normal distribution are identified. The number of rainfall simulation is one thousand cases. Hydrographs are computed through Instantaneous Unit-hydrograph Method for the simulated hyetographs. In TABLE II, the matrix means the conditional probability distribution at time stage 20 between sub-basin 7 and 8 where the one unit of discretized discharge is 30 m^3/sec and the dividing number is ten points. If one divided area of sub-basins are very different from neighbouring sub-basin, it tends to become the one-sided matrix.

TABLE I. Information data on sub-basin area

Sub-basin	Dam	Area (km³)
1	1	85.7
2	2	264.3
3		77.5
4	3	142.5
5	4	41.6
6	5	130.7
7		67.7
8		170.0
9	6	45.0
10		325.0

TABLE II. Conditional probability of rainfall

		Sub-basin 8									
		1	2	3	4	5	6	7	8	9	10
Sub-basin 7	1	0.0	0.0	0.0	0.0	0.0	0.0	0.0	0.0	0.0	0.0
	2	0.0	0.0	0.0	0.1316	0.3947	0.2105	0.1579	0.0789	0.0	0.0263
	3	0.0	0.0	0.0164	0.2131	0.3508	0.2492	0.1246	0.0230	0.0197	0.0033
	4	0.0	0.0	0.0147	0.1728	0.2941	0.3272	0.1397	0.0257	0.0184	0.0074
	5	0.0	0.0	0.0098	0.1275	0.3824	0.3039	0.0980	0.0784	0.0	0.0
	6	0.0	0.0	0.0303	0.2121	0.3333	0.2121	0.0909	0.1212	0.0	0.0
	7	0.0	0.0	0.0	0.0	0.3125	0.4375	0.0625	0.1250	0.0625	0.0
	8	0.0	0.0	0.0	0.1250	0.3750	0.2500	0.2500	0.0	0.0	0.0
	9	0.0	0.0	0.0	0.0	0.5000	0.5000	0.0	0.0	0.0	0.0
	10	0.0	0.0	0.0	0.0	0.0182	0.0227	0.0136	0.0045	0.0136	0.9273

6.2. Site and Scale Planning

The nummber of construction scale is three units, such as non-construction, small size and large size. As there are twelve projects, the number of experimental factors is twelve points and each factor has three level. Applying experimental Planning Method, we use $H_{3,36}$ table as the orthogonal layout. TABLE III shows the applied levels of dams and dikes. The construction costs of dam and dike are assumed to be two thousand yen per unit volume and zero yen. The constraint probability PFU_m assined to the river basin is 0.01. TABLE IV shows the superior solutions extracted from all of the feasible solutions. The better three alternatives have no flood inundation. As shown in the same table, dam 1 is important project to satisfy the constraints and dam 2 can control the flood discharge directly because of being constructed in the main

TABLE III. Information data on planning levels

Dam	Flood control capacity (m^3)			Estimate point	Allowable discharge (m^3/sec)		
	Level 0	Level 1	Level 2		Level 0	Level 1	Level 2
1	0	291600	680400	1	480	210	420
2	0	874800	1458000	2	750	450	900
3	0	291600	680400	3	1050	600	1200
4	0	2818800	3790800	4	1350	750	1350
5	0	291600	680400	5	1500	900	1650
6	0	291600	680400	6	1800	1050	1950
				7	1920(fixed)		

TABLE IV. Superior alternatives gained by Experimantal Planning

	Factor		Probability of flood inundation	Cost (x10^8 Yen)	Order
	Dam	Dike			
Experiment	1 1 1 2 2 1	1 2 1 0 0 1	0	124.3	
	1 1 1 1 0 1	0 1 0 1 1 2	0.0277		
	2 2 2 2 1 2	1 2 1 2 2 0	0	151.5	
	1 1 2 1 1 0	2 0 1 1 2 1	0.0317		
	2 2 0 2 2 1	0 1 2 2 0 2	0	137.9	
	2 2 1 1 2 0	2 2 0 0 2 2	0	118.5	3
	2 0 1 1 2 2	0 0 1 2 1 1	0.0106		
	1 2 2 0 2 0	0 1 1 0 1 0	0	62.1	2
	2 1 0 2 2 0	1 0 0 1 1 0	0.005		
	2 1 1 0 1 1	0 0 2 0 2 0	0	48.5	1

river. It is obvious for dam 3 to control the flood inundation probability because the flow discharge running off from sub-basin 6 has the high probability on the side of high discharge. Relating to the dike, the flow discharge of the downstream point increases according to the scale of upstream dike. So, it seems that the necessary scale of downstream dike depends on the scale of upstream dike.

6.3. Schedule Planning

The optimal sequence of construction schedule will be determined for the combination of projects which is selected as the cheapest cost in site and scale planning. Figure 4 shows the sequence of optimal construction schedule gained by Branch-bound Method. Figure 5 shows the illustration graphs of flood inundation probability and the system state composed of constructed projects at the time stages 0, 5, and 17. Dam 1 is constructed at the earlier stage because of its high effect on flood control. As the dike is firstly constructed at the lowest point of flood inundation, it is understood that it is important to keep the balance of flood inundation between all of defense points. Since, in the first half, the

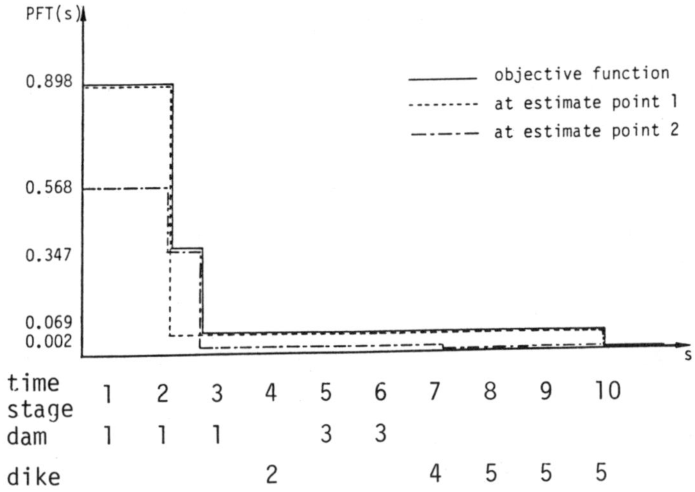

Figure 4. Flood inundation probability on shedule planning

flood inundation probability decreases in a great degree, the same result may be gained by introducing the objective function (iii).

7. CONCLUSION

In this paper, we tried to establish the universal and easy project planning of flood control system based on the economical and the physical criteria. These proposed procedure was applied to the real river basin for verification of our thesis. To sum up, the following results

i) Proposition of project planning procedure in the whole river basin

Project planning is divided into two parts composed of site and scale planning and schedule planning. Using screening model, simulation model and sequential model, each planning is formulated by LP or DP.

ii) Introduction of multi-objective programming

The problem is formulated by single optimization and vector optimization including several control objectives, such as minimization of construction cost, flood inundation probability, probability of duration time of flood.

iii) Modeling of flood control project in matrix equation

Flood control project is represented as the matrix in screening model and the flood inundation probability can be computed by Shift Operation Method.

iv) Introduction of Experimantal Planning Method

In order to reduce the computation time, Experimental Planning Method is introduced to extract the superior solutions. The optimal are gained.

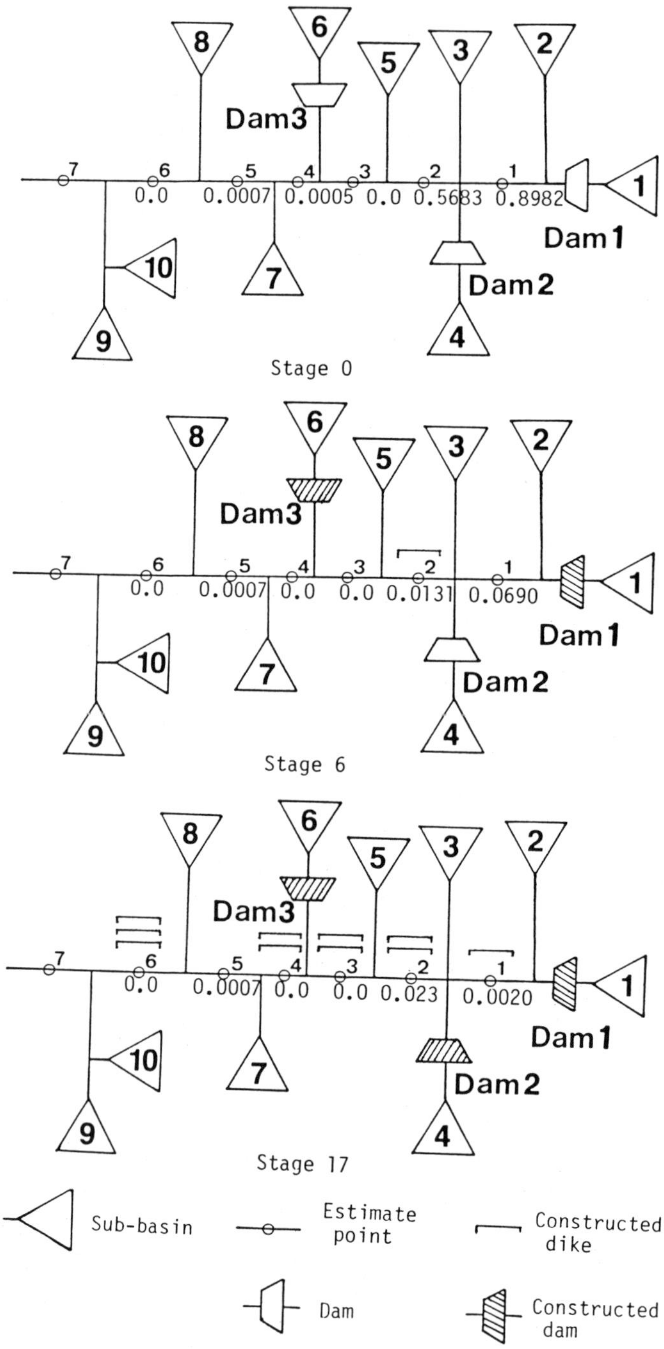

Figure 5. Sequence of construction schedule at time stage 0, 6, and 17

solution of site and scale planning is decided through Branch-bound Method.

v) Formulation of schedule planning considering the flood inundation probability

Schedule planning is formulated by LP and DP having no inconsistency with the objective of site and scale planning. These approches can be applied to not only the case that the single project is constructed at one time stage but also the case that several projects are constructed at the same time stage.

8. REFERENCES

8.1. Journal and Periodicals

Conover, W.J., 1965. The Distribution of $\mathfrak{F}(y_t)$ where $(y_0, y_1, \ldots)$ is Realization of a Non-Homogeneous Finite State Markov Chain. Biometrika 52, **No.1** pp. 277-279.

Ikebuchi, S., Kojiri, T. and Hori, T., 1985. A Study on Optimal Scheduling of Construction Works for Flood Control System. Annuals of Disaster Prevention Research Institute, Kyoto University, **No.28 B-2**, Japan, pp. 237-251.

Kojiri, T., Hori, T. and Ikebuchi, S., 1984. A Study of Optimal Flood Control System on Screening Model. Annuals of Disaster Prevention Research Institute, Kyoto University, **No.27 B-2**, Japan, pp. 241-254.

Takasao, T., Ikebuchi, S. and Kojiri, T., 1982 A Study of Optimal Flood Control System Considering the Flood Inundation Probability in Time and Space. Annuals of Disaster Prevention Research Institute, Kyoto University, **No.25 B-2**, Japan, pp. 285-296.

Nash, J. E., 1959. Systematic Determination of Unit Hydrograph Parameters. Jour. of Geophysical Research, **Vol.64 No.1**, pp. 179-194.

8.2. Edited Symposia and etc.

Chow, V. T., 1971. Stochastic Hydrologic Systems. Proceeding of 1st Bilateral U.S.-Japan Seminor in Hydrology, pp. 1.1-1.23.

Ikebuchi, S., 1981. Design Rainfall. Text Book of Summer Seminar, JSCE Kansai-branch, Japan, pp. 49-66.

Kojiri, T., Ikebuchi, S. and Hori, T., 1985. Stochastic Control of Dam Reservoirs Considering Flood Inundation Probability on Whole River Basin. Proceeding of 4th International Hydrology Symposium -Multivariate Analysis of Hydrologic Process-, Fort Collins, USA (in publication).

Yoshikawa. K., Haruna, M., Iyama, A. and Matsubara, Y., 1983. Systems Approach to Flood Control Problem in Urbanization Area. A Summary of Annual Lectur Meeting, JASE Kansai-branch, Japan, pp. IV-7.1-IV-7.2.

APPLICATION OF SYSTEM RELIABILITY ANALYSIS TO FLOOD FORECASTING

Charles S. Melching, Graduate Research Assistant

Harry G. Wenzel, Jr., Professor

Ben C. Yen, Professor

Department of Civil Engineering
University of Illinois at Urbana-Champaign
208 North Romine Street
Urbana, Illinois 61801

ABSTRACT. The decision making process regarding the issuance of flood watches and warnings often relies on the use of real-time flood forecasting models for watersheds. The floods predicted by these models are subject to uncertainties which may be significant for efficient operation of flood warning and preparedness systems. System reliability analysis offers a means to account for the uncertainties in the flood forecast in the form of occurrence probabilities for any flood level of interest. In this paper the advanced first-order second moment (AFOSM) reliability analysis method is used to analyze the uncertainties in a rainfall-runoff flood frequency model proposed by Eagleson (1972) and extended by Wood (1976) for an analytical derivation of the expected flood exceedance probability considering parameter uncertainty. Comparison of the results shows that the AFOSM method is a viable, reasonable, and practical method useful for hydrologic problems subject to uncertainties.

1. INTRODUCTION

Real-time flood forecasting is one of the primary sources of information for decision making regarding the issuance of flood watches and warnings. When using a hydrologic model to provide such forecasts, the reliability of the predicted hydrograph is influenced by four sources of uncertainties: (1) natural, (2) data, (3) model parameters, and (4) model structure. Natural uncertainties refer to the random temporal and areal fluctuations inherent in natural processes. Data uncertainties include: (a) measurement accuracy and

V. P. Singh (ed.), Application of Frequency and Risk in Water Resources, 335–350.

errors; (b) the adequacy of the rain and/or stream gage network, soil data, vegetation data, etc.; and (c) data handling and transmission errors. Parameter uncertainty reflects variability in the determination of the proper parameter values to be used in modeling a given storm event. Model structure uncertainty refers to the ability of the model to accurately reflect the watershed's true physical runoff process. In addition to these uncertainties there may also be uncertainties related to human factors, such as the operation of reservoirs, gated levees, etc.; these uncertainties will not be dealt with here.

Uncertainties may greatly affect the accuracy of the predicted hydrograph. Dawdy, Lichty, and Bergmann (1972) found a standard error of prediction of 30 to 35 percent when using the U.S. Geological Survey Flood Hydrograph Simulation Model on three small watersheds representing different geographic and climatic regions of the U.S. Bras and Rodriguez-Iturbe (1976) found that, contrary to the usual assumption, the spatial variability of rainfall is not filtered out by the natural runoff process and must be accounted for in the modeling process. Furthermore, the rainfall uncertainty primarily affects the peak and rising limb of the hydrograph, which are the key hydrograph characteristics for a flood warning and preparedness system. Troutman (1982, 1983) concluded that, even if measured rainfall at a small number of gages is equal to the areal average rainfall in an expected-value sense (i.e., over a large number of storms), the variance of basin average precipitation is always less than the variance of point rainfall, and this difference in variability can result in serious biases in runoff prediction. Wood (1976) found that considering the uncertainty in a single parameter of a simple hydrologic model significantly changed flood frequency curves derived from use of this model and the joint probability distribution of rainfall intensity and duration relative to the case where the uncertain parameter was simply set to its mean value. Hence, the uncertainties affecting real-time flood forecasting should be accounted for in order to have reliable information for flood warning decision making.

The traditional way to circumvent the forecasting uncertainty problem is through the use of updating. Updating involves adjusting the values of the model parameters or the rainfall input such that the predicted hydrograph agrees with the available runoff measurements. In recent years, several authors have proposed complex adaptive filtering techniques to improve the "state-of-the-art" in updating (e.g., Kitanidis and Bras, 1980a and b; Wood, 1981; and Georgakos and Bras, 1982).

An alternative and complimentary approach for considering the uncertainties in real-time flood forecasting is to apply system reliability analysis to the hydrologic modeling process. System reliability analysis examines each of the individual sources of uncertainty and relates them to their total effect on the performance of the system (i.e., structure, watershed, etc.) in the form of a "failure" probability. System "failure" is generally defined as a case where the load on the system exceeds its capacity to resist that load. For the case of a flood warning system the load may be thought of as the predicted flood peak and the capacity as prespecified flood danger levels. Hence, system reliability analysis provides flood warning decision makers with an estimate of the probability that the peak discharge will exceed prespecified flood danger levels. Such information will undoubtedly provide aid to efficient operation of the flood warning system.

Thus, the potential benefits of a methodology that combines system reliability analysis with real-time flood forecasting include:

1. The enhancement of the forecast itself by providing a measure of its precision;

2. Information which could allow flood warnings to be issued in such a way as to promote a more positive response (i.e., an anticipatory response) from the flood plain dweller;

3. More complete information on flood potential which can reduce "false alarms" and unanticipated floods, leading to increased public confidence in the warning system and, hence, improved response to warnings;

4. A criterion for determining the optimal level of model complexity required for flood warning systems to achieve a prescribed level of reliability;

5. Determination of which hydrologic areas -- e.g., data, model parameters, and/or model structure -- that would provide the greatest forecast improvement;

6. Determination of the expected flood damage from the current storm or simulated storms, provided that the economic data on avoidable flood loss versus stage is available.

Despite its apparent usefulness for hydrologic problems, system reliability analysis has been rarely used in hydrology. Perhaps the primary reason for this is that the methods are relatively new to hydrologists. In addition, there are questions regarding the accuracy and validity of the different approximate techniques of system reliability analysis. This paper addresses these questions by examining one of the most promising system reliability analysis methods, the advanced first-order second moment (AFOSM) method, and comparing the results of this method to the analytical results obtained by Wood (1976) for the problem of deriving flood frequency curves from rainfall intensity probability distributions and a runoff model with one uncertain parameter.

2. DEFINITION OF SYSTEM RISK AND RELIABILITY

The risk associated with an engineering system is simply the probability that the system will fail to perform the function for which it was designed. For engineering systems, system failure occurs when the load, L, placed on the system exceeds the system's capacity to resist, R. For a flood warning system, the resistance may be considered as the critical flood stage (e.g., bankfull) whose exceedance leads to flooding of low lying areas where damages could be reduced via flood warning. The load is the peak discharge estimated by the real-time flood forecasting model. From the previous discussion, it is seen that the peak discharge estimate (load) is highly uncertain. Furthermore, since the critical flood stage (resistance) is generally converted to an equivalent critical flood discharge for comparison to the estimated peak discharge, the resistance is also uncertain due to errors in the stage-discharge relation. Thus, the flood risk (system risk), R_s, may then be defined as

$$R_s = 1 - R_\ell = P_r(L > R) \tag{1}$$

where $P_r(X)$ is the probability of event X occurring and R_ℓ is the system reliability.

A convenient way to evaluate system failure risk is to use a system performance function, Z, which relates the basic variables of the system to the load and the resistance. The performance function may take on several equivalent mathematical forms where a negative value indicates system failure, e.g., R-L, (R/L)-1, and $\ell n(R/L)$. Thus, the system risk can be defined as

$$R_s = P_r(Z < 0) \tag{2}$$

where $Z = g(x_1, x_2, \ldots, x_m) = g(\underline{x})$

$\underline{x}$ = the vector of the basic variables of the system.

In recent years a number of methods have been developed to evaluate the system risk (Yen, 1985). Among them the most promising is the AFOSM method which is adopted in this paper.

3. ADVANCED FIRST-ORDER SECOND MOMENT (AFOSM) METHOD

The essence of the AFOSM method is linearization of the performance function via Taylor series expansion at a likely failure point $(\underline{x}^*)$ on the failure surface, i.e., when the performance function, $g(\underline{x}^*)$, equals zero. Details on the AFOSM method are presented in Ang and Tang (1984, Chapter 6) and Yen, Cheng, and Melching (1986) and so only a brief synopsis will be presented here.

The expected value and variance (assuming the basic variables are uncorrelated) of the performance function as approximated by a first-order Taylor series expansion at point $\underline{x}^*$ are.

$$E[Z] \approx g(\underline{x}^*) + \sum_{i=1}^{m} C_i(\overline{x_i} - x_i^*) \tag{3}$$

$$VAR(Z) = \sigma_Z^2 \approx \sum_{i=1}^{m} C_i^2\, VAR(x_i) \tag{4}$$

$$\sigma_Z \approx \left(\sum_{i=1}^{m} (C_i\, \sigma_i)^2 \right)^{1/2} \tag{5}$$

where $\overline{x_i}$ = mean value of basic variable i,
σ_i = standard deviation of basic variable i,
C_i = $\partial g/\partial x_i$ evaluated at $\underline{x}^*$.

The inverse of the coefficient of variation for the system is defined as the system reliability index, β, i.e.,

$$\beta = E[Z]/\sigma_Z \tag{6}$$

For a linear failure surface where all the basic variables are normally distributed, the probability of failure (risk) is exactly

$$R_s = 1 - \Phi(\beta) \tag{7}$$

where $\Phi(.)$ is the standard integral of the normal distribution. While for a nonlinear convex failure surface where all the basic variables are normally distributed, the probability of failure (risk) is bounded as

$$1 - \Phi(\beta) \leq R_s \leq 1 - \chi_m^2(\beta^2) \tag{8}$$

where $\chi_m^2(.)$ is a chi-squared distribution with m degrees of freedom. However, for most real systems not all the basic variables are normally distributed. Thus, Rackwitz (1976) proposed a method for transforming non-normal variables into equivalent normally distributed variables. Rackwitz and Fiessler (1978) point out that this normal transformation is exact within the accuracy of the first-order theory under consideration, and so Eqs. 7 and 8 may be used for problems with transformed normal basic variables. Finally, Rackwitz (1976) points out that for practical nonlinear performance function structural design problems the lower bound of Eq. 8 has provided probability estimates which are acceptable within the accuracy of the first-order approximation. Ang and Tang (1984, p. 383) further note that such estimates have been found to be quite accurate for a variety of problems with nonlinear performance functions which are typical of those found in practical engineering problems.

The major difficulty involved in the AFOSM method is the determination of the appropriate point on the failure surface at which to perform the Taylor series expansion. This point may be determined by a standard iteration technique (see Ang and Tang, 1984, p. 361) or by constrained nonlinear optimization as proposed by Cheng, Yen and Tang (1982). The primary advantage of Cheng's constrained nonlinear optimization approach is that it uses the generalized reduced gradient algorithm (GRG) to solve for the failure point. Sandgren and Ragsdell (1980a and b) found GRG to be far more efficient than the other constrained nonlinear optimization methods and capable of solving a wide variety of complex nonlinear programming problems. Hence, the AFOSM method using GRG to solve for the failure point is a fairly robust approach capable of examining a wide variety of problems. Therefore, Cheng's constrained nonlinear optimization approach was employed in this paper.

The AFOSM method is relatively simple, robust, and accurate, requiring knowledge of only the mean and variance of the basic variables. These properties have led this method to become extremely popular for safety and reliability studies in structural engineering. Several agencies have encouraged the use of the AFOSM method as a scientific basis for structural safety codes (Rackwitz, 1976, and CIRIA, 1977).

The AFOSM method has also been used quite successfully in assessing the risk for several types of hydraulic structures. Kooman et al. (1978) and Vrijling (1982) used the method to check the reliability of the structural components of the Oosterschelde storm surge barrier in The Netherlands. Melching and Yen (1984) used the method to analyze the effects of construction errors (as they affect the pipe design slope) on storm sewer reliability. Cheng, Yen and Tang (1982) used the method to evaluate dam overtopping risk. Furthermore, they compared the various risk evaluation methods for some simple examples. They concluded considering accuracy, consistency, and computational cost, the AFOSM method with the GRG optimization technique is highly appropriate for dam safety evaluation or for any other system where risk levels less than 10^{-3} are important.

Despite the success achieved for structural engineering (including hydraulic structures), first-order system reliability analysis methods (as well as the other system reliability analysis methods) have been used sparingly in the study of hydrologic system uncertainties. In the following sections the AFOSM method will be applied to a simple example involving an analytical rainfall-runoff flood frequency model developed by Eagleson (1972) with an uncertain water loss parameter. This simple example application serves the dual purpose of familiarizing hydrologists with the AFOSM method and demonstrating this method's validity via comparison with Wood's (1976) analytical solution to the problem.

4. FLOOD FREQUENCY DERIVED FROM RUNOFF MODELING

Eagleson (1972) derived flood frequency curves by using a simple hydrologic model to transform a joint probability distribution of rainfall duration, t_d, and intensity, i, into an equivalent peak discharge distribution. Eagleson showed that in the intensity-duration plane there exists a curve of constant peak discharges, Q_{max}, such that all combinations of intensity and duration to the lower left of this boundary produce discharges less than Q_{max}. Figure 1 displays the general form of this curve. By integrating the area under this curve the probability of the peak discharge being less than or equal to Q_{max} may be evaluated. Thus, Eagleson found the exceedance probability of Q_{max} conditioned upon the mean value of the spatially and temporally averaged potential loss rate, ϕ (in mm/hr or in./hr) to be

$$G(Q_{max}|\bar{\phi}) = I_o \exp\left(-\frac{\eta Q_{max}}{k_1 A_r K} - \frac{\eta}{K}\bar{\phi}\right) \tag{9}$$

where Q_{max} = peak discharge whose exceedance probability is sought, m^3/sec (cfs);
A_r = direct runoff producing area, km^2 (sq. miles);
η = parameter of the (exponential) distribution of point rainfall intensity, hr/mm (hr/in.);
K = factor reducing point rainstorm depth to average over A_r for events of common probability;
k_1 = 0.278 for SI units (645 for English units);
I_o = $\exp(-\sigma/N)\ \sigma^{-\sigma+1}\ \Gamma(\sigma)$;
N = an exponent, taken as 1/2;
$\Gamma(.)$ = Gamma function of (.);
σ = $\left[k_2 \frac{\eta\lambda^2 A_r}{K\alpha_c L_s}\left(1 - \frac{k_3 \alpha_s^{4/3} A_r^2}{\alpha_c L_s^3 Q_{max}^{1/3}}\right)\right]^{1/3}$
λ = parameter of the (exponential) distribution of storm duration, hr^{-1};
α_c = parameter of overland flow, sec^{-1};
α_s = parameter of streamflow, sec^{-1};
L_s = average stream length, km (mi);
= $(3\ A_c)^{1/2}$ (Eagleson, 1970, p. 379);
A_c = catchment area, km^2 (sq. miles);
k_2 = 34.7 for SI units (2.21 for English units);
k_3 = 347 for SI units (1830 for English units).

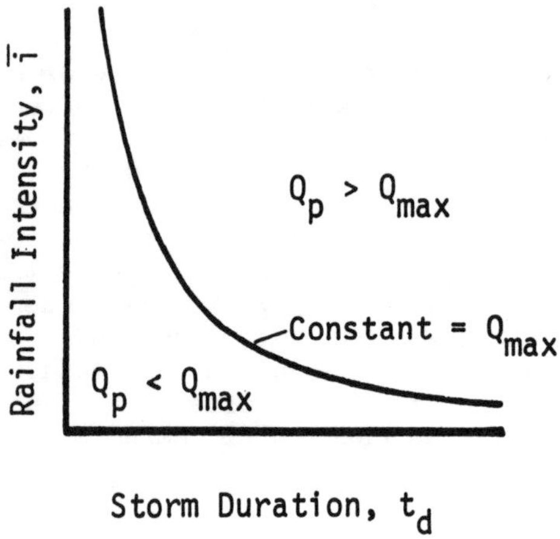

Figure 1. General relationship between peak discharge, Qp, and rainfall intensity and duration.

4.1. Wood's Consideration of Parameter Uncertainty

Wood (1976) extended Eagleson's (1972) work to consider uncertainty in the potential water loss rate, ϕ. He noted that if two random variables are functionally related, for example, $y = g(x)$, and if the function is monotonic and continuous, then the following relationships hold:

$$E[y^n] = \int_x g(x) \cdot f(x)\, dx$$
$$f(y) = f(x) \,.\, |dx/dy| \qquad (10)$$

where f(.) is the probability density function of (.). By defining y as $G(Q_{max}|\phi)$ these relationships were used to obtain the marginal probability density function, pdf, and the moments for the exceedance probability, given the peak discharge, which reflect the uncertainty in ϕ. For uniformly distributed ϕ the first and second moments of y are

$$E[y] = \frac{DK}{\eta(\phi^o - \phi_o)} \exp(-\phi_o \eta/K) \qquad (11)$$

$$E[y^2] = \frac{D^2K}{2\eta(\phi^o - \phi_o)} \left[\exp(-2\phi_o \eta/K) - \exp(-2\phi^o \eta/K)\right] \qquad (12)$$

where ϕ_o = lower bound for ϕ, mm/hr (in./hr);
ϕ^o = upper bound for ϕ, mm/hr (in./hr);
$D = I_o \cdot \exp[-\eta Q_{max}/(k_1 K A_r)]$

for gamma distributed ϕ the first and second moments of y are

$$E[y] = D\{\alpha/[\alpha + (\eta/K)]\}^r \tag{13}$$

$$E[y^2] = D^2\{\alpha/[\alpha + (2\eta/K)]\}^r \tag{14}$$

where α and r are parameters of the gamma distribution with the pdf

$$f(\phi) = \exp(-\alpha\phi)\ \phi^{r-1}\ \alpha^r/\Gamma(r). \tag{15}$$

Wood (1976) compared the flood frequency curve calculated using Eqn. 9 to those evaluated considering the random nature of ϕ to be represented by a uniform, gamma, or exponential (gamma with r = 1) distribution. He found that the flood exceedance probability significantly increased when the random nature of ϕ was considered. Thus, he demonstrated that E[g(x)] is not equal to g(E[x]), although the difference between them becomes acceptably small as the coefficient of variation of the random variable becomes small (say 0.1). Hence, when using a hydrologic model, the best estimates of the model parameters will not always give the best (truest) prediction and some consideration of the uncertainty in these parameters should be made.

Wood (1976) also plotted the one standard deviation confidence limits for the flood frequency curves derived considering the random nature of ϕ. As expected, the size of these limits is a function of the coefficient of variation of the random variable. Nevertheless, the size of the confidence limits for most of the cases Wood studied further displayed the need to consider uncertainty in hydrology.

4.2 AFOSM Analysis of the Flood Frequency Problem

The AFOSM method provides an interesting extension of the parameter uncertainty considerations made by Wood (1976). By identifying the most likely combination of ϕ, $\bar{i}$, and t_d which produces a peak discharge, Q_{max}, and the probability of this combination, the AFOSM method to some degree combines the expected value and confidence limits curves defined by Wood into an overall total exceedance probability curve. How reasonable this total exceedance probability curve is relative to the findings of Wood is discussed below.

According to Eagleson (1972) for Boston data the joint probability distribution of rainfall intensity and duration is reasonably approximated as

$$f(\bar{i},t_d) = \frac{\lambda\eta}{K} \exp(-\lambda t_d - \frac{\eta}{K}\bar{i}) \tag{16}$$

where η, λ, and K equal 1.18 hr/mm (30 hr/in.), 0.13 hr^{-1}, and 0.95, respectively. The relationship between rainfall duration, the overland flow time of concentration, t_c, and the total runoff producing area time of concentration, t_*, defines the effects of rainfall duration on peak discharge. The peak discharge is estimated according to a kinematic wave approximation by one of three different equations depending on whether $t_d < t_c$, $t_c \leq t_d < t_*$, or $t_d > t_*$. For the extreme levels of peak discharge considered here, the most likely combination of ϕ, $\bar{i}$, and t_d which produces the desired Q_{max} will be one where $t_d \geq t_*$ for which

$$Q_{max} = k_1 A_r(\bar{i} - \phi). \tag{17}$$

Hence, the rainfall duration need not be explicitly considered in the AFOSM analysis, and so only the pdf of intensity, $f(\bar{i})$, is considered, where

$$f(\bar{i}) = (\eta/K) \exp(-\eta\bar{i}/K) \tag{18}$$

It should be noted that the above approach is consistent with Eagleson's (1972) original approach since he found that his rainfall model implied the independence of rainfall intensity and duration.

The system performance function for the flood frequency problem simply becomes

$$Z = Q_{max} - k_1 A_r (\bar{i} - \phi) \tag{19}$$

In this analysis, Wood's (1976) example will be followed. Hence A_r is taken as one-third of the hypothetical catchment area of 259 km^2 (100 sq. miles) and α_s and α_c are taken as 0.1 sec^{-1} and 10 sec^{-1}, respectively. By varying Q_{max} and using Eq. 18 to describe the random rainfall intensity, curves of total flood exceedance probability, R_F, were derived for ϕ being uniformly distributed ($0 \leq \phi \leq 2.54$ mm/hr = 0.1 in./hr, cov = 0.577 where cov = mean/standard derivation of ϕ), exponentially distributed (α = 0.787 for ϕ in SI units and 20 for ϕ in English units, cov = 1), or gamma distributed (case A, r = 3, α = 2.36 for ϕ in SI units and 60 for ϕ in English units, cov = 0.577; and case B, r = 10, α = 7.87 for ϕ in SI units and 200 for ϕ in English units, cov = 0.316). Figures 2-5 display each of these curves of the total flood exceedance probability, R_F, along with the corresponding curves of the expected flood exceedance probability, E(y), and the upper one standard deviation confidence limit, $E(y) + \sigma_y$ (where $\sigma_y = (E[y^2] - E[y]^2)^{1/2}$ = standard deviation of the flood exceedance probability), derived by Wood (1976).

4.3 Comparison of AFOSM and Wood's Analyses

As shown in the figures, the total flood exceedance probability is approximately parallel to the semilog plot of the expected value of the flood exceedance probability and its magnitude is typically three times greater than the expected value (a detailed comparison is given in Table I). To understand the difference between these two curves

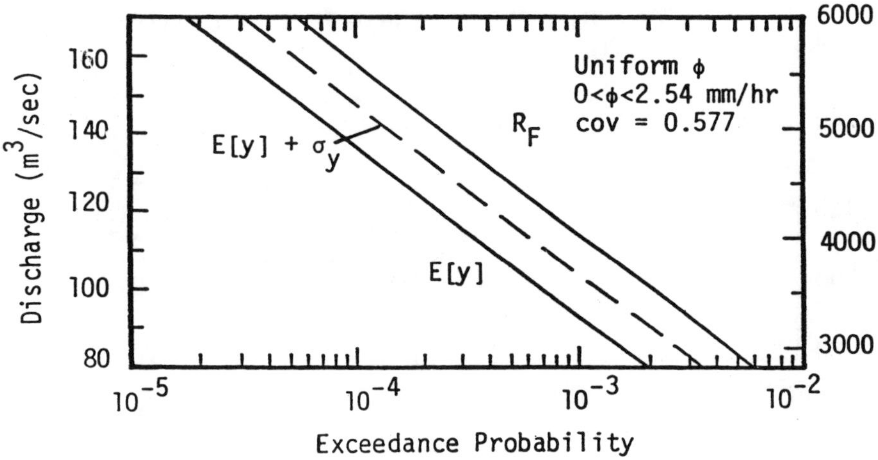

Figure 2. Comparison of R_F and E[y] for uniformly distributed loss rate.

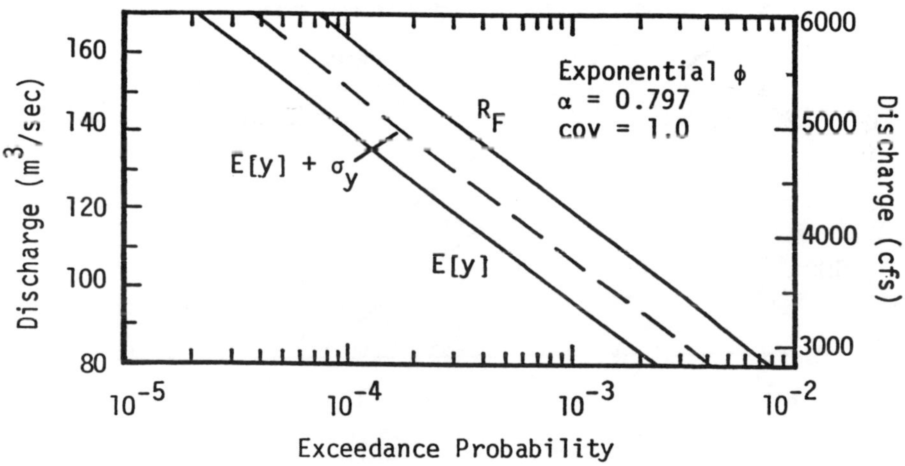

Figure 3. Comparison of R_F and E[y] for exponentially distributed loss rate.

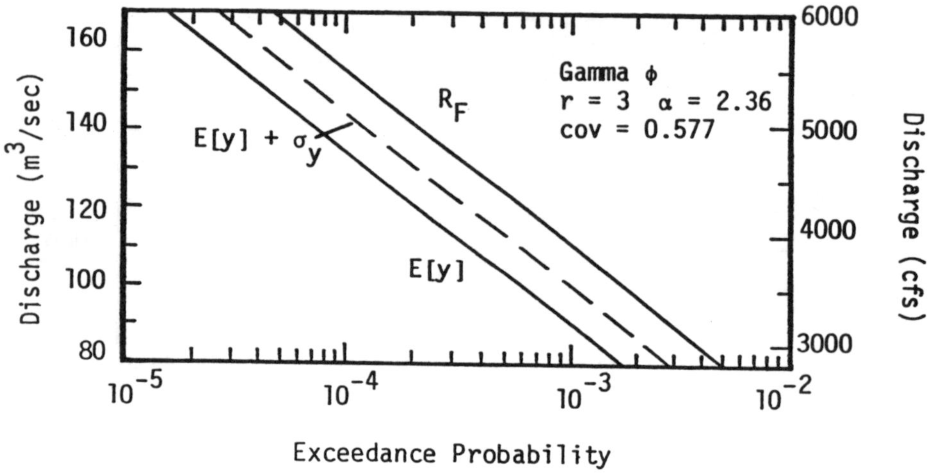

Figure 4. Comparison of R_F and E[y] for gamma distribution of loss rate, case A.

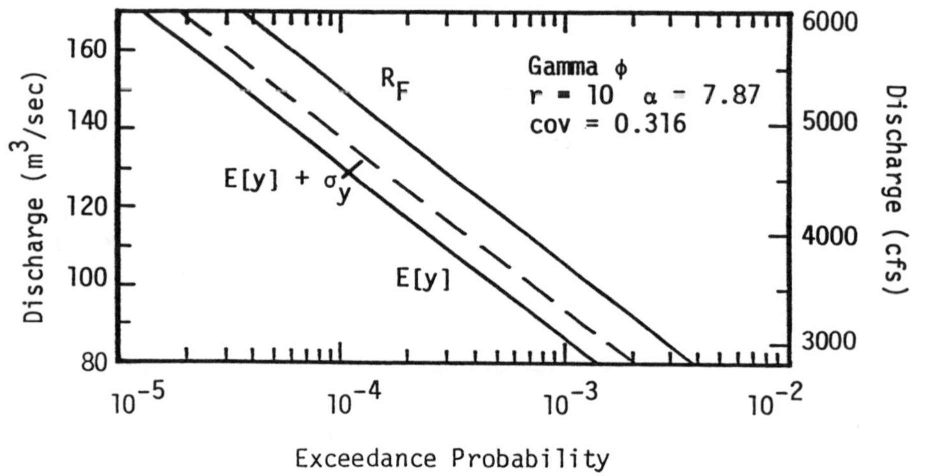

Figure 5. Comparison of R_F and E[y] for gamma distribution of loss rate, case B.

and their respective utility, it is necessary to consider the problem from a "forecaster's view" and a "designer's view." As a forecaster, one is interested in the most likely (expected) outcome from a system or process involving random variables, which in this case is provided by Wood's (1976) analysis. As a designer, one is interested in the probability of system failure and the most likely combination of the random variables which leads to failure. Such combinations and their probabilities are identified by the AFOSM method. Furthermore, for extreme discharges, such a combination will be at an extreme condition rather than the mean condition of the flood frequency relation. Hence, the large difference in magnitude was anticipated.

Table I. Comparison of results from AFOSM and Wood (1976)

Distribution	$R_F/E[y]$		$(R_F - E[y])/\sigma_y$		R_F as Percent Confidence Limit of $E[y]$	Relative Probability $\Phi[R_F-E[y]/\sigma_y]$
	Q_{max} = 85 m^3/sec	Q_{max} = 170 m^3/sec	Q_{max} = 85 m^3/sec	Q_{max} = 170 m^3/sec		
Uniform	2.96	3.05	2.59	2.71	99.18	0.00408
Exponential	3.33	3.42	3.02	3.13	99.79	0.00107
Gamma (A)	2.94	3.01	2.85	2.95	99.62	0.00189
Gamma (B)	2.73	2.78	3.81	3.91	99.99	0.00006

The extreme condition of the flood frequency relation represented by R_F found by the AFOSM method may be evaluated considering the difference between R_F and $E[y]$ in terms of standard deviations, as given in Table I. Furthermore, if it is assumed that the variations of the flood frequency relation about its expected value are normally distributed, R_F may be identified as a percent confidence limit on $E[y]$. As shown in Table I, the R_F curves identified by the AFOSM method correspond to 99 percent or greater confidence limits on $E[y]$. The range of $E[y]$ for the discharges considered is 10^{-3} to 10^{-5}. Hence, it follows that the total flood exceedance probability would be defined by a confidence limit corresponding to a similar probability of exceedance, and, as shown by the relative probability in Table I, this is the case for the R_F curves derived by the AFOSM method. Thus, the AFOSM method provides a reasonable estimate of the total flood exceedance probability relative to what might be anticipated based on Wood's (1976) work.

The fact that the R_F and $E[y]$ curves are nearly parallel on the semi-logarithmic plot also supports the contention that Rackwitz's (1976) normal transformation scheme for non-normal basic variables is exact within the accuracy of the first-order theory under consideration. The accuracy of the normal transformation further supports the contention that the AFOSM method provides valid results for hydrologic problems.

In summary, the results of this comparison not only show that AFOSM method provides a reasonable and valid analysis of hydrologic uncertainties, but it also demonstrates the AFOSM method's advantages for many practical problems. In practice, engineers tend to take a "designer's view," wishing to know the overall, total probability of occurrence of an undesirable (failure) event so that projects may be designed and/or operated to keep this probability at an acceptable level. Wood (1976) showed that the traditional method of using a model with the parameter "best estimates" without considering parameter uncertainty may greatly underestimate the flood exceedance probability. Similarly, it is shown here that even Wood's analytical solution does not provide an adequate representation of the true overall flood exceedance probability, as Wood's approach underestimates this probability by a factor of 3. Furthermore, similar results have been noted for structural design problems (e.g., Vrijling, 1982). Therefore, the AFOSM method provides reasonable estimates of the overall system failure probability which, in turn, is of great practical interest for design, planning, and operation of engineering systems.

5. CONCLUSIONS

The validity and practicality of the AFOSM method are demonstrated by a simple example wherein the uncertainties in an analytical rainfall-runoff model are evaluated. A simple kinematic wave runoff model is used to convert rainfall excess to runoff with the rainfall excess being a function of random variables $\bar{i}$ (average rainfall intensity) and ϕ (water loss rate). The AFOSM method provides a very reasonable estimate of the total flood exceedance probability in comparison to Wood's (1976) analytical derivation of the expected flood exceedance probability and corresponding confidence limits for this problem.

From a "designer's view" the large difference between the total and expected flood exceedance probabilities points out an important practical advantage of system reliability analysis, in general, and the AFOSM method, in particular. That is, current methods of estimating flood exceedance probabilities using rainfall-runoff models with mean parameter values greatly underestimate the true probability of exceedance. In fact, even analytical methods considering parameter uncertainty fail to give the true overall flood exceedance probability picture. Hence, system reliability analysis provides a much more useful estimate of the exceedance probability. Furthermore, similar results have been found for structural design problems.

Another key practical advantage of system reliability analysis methods is that they can consider all the sources of uncertainties -- natural, data, model parameters, and model structure -- while analytical methods can handle only a few sources. Thus, system reliability analysis methods should prove useful in analyzing the uncertainties in real-time flood forecasting, and hence provide useful information for flood warning decision making. Currently, at the

University of Illinois at Urbana-Champaign research is in progress to study the application of system reliability analysis to uncertainties in realistic, real-time flood forecasting models and scenarios.

In summary, this paper has demonstrated the validity and practicality of system reliability analysis for hydrologic problems. It is hoped that through this simple introduction, hydrologists will become aware of the potential utility of system reliability analysis and extend its use to hydrologic problems subject to uncertainties.

6. REFERENCES

Ang, A. H.-S., and Tang, W. H. 1984. Probability Concepts in Engineering Planning and Design, Vol. II: Decision, Risk, and Reliability. John Wiley & Sons Inc., New York.

Bras, R. L., and Rodriguez-Iturbe, I. 1976. 'Rainfall Network Design for Runoff Prediction,' Water Resources Research, 12(6):1197-1208.

Cheng, S.-T., Yen, B. C., and Tang, W. H. 1982. 'Overtopping Risk for an Existing Dam,' Civil Engineering Series Hydraulic Engineering Studies No. 37, University of Illinois at Urbana-Champaign.

CIRIA. 1977. 'Rationalization of Safety and Serviceability Factors in Structural Codes,' Report 63, London.

Dawdy, D. R., Lichty, R. W., and Bergmann, J. M. 1972. 'A Rainfall-Runoff Simulation Model for Estimation of Flood Peaks from Small Drainage Basins,' USGS Professional Paper 506-B, U.S. Geological Survey, Washington, D.C.

Eagleson, P. S. 1970. Dynamic Hydrology. McGraw-Hill Book Co., New York.

Eagleson, P. S. 1972. 'Dynamics of Flood Frequency,' Water Resources Research, 8(4):878-898.

Georgakokos, K. P. and Bras, R. L. 1982. 'Real-Time, Statistically Linearized, Adaptive Flood Routing,' Water Resources Research, 18(3):513-524.

Kitanidis, P. K., and Bras, R. L. 1980a. 'Real-Time Forecasting With a Conceptual Hydrologic Model 1. Analysis of Uncertainty,' Water Resources Research, 16(6):1025-1033.

Kitanidis, P. K., and Bras, R. L. 1980b. 'Real-Time Forecasting With a Conceptual Hydrologic Model 2. Applications and Results,' Water Resources Research, 16(6):1034-1044.

Kooman, D., Vrijling, J. K., Maldir, T., and de Quelirij, L. 1978. Probabilistic Approach to Determine Loads and Safety Factors. Symposium on Foundation Aspects of Coastal Structures, Delft.

Melching, C. S., and Yen, B. C. 1984. 'Slope Influence on Storm Sewer Risk,' Proceedings, 4th IAHR International Symposium on Stochastic Hydraulics, Urbana, Illinois, 277-284.

Rackwitz, R. 1976. 'Practical Probabilistic Approach to Design,' Bulletin 112, Comite European du Beton, Paris, France.

Rackwitz, R., and Fiessler, B. 1978. 'Non-normal Vectors in Structural Reliability,' SFB 96 Report 29, Technical University of Munich, 1-22.

Sandgren, E., and Ragsdell, K. M. 1980a. 'The Utility of Nonlinear Programming Algorithms: A Comparative Study - Part I,' Journal of Mechanical Design, ASME, **102**(3):540-546.

Sandgren, E., and Ragsdell, K. M. 1980b. 'The Utility of Nonlinear Programming Algorithms: A Comparative Study - Part II,' Journal of Mechanical Design, ASME, **102**(3):547-551.

Troutman, B. M. 1982. 'An Analysis of Input Errors in Precipitation-Runoff Models Using Regression with Errors in the Independent Variable,' Water Resources Research, **18**(4):947-964.

Troutman, B. M. 1983. 'Runoff Prediction Errors and Bias in Parameter Estimation Induced by Spatial Variability of Precipitation,' Water Resources Research, **19**(3), pp. 791-810.

Vrijling, J. K. 1982. 'Probabilistische Methodes in hit Ontwerp van de Stormvlocdkering,' Cement, **34**(11):721-728.

Wood, E. F. 1976. 'An Analysis of the Effects of Parameter Uncertainty in Deterministic Hydrologic Models,' Water Resources Research, **12**(5):925-932.

Wood, E. F. 1981. 'Filtering of Partitioned Large Scale Hydrologic Systems,' Hydrological Sciences Bulletin, **26**(1):33-46.

Yen, B. C. 1985. Reliability of Hydraulic Structures Possessing Random Loading and Resistance. Lecture notes, NATO Advanced Study Institute on Engineering Reliability and Risk in Water Resources, Tucson, Arizona.

Yen, B. C., Cheng, S.T., and Melching, C.S. 1986. 'First Order Reliability Analysis,' in Stochastic and Risk Analysis in Hydraulic Engineering (ed. by B. C. Yen) 1-36, Water Resources Publications, Littleton, Colorado.

A STOCHASTIC APPROACH TO FLOOD DAMAGE ESTIMATION

P. Ouellette
Researcher, Centre de Recherche et Développement en économique
Université de Montréal

N. El-Jabi
Associate Professor, School of Engineering
Université de Moncton

J. Rousselle
Professor, École Polytechnique de Montréal
P.O. Box 6079, Station "A"
Montréal, Québec H3C 3A7
Canada

ABSTRACT. Effective flood plain management requires estimation of the costs and benefits of all contemplated projects. In this study the focus is on estimating the benefits of such schemes. Starting from the similarity between flood flow and flood damage time series, the authors take a probabilistic approach to flood damage estimation. They first develop a hydroeconomic model to assess flood-related damages and then derive a damage distribution function by applying the theory of extreme values and the sum of the random variables to the estimated damage series. It is assumed that the values of extreme damages are independent and identically distributed over the time interval $(0,t]$ of one year and one season simultaneously. The distribution function can then be used to estimate the benefits of flood plain management projects. The Richelieu River basin has been used for a numerical application because of its combined rural and urban characteristics and the fairly large amount of information acquired through previous studies.

1. INTRODUCTION

Floods are a source of physical and material damage and a hindrance to human activity. In some areas of the world, flooding is a major problem, with large sums of money being spent to protect potential flood victims and rehabilitate devasted areas. In seeking to rationalize their interventions, authorities have been prompted to

V. P. Singh (ed.), Application of Frequency and Risk in Water Resources, 351–360.

evaluate the impact of their projects on the target areas. The benefits of such intervention, although many, center on reducing the damages sustained by the different economic sectors, namely the residential, industrial, commercial, agricultural and public works sectors.

The evaluation of flood damage reduction is a twofold process calling for (a) estimation of flood damages prior to intervention and (b) estimation of the proportion of damages eliminated by such action. The second preoccupation lies in the area of civil engineering and will not be broached in this paper, which proposes a methodology for flood damage estimation. The thrust of this paper lies in setting up a comprehensive management information system which incorporates the hydrologic characteristics of flooding and the economic characteristics of the study area. The chief outcome of this approach is the estimation of mean annual damages, a key variable in the cost-benefit analysis of flood plain management. It differs from other probabilistic approaches in that it does not estimate a damage distribution function solely from the maximum damage occurrence within a given interval of time, such as one year or one season. It considers instead the distribution of the number of damage events and of the damage value within a given time interval, resulting in a broader information base. In sum, the proposed method takes account of all damage occurrences, not simply the maximum occurrence.

However, this approach requires a fairly long flood damage series. As such data are usually non-existent, we will present a simple method whereby a damage estimate can be derived for each individual flood.

2. THEORETICAL CONSIDERATIONS

2.1 Hydroeconomic Model

The damages associated with each flood comprise two types of variables: natural random variables, such as flood depth of submersion, and non-natural determinist or random variables, i.e. the economic development of the study area. The damage function can be written as

$$d = f(K,Y) \tag{1}$$

where

- d = a variable describing various possible types of physical and non-physical damage;
- K = a variable describing all the physical capital such as residences, commercial and industrial buildings, stocks, and associated activities such as production flow and domestic services;
- Y = a vector of the elements that describe the flood characteristics such as water depth and velocity and duration of submersion.

For the estimation of the equation (1) we have transformed it as follows:

$$d = \theta(Y)K \tag{2}$$

where $\theta(Y)$ is the part of K destroyed by a flood having the characteristics Y.

Given the characteristics of flood damages, the function $\theta(Y)$ is a monotonic, continuous, non-decreasing function that takes the form of an S curve (Dantzig, 1956). The Gompertz curve complies with these characteristics and was thus the one used. If, in addition, we take only the depth of submersion (z) as the hydrologic variable of damage, the damage function is expressed as

$$\frac{d}{K} = \frac{1}{e^{e^{-\alpha}}-1} \{e^{[1-e^{-\gamma^{z}}]e^{-\alpha}}-1\} \tag{3}$$

where z = the depth of submersion;
α and γ = parameters to be estimated.

In equation (3), the damage variables are regarded as unitary for the various economic sectors ($/unit). Simply knowing the depth of submersion of any given economic unit is therefore assumed to be sufficient to calculate the damage which it sustains.

Note that equation (3) considers a constant depth of submersion for the entire stock of physical capital. It is highly unlikely, however, that all of the economic units are situated on the same level and thereby subjected to the same submersion depth. The enormous task of calculating the depth of submersion for each individual unit is alleviated by determining the mean unit damage corresponding to a particular maximum depth. This mean unit damage corresponds to the damage which would occur if the economic units were uniformly distributed within a sector lying between the base level ($z = 0$) and the elevation considered ($z = z_1$). This hypothesis is valid for certain economic units (residences) but does not apply across the board. Industrial sectors, for instance, must be dealt with on a per unit basis. The mean unit damage becomes

$$\overline{(d/K)_{z_1}} = \int_0^{z_1} \frac{G(z)}{z_1}\, dz \tag{4}$$

where G(z) represents the unit damage function (equation (3)).

2.2 Model of Extreme Values

Let us examine a hydrograph which plots the instant river flow at a given station for a time interval (0,t] (Fig. 1):

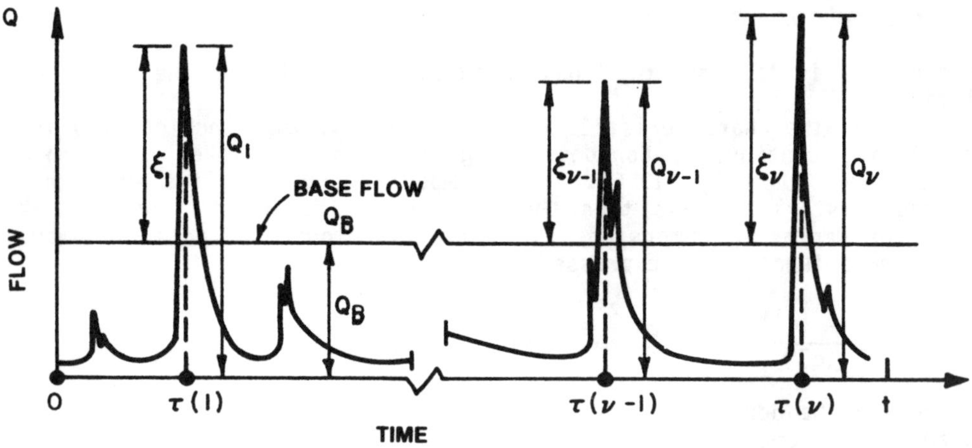

Figure 1. Hydrograph of instantaneous flow of a river at a given station.

Given the discontinuity of the flood phenomenon over time, the flood hydrograph can be plotted by means of the following model (Fig. 2):

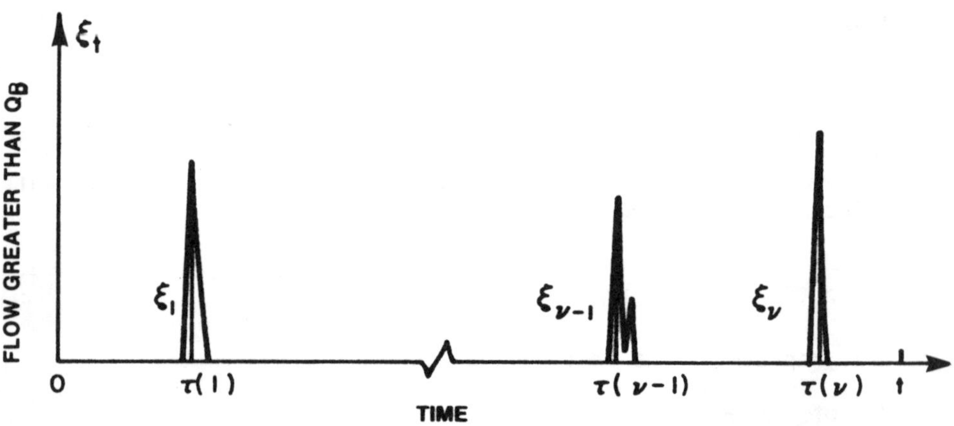

Figure 2. Hydrograph of flood flow.

$$\xi_\nu = \begin{cases} 0 & ; Q_\nu \leq Q_B \\ Q_\nu - Q_B & ; Q_\nu > Q_B \end{cases} \tag{5}$$

Considering only the maximum occurrence of the intermittent series yields a discrete, non-negative stochastic process for the exceedances in time interval (0,t]. Thus let ξ_ν be the νth exceedance at time $\tau(\nu)$ (Fig. 3).

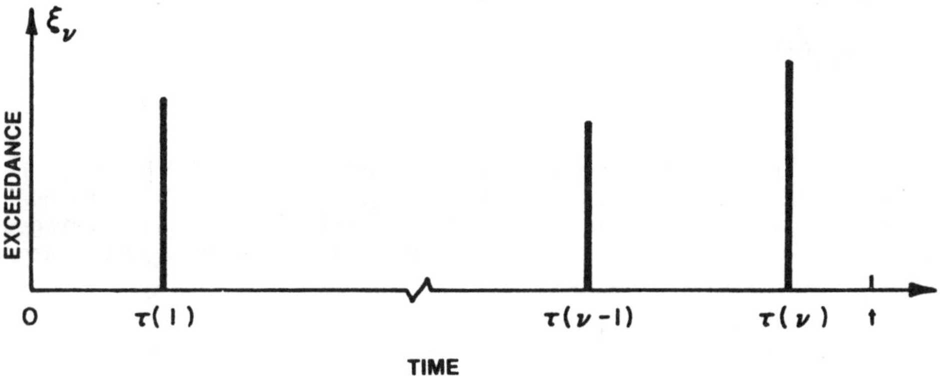

Figure 3. Stochastic process of exceedances over an interval (0,t].

The flood phenomenon will be rendered by

$$\{\ \xi_\nu;\ \nu = 0,1,2,\ldots,\ \tau \geq 0\} \tag{6}$$

The hydroeconomic model presented here associates a mean unit damage MUD_ν with each exceedance ξ_ν. For this association the mean unit damage is preferred over the total damage since the total damage series is non-stationary, a characteristic that would distort the results. The problem stems from the fact that the residence stock is not constant but displays a trend which alters the damage expectation form one period to another. This phenomenon is not encountered with the unit damage series. The non stationarity of the residence stock is taken into account by the introduction of one variable representing the growth rate of the value of this stock, see equation (14). These series can be used to derive a family of discrete stochastic processes serving as the basis for the flood damage evaluation model. The evaluation comprises two types of estimates:

- the distribution of the number of damage events estimated by (Zelenhasic, 1970) and (Todorovic and Yevjevich, 1969) as:

$$P(E_\nu^t) = \exp\{-\Lambda(t)\} \frac{\{\Lambda(t)\}^\nu}{\nu!} \tag{7}$$

where $P(E_\nu^t)$ = the probability of occurrence of ν damage $\varepsilon(0,t]$
$\Lambda(t)$ = mean number of damages in unit time $(0,t]$

- the distribution of the annual maximum unit damage, $F_t(x)$ which when estimated on an annual time base yields:

$$F_t(x) = \exp\{-\Lambda(t)\, e^{-\beta x}\} \tag{8}$$

with $\beta = \{E[MUD]\}^{-1} > 0$ where the mean unit damages (MUD) are independent and identically distributed over a one-year interval (IIDA) (Zelenhasic, 1970).

When estimated on a seasonal time base this yields:

$$F_{t_4}(x) = \exp\{-\Lambda(T_1)\, e^{-\beta_1 x} - [\Lambda(T_2) - \Lambda(T_1)]e^{-\beta_2 x}$$
$$- [\Lambda(T_3) - \Lambda(T_2)]e^{-\beta_3 x} - [\Lambda(T_4) - \Lambda(T_3)]e^{-\beta_4 x}\} \tag{9}$$

with $\Lambda(T_1)$, $[\Lambda(T_2)-\Lambda(T_1)]$, $[\Lambda(T_3)-\Lambda(T_2)]$ and $[\Lambda(T_4)-\Lambda(T_3)]$ being the mean number of damages per season and β_1, β_2, β_3, β_4 being the exponential function parameters for each season, where the mean unit damages (MUD) are independent and identically distributed over a one-season interval (IIDS) (Rousselle, 1972).

2.3 Damage/Probability Relation

The damage/probability relation is expressed here as

$$\rho(x) = 1 - F_t(x) \tag{10}$$

The damage/probability relation is in an important one, being the focus, as it were, of the theory of extreme values. Based on the assumption of independence of the number and value of flood damages, this model links flood damages with the probability of occurrence of such damages. This is of particular importance in that this relation yields the mean annual unit damage, a key variable in the profitability analysis of flood plain management programs. The mean annual unit damage (MAUD) is calculated using the formula below:

$$MAUD = \int_0^\infty \rho(x)\,dx \tag{11}$$

3. NUMERICAL APPLICATION

This section presents a numerical illustration, beginning with the development of the intermittent series of flood damages in the town of Saint-Jean on the Richelieu River over the period 1938-1981. Secondly, this series is analyzed by means of the theory of extreme values. Lastly, we comment on the interest of this approach. The application is confined to the residential sector. Note, however, that with slight modification of the application, this methodology is valid for any other economic activity.

3.1 Estimation of Mean Unit Flood Damage Series

Mean unit damages (MUD) are estimated from equation (3) as follows:

$$\text{MUD} = \frac{1}{z_1} \int_0^{z_1} \frac{1}{e^{e^{-\alpha}}-1} \{e^{[1-e^{-\gamma z}]e^{-\alpha}} -1\} \, dz \qquad (12)$$

The z_1 values for floods having occurred during the period 1938-1981 were supplied by the Canadian Department of the Environment. The parameters α and γ were estimated from data taken from the 1976 post-flood survey conducted by CRAR (1977) in the Richelieu River area.

Given the variable MUD ($=\overline{d/K}$) and z for the 1976 flood, the parameters α and γ were estimated by the method of least squares for non-linear functions (Draper and Smith, 1966). The results are: α = 0.953 and γ = 0.108.

Knowing α, γ and z allows for estimation of the mean unit damage associated with each flood. The mean unit flood damage series is given in Table I.

3.2 Application of the Extreme Value Theory

The series of mean unit flood damages sustained by Saint-Jean comprises 66 events over a 44-year period, for an average 1.5 event par year. The seasonal distribution of the damages is given in Table II.

The distribution function of the mean unit damage value is required to derive the damage distribution function, $F_t(x)$, in the case of annual damage distribution, assuming an exponential distribution of the mean unit damage values gives the parameter β = 21.671. The parameters of the exponential function for seasonal distribution of the mean unit damages are given in Table III. Using the Kolmogorov-Smirnov test, the goodness-of-fit of the observed values to the exponential function was verified at a confidence level of 5 percent.

TABLE I Mean unit Damage (MUD)

Year	MUD	Year	MUD
1938	0.074 000	1955	0.106 000
1938	0.004 860	1956	0.055 200
1938	0.003 070	1958	0.099 800
1938	0.031 200	1959	0.066 800
1938	0.011 000	1960	0.086 300
1939	0.115 000	1961	0.077 900
1940	0.088 300	1961	0.003 530
1940	0.004 830	1961	0.014 900
1942	0.055 200	1962	0.039 700
1943	0.003 070	1963	0.090 700
1943	0.093 100	1968	0.038 000
1944	0.005 270	1969	0.006 590
1944	0.068 000	1969	0.098 600
1944	0.011 800	1970	0.097 800
1945	0.074 200	1971	0.104 000
1945	0.001 750	1972	0.113 000
1945	0.070 500	1972	0.009 650
1946	0.010 990	1972	0.005 280
1946	0.007 640	1973	0.085 100
1947	0.122 000	1973	0.039 700
1948	0.034 600	1973	0.007 030
1948	0.014 900	1974	0.015 300
1948	0.007 030	1974	0.004 840
1950	0.012 700	1974	0.015 300
1950	0.008 780	1974	0.077 400
1950	0.027 000	1975	0.016 200
1950	0.039 700	1976	0.111 000
1951	0.095 800	1977	0.077 400
1952	0.062 700	1977	0.007 030
1952	0.011 000	1978	0.089 900
1953	0.035 500	1979	0.072 900
1953	0.055 200	1981	0.020 500
1954	0.086 270	1981	0.015 300

TABLE II Seasonal and Annual Values of $\Lambda(t)$

Season	Period	Interval	$\Lambda(t)$
Autumn	Sept. 21 - Dec. 20	$(0, T_1]$	$\Lambda(T_1) = 0.023$
Winter	Dec. 21 - March 20	$(T_1, T_2]$	$\Lambda(T_2) - \Lambda(T_1) = 0.045$
Spring	March 21 - June 20	$(T_2, T_3]$	$\Lambda(T_3) - \Lambda(T_2) = 1.386$
Summer	June 21 - Sept. 20	$(T_3, T_4]$	$\Lambda(T_4) - \Lambda(T_3) = 0.045$
Year	Sept. 21 - Sept. 20	$(0, T_4]$	$\Lambda(t) = 1.500$

TABLE III Seasonal and Annual Values of β

Season	Average of Mean Unit Damages, E[MUD]	$\beta = \{E[MUD]\}^{-1}$
Autumn	0.0225	44.444
Winter	0.0150	66.667
Spring	0.0488	20.487
Summer	0.0075	133.333
Year	0.0461	21.671

Given the values of the parameters $\Lambda(t)$ and β, the damage distribution function $F_t(x)$ can be determined. Using the Kolmogorov-Smirnov test, adjustment of the MUD distribution to equation (8) or (9), as appropriate, was verified at a confidence level of 1 percent. Lastly, $F_t(x)$ yields the mean unit damage/probability relation (equation (10)), which is used to estimate the mean annual unit damage (MAUD) by calculating the area under curve $\rho(x)$. For Saint-Jean, the mean annual unit damage value is 0.116 87 or 0.115 43, depending on whether flooding was assumed to be identically and independently distributed over a period of one year or one season.

Knowing the mean annual unit damage, we can estimate the actualized mean annual damage by the following relation:

$$\sum_{i=1}^{n} \text{MAD } (1+r)^{-i} = \sum_{i=1}^{n} (\text{MAUD} \cdot K_i)\ (1+r)^{-i} \tag{13}$$

where K_i is the mean annual value of the residences affected by flooding.

Supposing that K_i represents a certain proportion $\emptyset$ if the total value of the residences in the town (KT) and that value KT changes at a constant rate δ, we obtain

$$\sum_{i=1}^{n} \text{MAD } (1+r)^{-i} = \text{MAUD} \cdot \text{KT}_0 \cdot \overline{\emptyset} \cdot \sum_{i=1}^{n} ((1+\delta)/(1+r))^{i} \tag{14}$$

where $\overline{\emptyset} = E(\emptyset)$

The parameter $\emptyset$ was estimated on the assumption of a monotonic, non-decreasing, non-negative relation between the depth of submersion and the proportion of KT affected by the flood. The following function was used:

$$\emptyset = x/(x+b) \tag{15}$$

The parameter b was estimated from CRAR (1977) data. In our case, b = 656 400. The mean proportion is $\overline{\emptyset}$ = 0.009 37.

Using 1981 as the base year, KT_0 = \$3.72 million. By way of example, if the actualization factor is assigned the values r = 0.10, δ = 0.03 and n = 30, the actualized damage is

$$\sum_{i=1}^{30} \text{MAD } (1+r)^{-i} = \begin{cases} \$5.21 \text{ million (IIDA)} \\ \\ \$5.15 \text{ million (IIDS)} \end{cases}$$

4. CONCLUSIONS

This paper has presented a methodology for incorporating the stochastic nature of flooding in the process of flood damage estimation, one element of the cost-benefit analysis of flood plain management programs. We have sought here to emphasize the ease of application of this approach. However, certain flood characteristics prompt a number of comments on some aspects of such programs. The cost-benefit analysis should cover the entire drainage basin of the area in question, for most interventions designed to counter the floods sustained by a given town will have some impact on the towns located downstream and/or upstream. Moreover, unlike what was supposed here for the puposes of this paper, flood damage reduction will not be the only outcome of a given intervention. A comprehensive study should also estimate the impact which an intervention will have on the growth rate of the study area and on the relationships between the study area and areas beyond.

5. ACKNOWLEDGMENTS

This study was jointly supported by the Federal Department of the Environment and the Natural Sciences and Engineering Research Council of Canada.

6. REFERENCES

CRAR (Centre de Recherche en Aménagement Régional). 1977. Évaluation des dommages causés par les inondations en 1976 à divers niveaux d'élévation des eaux de la rivière Richelieu et de la baie Missisquoi, Haut-Richelieu, Québec, Université de Sherbrooke, 210 pages.

Dantzig, D.V. 1956. "Economic Decision Problems for Flood Prevention". Econometrica, Vol. **24**, No. 3, pp. 276-287.

Draper, N.R. and Smith, H. 1966. Applied Regression Analysis, Wiley, 407 pages.

Rousselle, J. 1972. On Some Problems of Flood Analysis, Ph.D. Thesis, Colorado State University, 226 pages.

Todorovic, P. and Yevjevich, V. 1969. Stochastic process of precipitation, Colorado State University, Hydrology Paper No. 35, 61 pages.

Zelenhasic, E. 1970. Theoritical probability distributions for flood peaks, Colorado State University, Hydrology paper, No. 42, 35 pages.

FLOOD HAZARD ASSESSMENT ON ALLUVIAL FANS: AN EXAMINATION OF THE METHODOLOGY

Richard H. French, Ph.D., P.E.
Water Resources Center
Desert Research Institute
2505 E. Chandler, Suite #1
Las Vegas, Nevada 89120

ABSTRACT. The recognition and assessment of the hazard that flash floods can pose to structures located on alluvial fans is seriously deficient relative to the ability of the engineer to assess flood risk to structures located in the vicinity of perennial rivers. With the development of major urban areas and hazardous, radioactive, and mixed waste sites on alluvial fans in the southwestern United States, there is both an increasing need and interest in developing rational and reliable techniques for assessing the hazards floods pose to these developments and facilities.

In this discussion, the three primary assumptions on which the assessment of flood hazard on alluvial fans are usually based are critically examined. Of these three assumptions, two are demonstrated to be reasonable based on a survey of the pertinent geologic and engineering literature and available field data while the third assumption which quantifies alluvial channel geometry during extreme flow events is shown to be invalid. A number of methods which may be used to predict alluvial channel geometry are discussed along with critical research needs.

1. INTRODUCTION

Although alluvial fans are common features of the landscape throughout the world, especially in arid and semi-arid regions, the hydraulic processes which formed these features and continue to modify them are poorly understood from the viewpoint of hydraulic/water resources engineering. An improved, quantitative understanding of hydraulic processes on alluvial fans is crucial for two reasons. First, the recognition and assessment of the hazard that flash floods can pose to structures located on alluvial fans is seriously deficient relative to the ability of the engineer to assess the flood risk to structures located in the vicinity of perennial rivers. To a large extent, this situation exists because until recently there has been a general lack of economic development in desert areas; and therefore, there is a relatively short historical record of flood damage to structures on

V. P. Singh (ed.), Application of Frequency and Risk in Water Resources, 361–375.

alluvial fans. With the development of major urban areas and the location of hazardous and radioactive waste management sites on alluvial fans in the southwestern United States, there is both an increasing need and interest in developing rational and reliable techniques for assessing the hazards floods pose to such developments and facilities. Second, in many arid and desert areas alluvial fans are an important source of groundwater; and further, recharge to many groundwater basins is through alluvial fan deposits. A better understanding of the hydraulic processes that form alluvial fans would aid the water resources engineer and hydrogeologist in estimating aquifer parameters and/or interpreting well-logs and aquifer tests in arid areas.

The study of hydraulic processes on alluvial fans is difficult both because of a lack of data and because it requires an understanding of geology, geomorphology, and hydraulic engineering. Further, much of the geological literature regarding alluvial fans is qualitative rather than quantitative; and much of the literature of hydraulic engineering regarding the behavior of alluvial channels is not applicable to the types of flows and alluvial channels found on fans. Finally, it should be mentioned that the seriousness of flooding on alluvial fans is less related to the absolute magnitude of the flood than is the case with the flooding of perennial rivers and more related to the quickness and ferocity of the event.

The purpose of this discussion is to present general descriptive information regarding alluvial fans from the geologic literature and some quantitative data from alluvial fans on the Nevada Test Site in southern Nevada, which both lends credence to and casts doubt on many of the assumptions engineers and geologists have made about hydraulic processes on alluvial fans. This paper focuses specifically on concepts important in flood hazard evaluation as delineated by the Federal Emergency Management Agency, Anon. (1982). It should be noted that other methodologies of flood hazard analysis have been used by other agencies; see for example Magura and Wood (1980), Christensen and Spahr (1980), and Squires and Young (1984).

2. ALLUVIAL FANS: GEOLOGIC DESCRIPTION

Although many definitions of alluvial fans can be found in the literature; see for example Anstey (1965), Bull (1977), and Rachocki (1981), there is general agreement that an alluvial fan is a fan or cone-shaped deposit of sediment found at the base of some mountain fronts. While alluvial fans are found in humid areas, they are primarily features associated with arid climates. Rachocki (1981) attributed the abundance of fans in arid regions to excellent preservation rather than unusually favorable conditions for formation.

Alluvial fans are formed by water transporting debris from intermountain canyons into adjacent valleys. A reasonable scenario for alluvial fan formation might be as follows. Debris accumulates along the flanks of mountains due to weathering; and when an intense precipitation event occurs, the accumulated debris is transported downslope in an intermountain canyon. At the point where the canyon enters the valley - the apex of the fan - the widening of the flow results in a decrease of its debris carrying capacity, and the debris is deposited at the apex and downslope from it. Through time, a series of depositional events cause the fan to aggrade and give it its characteristic shape. It should be noted that fans form because of base-level fall of the depositional area relative to the source area. Erosional base-level falls tend to result in temporary, thin fans, while tectonic base-level falls tend to result in the prolonged accumulation of thick fans.

The transport of material on alluvial fans can take place as either stream or debris flow. Streamflow is believed to be the more important mechanism in areas where annual precipitation is high, Hooke (1967); and Blissenbach (1954), based on an examination of field observations, also suggested that the relative importance of debris flow to streamflow deposition increased with decreasing average precipitation. However, it must be recognized that debris flow deposits are commonly sorted and stratified by subsequent streamflow; that is, the evidence of debris flow is continuously destroyed. Thus, one explanation of the observations of Blissenbach is that on fans in areas of higher precipitation there is less evidence of debris flows because the evidence of past debris flows has been modified beyond recognition. Hooke (1967) attributed the formation of debris flows to intense episodic rainfall, unconsolidated fine material, sparse vegetative cover, and reasonably steep slopes. This view was supported by Bull (1977) and Beaty (1974); however, none of these investigators provide quantitative threshold limits on these causative factors. Beaty (1974) and Hooke (1967) have asserted that in the White Mountains of Nevada and California spectacular episodes of debris flow deposition interspersed with periods of quiescence have been primarily responsible for building the fans.

A common characteristic of alluvial fans is the entrenchment of a channel near the fan apex - a geomorphologic feature termed the fanhead trench. Channel entrenchment occurs when erosion rather than deposition occurs near the apex. Bull (1977) noted that the fanhead trench is not always a permanent feature and distinguished between temporary and permanently entrenched channels. Laboratory observations, Hooke (1967), suggest that channel incision at the fanhead is the result of the alteration of debris and stream flows. Further, even when a fanhead channel is deeply entrenched, debris flows may either exceed the channel depth, resulting in the deposition of material on the fan surface above the intersection point or blocking the original channel and causing a new channel to be formed.

A second characteristic of alluvial fans which should be noted is that the slope of major channels on many fans is less than the slope of the adjacent fan surface. This results in channels which are deeper upstream and grow shallower in the downstream direction until they merge with the fan surface at a place known as the intersection point. This point is also a locus of deposition, and the material found in this location is usually coarser than the average material found in the channel. It is also in the vicinity of this point that most debris flows will terminate. Field observations indicate that in general most deposition near the fanhead was caused by debris flows while at the toe of the fan deposition is the result of streamflow.

Alluvial fans are depositional features which can be active for long periods of time. Many of the large alluvial fans in the southwestern United States are believed to be of Pleistocene age. Bull (1964) estimated that the Arroyo Ciervo fan in the San Joaquin Valley of California has an average vertical accretion rate of 0.03 - 0.2 m per decade; Beaty (1974) estimated the average vertical accretion rate of the Milner Creek Fan in the White Mountains of Nevada and California to be 0.08 - 0.15 m per thousand years; and French and Lombardo (1984) estimated the average vertical accretion of fans located on Frenchman Flat of the Nevada Test Site to be 0.07 m per thousand years.

Given the rather low estimated average rates of vertical fan growth, it would appear that structures sited on fans are reasonably safe. This is far from the truth. Blissenbach (1954) and Bull (1964) have measured debris flows with apex to toe thicknesses of 6 m. Alluvium deposited by streamflow may range from a fraction of a foot to several feet. Anstey (1965) reported on the destructive nature of an alluvial fan flash flood in Death Valley which, while depositing as much as 1.2 m of alluvium in some areas, eroded channels up to 1.8 m deep in other parts of the fan. During this event, boulders up to 1.8 m in diameter were moved. Although the average rate of fan growth in geologic time is slow, it cannot be assumed that catastrophic flow events do not occur. The engineer must further realize that he and the geologist have time scales that are orders of magnitude apart. While the engineer considers time scales of hundreds of years long, the geologist commonly uses time scales on the order of thousands to millions of years. This difference in time scales can be a significant impediment to understanding and effective communication when flood hazard and the average rate of fan growth are discussed.

3. ALLUVIAL FANS: HYDRAULIC PROCESSES

From the viewpoint of hydraulic engineering and flood hazard evaluation, the following primary assumptions are usually made regarding alluvial fans; see for example Dawdy (1979) and Anon. (1982):

1. Flows only rarely spread evenly across the surface of a fan. In general, a flood flow across a fan will initially be concentrated in an identifiable, temporary channel or will be confined to a specific portion of the fan. These initial channels are prone to lateral migration and/or sudden relocation to almost any other portion of the fan during a single extreme flow event.

2. For a majority of fans, critical slope; and hence critical flow is the norm.

3. Channels formed on the face of the fan are shaped by the flow

itself. If supercritical flow occurs, the channel banks will erode so that a wider channel is formed and the flow will return to a critical state.

Two crucial assumptions are also tacitly implied. First, given a specific location, a flood-discharge frequency distribution can be determined using a rational and accurate method; see for example, Anon. (1977). Second, debris flows do not present a flood hazard to facilities on alluvial fans.

At this point, these basic assumptions will be examined from the viewpoint of determining their validity since they are crucial to any flood hazard analysis. The first assumption implies that fans are created and modified by what might be termed a random process. In fact the U.S. Federal Emergency Management Agency, Anon. (1982), assumes that a channel caused by a flood event is equally likely to cross a contour of the fan at any point or

$$p(x|f) = T/W_c \tag{1}$$

where $p(x|f)$ = conditional probability that given flood f occurs, point x on a specified contour will be hit; T = channel width, and W_c = alluvial fan width at point x. To some degree, the validity of this assumption depends on the time scale considered. Over short periods of time, flow events on both laboratory and natural fans are localized, Hooke (1967). However, over longer periods of time, the locus of deposition must shift to yield uniform deposition over the entire fan surface; see for example Price (1974).

The second primary assumption is that the slope of the fan is such that critical flow will occur. While it is true that the qualitative literature available demonstrates that alluvial fans have steep slopes in comparison to perennial streams and rivers, there has been in the past no quantitative justification of this assumption. Recall that, by definition, the critical slope is one on which uniform or normal flow occurs at critical depth. An equation for the critical slope can be obtained from the Manning equation.

$$S_c = \frac{12.6n^2}{q^{2/9}} \tag{2}$$

where S_c = critical slope; q = flow per unit width and a wide rectangular channel has been assumed. When $S > S_c$, supercritical flow occurs; and when $S < S_c$, subcritical flow occurs. There are a number of semi-empirical methods of estimating a value of n based on an analysis of the size of the materials composing the bed of the channel. In the material which follows, the equation suggested by Lane and Carlson (1953) for channels whose beds are paved with cobbles or

$$n = 0.047\ (d_{75})^{1/6} \tag{3}$$

and the equation suggested by Meyer-Peter and Muller (1948) for channels whose beds are not paved with cobbles or

$$n = 0.038\ (d_{90})^{1/6} \qquad (4)$$

are used. In Equations (3) and (4), d_{75} = diameter of the bed material in meters such that 75% of the material, by weight, is smaller and d_{90} = diameter of the bed material in meters such that 90% of the material, by weight, is smaller.

In Area 5 of the Nevada Test Site a detailed study of the hazard floods present to a facility, including the estimation of peak flood flows and an intensive soil sampling program, was performed. In Tables 1, 2, and 3 a limited portion of the results of this study are summarized for the Scarp Canyon Fan. The peak flood flows and their associated return periods, Columns (1) and (2) in Table 3, were estimated by regional analyses of the available hydrological data; see for example French and Lombardo (1984), Christensen and Spahr (1980), and Squires and Young (1984). The channel top width and depth of flow, Columns (3) and (4) in Table 1, were estimated by hydraulic geometry hypothesis which is discussed in the next section of this paper; and the flow per unit width, Column (5) in Table 1, is the result of dividing Column (2) by Column (3). The slopes in Table 2 were estimated from the topographic maps available for the area. The average values of d_{75} and d_{90}, Columns (2) and (6) in Table 3, and the standard deviations associated with these values, Columns (3) and (7), were obtained from sieve analyses of material from sixty sampling sites on the Scarp Canyon Fan. The expected values of the Manning resistance coefficient, E(n), and the variances associated with these values, Var(n), were estimated by the method of first order analysis using Equations (3) and (4); see for example Benjamin and Cornell (1970).

TABLE 1. SUMMARY OF PEAK FLOOD FLOWS AND OTHER VARIABLES FOR THE SCARP CANYON FAN

Return period R (years)	Peak flood flow Q_R (m^3/s)	Channel top width T (m)	Depth of flow y (m)	Flow per unit width q ($m^3/s/m$)
(1)	(2)	(3)	(4)	(5)
10	22	34	0.30	0.65
25	57	61	0.46	0.93
50	100	76	0.55	1.3
100	170	94	0.70	1.8
500	450	137	1.0	3.3

TABLE 2. SUMMARY OF AVERAGE SLOPE DATA FOR THE SCARP CANYON FAN

Elevation (m) (1)	Distance between elevations (m) (2)	Average slope (3)
1158		
	1547	0.0194
1128		
	2118	0.0146
1097		
	1311	0.0229
1067		
	1265	0.0245
1036		
	1372	0.0219
1006		
	1585	0.0196
975		
	2256	0.0133
945		
	1006	0.0060
939		

TABLE 3. SUMMARY OF SOIL SAMPLE AND SIEVE ANALYSIS DATA FOR SITES ON THE SCARP CANYON FAN

Sample site location (1)	d_{75} (m) (2)	$_{75}$ (m) (3)	E(n) by Eq (7) (4)	Var(n) (5)	d_{90} (m) (6)	$_{90}$ (m) (7)	E(n) by Eq (8) (8)	Var(n) (9)
Current channels	0.0056	0.0038	0.020	5.0×10^{6}	0.012	0.0069	0.0017	3.0×10^{-6}
Non-channels	0.0028	0.0038	0.015	1.6×10^{-5}	0.0074	0.0063	0.014	5.7×10^{-6}

In Figure 1, the critical slope, S_c, is plotted as a function of the flow per unit width for the extreme values of the expected value of n, Table 3. Also plotted in this figure are the extreme values of the slope of the fan from Table 2. At the bottom of Figure 1, the top width of a rectangular channel, T_R, required so that the product of T_R and q equals the peak flood flow, Q_R in Table 1, with a return period of R years is specified.

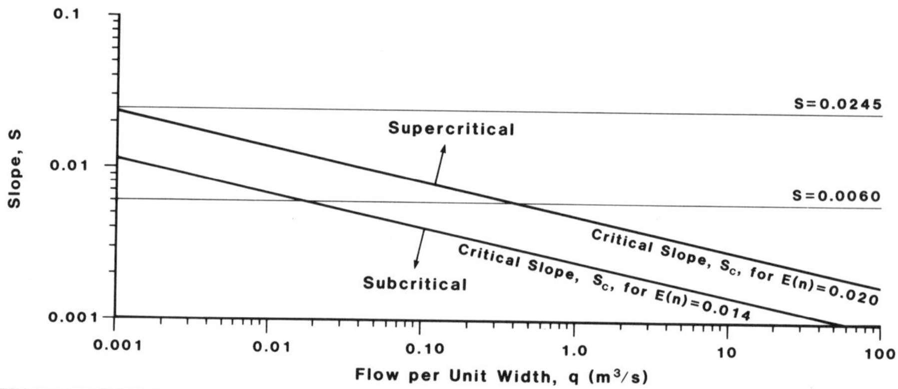

Top Width, T_R (m)					
T_5	22,000	2,200	220	22	0.22
T_{25}	57,000	5,700	570	57	0.57
T_{50}	100,000	10,000	1000	100	1.0
T_{100}	170,000	17,000	1,700	170	1.7
T_{500}	450,000	45,000	4,500	450	4.5

Top Width T_R (m) Corresponding to a Flow Rate Q_R (m^3/s) Having a Return Period of R Years With a Flow per Unit Width of q (m^3/s).

Figure 1. Critical slope as a function of flow per unit width for floods of various and extreme values of n.

Figure 1 demonstrates that under most conditions, flood flows on the Scarp Canyon Fan will be critical or supercritical. Although the state of flow is strongly dependent on the width of the channel, the assumptions inherent in constructing this figure are somewhat independent of any specific assumption regarding how channel width should be estimated.

The third assumption is perhaps the most difficult to evaluate. The assessment of flood risk by the conditional probabilities given in Equation (1) requires that the width of the flood channel be estimated. On many alluvial fans below the intersection point there are no well-defined, stable channels; and thus, some methodology must be used to estimate the geometry of the channel which will be formed by the flood flow. Dawdy (1979), and Anon. (1982) assumed that the width, depth, and velocity of flow can be written in terms of flow rate Q or

$$u = C_1Q^m \tag{5}$$

$$y = C_2Q^f \tag{6}$$

$$T = C_3Q^b \tag{7}$$

where Q = flow rate, y = depth of flow, and C_1, C_2 C_3, m, f, and b = undetermined coefficients and exponents. Note, to'maintain the validity of the continuity equation, it is usually assumed that both the sum of the exponents; and the product of the leading coefficients equal 1. In the terminology of hydraulic engineering, Equations (5) (7) have been previously termed the regime theory. In fact, this is not a theory but an empirical correlation of data from stable canals and rivers. The theory implies that a channel will adjust its slope and channel section so that the rate of sediment transport equals the rate of sediment supply. In the geologic literature, the hypothesis quantified by Equations (5) - (7) is commonly referred to as the theory of hydraulic geometry. It is obvious that these equations present an easy and explicit solution to the problem of estimating channel width and depth and the velocity of flow across an alluvial fan; and hence a method of quantifying flood hazard. The crucial question is whether these equations are valid; and if they are valid, then what are the true values of the coefficients?

Numerous investigators have made detailed studies of alluvial channels in attempts to determine an average set of coefficient values, either for a group of streams in a particular physiographic setting or on a global basis. There have also been a number of attempts to use theoretical methods to determine the hydraulic geometry. Some of the field and theoretical results that are available are summarized and discussed in French (1987).

For the moment, assume that the basic tenets of the hydraulic geometry hypothesis are valid in the general case, we must still consider the validity of this hypothesis in relation to hydraulic processes on alluvial fans. Two points should be noted. First, almost without exception all previous hydraulic geometry studies, both field and theoretical, have examined perennial alluvial streams which according to the hypothesis are stable channels. Further, these previous investigations have considered only steady flow. Thus, if the validity of the hydraulic geometry hypothesis is questioned in perennial alluvial channels with reasonably stable geometry and steady flow, it must certainly be questioned in ephemeral channels composed of non-cohesive materials and under very unsteady conditions. Second, the current hydraulic geometry hypothesis does not explicitly mention the Froude number. In the foregoing material, it was demonstrated that flood flow across one particular alluvial fan can reasonably be assumed to be critical or supercritical flow. Field experience with the regime/hydraulic geometry theory in the India-Pakistan Canals has demonstrated that the stability of canals will be maintained if they are in the lower flow regime but that meandering and bank erosion often develop as the upper flow regime is reached. Canal designers often use a Froude number criterion for regime flow; and Chang (1982b) noted that the value of the Froude number has usually been kept at about 0.2 and has never been allowed to exceed 0.3.

Again using the data from the Scarp Canyon Fan on the Nevada Test Site, the validity of the regime theory in the case of flood flows across alluvial fans can be examined. In Table 4, the average slopes of the Scarp Canyon Fan between elevations 1160 m and 945 m are summarized along with the peak flood flows of both a ten and five hundred year return period (Table 1). Subsequent columns of this table provide the following information; Column (3) is the depth of flow estimated by

$$y = 0.07Q^{0.4} \tag{8}$$

when Q has units of (ft^3/s) and y has units of (ft) which is the formulation suggested by Dawdy (1979); Column (4) is the shear velocity estimated by

$$u_* = \sqrt{gyS} \tag{9}$$

where y is obtained from Column (3), S from Column (1), and a wide channel is assumed; Column (5) is the average particle size estimated from soil samples; Column (6) is the parameter F_s (the entrainment function) where

$$F_s = \frac{u_*^2}{(S_s - 1)gd} \tag{10}$$

and S_s = specific gravity of the soil (S_s = 2.65); Column (7) is the parameter R_* (the particle Reynolds number) where

$$R_* = \frac{u_* d}{\nu} \tag{11}$$

and $\nu = 0.984 \times 10^{-6}\ m^2/s$; and Column (7) identifies the point in Figure 2 which represents these data. Figure 2 is a plot of R_* versus F_s demonstrating the dependence of the bed forms identified by Simons and Richardson (1961) on these parameters. Thus, on the basis of the assumption made by Dawdy (1979), these flows will result in antidunes in the upper regime of flow. Based on these data, it is concluded that the regime/hydraulic geometry theory is not a valid methodology for predicting channel widths and depths for flood flows across the Scarp Canyon Fan on the Nevada Test Site.

If the regime hypothesis is not valid for use in evaluating hydraulic processes on alluvial fans during flood flows, then alternative techniques for estimating channel geometry must be sought. The primary problem of fluvial hydraulics is that there are more unknowns than there are equations. The concepts of minimum unit stream power

TABLE 4. SUMMARY OF DATA REQUIRED TO TEST THE VALIDITY OF THE HYDRAULIC GEOMETRY THEORY FOR THE SCARP CANYON FAN

Slope	Peak flood flow Q_R (m^3/s)	Depth of flow y (m)	Shear velocity u_* (m/s)	Average particle size (m)	F_s	R_*	ID number
(1)	(2)	(3)	(4)	(5)	(6)	(7)	(8)
0.0194	22	0.306	0.241	3.05×10^{-3}	1.2	750	1
				3.35×10^{-4}	11	80	2
	450	1.02	0.441	3.05×10^{-3}	3.9	1370	3
				3.35×10^{-4}	36	150	4
0.0146	22	0.306	0.209	3.05×10^{-3}	0.86	648	5
				9.45×10^{-4}	2.9	200	6
	450	1.02	0.382	3.05×10^{-3}	3.0	1180	7
				9.45×10^{-4}	9.5	368	8
0.0229	22	0.306	0.262	4.27×10^{-3}	0.99	1140	9
	450	1.02	0.478	4.27×10^{-3}	3.3	2070	10
0.0245	22	0.306	0.271	2.41×10^{-3}	1.9	664	11
				5.18×10^{-4}	8.8	143	12
	450	1.02	0.495	2.41×10^{-3}	6.3	1210	13
				5.18×10^{-4}	29	261	14
0.0219	22	0.306	0.256	1.10×10^{-3}	3.7	286	15
				1.40×10^{-3}	2.9	364	16
	450	1.02	0.468	1.10×10^{-3}	12	523	17
				1.40×10^{-3}	9.7	666	18
0.0196	22	0.306	0.242	4.27×10^{-4}	8.5	105	19
				3.05×10^{-4}	12	75	20
	450	1.02	0.443	4.27×10^{-4}	28	192	21
				3.05×10^{-4}	40	137	22

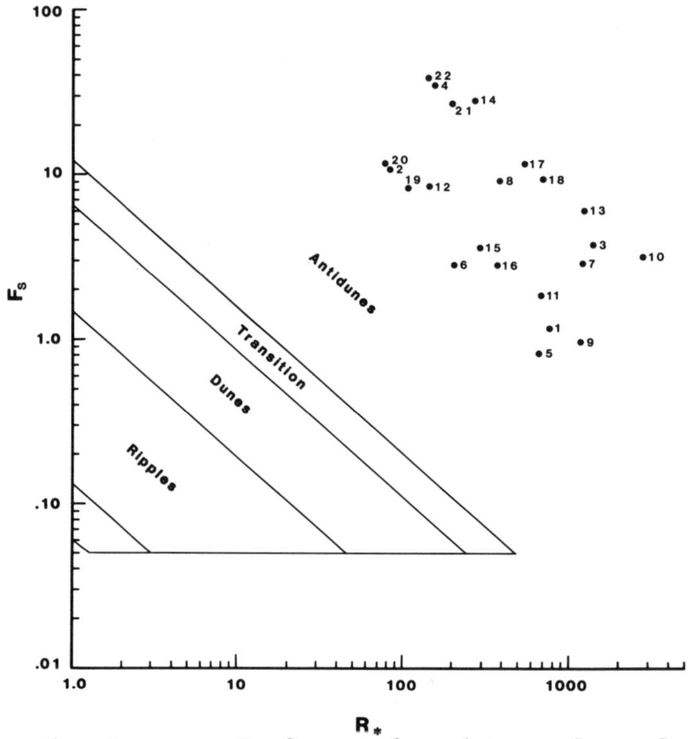

Figure 2. F_s versus R_* for sample points on Scarp Canyon Fan.

and minimum stream power are attempts to solve the indicated closure problem. The minimum unit stream power hypothesis asserts that an alluvial channel tends to adjust its velocity, slope, roughness, and channel geometry such that minimum unit stream power is used to transport a given sediment concentration and water discharge; see for example, Yang (1976). The minimum stream power hypothesis asserts that an alluvial stream in response to changes in the environment adjusts itself so that the total stream power of the channel reach is minimized; see for example, Chang (1982a). The minimum unit stream hypothesis, in its current form, is not applicable to flood flows across alluvial fans because this hypothesis was developed for subcritical flows in the lower flow regime, Yang (1976). The applicability of the minimum stream power hypothesis to the type of alluvial channel formed on a fan is less clear.

Thus, the conclusion regarding the third primary hypothesis is that it is not valid and must be reconsidered. However, it must be admitted that whatever the inadequacies of the Dawdy (1979) theory regarding the estimation of channel width and depth, there is not presently a theory which is superior to it.

At this point, attention must be given to the two assumptions which were not explicitly stated by Dawdy (1979) and Anon. (1982) but were implied. First, flow rates in many arid areas cannot be estimated by the standard methods recommended in Anon. (1977) because there is not a sufficient period of record. French and Lombardo (1984) discussed this problem and concluded that for the present in the southern Nevada area peak flood flow rates should be estimated by either regional regression models or envelope curve methods. It should be noted that in many arid areas both of these methods have a number of severe limitations; see for example, French and Lombardo (1984). Second, the validity of the implied assumption that debris flows do not present a significant hazard to facilities on alluvial fans depends to a large extent on the location of the facility on the fan. For facilities located in the vicinity of the intersection point, debris flows must be considered. For facilities located in the vicinity of the toe of the fan, debris flows may not present a severe hazard. The analytic methods for assessing the debris flow hazard can best be described as a developing technology; see for example DeLeon and Jeppson (1982). For facilities located away from the intersection point, debris flows present less of a hazard. It is clear that more consideration should be given to this implied assumption.

4. CONCLUSION

In conclusion, a number of comments can be made and a number of questions must be raised. First, the assumption that a flow on an alluvial fan has an equal probability of crossing any point on a given contour seems to be a very conservative assumption; however, from the viewpoint of flood hazard analysis this assumption is for the present acceptable. Second, given the data from the Nevada Test Site, it would appear that the assumption that fans have critical to supercritical slopes is acceptable. However, validity of this assumption is dependent on the third assumption regarding channel width and depth. Third, the present methods of estimating channel width and depth on alluvial fans seem to be invalid. However, it must be noted that currently there does not seem to be a superior method available. Therefore, for the present, the assumptions of Dawdy (1979) and Anon. (1982) should be accepted, but the unquantifiable errors present in these assumptions must be recognized. Fourth, the specific flood hazard evaluation procedures described by Dawdy (1979) are not valid because they are based on the assumption that sufficient records exist to do a peak flow analysis along the lines recommended by Anon. (1977). It is not obvious that for much of the Southwest this is a valid assumption since sufficient data are not available. Fifth, the implied assumption regarding debris flows is only valid once a location on a fan relative to the intersection point has been established. It must also be recognized that the location of the intersection point may change dramatically even on an engineering time scale. Thus, it is concluded that the current methods of flood hazard assessment on alluvial fans are not adequate given the current and projected economic value and potential environmental hazard of structures and development on alluvial fans in the southwestern United States.

Given the above comments, at least three areas of basic and applied research can be identified. First, there is a need to understand how alluvial fans develop on a geologic time scale. The initial laboratory studies of Hooke (1967) and the model developed by Price (1974) should be carefully studied and a new program incorporating coordinated physical and numeric model studies with field verification of the results must be designed.

Second, numeric models capable of estimating the location and size of channels formed by unsteady, high Froude number flows in alluvial fill must be developed. Although it may be possible to modify currently available models to accomplish this goal, field and laboratory verification of the results will be required.

Third, in areas where there are not adequate stream gaging records, techniques which are superior to the regional method of peak flood flow analysis and the envelope curve method must be developed. If this is not possible, then error limits on these methodologies must be developed and the error estimates carried through the subsequent calculations. Further, if the regional or envelope curve methodologies are used, then a technique of estimating of hydrograph shape must be developed for use in routing the flow across the fan and estimating sediment transport.

ACKNOWLEDGMENTS

The author acknowledges the financial support of the U.S. Department of Energy under Contract No. DE-AC08-81NV10162 and the University of Nevada System Board of Regents.

REFERENCES

Anon. 1982. "Alluvial Fan Studies," Flood Insurance Study: Guidelines and Specifications for Study Contractors, FEMA-37/July 1983, Federal Emergency Management Agency, Washington, D.C., pp. A6-1:A6-10.

Anon. 1977. "Guidelines for Determining Flood Flow Frequency," Bulletin 17A, Water Resources Council, Washington, D.C.

Anstey, R.L. 1965. "Physical Characteristics of Alluvial Fans," Technical Report ES-20, U.S. Army Natick Laboratories, Natick Laboratories, Natick, Mass.

Beaty, C.B. 1974. "Debris Flows, Alluvial Fans, and a Revitalized Catastrophism," Zeitschrift fur Geomophologie N.F., Suppl. Bd. 21, pp. 39-51.

Benjamin, J.R. and Cornell, C.A. 1970. Probability, Statistics, and Decision for Civil Engineers, McGraw-Hill Book Company.

Blissenbach, E. 1954. "Geology of Alluvial Fans in Semi-Arid Regions," Geological Society of America Bulletin, Vol. 65, pp. 175-189.

Bull, W.B. 1977. "The Alluvial Fan Environment," Progress in Physical Geography, Vol. 1, pp. 222-270.

Bull, W.B. 1964. "Geomorphology of Segmented Alluvial Fans in Western Fresno County, California," Professional Paper 352-E, U.S. Geological Survey, Washington, D.C.

Chang, H.H. 1982a. "Fluvial Hydraulics of Deltas and Alluvial Fans," ASCE, Journal of the Hydraulics Division, Vol. 108, No. HY11, pp. 1282-1295.

Chang, H.H. 1982b. "Stable Alluvial Canal Design," ASCE, Journal of the Hydraulics Division, Vol. 106, No. HY5, pp. 873-891.

Christensen, R.C. and Spahr, N.E. 1980. "Flood Potential of Topopah Wash and Tributaries, Eastern Part of Jackass Flats, Nevada Test Site, Southern Nevada," Open File Report 80-963, U.S. Geological Survey, Lakewood, Colorado.

Dawdy, D.R. 1979. "Flood Frequency Estimates on Alluvial Fans," ASCE, Journal of the Hydraulics Division, Vol. 105, No. HY11, pp. 1407-1413.

DeLeon, A.A. and Jeppson, R.W. 1982. "Hydraulics and Numerical Solutions of Steady-State but Spatially Varied Debris Flow," UWRL/H-82/02, Utah Water Research Laboratory, Logan, Utah.

French, R.H. and Lombardo, W.S. 1984. "Assessment of Flood Hazard at the Radioactive Waste Management Site in Area 5 of the Nevada Test Site," DOE/NV/10162-15, Water Resources Center, Desert Research Institute, Las Vegas, Nevada.

French, R.H., 1987. Hydraulic Processes on Alluvial Fans, Elsevier Scientific Publishers, Amsterdam.

Hooke, R.L. 1967. "Processes on Arid-Region Alluvial Fans," Journal of Geology, Vol. 75, pp. 438-460.

Lane, E.W. and Carlson, E.J. 1953. "Some Factors Affecting the Stability of Canals Constructed in Coarse Granular Materials," Proceedings of the Minnesota International Hydraulics Convention.

Magura, L.M. and Wood, D.E. 1980. "Flood Hazard Identification and Flood Plain Management on Alluvial Fans," AWRA, Water Resources Bulletin, Vol. 16, No. 1, pp. 56-62.

Meyer-Peter, E. and Muller, R. 1948. "Formulas for Bed Load Transport," Proceedings of the 3rd Meeting of the IAHR, Stockholm, pp. 39-64.

Price, W.E., Jr. 1974. "Simulation of Alluvial Fan Deposition by a Random Walk Model," Water Resources Research, Vol. 10, No. 2, pp. 263-274.

Rachocki, A. 1981. Alluvial Fans, An Attempt at an Empirical Approach, John Wiley and Sons, Inc., New York.

Simons, D.B. and Richardson, E.V. 1961. "Forms of Bed Roughness in Alluvial Channels," ASCE, Journal of the Hydraulics Division, Vol. 87, No., HY3, pp. 87-105.

Squires, R.R. and Young, R.L. 1984. "Flood Potential of Forty-Mile Wash and Its Principal Southwestern Tributaries, Nevada Test Site, Southern Nevada," Water Resources Investigations Report 83-4001, U.S. Geological Survey, Carson City, Nevada.

Yang, C.T. 1976. "Minimum Unit Stream Power and Fluvial Hydraulics," ASCE, Journal of the Hydraulics Division, Vol. 102, No. HY7, pp. 919-934.

FLOOD DAMAGE-FREQUENCY ANALYSIS OF AN URBANIZING WATERSHED

Nageshwar R. Bhaskar
Assistant Professor
Department of Civil Engineering
University of Louisville
Louisville, Kentucky 40292

ABSTRACT. The impact of urbanization resulted in serious flooding problems in the Beargrass Creek basin, which lies principally in the City of Louisville, Kentucky. The watershed is 60.9 square miles in size with the South Fork and Middle Fork as the two principal streams draining into the Ohio River. In order to mitigate the damage due to flooding, the U.S. Army Corps of Engineers, Louisville District undertook a study in 1973 to evaluate the existing and future flood damage potential in the watershed. Using the Clarks Instantaneous Unit Hydrograph approach and the rainfall events associated with the maximum annual flood events over the period of record (1941-1973), a flood-frequency analysis was carried out for both the present (1973) and future conditions (1990). Future flood conditions were simulated using projected Unit Hydrograph parameters and the historical rainfall record. The year 1990 was selected since the watershed is expected to be completely urbanized by that time. Backwater profiles were developed using Army Corps of Engineers HEC-2 program in order to establish flood elevation-frequency curves in selected reaches. Using this information, flood damage-frequency curves, both under existing and projected conditions, were derived.

In order to provide flood protection in the Beargrass Creek basin, several structural and non-structural measures of flood control were examined in the study. A dry-bed reservoir located upstream on the South Fork, was recommended as the best alternative using a combination of economic, environmental, technical and social criteria. A comparison of the flood damage-frequency curves under modified conditions along the South Fork due to the dry-bed reservoir with similar curves under pre-development conditions is presented.

V. P. Singh (ed.), Application of Frequency and Risk in Water Resources, 377–394.

INTRODUCTION

Changes in land use, as a major component of urbanization, has had a significant impact on flood levels in most metropolitan areas across the United States. In this respect, the City of Louisville in Western Kentucky has been no exception. An increase in the degree of imperviousness in the watershed and the provision of improved conveyance has resulted in major changes in the timing and volume of peak discharges resulting in serious problems of flooding. After the Flood Control Act of 1936 and the National Flood Insurance Act of 1968 were enacted by the Congress, efforts have been underway by various researchers and agencies to quantify the relationship between flooding and urbanization. (Epsey et.al., 1965; Leopold, 1968; Rao et.al., 1972; Bras and Perkins, 1975; Delleur et.al., 1976; Hall, 1977; Beard and Chang, 1979; McCorkle and Halff, 1979; Bost et.al., 1980; Kibler et.al., 1981; and Ellis et.al., 1983). An article by Foster (1973) reviews very clearly the various flood management techniques, both in and off the flood plain, currently available in mitigating the adverse impacts of flooding. The methods outlined in his paper can be broadly classified as structural and non-structural, the former involving flood control structures such as reservoirs, detention/retention basins, channel modification and levees while the latter includes land use, flood plain zoning, and flood insurance.

An important aspect of studying the impact of flooding in urbanized areas involves the estimation of flood damage-frequency curves (Golden, 1977; Walesh and Videkovich, 1978; and Debo, 1981, 1982). In order to accomplish this, a detailed evaluation and mapping of the flood plain for various flood frequency levels must be done. As pointed out by Yomtovian and Ford (1982), the use of sophisticated and complex computer programs, such as HEC-1 and HEC-2, can be justified where accurate flood elevations are required. This is often the case, since the delineation of the flood plain is not only important in flood insurance studies, but is required in assessing various flood control strategies (Ball et.al, 1978; and Bialas and Loucks, 1978).

In order to alleviate serious flooding problems brought upon by rapid urbanization in the Beargrass Creek basin (which includes a major portion of the City of Louisville), the U.S. Army Corps of Engineers initiated a detailed flood study in 1973. (U.S. Army Corps of Engineers, 1975). The problems and solutions examined in the study were similar in nature to those discussed above. The principal objective of this study was to perform a detailed evaluation of the existing and future flooding potential and damage in basin and to provide suitable flood control measures. This paper presents the results of this study. Using the Clark Instantaneous Unit Hydrograph approach and rainfall events associated with the maximum annual flood events, a flood-frequency analysis was carried out for both present (1973) and future (1990) conditions. To predict the future flooding potential, projected unit hydrograph parameters were used in the analysis. Since most of the flood damage in the Beargrass Creek basin occurred along the South fork (a principal stream draining the basin), most of the efforts of the flood study focussed on a detailed analysis of the flood characteristics of this stream. Based upon selected design floods and results of the flood-frequency analysis, flood damage-frequency curves were developed along selected reaches of the South Fork under existing and future conditions.

To provide flood mitigation measures, several structural and non-structural plan components were examined either individually or in combination. After a careful evaluation of all the alternatives based upon economic, environmental, technical and social criteria, a dry-bed reservoir, to be located in the upstream reaches of the South fork, was selected as the best alternative (Bhaskar, 1985).

STUDY AREA

Basin Location

The Beargrass Creek basin lies entirely in Jefferson County in Western Kentucky (Figure 1) and incorporates over half of the heavily urbanized City of Louisville in its 60.9 square miles (157.7 km^2) drainage area. The South Fork, which is 15.4 miles (24.6 km) in length, is the principal stream and has its confluence with the Ohio River near downtown Louisville. The Middle and Muddy Forks are tributaries to the South Fork having lengths of 16 and 7 miles (25.6 and 11.2 km), respectively. The Muddy Fork drains the northern sections of the basin which lies primarily in the uplands near the Ohio River flood plain. An initial evaluation of the Muddy Fork subbasin indicated no serious flooding problems and, consequently, a detailed investigation was not included in the flood study. The Middle Fork originates in and drains the easternmost and central portions of the basin, while the South fork drains the rest of the basin lying to the south.

Physical, Climatic and Hydrological Characteristics

The topography of the Beargrass Creek basin is composed of the flat alluvial flood plain and terraces of the Ohio River and plateaus of the uplands. The geology of the basin consists of an exposed bed rock of the Silurian and Devonian age overlain with sediment strata, mainly shale and limestone, of the Paleozoic times. The two principal soil associations in the basin are the Wheeling-Weinbach-Huntington series and the Crider-Corydon series. These soils generally have low permeability. The climate is quite variable due to its location in the Ohio River Valley. Mean annual temperature is 56°F (13.3^{o}C) while the mean annual precipitation is 41 inches (1.04 m). All the three streams draining the basin have slopes of 10-12 ft/mile (1.89-2.27 m/km) and less than 5 ft/mile (0.95 m/km) near the confluence with the Ohio River. Hydrologic characteristics of the Middle Fork and South Fork are summarized in Table 1. Data on the Muddy Fork is not available.

Land Use

Beargrass Creek basin consists of land use which is typical of most urban areas. Areas along the South Fork are heavily developed with a higher degree of imperviousness than the Middle Fork. Hence, flooding has been a more serious problem in portions of the basin drained by the South Fork. Table 2 summarizes data on existing (prior to 1973) and projected (1990) land use expected in the Jefferson County (which

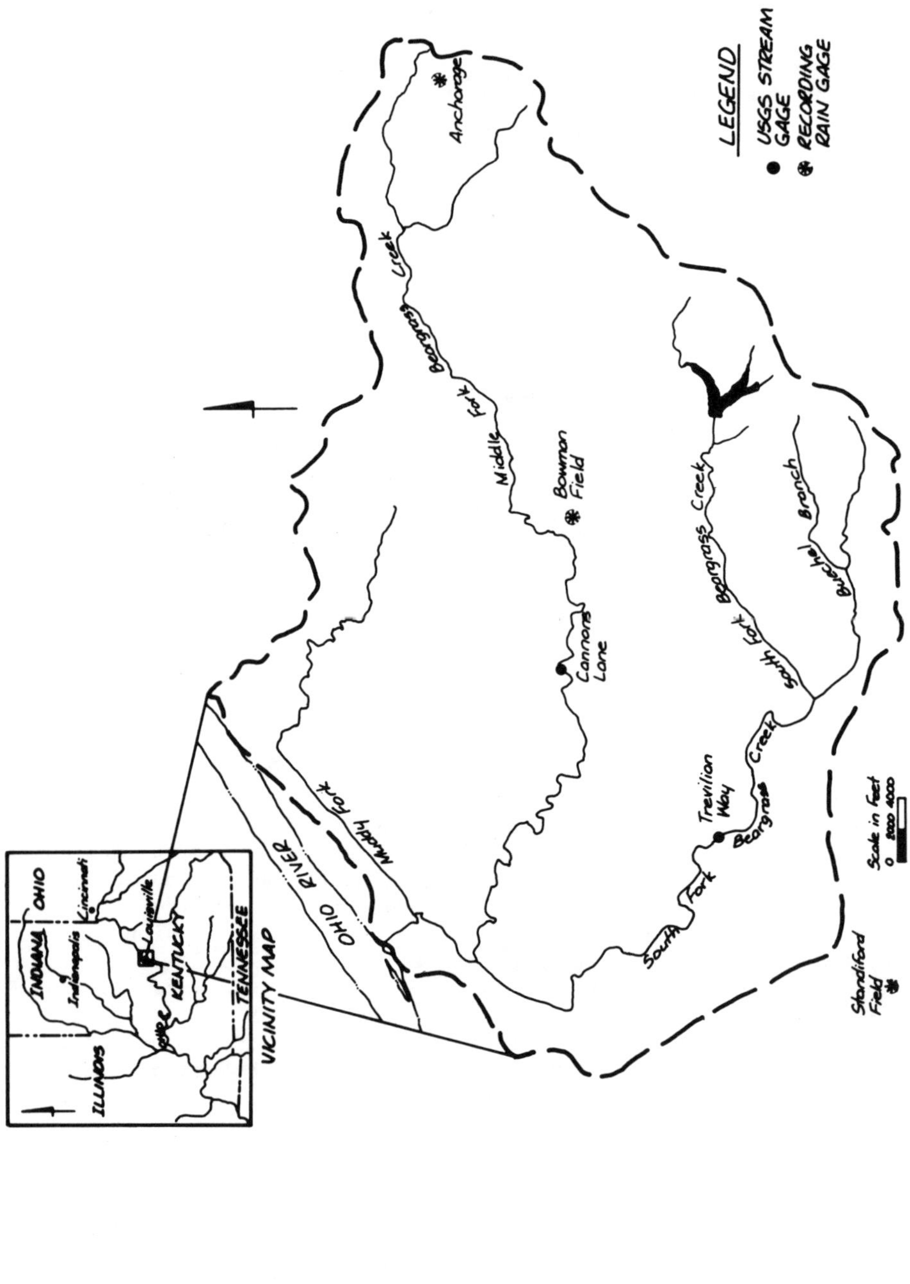

Figure 1. Location Map of Beargrass Creek Basin, Louisivlle, Kentucky

Table 1. Hydrologic Characteristics of Principal Streams In the Beargrass Creek Basin

Item	South Fork	Middle Fork
Channel Length (miles)	15.4 (24.8 Km)	16.4 (26.4 Km)
Drainage Area (square miles)	17.2 (44.6 Km^2)	18.9 (48.9 Km^2)
Area of Lakes and Ponds (Percent of drainage area)	0.12	0.0
Forest Cover (Percent of drainage area)	3.0	3.0
Noncontributing Area (Percent of drainage area)	1.0	3.0
Slope of Stream (feet/mile above gage)	19.4 (3.7 m/Km)	18.5 (3.5 m/Km)
1973 Average Discharge (cfs)	21.4 (0.61 m^3/s)*	24.8 (0.70 m^3/s)**
1973 Average Rainfall (inches)	43.0 (1.09 m)	43.0 (1.09 m)
1973 Average Runoff (inches)	16.9 (0.43 m)	17.8 (0.45 m)
Magnitude of Flood of Record (cfs)	4940.0 (139.8 m^3/s)	5200.0 (147.2 m^3/s)

*As measured at USGS Trevilian Way streamflow gage.
**As measured at USGS Cannons Lane streamflow gage.

Table 2. Land Use Acreages, Jefferson County, Kentucky*

Categories	1968 Acres	1968 Percent of total	1968 Percent urbanized	1990 Acres	1990 Percent of total	1990 Percent urbanized
Residential	45,424	18.4	50.2	81,355	33.0	45.2
Commercial	2,659	1.1	2.9	4,432	1.8	2.5
Industrial	8,554	3.5	9.4	19,829	8.1	11.1
Open Space and Recreational	9,732	4.0	10.7	24,868	10.1	13.8
Community Facilities	7,100	2.9	7.8	13,500	5.4	7.4
Public Right of Way	17,066	6.9	19.0	35,844	14.6	20.0
Agriculture and Vacant	155,733	63.2	0.0	66,480	27.0	0.0
TOTAL	246,308	100.0	100.0	246,308	100.0	100.0

*1 acre = 4.047×10^{-3} Km^2; or 1 Km^2 = 247.1 acres

includes the Beargrass Creek basin). As shown in Table 2, only 27% of the county will be undeveloped by the year 1990. Most of this area lies in Southwestern portions of the basin where the terrain features makes it unsuitable for future development. Consequently, the Beargrass Creek basin can be assumed to be 100% urbanized by the year 1990.

URBANIZATION EFFECTS ON RUNOFF

The Clark Instantaneous Unit Hydrograph (IUH) method (Clark, 1945) was used to simulate 1973 and 1990 flood conditions along the Middle and South Fork sections of the Beargrass Creek basin. Using historical rainfall-runoff records of the maximum annual floods observed at Cannons Lane gage on the Middle Fork (period of record 1944-1973) and at Trevilian Way gage on the South Fork (period of record 1940-1973), Clark Instantaneous Unit Hydrographs are derived for each year over the period record. Trends in the two important parameters of these unit hydrographs, namely, the time of concentration (T_c) and the storage attenuation constant (K), were used to obtain the unit hydrograph for the year 1990 (Bhaskar, 1985). Results show that the unit hydrograph peaks had more than doubled with a corresponding reduction in the time of concentrations in excess of 70%. Consequently, the effects of urbanization in the Beargrass Creek basin resulted in a dramatic increase in the flooding potential since 1940. A discharge-frequency analysis (assuming log-Pearson Type-3 distribution) of simulated annual maximum flood peak flows using the historical rainfall record (each of the rainfall events corresponds to the observed maximum annual flood event selected in the study) and the unit hydrographs, reflecting present (1973) and future (1990) runoff potential in the basin, indicates a significant increase in the flood levels due to urbanization (Bhaskar, 1985). Implicit in this procedure is the assumption that similar meteorological conditions will be experienced in the future. Although this is incorrect, it is a reasonable assumption in the absence of future rainfall data. Table 3 shows the flood peak discharges at selected return periods (or frequencies) for the Middle and South Forks, respectively. Also, included in this table are historical floods of record. The maximum flood of record has a magnitude of 4940 cfs (139.8 cumecs) for the South Fork and 5200 cfs (147.2 cumecs) for the Middle Fork. Based on 1973 discharge-frequency curves; these two flood events have return periods of 25 years and 50 years, respectively.

EXTENT AND CHARACTER OF THE FLOOD PLAIN

The following information was compiled by the U.S. Army Corps of Engineers in 1973 to define the character of the flood plain prior to an assessment of flood damage. A detailed computation of the flood plain, under existing (1973) and future (1990) conditions, was performed using the U.S. Army Corps of Engineers computer program HEC-2 (U.S. Army Corps of Engineers, 1981) in conjunction with simulated and observed maximum annual flood events. By taking several reaches along the South and Middle Fork, the flood plain for the SPF and the 100-year floods was mapped. Elevation-frequency curves similar to the one in Figure 2 were used for this purpose. The area within the 100-year flood plain as of

Table 3. Annual Maximum Flood Peak Levels for Selected Return Periods[@]

Return period (years)	Simulated flood peaks* (cfs)					
	South Fork 1973	South Fork 1990	Increase	Middle Fork 1973	Middle Fork 1990	Increase
100	6950	10900	3850	5950	8250	2300
50	5800	8950	3150	5200	7200	2000
20	4350	6700	2350	4150	5900	1750
10	3300	5200	1900	3400	4800	1400

Date	Historical flood peaks (cfs) South Fork	Middle Fork
March 9, 1964	4940**	3920
April 2, 1970	4610	5200**
July 22, 1973	4110	2150

@ Data represents simulated maximum annual flood peaks at Trevilian Way gage on South Fork and Cannons Lane gage on Middle Fork

* As obtained from flood-discharge frequency curves derived using simulated flood conditions (1.0 cfs = 0.0283 cumecs)

** Maximum flood peak of record

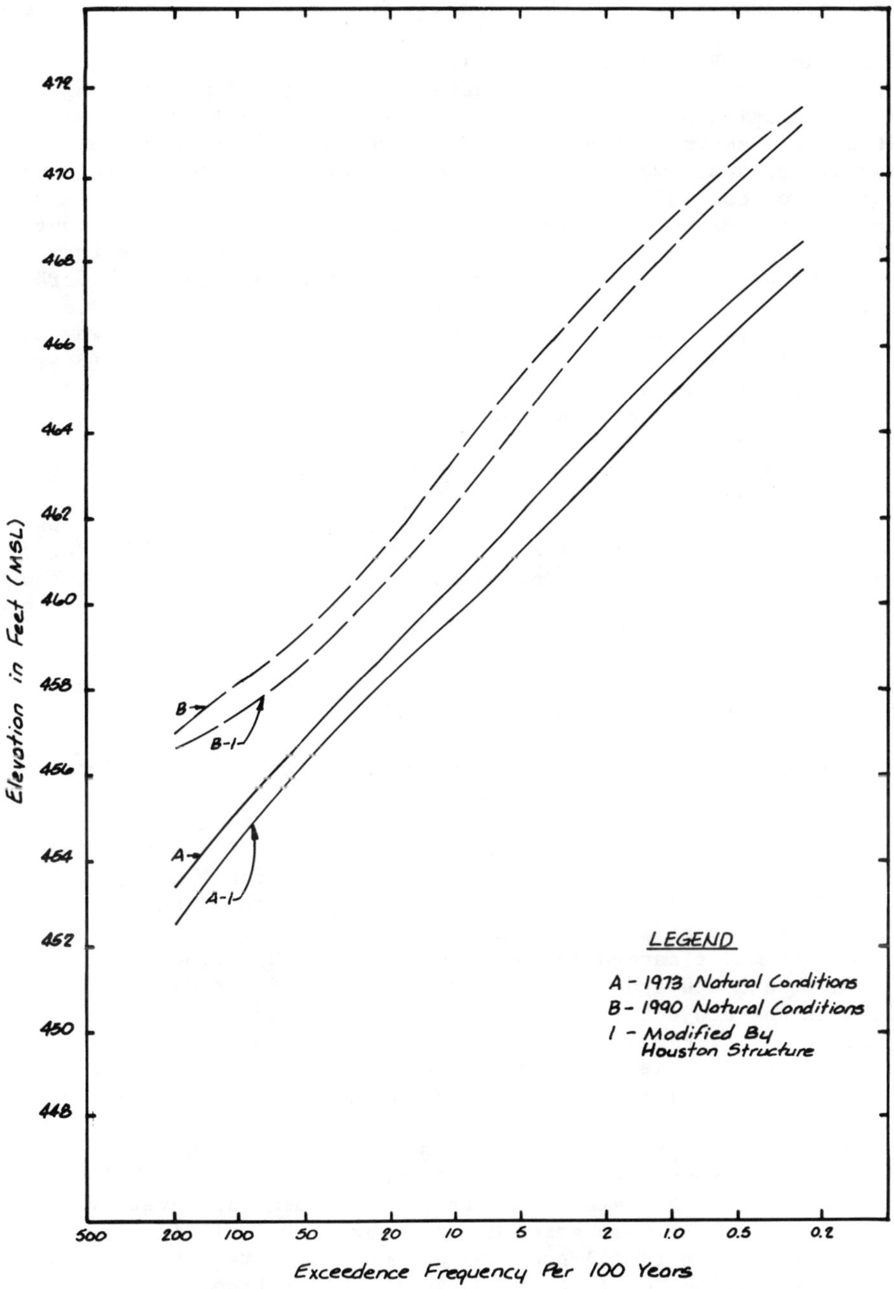

Figure 2. Elevation - Frequency curves for present (1973) and future (1990) 100 - year flood at Trevilian Way Gage on South Fork

1973 natural conditions is 800 acres and 400 acres along the South and Middle Forks, respectively. The flood plain is approximately 50% more for the Standard Project Flood (SPF).

For the purpose of illustration a 100-year flood plain along a particular section of the South Fork, under 1973 and 1990 natural conditions, is shown in Figure 3. The word natural implies that the flood plain boundary was computed prior to the recommendation of any suitable flood control measure. This would, therefore, include any backwater effects resulting from existing (1973) channel conditions or modifications. Since most of the severe flooding problems in the Beargrass Creek basin were determined to exist along the South Fork, a dry-bed reservoir (Figure 1) was recommended as a suitable device to reduce flood damage along this stream. Such a recommendation was made after a detailed evaluation of several plans and alternatives (Bhaskar and Singh, 1985). Figure 3 also illustrates the modified flood plain with the dry-bed reservoir in place. A comparison of the modified 1973 and 1990 flood plain indicates that the future (1990) flood plain continues to be wider than the present (1973) flood plain. This is primarily due to the anticipated increase in the runoff due to urbanization. As presented in the detailed report (U.S. Army Corps of Engineers, 1975), prohibitive costs in excess of benefits, prevented the implementation of any structural measures of flood control along the Middle Fork. Furthermore, flood damages along the Middle Fork are small in comparison to the South Fork.

EVALUATION OF FLOOD DAMAGE – PRESENT AND FUTURE

Having defined the flood plain, the next step was to assess flood damage for floods of selected return periods or frequency. In order to accomplish this, a detailed evaluation of the development in the Standard Project Flood (SPF) flood plain was made. Since the Standard Project Flood has the highest flood level (apart from the maximum probable flood (MPF)), it is reasonable to assume that the development will be maximum in this flood plain. Table 4 summarizes the type type and extent of major categories of development in SPF flood plain as of 1973 along the South and Middle Forks. No significant damagable development was projected in the flood plain by 1990, due to the enforcement of the Flood Insurance Act of 1973. As of 1973, the total property value of these developments (based on 1973 prices) was approximately $145 million for South Fork and $30 million for the Middle Fork.

Damages to the property in the flood plain is evaluated by considering only tangible losses due to physical damage to buildings and to their contents. For example, losses to residential development include damage to the main structure and auxiliary buildings, equipment, clothing, items of personal property and cost of alternative accommodation during periods of evacuation and rehabilitation. Table 5 shows the total value (based on 1973 prices) of each type of property lying in the SPF flood plain. Estimated existing (1973) damages to this or part of the property due to the SPF, 100-year and the observed flood of 1970 are also included in the table. An examination of Table 5 indicates that the value of the property for exceeds the actual damage associated with flooding. The underlying assumption is that flooding will not

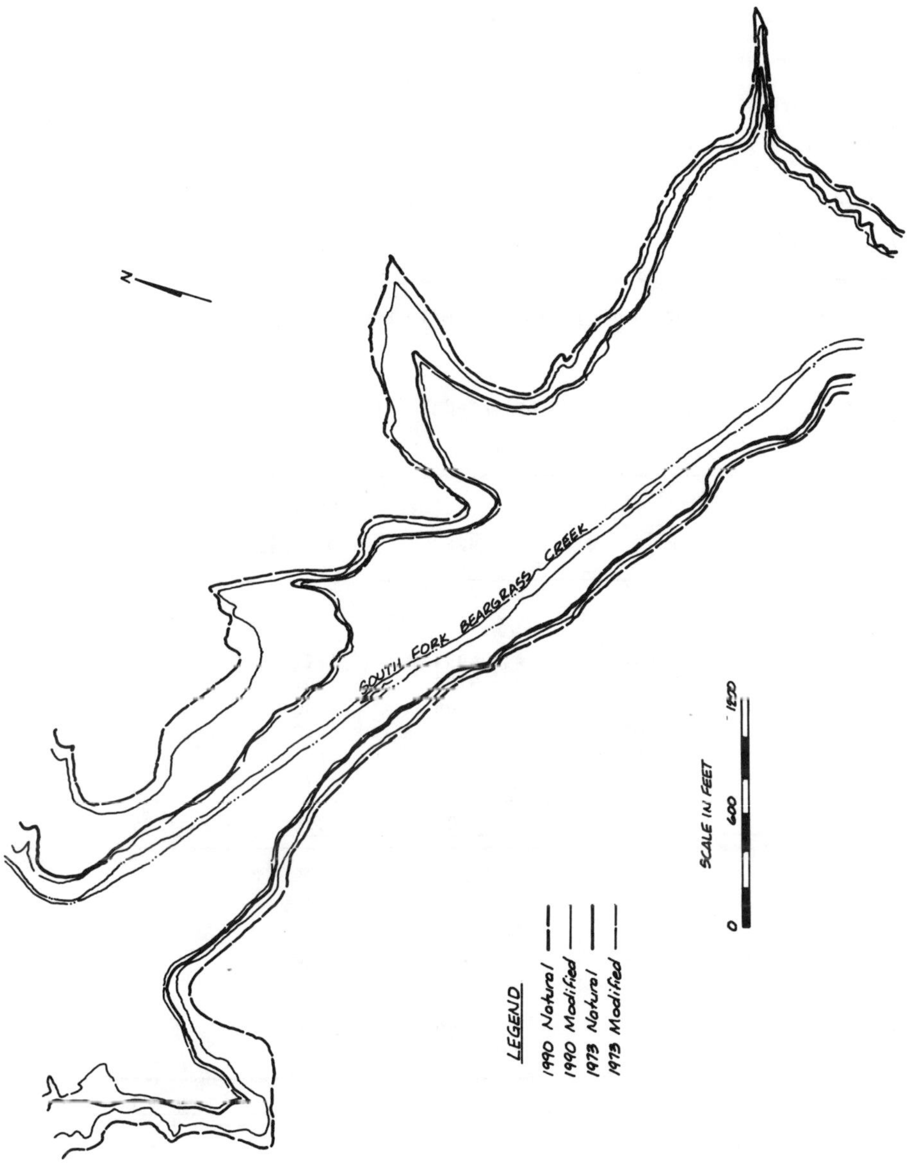

Figure 3. Natural and Modified 100 - year flood limits
Reach SF-1 on South Fork

Table 4. Existing Nature of Development Along the SPF Flood Plain (as of 1973)*

Land use	South Fork	Middle Fork
Residential (single-family)	1300 units	4 units
Apartments (multi-family)	1700 units	300 units
Commercial Buildings	130 units	15 units
Public Buildings	15 units	2 units
Transportation Routes	23 miles	13 miles

*Projected development in the flood plain to remain about the same level by 1990.

Table 5. Property Value and Damage for Selected Existing (1973) Flood Levels[@]

Property Type	Total property* value (x 1000 $)	Damages (x 1000 $) SPF	100-year	1970**
South Fork				
Residential	40092	4955	1927	632
Apartments	33612	3190	1207	360
Commercial	58987	8906	3790	48
Public	9636	370	50	3
Transportation	3094	257	193	50
Utilities	-	129	91	42
Totals	145421	17807	7258	1135
Middle Fork				
Residential	267	14	4	2
Apartments	5723	621	128	12
Commercial	16893	1235	163	3
Public	3000	33	5	1
Transportation	4440	99	38	4
Utilities	-	9	4	1
Totals	30323	2011	342	23

@ Based on seven reaches along South Fork and two reaches along Middle Fork using 1973 prices

* Based on the development in the Standard Project Flood Plain as of 1972

** See Table 3 for Flood magnitude. Estimated return period - 25 year event for South Fork and 50-year event for Middle Fork (based on 1973 simulated discharge-frequency curves) result in a complete loss of the property. Intangibles such as loss of human life and adverse impact on the environment have not been included in assessing flood damage. Damage at other flood levels can be evaluated by using elevation-damage data, developed for selected reaches, together with elevation-frequency data (refer to Figure 2). For the purpose of illustration, Figure 4 shows the elevation-value-damage curves for the reach shown in Figure 3. The combination of information contained in Figure 2 and Figure 4 yields frequency-damage curves similar to Figure 5. Since future development in the flood plain by the year 1990 is expected to be near 1973 levels, any increase in flood damages would be primarily due to higher flood levels as a result of urbanization in the Beargrass Creek basin. This is well illustrated in Figure 5 where future damages increase dramatically over existing values. It is interesting to note that although no damages from more frequent flood events (example 20 year flood) are expected under existing (1973) conditions, such is not the case in 1990. Figure 5 also includes the damage-frequency curves with the recommended dry-bed reservoir on South Fork. The effect of this reservoir is to reduce flood damages under existing and future conditions. Furthermore, the reservoir is more effective in reducing damages in the upstream reaches of the South Fork.

SUMMARY

The impact of urbanization caused severe flooding problems in the Beargrass Creek basin which incorporates a major portion of the City of Louisville in North Central Kentucky. The South Fork and Middle Fork are the two principal streams draining this 60.9 square miles (157.7 km^2) basin. In order to alleviate these problems the U.S. Army Corps of Engineers undertook a detailed study to evaluate the existing (1973) and future (1990) flooding potential in the basin. A discharge-frequency analysis of simulated maximum annual flood peaks generated by using 1973 and 1990 Clark Instantaneous Unit Hydrographs in conjunction with historical rainfall records of observed maximum annual flood events, shows a substantial increase in the flood levels. Using simulated and observed maximum annual floods, flood elevations-frequency curves are evaluated for selected reaches along the South and Middle Forks using the U.S. Army Corps of Engineers computer program HEC-2. In order to assess flood damage, the development along the Standard Project Flood (SPF) is determined since such a flood plain would incorporate development associated with other frequency floods. Flood damages are evaluated for existing (1973) and future (1990) flood conditions. Integrating the discharge-elevation-frequency curves with elevation-damage curves, damage-frequency are developed for the selected reaches along the South Fork and Middle Fork. Results show that although significant development is not expected to occur by 1990 over the existing (1973) levels, damages increase substantially due to increase in flood levels as a result of urbanization.

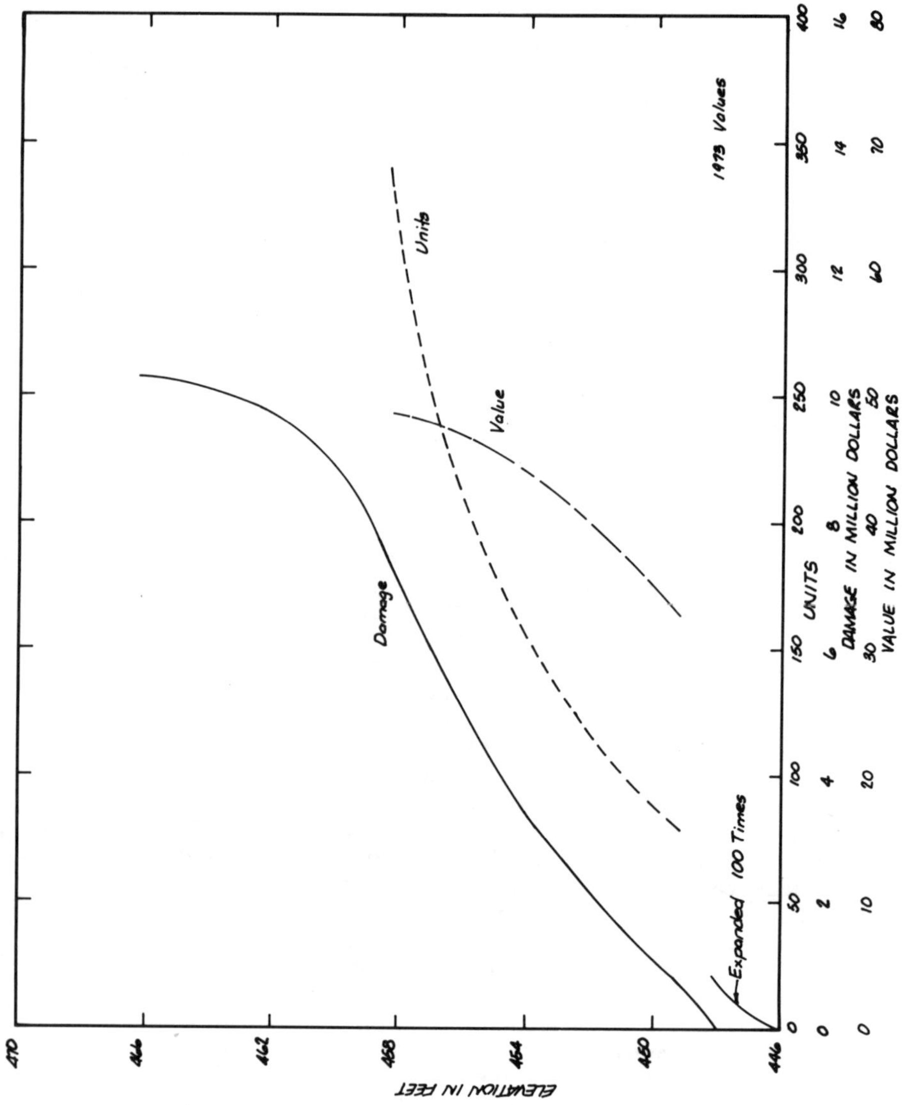

Figure 4. Elevation - Unit, Value and Damage Curves for Reach SF-1 on South Fork

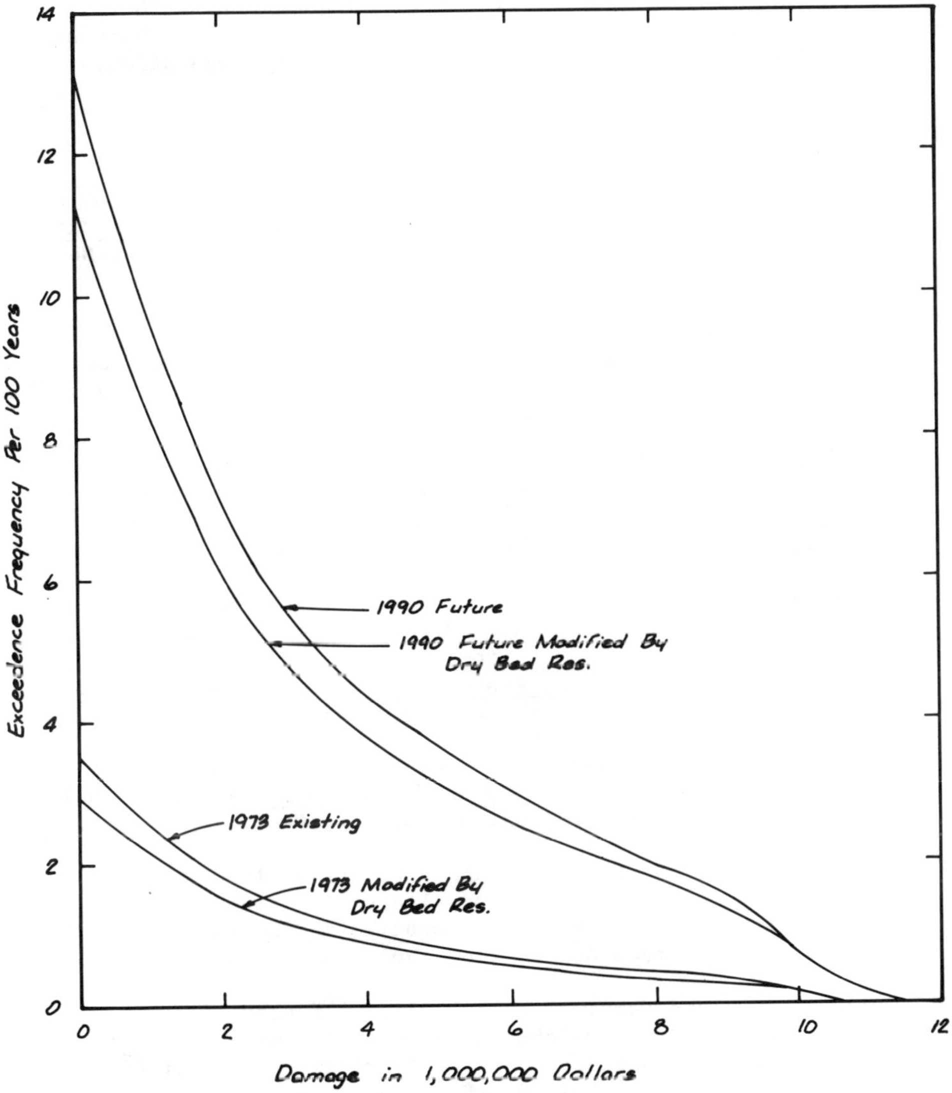

Figure 5. Natural and Modified Frequency - Damage Curves for Reach SF-1 on South Fork under present (1973) and future (1990) conditions

ACKNOWLEDGEMENTS

The author extends his sincere appreciation to the U.S. Army Corps of Engineers, Louisville District for allowing the use of the contents of their feasibility in putting together this paper. The author is greatly indebted to Mr. Robert L. Mullins, who is presently the Chief of the Hydrology Section in the District Office, for his cooperation.

REFERENCES

1. Ball, M. O., Bialas, W. F., and Loucks, D. P. 1978. 'Structural Flood Control Planning,' Water Resources Research, Journal of the American Geophysical union, **Vol. 14, No. 1**, pp. 62-66.

2. Beard, L. R., and Shin Chang 1979. 'Urbanization Impact on Streamflow,' Journal of Hydraulics Division, Proceedings of the American Society of Civil Engineers, **Vol. 105, No. HY6**, pp. 646-659.

3. Bhaskar, N. R. 1985. 'Evaluation of Urbanization Effects on Runoff Using Clark Unit Hydrograph Method - A Case Study,' paper submitted to Water Resources Bulletin, Journal of American Water Resources Association.

4. Bhaskar, N. R., and Singh, V. P. 1985. 'Environmental Issues for Planning Water Resources Projects in Urbanizing Areas,' Proceedings of the International Seminar on Environmental Impact Assessment of Water Resources Projects, University of Roorkee, Roorkee, India, pp. 231-249.

5. Bialas, W. F., and Loucks, D. P. 1978. 'Non-structural Floodplain Planning,' Water Resources Research, Journal of the American Geophysical Union, **Vol. 14, No. 1**, pp. 67-72.

6. Bost, R. C., Bedient, P. B., and Rowe, P. G. 1980. 'Effect of Urbanization on Alternative Flood Control Strategies,' Water Resources Bulletin, Journal of the American Water Resources Association, **Vol. 16, No. 4**, pp. 710-716.

7. Bras, R. L., and Perkins, F. E. 1975. 'Effects of Urbanization on Catchment Response,' Journal of Hydraulics Division, Proceedings of the American Society of Civil Engineers, **Vol. 101, No. HY3**, pp. 451-466.

8. Clark, C. O. 1945. 'Storage and the Unit Hydrograph,' Transactions of the American Society of Civil Engineers, **Vol. 110**, pp. 1419-1446.

9. Debo, T. N. 1981. 'Estimating Flood Damages in the State of Georgia,' Environmental Resources Report ERC01-81, Georgia Institute of Technology, Atlanta, Georgia.

10. Debo, T. N. 1982. 'Urban Flood Damage-Estimating Curves,' Journal of the Hydraulics Division, Proceedings of the American Society of Civil Engineers, **Vol. 108, No. HY10**, pp. 1059-1069.

11. Delleur, J. W., Miller, W. L., and Potter, H. R. 1976. 'Interactions Between Land Use and Urban Water Resources Planning,' Water Resources Bulletin, Journal of the American Water Resources Assc., **Vol. 12, No. 4**, pp. 759-777.

12. Ellis, S. R., Linder, J. B., and Patterson, D. A. 1983. 'Effects of Increased Urbanization on the Storm Runoff from a Small Urban Basin in Metropolitan Denver, Colorado,' Proceedings of the International Symposium on Urban Hydrology, Hydraulics and Sediment Control, University of Kentucky, Lexington, Kentucky, pp. 79-86.

13. Epsey, W. H., Morgan, C. W., and Masch, F. D. 1965. A Study of Some Effects of Urbanization on Storm Runoff from a Small Watershed, Technical Report No. HYD07-6501, CRWR-2, Center for Research in Water Resources, Univ. of Texas, Austin, Texas.

14. Foster, J. H. 1976. 'Flood Management: Who Benefits and Who Pays,' Water Resources Bulletin, Journal of the American Water Resources Association, **Vol. 12, No. 5**, pp. 1029-1039.

15. Golden, H. G. 1977. Preliminary Flood-Frequency Relations in Urban Streams, Water Resources Investigations 75-77, United States Geological Survey, Doraville, Georgia.

16. Hall, M. J. 1977. 'The Effect of Urbanization on Storm Runoff from Two Catchment Areas in North London,' Proceedings of Amsterdam Symposium on Effects of Urbanization and Industrialization on the Hydrological Regime and on Water Quality, Convened by UNESCO and the IAHS.

17. Kibler, D. F., Froelich, D. C., and Aron, G. 1981. 'Analyzing Urbanization Impacts on Pennsylvania Flood Peaks,' Water Resources Bulletin, Journal of the American Water Resources Assc., **Vol. 17, No.2**, pp. 270-274.

18. Leopold, L. B. 1968. Hydrology for Urban Planning - A Guidebook on the Hydrologic Effects of Urban Land Use, Circular 554, United States Geological Survey.

19. McCorkle, M., and Halff, A. H. 1979. 'Flood-Plain Management in Dallas, Texas: 1908-1978,' Journal of Hydraulics Division, Proceedings of the American Society of Civil Engineers, **Vol. 105, HY2**, pp. 125-137.

20. Rao, R. A., J. W. Delleur and Sarma, B. S. P. 1972. 'Conceptual Hydrologic Models for Urbanizing Basins,' Journal of Hydraulics Division, Proceedings of American Society of Civil Engineers, **Vol. 98, No. HY7**, pp. 1205-1220.

21. U.S. Army Corps of Engineers 1975. Feasibility Report for Water Resources Development, Beargrass Creek Basin, Jefferson County, Kentucky, Report Prepared by the U.S. Army Corps of Engineers, Louisville District, Louisville, Kentucky, 47 pp.

22. U.S. Army Corps of Engineers 1981. HEC-2-Water Surface Profiles. The Hydrologic Engineering Center, U.S. Army Corps of Engineers, Davis, California, 39 pp.

23. Walsh, S. G., and Videkovich, R. M. 1978. 'Urbanization: Hydrologic-Hydraulic Damage Effects,' Journal of the Hydraulics Division, Proceedings of the American Society of Civil Engineers, **Vol. 104, No. HY2**, pp. 141-155.

24. Yomtovian, I., and Ford, D. 1982. 'Floodplain Analysis, Simplified vs. Complex Methods: A Case Study,' Water Resources Bulletin, Journal of the American Water Resources Association, **Vol. 18, No. 2**, pp. 259-263.

FLOOD RISK BIAS ANALYSED THROUGH A MULTI-STATE FLOOD INSURANCE MODEL

K.C.TAI
Department of Water Resources Victoria
590 Orrong Road
ARMADALE 3143
AUSTRALIA

ABSTRACT. Flood risk bias introduced through the traditional method of flood frequency analysis can distort the assessment of flood insurance premium, the fair actuarial rate of flood insurance, even though the insurance payout is maintained the same. The extent of this statistical bias is examined through the concept of expected probability analysis.

To assist in the analysis of flood risk bias, a multi-state insurance is developed, based on some of the basic assumptions of the two-state insurance model developed by Lippman and McCall.

INTRODUCTION. The probability of exceedance of a given flood peak obtained from a mathematically fitted flood frequency curve is a biased estimate of flood risk.

This estimate is due to time-sampling error of a T-year peak estimated from a finite series of annual peaks, where the population distribution was assumed to be normal or log-normal.

However, this bias can be estimated by incorporating the standard error of estimated T-year peak flows, resulting in a flood risk curve significantly different from the flood frequency curve. The flood risk curve rather than the flood frequency curve should be used to compute the average annual cost of flood damage or the actuarial rates of flood insurance. (Hardison and Jennings, 1972).

With perfect information on the random variable of annual flood peak X, the Pth quantile X_p, being the true flood event value, would be exceeded with probability (1-p).

However, the distribution of the annual flood peak X is known with limited precision. Therefore, the best estimate of the Pth quantile of X_p should be used in the flood risk analysis and that best estimate is denoted as Xp^o, which is derived either through the Bayesian or the Classical approach. (Stedinger,1983).

Xp^o, the flood event that is exceeded with probability (1-p), is expressed as:

$$Xp^o = \overline{X} + S(1 - (1/n))^{\frac{1}{2}} \cdot t_{p,\nu}$$

where $t_{p,\nu}$ is the Pth quantile of the Student's t distribution.

V. P. Singh (ed.), Application of Frequency and Risk in Water Resources, 395–404.

$\bar{X}$ and S are viewed as random variables of sample mean and sample standard deviation of different samples n, with $\nu = n-1$ degrees of freedom.

Hence if one samples many values of Xp^o, the probability of its exceedance will be (1-p) on the average. This interpretation supports Beard's (1974) denoting of (1-p) as an exceedance probability and Hardison and Jenning's use of the term, the 'average exceedance probability'.

On the other hand, the traditional flood frequency uses the best estimate of the pth quantile of Xp as $\hat{X}p$:

$$\hat{X}p = \bar{X} + S.Zp$$

where Zp is the pth quantile of the standard normal distribution. In this approach no allowance is made for repeated sampling.

Both the estimate Xp^o and $\hat{X}p$ will converge with probability one to the pth quantile Xp as sample size n tends towards infinity. Therefore both the estimates are consistent estimator of Xp asymptotically. The large difference in values of estimates when n is small is analysed in terms of insurance premiums.

ANALYSIS. The Multi-State Insurance Model.

In developing the multi-state insurance model, the following assumptions are made:

(a) each interval of flood stage j (j=1,2,......,n), has a probability flooding segment Pj, with a corresponding mean flood damage $\bar{D}j$, for a house on a flood plain;

(b) each interval of flood stage j is subject to two states of nature; either non-flooding with probability q_j or flooding with probability p_j, such that $p_j + q_j = 1$, as is the case in the two state insurance model;

(c) the individual is endowed with a wealth w in the non-flooding state and in the flooding state, his wealth is reduced to Y_j with Y_j less than w for each flood stage interval j;

(d) before the state of nature is known, the individual can insure against flooding by purchasing flood insurance. The actuarial rate between flooding and non-flooding is π_j for the interval of flood stage j. π_j is known as the cost of insurance, i.e. an insurance S_j in the flood state at stage interval j can be purchased by a reduction of $\pi_j S_j$ in the non-flooding state;

(e) the individual possesses an increasing and strictly concave utility function and his objective is to select the optimal amount of insurance S_j^* by maximising the expected utility function of his wealth; and

(f) a proportion θ_j of flood losses in the stage interval j is covered by insurance S_j such that $S_j \leq \theta_j \leq D_j$ where

$0 \leq \theta_j \leq 1.$

In selecting the optimal amount of insurance S* by maximising the expected utility function of his wealth, the individual would purchase the insurance cover S^*_j for each stage interval at the same time.

Hence the optimal level of insurance would be determined as follows:

$$U(\tilde{S}^*) = \max_{\tilde{S} \geq 0} U(\tilde{S}) \text{ where } \tilde{S} = \begin{bmatrix} S \\ \cdot \\ \cdot \\ \cdot \\ Sn \end{bmatrix}$$

and the expected utility is:

$$U(\tilde{S}) = q.u(w - \tilde{\pi}^T \tilde{S}) + \sum_{j=1}^{n} p_j \ .u(Y_j + S_j) \qquad (1)$$

where $u(w - \tilde{\pi}^T \tilde{S})$ is the expected no-flood utility and $u(y_j + S_j)$ is the expected utility if flood occurs at stage interval j.

P_j is the probability of flood in stage interval j occuring and q is the total probability of no-flooding.

It can be shown from a subsequent analysis using the tree diagram that for a given stage interval j, (j=1.2,......n):

$$p_j + q'_j + q_j = 1$$

and

$$q'_j + q_j = (1 - p_j).$$

Therefore

$$q = \sum_{j=1}^{n} (q'_j + q_j) = \sum_{j=1}^{n} (1 - p_j) \qquad (2)$$

It is also shown in a subsequent analysis that q_j is propogated in that state with a constant probability q_o of no-flooding.

Threfore with $q_j = q_o$, then:

$$q = \sum_{j=1}^{n} (q'_j + q_o) = \sum_{j=1}^{n} (1 - p_j) \qquad (3)$$

Substituting equation (3) in the expected utility gives

$$U(\tilde{S}) = \sum_{j=1}^{n} (q'_j + q_j).u(w - \tilde{\pi}^T \tilde{s}) + \sum_{j=1}^{n} p_j \ .u(y_j + s_j) \qquad (4)$$

where $\tilde{\pi}^T = (\pi_1, \pi_2, \ldots\ldots \pi_n)$, and

$\tilde{\pi}^T . \tilde{S} = \sum_{j=1}^{n} \pi_j S_j$ which is the annual payment to the insurer.

The first order equilibrium condition gives:

$$\frac{\partial U(S)}{\partial S_j} = -\pi_j (q'_j + q_j).u'(w - \tilde{\pi}^T \tilde{S}) + p_j.u'(y_j + s_j) = 0$$

for $j = 1,2,......,n.$

This furnishes n simultaneous equations:

$$\pi_j = \frac{P_j}{(q_j' + q_o)} \cdot \frac{u'(Y_j + S_j)}{u'(w - \pi^T S)} = \frac{P_j}{(1-p_j)} \cdot \frac{u'(Y_j + S_j)}{u'(w - \pi^T S)} \tag{5}$$

$$= \frac{P_j}{(q_j' + q_o)} = \frac{P_j}{(1-p_j)},$$ is the actuarial fair rate of insurance (see Lippman and McCall) because the expected insurance payout balances the expected premium for a given stage interval j: $P_j S_j = (1-p_j)\pi_j S_j$

It follows that:

$$u'(y_j + S_j^*) = u'(w - \pi^T S), \text{ or}$$

$$(y_j + S_j^*) = (w - \pi^T S), \text{ or}$$

$$S_j^* = (w - y_j) - \sum_{j=1}^{n} \pi_j S_j^*, \quad \text{where } j=1,2,.....,n. \tag{6}$$

where S_j^* is the optimal insurance cover for stage interval j; $(w-Y_j)$ is the reduction in wealth or damage to wealth due to segment j flood, and

$\sum_{j=1}^{n} \pi_j \cdot S_j$ is the annual payment to insurer to cover against flood damage.

Tree Diagram Analysis for Fair Actuarial Rate

Actuarial rate may be assessed on the basis of a tree diagram (fig.1A) showing at the first 'chance fork', the occurrence of flooding F_o at the stage interval j=0 with probability P_o, and the occurrence of no flooding NF_o at the floor level with probability q_o. This stage interval would correspond to almost the floor level interval.

At the second 'chance fork'. at stage interval j=1, the joint probabilities are recorded in the respective cells of figure 1B as shown, and the marginal probabilities for stage intervals j=0 and j=1 are indicated at the margins.

The actuarial fair rate for stage interval j=1 is therefore:

$$\pi_1 = \frac{P(F_1)}{P(NF_1)} = \frac{P_1}{(q_1' + q_o)} = \frac{P_1}{(1-p_1)},$$ which is the ratio of the marginal probability of flooding to the marginal probability of no-flooding.

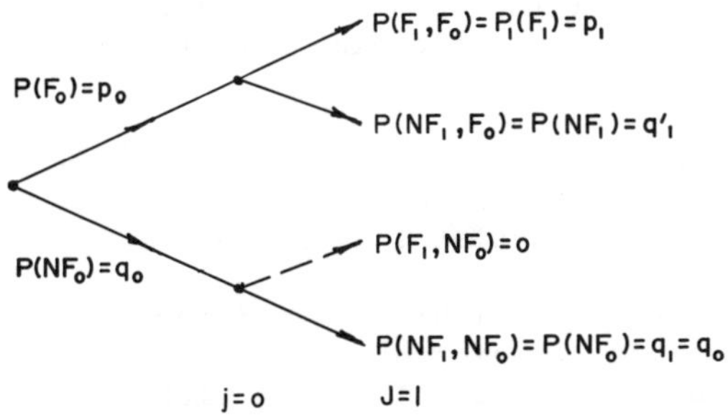

FIG. 1A. TREE DIAGRAM ANALYSIS

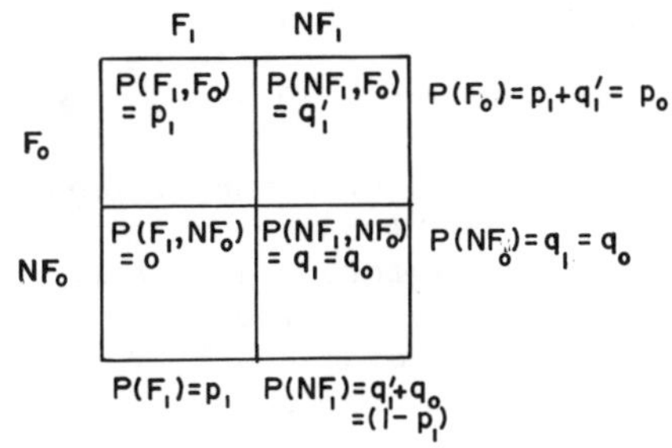

FIG. 1B. JOINT AND MARGINAL PROBABILITIES

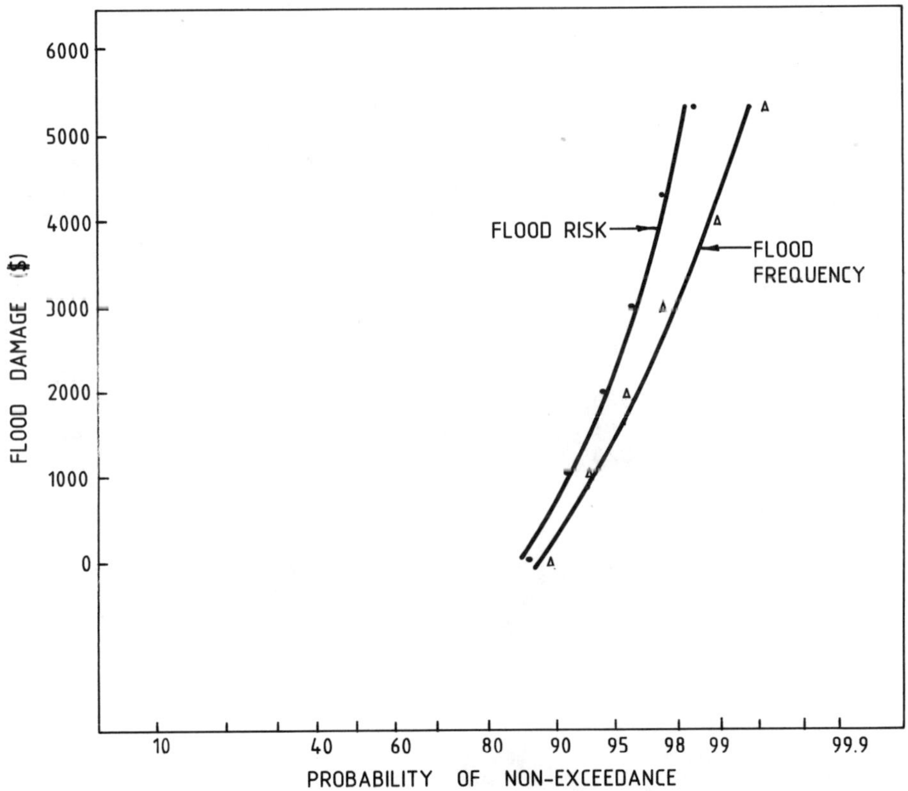

FIG. 1C. COMPARISON OF FLOOD FREQUENCY WITH FLOOD RISK CURVES

It can be seen from the tree diagram that the conditional probability of flooding at stage interval j=1, given that no flooding has occurred at stage interval j=0 zero, i.e. $P(F_1/NF_0)=0$.

With this observation, two results could be deduced:

(a) the no-flooding state from the floor level corresponding to the stage interval j=0 cannot propogate a state of flooding to higher stage intervals, and

(b) the flooding state from the floor level within stage interval j=0 can propogate either a state of flooding or a state of non-flooding but the latter would be a terminal state with a probability q_j' corresponding to stage interval j.

RESULTS. A biased estimate of flood risk as displayed by the flood frequency curve shown in figure 1c results in a biased estimate of average annual flood damage. Taking the flood damage example from Hardison and Jennings (1972) as an example, annual expected flood damage is estimated to be $202 as shown in Table 1.

TABLE 1: COMPUTATION OF AVERAGE ANNUAL FLOOD DAMAGE FOR FLOOD FREQUENCY CURVE IN FIGURE 1.

Damage Stage Interval j	Stage in metres	Exceedance Probability	Segmented Probability Pj	Damage Dj ($)	Average Damage Range ($)	Average Annual Damage ($)
	3.81	0.105		0		
1			0.038		500	19
	3.96	0.067		1000		
2			0.027		1500	40
	4.11	0.040		2000		
3			0.017		2500	42
	4.27	0.023		3000		
4			0.013		3650	48
	4.47	0.010		4000	*	
5			0.010		5300	53
	4.62	0.005		5300		

Total average annual flood damage: $202.

The flood risk curve shown against the flood frequency curve in figure 1c is derived on the basis of the average exceedance probability concept based on an assumed normal distribution with zero skew for a sample size of 10 annual flood peaks. (see Table 2 of Hardison and Jennings (1972). This is a very small sample size to indicate the extent of the bias in flood risk.

The table below shows the unbiased estimate of average annual flood estimate.

TABLE 2: COMPUTATION OF AVERAGE ANNUAL FLOOD DAMAGE FOR THE FLOOD RISK CURVE SHOWN IN FIGURE 1.

Damage Stage j	Stage in metres	Exceedance Probability	Segmented Probability Pj	Damage ($) Dj	Average Damage ($)	Average Annual Damage ($)
	3.81	0.132		0		
1			0.046		500	23
	3.96	0.086		1000		
2			0.030		1500	45
	4.11	0.056		2000		
3			0.020		2500	50
	4.27	0.036		3000		
4			0.013		3650	47.5
	4.47	0.026		4300		
5			0.009		4800	43.2
	4.42	0.014		5300		
6			0.006		5300	31.8
	4.62	0.008		5300		
7			0.005		5300	26.5
	4.62	0.003		5300		

Total average annual flood damage: $267.00

The bias in flood risk has resulted in an under-estimate of average annual flood damage by $65.00 or 32%.

The actuarial fair rate of flood insurance is also assessed as the result of the flood risk bias introduced by the flood frequency procedure of analysis. This biased actuarial rate is compared with the unbiased rate obtained through the average probability exceedance procedure.

TABLE 3: ACTUARIAL FAIR RATE OF INSURANCE OBTAINED THROUGH TRADITIONAL FLOOD FREQUECY ANALYSIS

Damage Stage Interval j	Segmented Probability of Flooding	Segmented Probability of Non Flooding	Incremental Probability of Non Flooding	Total Segmented Probability of Non Flooding	Actuarial Fair Rate of Insurance
	$P(F_j)=P_j$	$P(NF_j)=q_j$	$P(NF_j, F_{j-1})=q'_j$	$(q_j + q'_j)$	$\pi_j = p_j/(1-p_j)$
(1)	(2)	(3)	(4)	(5)=(3+4)	(6)=(2)/(5)
0	0.105	0.895	-	0.895	-
1	0.038	0.895	0.067	0.962	0.040
2	0.027	0.962	0.011	0.973	0.028
3	0.017	0.973	0.010	0.983	0.017
4	0.013	0.983	0.004	0.987	0.013
5	0.010	0.987	0.003	0.990	0.010

TABLE 4: ACTUARIAL FAIR RATE OF INSURANCE OBTAINED THROUGH THE AVERAGE EXCEEDANCE PROBABILITY CONCEPT

Stage Interval j	Segmented Probability of Flooding	Segmented Probability of Non Flooding	Incremental Probability of Non Flooding	Total Probability of Non Flooding	Actuarial Fair Rate of Insurance $\pi_j = Pj/1-Pj$
(1)	(2)	(3)	(4)	(5)	(6)
	Pj	q_j	q'_j	$(q_j + q'_j)$	
0	0.132	0.868	-	0.868	-
1	0.046	0.868	0.086	0.954	0.048
2	0.030	0.954	0.016	0.970	0.031
3	0.020	0.970	0.010	0.980	0.020
4	0.013	0.980	0.007	0.987	0.013
5	0.009	0.987	0.004	0.991	0.009
6	0.006	0.991	0.003	0.994	0.006
7	0.005	0.994	0.001	0.995	0.005

The bias in flood risk distorts the actuarial fair rate of insurance and the distortion in terms of the insurance Sj is displayedas follows:

TABLE 5: ACTUARIAL FAIR RATE OF INSURANCE AND INSURANCE PREMIUMS COMPUTED BY THE FLOOD FREQUENCY AND AVERAGE EXCEEDANCE PROBABILITY PROCEDURES.

	Flood Frequency Procedure			Average Exceedance Probability Procedure		
Stage Interv j	Average Damage ($)=Sj	Ratio	($)	Average Damage=Sj ($)	Ratio	Sj ($)
1	500	0.040	20	500	0.048	24.0
2	1500	0.028	42	1500	0.031	46.5
3	2500	0.017	42.5	2500	0.020	50
4	3650	0.013	47.5	3650	0.013	47.5
5	5300	0.010	53.0	4800	0.009	43.2
6				5300	0.006	31.8
6=7				5300	0.005	26.5
		Total:	$205.00			$269.50

The difference in bias is confirmed with significant under-estimation of pure insurance premiums based on the traditional flood frequency estimation procedure.

These results confirm the findings of Beard (1978) that the bias in flood frequency estimates would lead to intolerable under-estimates of flood insurance premiums.

The procedure of expected probability or average exceedance probability, which removes the bias, should be used in flood risk estimates for setting fair actuarial rates.

It comes as no surprise that the fair rate of insurance premium almost equals the expected annual flood damages, for these reasons:

(a) since the actuarial fair rate for stage interval j is $\pi_j = Pj/(1-pj)$, and with pj being small, $\pi_j \doteq pj$ (see Tables 3 and 4), and

(b) average damage $\bar{D}_j$ is set equal to S_j with $\theta_j = 1$ (see Table 5).

Note however that Sj is not the optimal insurance S_j^* previously established in the mult-state insurance model. The use of pj and $\bar{D}j$ to estimate insurance premium has been widely used in practice and this analysis has been influenced by that consideration.

CONCLUSIONS. Flood risk bias introduced through the traditional method of flood frequency analysis can distort the assessment of flood insurance premium, expressed as the annual flood damage value. The flood risk bias has also the distorting effect on the actuarial fair rate of flood insurance, even though the insurance payout is maintained equal to the flood damage.

The extent of the bias is examined through the concept of expected probability or average exceedance probability analysis.

A multi-state flood insurance model is developed which is unique to the multi-stage analysis of segmental probability for flooding and non-flooding at each given stage interval. An actuarial fair rate of flood insurance for each stage interval can be established, based on the ratio of the segmental probability for flooding and non-flooding. This actuarial fair rate is a pure rate of insurance.

ACKNOWLEDGMENT. Acknowledgment is made to Dr. George Kuczera, of the Civil Engineering Department of the University of Newcastle, New South Wales, for his critical comments and useful suggestions.

REFERENCES. Beard, L.Rl, 1978. Impact of hydrologic uncertainties on flood insurance.
Journal of Hydrau. Div., Am. Soc. Civil Eng., V.104, No.HY11, Nov. pp. 1473-1484.

Hardison, C.H. and M.E. Jennings, 1972. Bias in computed flood risk. Journal of Hydrau.Div., Am. Soc. Civil Eng., V. 98, No. HY3, March, pp. 415-427.

Lippman, S.A. and J.J. McCall, 1981. The economics of uncertainty, ch:6 in Ed. K.J. Arrow and M.D. Intriligator, "Handbook of Mathematical Economics", North Holland Publishing CO., N.Y., pp230-232.

Stedinger, J.R., 1983. Design events with specified flood risk. Water Resources Research, V.19, No.2. pp. 511-522.

CORPS OF ENGINEERS PROCEDURES FOR EVALUATING AND MODIFYING EXISTING DAMS TO SAFELY ACCOMMODATE THE CURRENT INFLOW DESIGN FLOOD

Roy G. Huffman, P.E.
Chief, Hydrologic Engineering Branch
Directorate of Civil Works
Office, Chief of Engineers
Department of the Army
Washington, D. C. 20314-1000

ABSTRACT. It is the Corps practice to design dams above populated areas to safely accommodate the largest flood considered physically possible. This magnitude of flood is refered to as the Probable Maximum Flood (PMF). The Corps existing dams are being evaluated to determine how safely they could accommodate inflow floods up to a PMF developed in accordance with todays standards. If it is found that failure of a dam could endanger peoples lives, a modification of the dam will be developed that would remove the threat to human life. Alternatives will be studied to determine the least costly modification that will meet the requirement, to not endanger lives. Economic analysis, that is comparing average annual benefits to average annual cost, will not be used as a basis for selecting the dam modification. Why economics is not considered to be an appropriate decision criteria is explained in some detail.

The U.S. Army Corps of Engineers is the Federal Government's largest water resources management and development agency. Throughout our history, we have adapted to changing national priorities. We continue to improve navigation and reduce flood damage while we provide a recreation/resource management program, generate hydroelectric power, provide water supply storage to cities and industry, regulate development in navigable waters and wetlands, and support the nation in many other ways. The Corps has constructed and operates about 600 dams and 270 navigation locks, built thousands of miles of levees and many pumping stations. Hydroelectric power is produced at 104 Corps dams, representing 4% of the total U.S. energy production. Also, the Corps maintains 49 major ports and 301 small boat harbors. The Corps employs full-time about 40,000 people, 1,000 of which are military.

The Corps of Engineers has been in the process for several years of reviewing its dams to determine how well they meet todays inflow design flood criteria. Policies and procedures for conducting this review, and formulating remedial actions, have been evolving and will continue to be perfected. There are many details that are being

V. P. Singh (ed.), Application of Frequency and Risk in Water Resources, 405–409.

handled on an ad-hoc basis and will require both research and experience to establish general policies. Basic policies used to make decisions on the merits of dam safety modifications related to hydrologic deficiencies were issued by letter: Policy for Evaluating Modifications of Existing Dams Related to Hydrologic Deficiencies, 8 April 1985. The origin of this dam safety modification policy can be found in the Corps traditional practices for new dams.

The Corps of Engineers has a long established policy for the design, construction and operation of new dams. It has been the Corps intent that new dams impounding large volumes of water located above populated areas be designed to safely accommodate the Probable Maximum Flood (PMF). Because of ethical and public welfare considerations, the Corps has not deliberately accepted recognizable risk simply to reduce the cost of a dam if failure of the dam would imperil the lives or property of the downstream populace. Preventing the loss of human life due to a dam failure is the dominant consideration. This fundamental attitude has not changed and is considered appropriate for the analysis and design of modifications to existing dams.

The Corps program to evaluate existing dams for hydrologic safety has been carried out in two phases. Phase No. 1 was a reconnaissance study to determine how well all Corps dams could accommodate the present PMF. Dams that can handle today's PMF were declared safe and no further study is required. Dams that can accommodate the present PMF with some encroachment into the freeboard were deferred for later study. Those dams that would be overtopped by today's PMF were considered potentially hazardous and were carried into the second phase.

Phase No. 2 moves into the determination of failure consequences, the analysis of alternatives, and the selection of an appropriate dam safety modification. This Phase No. 2 analysis may be envisioned as a four step process as follows:

Step I: This step selects an appropriate inflow design flood; designated the "Base Safety Condition" (BSC). Inflow floods from 10% to 100% of the PMF are routed through the system. A failure of the dam is postulated for those floods which would encroach on the freeboard or overtop the dam. The flood magnitude that would just start to encroach on the dam freeboard is designated as the "Threshold Flood". Downstream impacts on people and property are determined for all floods from 10% to 100% of the PMF without dam failure. Similar downstream impacts are determined with dam failure for floods of threshold and larger magnitude. The consequences of dam failure are then compared to the flood impacts without failure. The flood magnitude where the differences converge and become insignificant is selected as the BSC. When the differences between without and with dam failure consequences remain significant up to the PMF then the BSC becomes the PMF.

Step 2: A cost effective modification to accommodate the BSC flood is selected in this step. The possible impacts on both upstream and downstream areas must be carefully evaluated. Increased spillway capacity could increase downstream flooding. Raising the dam could raise upstream flood levels and place people at risk who were not previously subject to flooding. Life cycle cost, including operation, maintenance and replacement, are used. All types of modifications including structural, nonstructural, and relocation of people are open for consideration.

Step 3: In this step, the cost of modification is compared to the project benefits. The basic criterion is that the benefits of continued operation of the project should exceed the modification cost of the project. If modification cost exceed project benefits then breaching the dam and the accompanying loss of national economic development (NED) benefits will be considered. It is not clear at this time what consideration will be given regional and local concerns such as lower property values, lost taxes, lost income and jobs, etc.

Step 4: In those cases where the BSC is less than the PMF, consideration will be given to increasing the inflow design flood to something greater than the BSC. Conventional economic analysis may be a basis for this decision.

There are several technical questions and policy areas that have yet to be resolved. Briefly, some of the more obvious are as follows:

Freeboard: The assumption that failure could occur at the bottom of the freeboard may not produce the most critical failure flood consequences.

Flood Warning: A method to estimate any reduction in loss of life due to people reacting to a flood warning is under investigation. Flood warning has many social and other problems that need to be thoroughly thought through. It should be noted that the Corps requires a flood warning and evacuation plan for all its dams. It is therefore possible that flood warning is a base condition for all dam safety studies.

Antecedent Flood: The reservoir level at the start of the PMF, or some fraction of, is under consideration.

Failure: More objective criteria for breach dimensions and rate of failure are needed.

Health: Trauma related symptoms are real and there costs are real. Procedures for quantifying mental and physical health cost due to flood disasters are needed.

Cultural: Evaluation of the impacts with and without a proposed dam safety modification on cultural and environmental assets should be carried out. However, uniform procedures are needed. For example: How and at what cost should a historic landmark be valued?

Regional and Local: What weight should be given to regional and local cost, benefits, and opinions?

A probabilistic based economic analysis may be used in Step 4 to capture NED impacts, however, it should be apparent that most of the effort will be expended in Step 1 through Step 3 on a deterministic based analysis. There are a number of reasons why a deterministic analysis is considered more appropriate than a probabilistic analysis for establishing the BSC inflow design flood.

Doing a probabilistic risk based analysis for dam safety requires reasonable estimates of probability out to the region of the PMF. The problems and controversy associated with estimating the frequency of extremely large flood peaks should be well known and will not be repeated. It should be noted, however, that for dam safety analysis purposes, the peak inflow is usually of little concern. Instead, it is the maximum pool elevation, maximum discharge, and downstream water surface elevation, velocity etc., that are of interest. The pool elevation is generally more sensitive to flood volume and gate operation, or misoperation, then peak inflow. Downstream flood conditions are sensitive to many assumptions such as, coincident flooding on tributaries, dam breach formation, debris on bridges, etc. A probabilistic-based analysis would therefore become a multiple parameter problem of immense complexity, too difficult and uncertain for practical use.

As difficult and uncertain as probabilistic risk based analysis is, some may still be inclined to use it. Before undertaking probabilistic risk based procedures for analysis of dam safety, one question should be answered. The question very simply is: if the probability (exceedance interval) of the PMF could be known exactly would this knowledge be of any use? My contention is that it would not. Why it would not be of any practical use is because of the way economic analysis is conducted and the way random numbers behave.

Economic analysis by definition deals with monetary values. This may lead to attempts to place dollar values on intangibles, such as human life, or develop other devices to include intangible values in the analysis. Placing monetary values on intangibles is a highly subjective and questionable process. In the process, intangibles of considerable value may be obscured. Also, economic analysis discounts the future and focuses on the present. At a 6% discount rate, the present value of one dollar 20 years from now is worth 31 cents and 50 years from now only 5 cents. Therefore, inherent to any economic analysis is a short time horizon. Anything very far in the future is not worth much today. Agencies entrusted with the public welfare should be as concerned with 100 or 200 years from now as 10 or 20 years into the future.

Floods are assumed to be random independent events and this random nature of floods appears to be a good working assumption. It can be shown by random number experiments that short data sets can depart significantly from the true parent population. Also, even for high probability events (like 0.10) it might take a thousand or more repetitions before the average ratio of successes to failures is about one in ten. The same phenomenon can be observed using real flood data by comparing a long record to the same data divided into shorter subsets. The point in relation to risk assessment is, that the expected average annual benefits of many flood damage reduction projects must be averaged together before the expected average rate of return will approximately equal the actual rate of return. This is true even for projects that depend on reducing the damage from frequent flood events for the bulk of their benefits. As the exceedance intervals of the flood events where the bulk of the benefits are expected become longer, there must be more projects and/or a longer time to average benefits over. There is essentially no possibility that the actual benefits from one dam safety modification would be close to the expected average annual benefits for that one project. It is unlikely that the actual rate of return would approximate the expected rate of return even when averaged over several hundred projects because the economic time horizon is short (20 to 30 years) and flood exceedance intervals are long (about 10^5 to 10^8 years). This is true if the damage reduction benefits are in dollars, lives saved or any other measure.

To bring this discussion together, a probabilistic risk analysis would be based on flood frequency estimates with very wide confidence bands that cannot be defined. Economic projections are uncertain, intangibles may be obscured, and time horizons are short. Using average annual values tends to average out and makes large damages from infrequent events seem less significant. Lastly, the estimated average annual benefits will not be realized because flood events, especially large floods, over the next few years will not conform to the expected long term norm. The objective of economic analysis is to maximize net benefits or to put it another way, to make the very best decision. It should be apparent that the best decision is not possible and a more appropriate decision criterion is to try to avoid making bad decisions. Designing a dam so it will not endanger lives in the event a large flood up to PMF magnitude should occur is such a decision criteria.

In conclusion, the Corps dam safety program is progressing. Studies are underway and research is in progress. A more complete statement of policy and procedures is contemplated by the fall of 1987.

BUREAU OF RECLAMATION USE OF RISK ANALYSIS

Neil F. Parrett
Chief, Division of Dam and Waterway Design
Bureau of Reclamation
U.S. Department of the Interior
P.O. Box 25007
Denver, Colorado 80225-0007

ABSTRACT. This paper briefly presents the Bureau of Reclamation's philosophy and methodology in using risk-based analysis to select appropriate actions in its safety of dams program.

Before presenting USBR (United States Bureau of Reclamation) practice and experience using risk analysis techniques in assessments for management decisions, I want to comment on causes for dam failures. While we have discussed risk associated with large remote occurring floods, actual failures of modern major dams result most often from causes other than floods, usually "piping," and usually resulting from a condition not detected by the designers before the failure.

If dams that fail due to "piping" during intial filling and early operations, are removed from the historic data base on failures, major dams have a low risk (10^{-7}) for failure. One recommendation is not to spend resources needed for geotechnical design, construction, and performance monitoring against the occurrance of "piping" on risk analysis of hydrologic events.

The typical applications by the Bureau of risk-based decision analyses are shown in table I. The Bureau's programs for design of dams and assessing existing dams strive for balance in assessing the several loading events that may threaten a dam. Having said that, I will focus my remaining comments on the Bureau's assessments of existing dams for large, infrequent flood events.

Figure 1 shows number of recommentations from examinations of 362 dams by subject area, and the number that had been analyzed by the end of 1984. Due to additional stream flow data in most basins and revised procedures for developing the PMF (probable maximum flood), virtually all dams require a new PMF to be prepared.

V. P. Singh (ed.), Application of Frequency and Risk in Water Resources, 411–428.

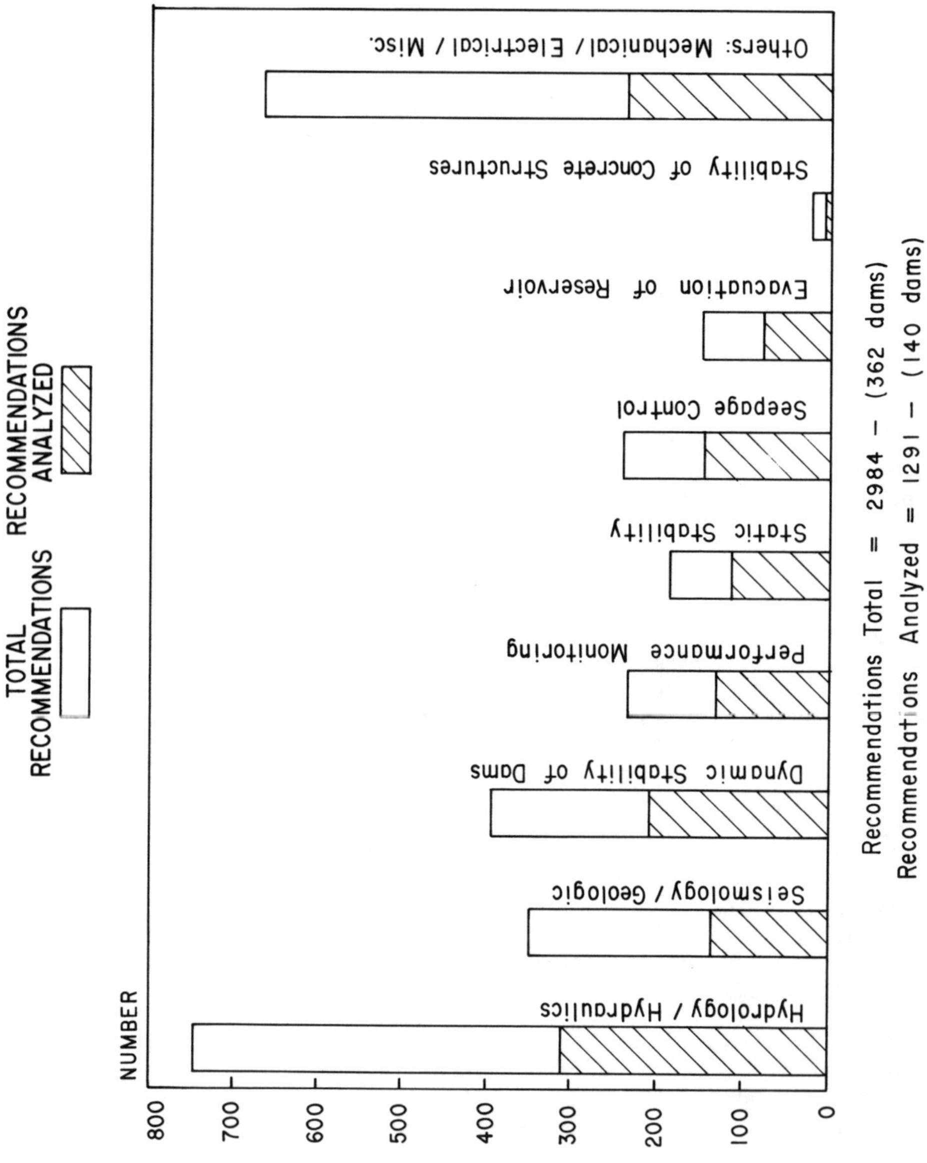

Figure 1. - Distribution of SEED recommendations by topic.

TABLE I. - USBR DAM SAFETY MODIFICATION STUDIES APPLICATION OF RISK-BASED DECISION ANALYSIS

1. HAZARD ANALYSIS
   - Evaluation Problem
   - Select Loading

2. RISK ANALYSIS
   - Present Alternatives
   - Compare Alternatives

3. DECISION ON MODIFICATION
   - Input Items 1, 2, and Other Factors

Before I share our approach and some case histories on risk analysis input to decisions on modification of dam decisions, I want to share three conclusions about how people, in general, deal with risk. I have lifted these conclusions from an article by William Allman, titled 'Staying Alive in the 20th Century,' which was published in the October 1985 issue of SCIENCE Magazine.

1. People's wisdom depends less on understanding the details of scientific risk assessment than understanding the limits of what science can do. "People are often more scientific in watching the processes of science than scientists themselves."

2. A feeling of control can actually make a risky technology even more dangerous. Our purpose for use of risk analysis is to counter our inherent overconfidences as we seek and compare information so we better realize the risks we face.

3. There is no neutral way to present information .... the best way to inform people about risks is to give them information in several different and understandable ways. How data are presented does influence. An engineer presenting the many low probability events that must occur in succession to result in a mishap will probably communicate to a public listener, "I didn't realize there were so many things that could go wrong."

Going back to the first conclusion on science and people's wisdom ... More than a few engineers, people, and politicians in management, look at PMF's and especially the apparent periodic quantum increases in PMF magnitudes with jaundice judgment. Reclamation, along with most Federal agencies, calculates PMF's using PMP (probable maximum precipitation) estimates obtained using the hydrometeorological reports prepared by the National Weather Service. The adopted use of these reports will tend to stabilize PMF estimates. Future changes, if needed, are expected to be reduced in frequency and magnitude. The Bureau's efforts to deal with increasing PMF's included the study of actual flood events against PMF's for the same basins. Because stream gauging stations are located to monitor low flows, and do not usually record data or even physically survive large floods, we could only compare peak flows determined using slope area techniques. No actual flood volume data were available to compare to PMF volumes. The large floods data the Bureau has obtained within the United States are shown on figure 2.

The Bureau uses risk analysis as a method of assessing selected elements believed important to a decision we need to make. The risk analysis hopefully influences, but it does not make the decision. The decision is consciously made by the actions or unconsciously made by the inactions of people. Like other technical analyses tools, if accurate data and objectivity are not employed in a risk assessment, the results will be misleading. The benefits from risk assessments are:

- To increase comprehension as to interrelations of factors impacting on decisions
- To array information in a manner that will provide a good overall understanding of the problem
- To provide a concise presentation on comparitive benefits of alternative actions
- To document the basis of decision

What I want to focus on is our decision process to 1) select level of modification for an existing dam, 2) select an alternative design that provides that level of protection, and 3) how we have employed risk-based or probability analysis in our decisions.

The goals in the Bureau's modification of existing dams activities are to reduce risk for potential loss of life and reduce adverse effects of loss of reservoir. An example of a decision sequence is shown in table II.

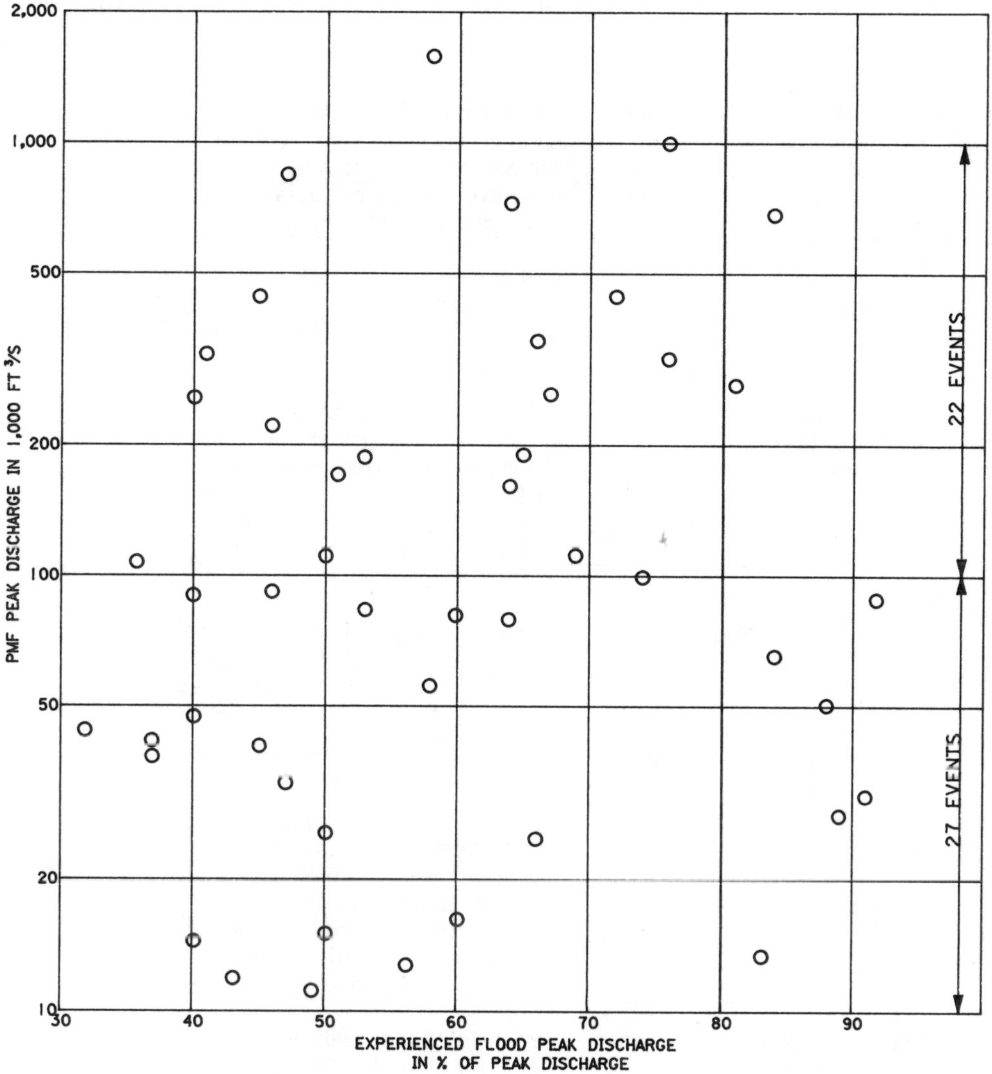

Figure 2. - Comparison of peak discharges experienced vs. probable maximum floods. (U.S. - 1862 through 1982)

TABLE II. - DECISION QUESTIONS FOR SAFETY OF DAMS MODIFICATION

Question	Answer	Action
A. Will the dam fail if the loading event selected for safe design occurs:	No	Stop modification decision process, continue periodic assessment of dam for adequate performance.
	Yes	Describe modes of potential failure and continue to step B.
B. Is failure possible from events likely to occur during normal anticipated operation?	No	Continue to step C.
	Yes	Define interim actions required and define priority to continued work. Continue to step C.
C. Will the physical consequences of downstream flooding be worse because of dam failure?	No	This question needs to be addressed to interim level events if the answer to the maximum magnitude event is No. If No to all level of an event, stop modification decision process.
	Yes	Continue to step D.
D. Are human lives likely to be lost as a result of dam failure?	No	Define most cost-effective level of protection and least cost modification alternative for that level of protection. Consider use of a probability-based analysis. Stop modification decision analysis process.
	Yes	Continue to step E.
E. Is level of protection required to minimize loss of lives the maximum magnitude event?	No	Determine if higher level of protection alternative is more cost-effective. Consider use of a probability-based analysis Continue to step F.
	Yes	Design least cost modification alternative for the maximum magnitude event. Stop modification decision analysis process.
F. Is most cost-effective level of protection above level of protection required to minimize loss of lives?	No	Design least cost modification alternative for level of protection required to minimize loss of lives. Stop modification decision analysis process
	Yes	Design at higher level of protection the least cost modification alternative. Stop modification decision analysis process.

The application of risk analysis philosophies has been in the areas of hazard assessment, impacts of alternative actions and comparisons of hazard, impacts and other identified factors for a risk management decision. In a hazard assessment, separate relationships are prepared to illustrate the potential loss of life and property damage versus flood discharge.

Figures 3a and 3b illustrate a hypothetical case indicating a small incremental difference in population at risk and estimated loss of life for inflow flood levels above about 60 percent of the maximum discharge, and significant incremental property damages for all flood levels with a failure of dam potential. A case can be established for a Safety of Dams Modification above 60 percent of PMF being determined solely upon economic analyses. Figures 3c and 3d illustrate a case where the incremental estimated loss of life is significant for all flood levels up to the PMF. For some damsites, the nature of the incremental hazard posed by potential dam failure can be discerned readily with a rather cursory assessment of the damages and population exposure. Other damsites require expensive detailed assessments. Attention must be given to inundation mapping, location of affected population, method of flood wave detection, warning dissemination, route and time for escape, and other site-related factors.

Catagories and examples of application of risk analyses between alternatives and various givens are illustrated on figure 4. Case C may lead to selection of other than the least total cost alternative if a higher level of protection is mandated.

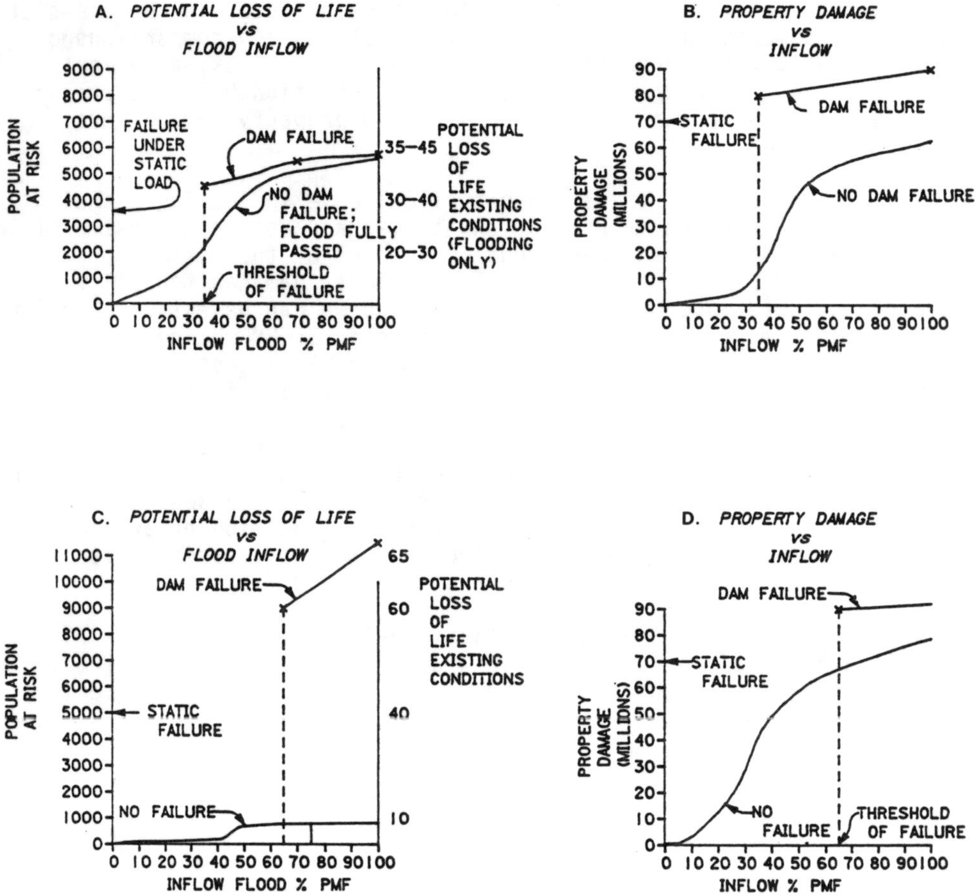

NOTES:

1. The incremental effects for dam failure due to static or earthquake loading are indicated at 0 discharge.
2. The relations between population at risk and estimated loss of life for the examples are purely conjecture and should not be considered an average or typical relationship.

Figure 3. - Typical results from a hazard assessment.

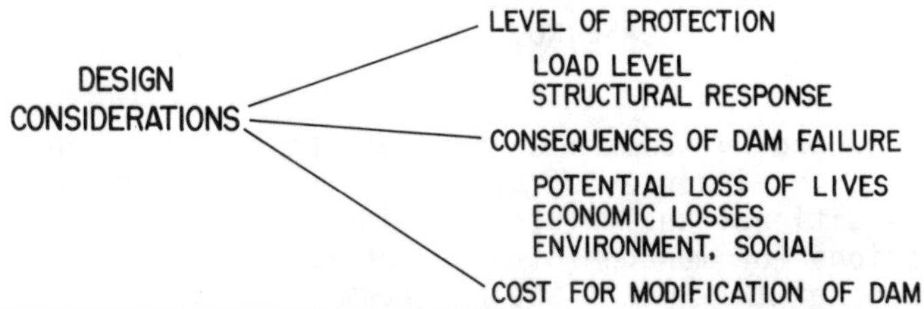

APPLICATION CATEGORY \ DESIGN FACTOR	LEVEL OF PROTECTION	CONSEQUENCES FROM DAM FAILURE	COST FOR MODIFICATION OF DAM
A	FIXED	FIXED	VARIABLE FOR ALTERNATIVE SELECTED
B	FIXED	VARIABLE FOR ALTERNATIVE SELECTED	VARIABLE FOR ALTERNATIVE SELECTED
C	VARIABLE FOR ALTERNATIVE SELECTED	VARIABLE FOR ALTERNATIVE SELECTED	VARIABLE FOR ALTERNATIVE SELECTED

CASE A - PROBABILITY BASED ANALYSIS USUALLY NOT NECESSARY WHEN SELECTING A MODIFICATION ALTERNATIVE

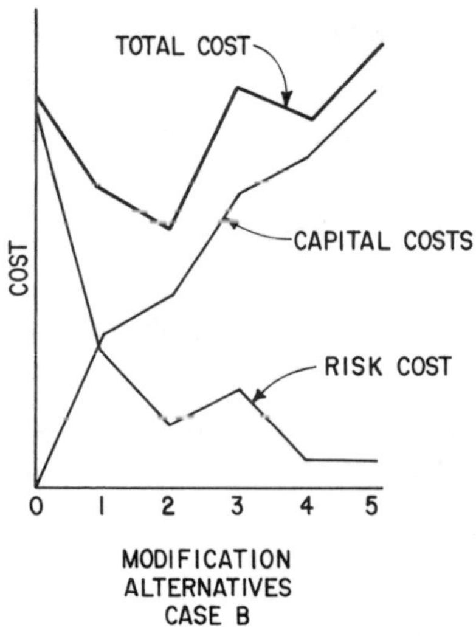

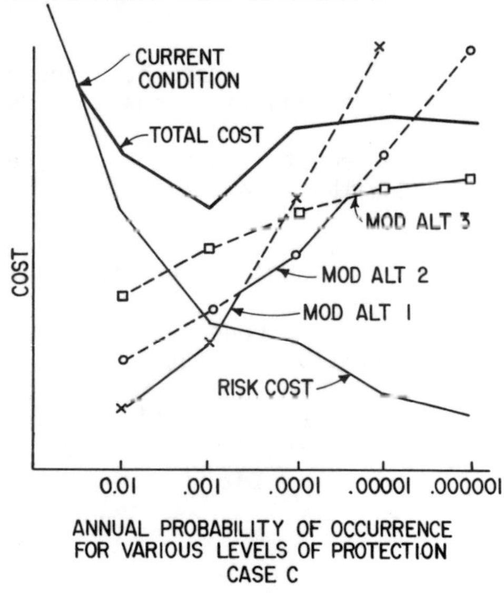

Figure 4. - Application categories of probability-based decision analysis.

CASE NO. 1 DAM

Figures 1-1, -2, -3, and -4 summarize, for Case No. 1 dam, a hazard assessment conducted for a series of dams upstream from a major population center. An additional analysis, not illustrated, was a comparison of alternative actions for management of a PMF through the system which selected the storage of the PMF at the upstream dam as much less expensive than modifying all spillways to pass the PMF through the series of dams. The hazard assessment included the following steps:

1. Flood routings to determine impact of dam as a function of flood level and establish outflow with no dam failure.

2. Identify failure modes and failure threshold for each.

3. Route flood with dam failure to obtain outflow.

4. Determine consequences as a function of flow.

5. Make hazard evaluation based on incremental consequences.

To summarize, the Bureau has used risk analyses in decision making on safety of dam modifications to assess:

- incremental consequences (human lives and property damage.) with and without dam failure

- determination of likelihood of failure for a particular loading level (flood, earthquake)

- comparing consequences, benefits, and costs between an array of alternative actions

- comparing qualitative factors between alternative actions (social, environmental, political, likelihood of success)

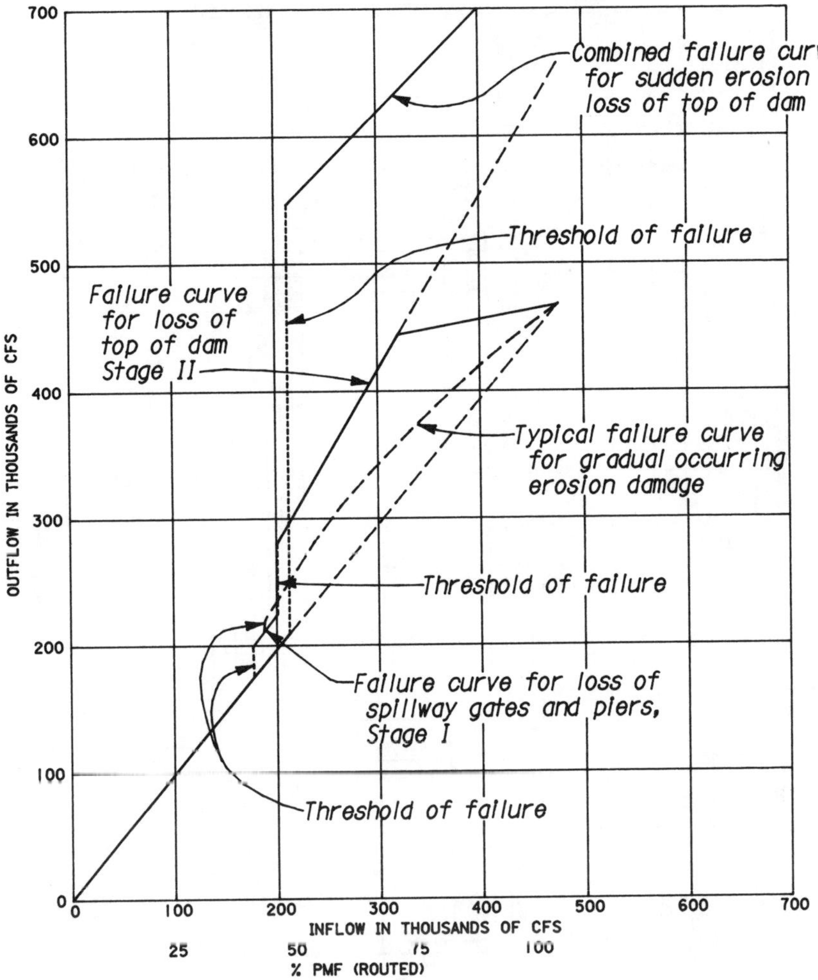

PMF — Probable Maximum Flood
CFS — Cubic Feet Per Second

Figure 1-1. - Discharge curves for various stages of failure.

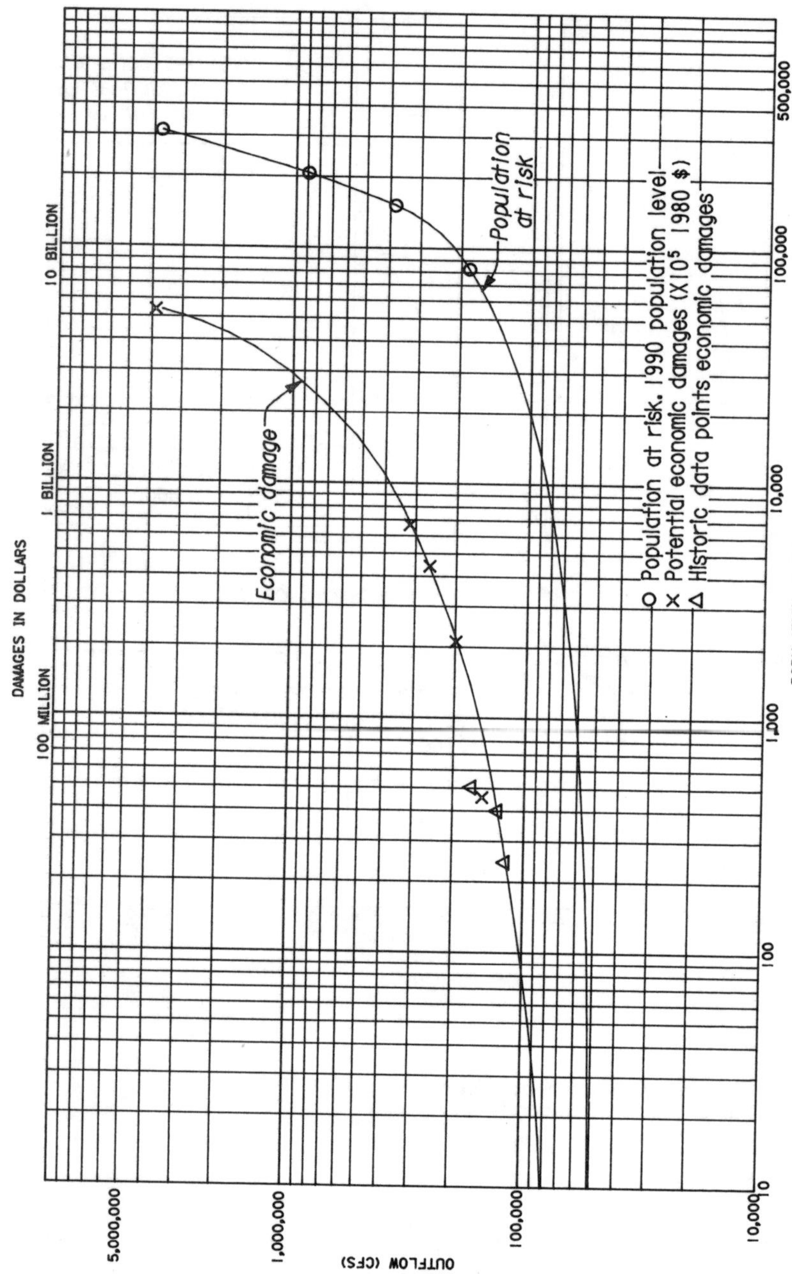

Figure 1-2. - Downstream consequences for various discharge levels following dam failure.

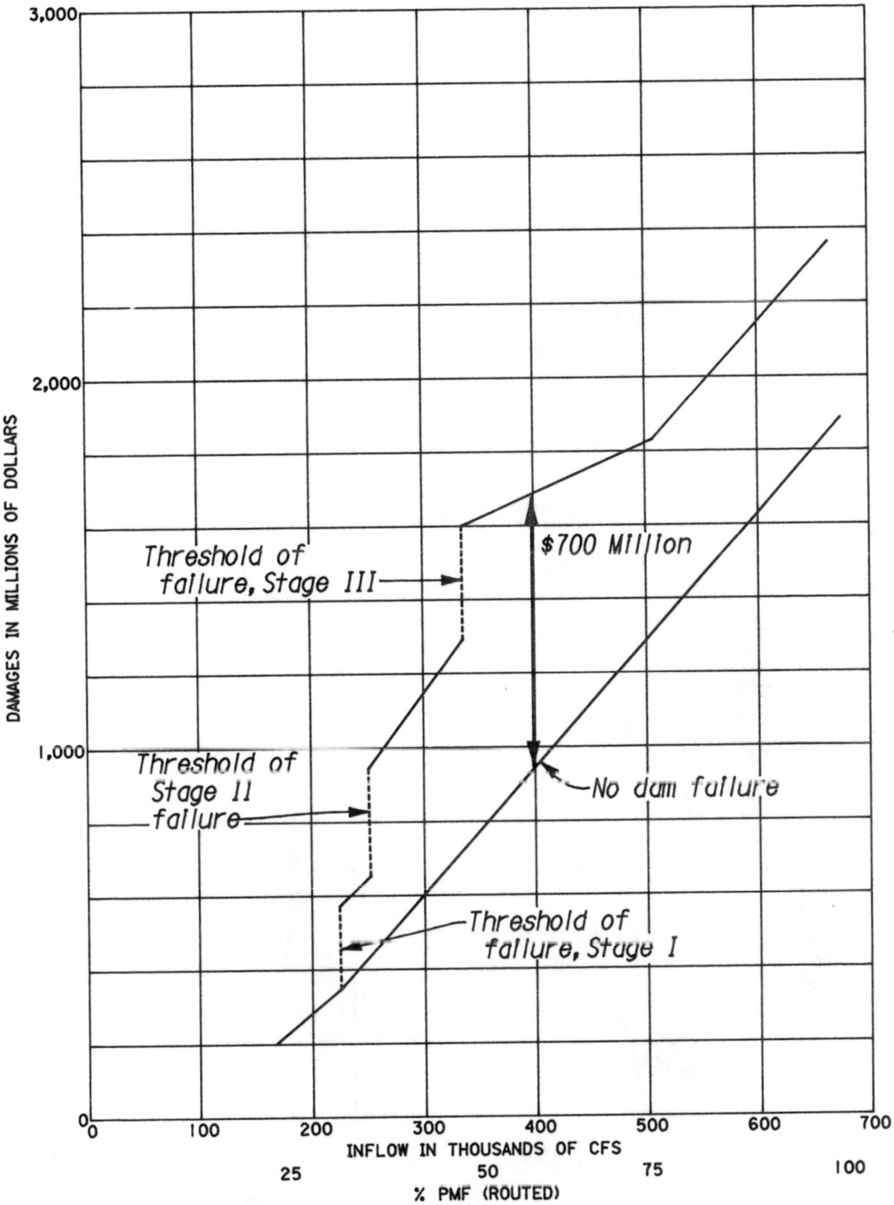

Figure 1-3. - Incremental economic loss due to potential dam failure.

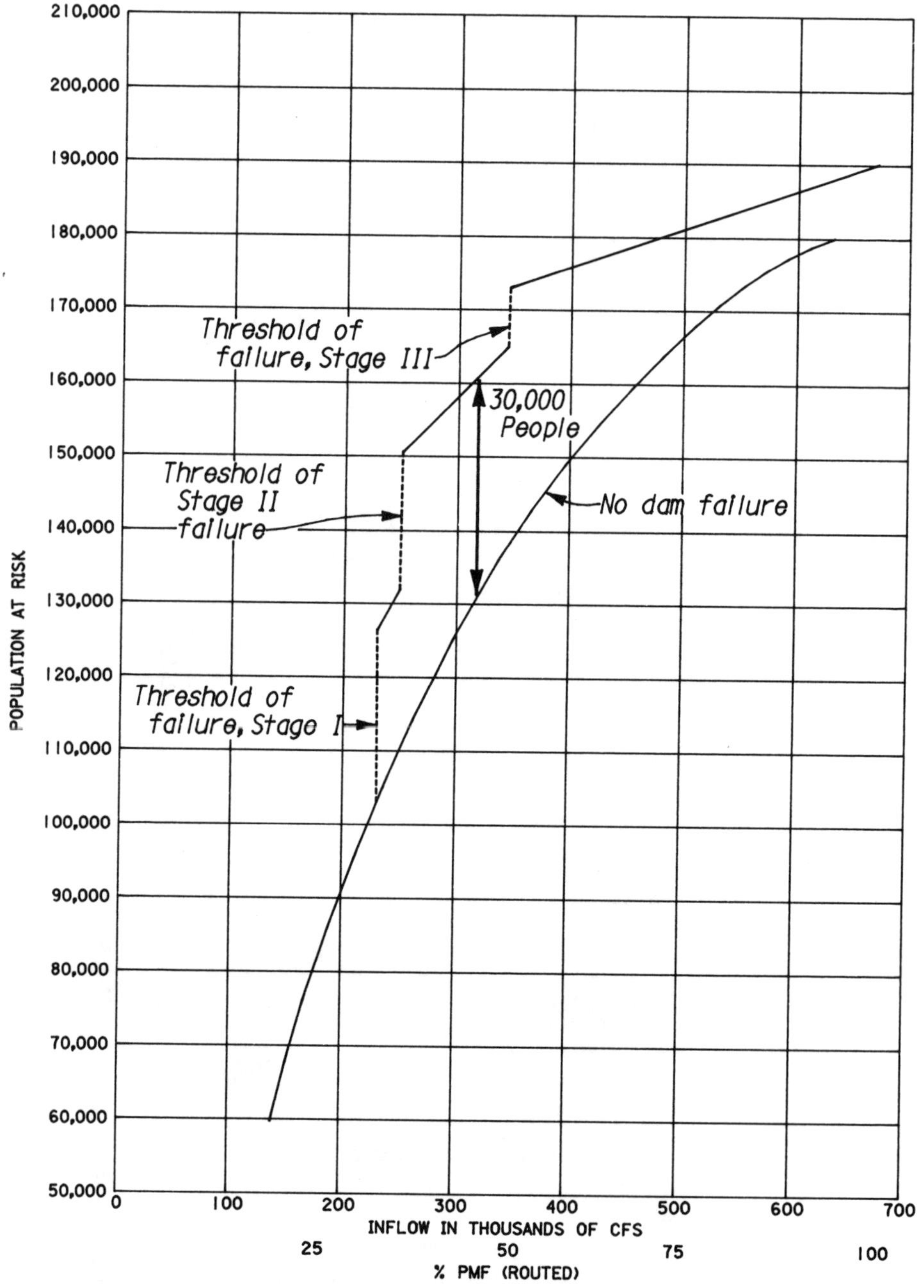

Figure 1-4. - Incremental population at risk due to potential dam failure.

CASE NO. 2 DAM

The second case shows a risk-cost analysis, in which the apparent recommended level of modification would be to balance protection for remote events to that which could be expected from extreme operational requirements, and also the dependence of such a logic upon the frequency of reoccurrence selected for large volume, remotely occurring floods. Refer to figures 2-1, -2, and -3.

Irrigation Project

Earth dam 40 feet height

Gated spillway 22,5000 ft^3/s capacity

Small reservoir outflow = Inflow

PMF (probable maximum flood) (revised) 84,000 ft^3/s

Condition of structures excellent

Existing single spillway may clog due to debris at about 10,000-ft^3/s flow

ALTERNATIVES

1.	Add second spillway of 20,000 ft^3/s and roller-compacted concrete to crest and downstream slope of embankment	\$2.8 million
2.	Add spillway of 40,000 ft^3/s and raise dam (no overtopping)	\$4.8 million
3.	Add spillway of 60,000 ft^3/s	\$6.0 million
4.	Add spillway of 84,000 ft^3/s	\$7.1 million
5.	Establish and maintain upstream and downstream alert	
6.	No action	Reputation

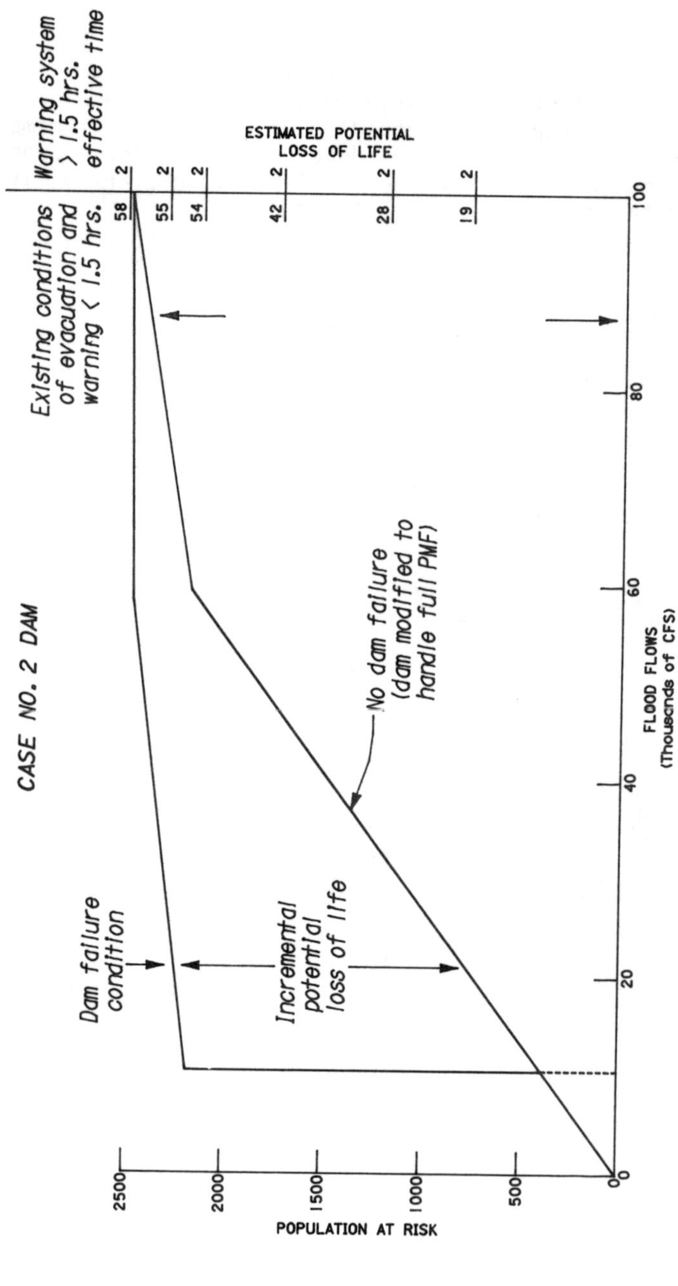

Figure 2-1. - Effects of various options upon results of potential dam failure.

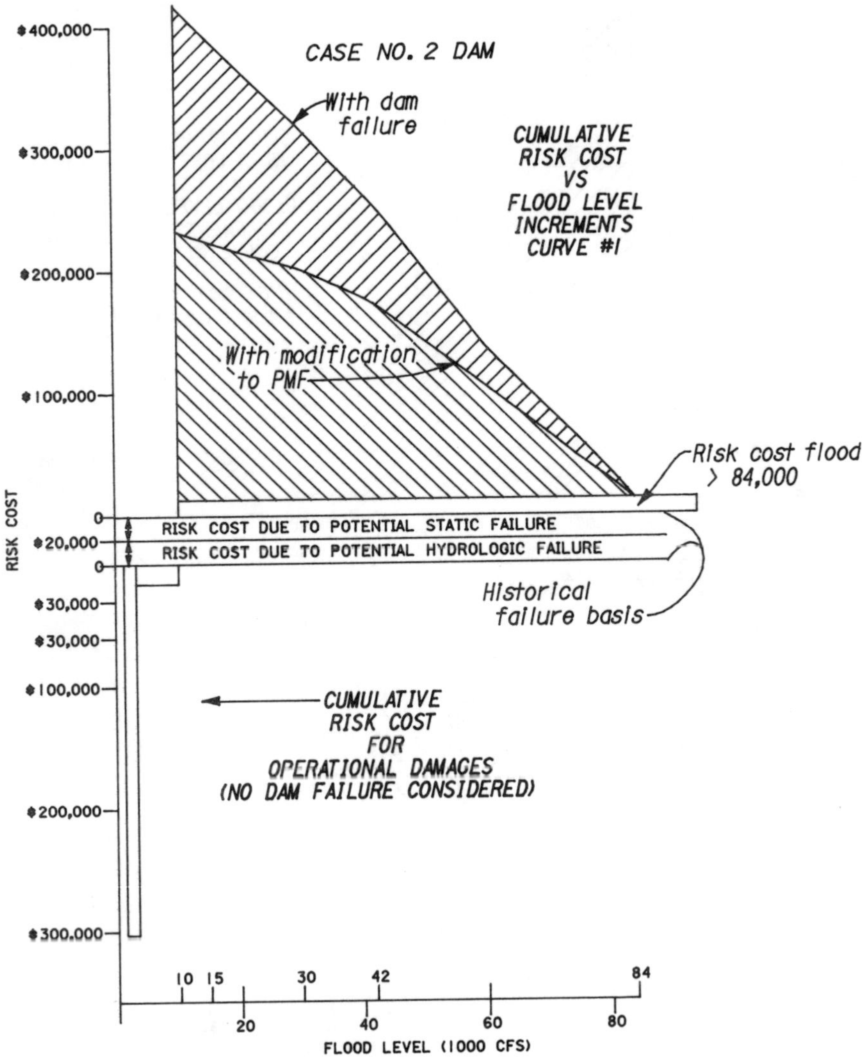

Figure 2-2. - Cumulative risk costs for various flood level increments. (No dam or operational modifications)

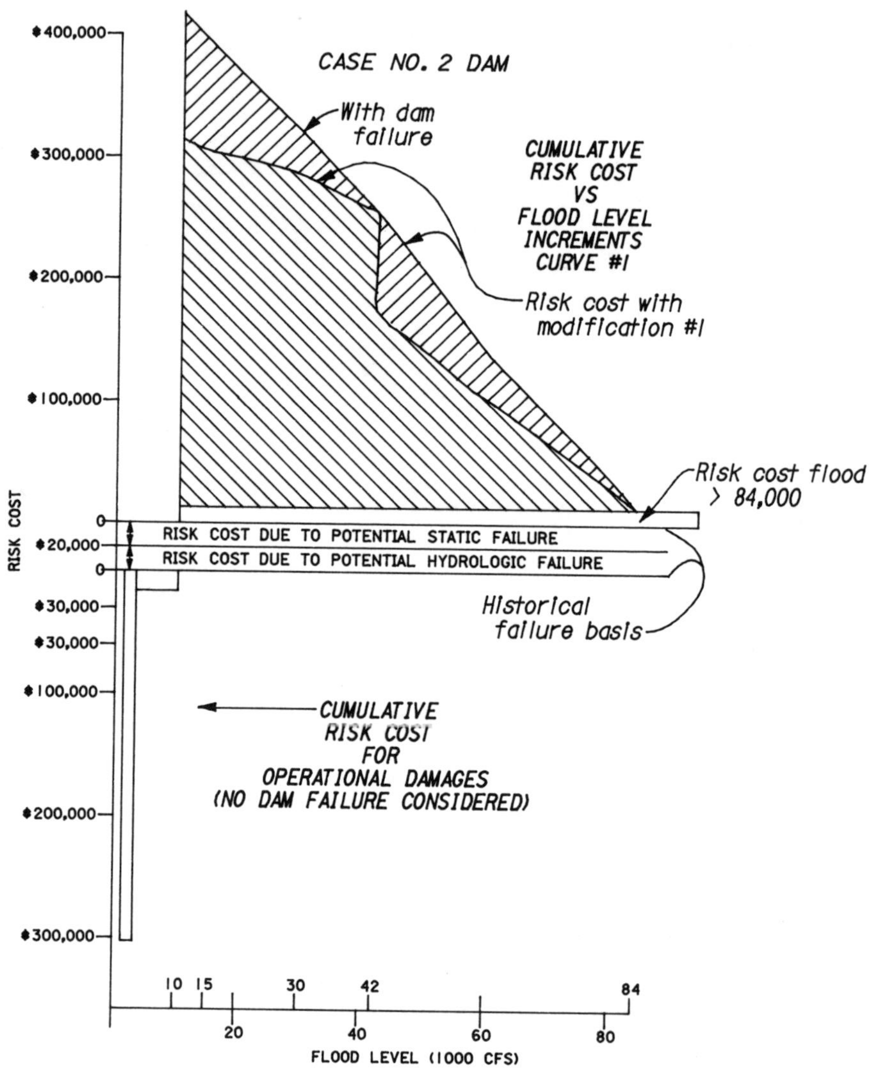

Figure 2-3. - Cumaltive risk costs for various flood level increments. (Modification option No. 1 included)

TVA PRACTICE IN FLOOD FREQUENCY AND RISK ANALYSIS

Donald W. Newton, Supervisor
Flood Protection Branch
Tennessee Valley Authority
200 Liberty Building
415 Walnut Street
Knoxville, Tennessee 37902

ABSTRACT. The Tennessee Valley Authority (TVA) is a government corporation established in 1933 with broad regional resource development responsibilities. In carrying out its responsibilities TVA has constructed dams and fossil and nuclear power plants, operates and maintains its dams including their safety evaluation, and maintains a program of floodplain management activities with communities, industries, and individuals. This paper describes the flood frequency and risk analysis procedures used by TVA in these activities. Discussed are (1) TVA's conclusions about the most accurate methods to compute flood frequency at ungaged locations including use of limited site historic flood data obtained through TVA's flood documenting activities, (2) the procedures and data used by TVA to estimate the PMF to ensure estimates are realistic and compatible with meteorological experience, (3) approaches TVA has used to define flood frequencies of extreme events up to and including the PMF, and (4) risk analysis as applied by TVA in community planning and dam safety evaluations. Recommendations as to additional procedures needed to enhance the engineer's capability for improved flood frequency and risk analysis are made.

INTRODUCTION

TVA is a government corporation established in 1933 with broad regional resource development responsibilities. Those responsibilities include agricultural and industrial development, navigation, and flood control. It has the authority to construct dams and to produce, distribute, and sell electric power. These broad responsibilities have resulted in a variety of activities which involve flood frequency and risk analysis. These include (1) the construction, operation, maintenance, and safety evaluation of dams, (2) the construction of fossil and nuclear power plants, (3) flood control operations, and (4) floodplain management activities with communities, industries, and individuals.

The Tennessee River watershed encompasses an area of 40,910 square miles extending from its mouth at the Ohio River near Paducah, Kentucky, upstream to its headwaters in the Blue Ridge and Great Smoky Mountains of western Virginia, western North Carolina, and northern Georgia. Flooding results from storm rainfall rather than snowmelt, although snow has been a contributor to flooding on at least one occasion. Mean annual rainfall

V. P. Singh (ed.), Application of Frequency and Risk in Water Resources, 429–438.

varies from 40 inches in sheltered portions of the Appalachian Mountains to 90 inches along the Blue Ridge Mountains of northern Georgia and western North Carolina with a basinwide average of 52 inches.

TVA programs and activities which require flood frequency and risk analysis include (1) the design, safety evaluation, and operation of its dams, (2) the siting of its fossil and nuclear power plants, and (3) its community flood damage reduction program. TVA has constructed or acquired 47 dams, 36 of which are operated as part of the Tennessee Valley's integrated water control system; one is outside the watershed and 10 are multipurpose nonpower dams. TVA has built or acquired 11 coal-fired plants and 4 nuclear plants and has designed 3 additional nuclear plants. TVA has worked with over 350 communities to provide technical assistance to encourage adoption of floodplain regulations and other flood damage prevention measures.

The following sections describe the data base available for estimating flood frequencies in the Tennessee Valley region, the flood-frequency procedures used for community studies and site evaluation, and those used for dam design and safety evaluation, nuclear plant siting, and risk assessment.

DATA BASE

At its inception TVA recognized that basic planning for multipurpose river basin development required a thorough familiarity with the flood-producing characteristics of the region. To this end they instituted a program of (1) rain and stream gaging, in cooperation with the U.S. Geological Survey (USGS) and National Weather Service (NWS), (2) collection and evaluation of flood history, (3) documentation of major floods immediately after they occur, and (4) definition of probable maximum precipitation (PMP).

TVA's gaging program was designed to provide information at potential dam building sites as well as broad regional coverage with special emphasis in the mountainous east. By 1974 there were 530 rain gages and 287 stream gages located within the Tennessee Valley watershed. Today TVA continues an active stream and rain gaging program in cooperation with the USGS and NWS but at a reduced scale.

TVA initiated a search for historical information on floods in 1933 (TVA, 1961, chapter 3) which included review of (1) published records and reports, including newspapers; (2) unpublished records and reports; and (3) information obtained through interviews with local people. Flood history reports were prepared by TVA (1961, chapter 4) for all major streams with information extending back to the early 1800s.

TVA documents major floods that occur in the Valley, particularly at locations where flood information is needed for TVA's flood damage abatement programs. The flood documenting program includes a field investigation to mark flood profiles and obtain data for indirect flow measurements and the collection of rain information. This documentation provides site-specific flood information for many locations where otherwise there would be no flood information. It serves two purposes: (1) provides information useful in flood-frequency analysis and (2) provides flood profile data for calibrating hydraulic models.

TVA initiated PMP studies for the Valley region in the early 1940s. Beginning in 1963, TVA contracted with the National Weather Bureau (now NWS) to define PMP and antecedent rain criteria for the Tennessee Valley region as needed for dam design and safety evaluation of existing dams (Schwarz, 1965, 1973; Schwarz and Helfert, 1973).

COMMUNITY FLOOD STUDIES

Flood-frequency analysis for community flood studies requires both an evaluation of systematic stream gaging records and estimates at ungaged locations. In both cases incorporation of available historic flood information is used to improve the analysis.

Gaged Site Analysis

Where there is a systematic stream gage record of more than 10 years, TVA uses the procedures outlined in Bulletin 17B (U.S. Water Resources Council, 1981a) including the skew map of Plate 1 and adjustments for historic information where available. TVA has used probable maximum flood (PMF) estimates or an intermediate flood (the TVA maximum probable flood) to guide extrapolation of flood-frequency estimates to the more extreme flood levels. Where systematic records are less than 10 years long, this short-record information is incorporated into the estimates based on regional analysis as discussed in the next section.

Ungaged Site Analysis

A common problem for hydrologists is to find a reliable means to estimate flood frequencies at ungaged locations. TVA had found that the different procedures in common use often gave significantly different results at a site. Furthermore, estimates made by different hydrologists using the same procedure often varied. The most comprehensive, objective evaluation of the performance of commonly used procedures is the pilot test conducted by the U.S. Water Resources Council (1981b). Based upon an analysis of this pilot test, Newton and Herrin (1982, 1983) concluded that the most accurate, reproducible, and cost-effective procedures for estimating peak flow frequencies at ungaged locations are prediction equations based upon regression analysis of flood-frequency determinations at gaged locations.

As a result, TVA uses regression-based procedures to estimate flood frequencies at ungaged sites. Because of the experience gained in working in a limited region and the large amount of available flood data, TVA develops its own flood prediction equations focused on relatively small subregions of the Valley. TVA is evaluating means to improve these flood peak estimating procedures for the Tennessee Valley watershed. It is clear, based upon the pilot test data, that improved procedures are needed. Even with the best performing procedures, only about 50 percent of the estimates were with ± 25 percent of the gaged estimate (Newton and Herrin, 1982).

In many cases TVA will have some site-specific flood information. This includes information about large past floods (historic information acquired as a result of TVA's flood documenting program) and/or less than 10 years of gage record (short-record flood information). Procedures have been developed by Newton, et al (1984) for incorporating this information into the analysis to improve results obtained by regression alone.

Two Flooding Sources

Occasionally a community or a parcel of land will be subject to flooding from two sources. Where small streams enter large rivers, the flooding by the two sources is essentially independent and the frequency of flooding is defined by the sum of the probabilities of flooding by the independent sources. Where flooding is concurrent, flood levels are defined by backwater analysis assuming flood peaks of a given frequency coincide.

Risk Analysis

A formal risk analysis is not used by TVA in its community studies or when advising individuals of potential flood problems. TVA encourages communities to weigh the probability of flooding with the consequences of flooding when making a land-use decision. Floods with annual exceedance probabilities of 1/100 are commonly used because they are the minimum required by the National Flood Insurance Program and Executive Order 11988. However, communities and individual developers are encouraged to consider floods with annual exceedance probabilities of 1/500 and several communities use this standard as a basis for floodplain regulations. In presenting flood data TVA identifies the probability of flood exceedances during a 30-year period as well as annually. The 30-year period was selected as representing a typical mortgage period.

For the past 5 years TVA has been working cooperatively with the city of Huntsville, Alabama, to develop a flood drainage plan that incorporates the impacts of future development (Newton and Black, 1983; Newton, Cripe, and Hughes, 1984). The information developed included potential flood damages in floods with annual exceedance probability of 1/10 and 1/100 and average annual damages for both present conditions and for projected future development with and without selected mitigation measures. Changes in potential flood damages provided the information that city executives and council used to select a strategy to be adopted for reducing flood damages expected from development. This same approach is being used at Maryville, Tennessee.

DAM DESIGN, NUCLEAR PLANT SITING

The design and safety evaluation of dams and the siting of nuclear plants involve consideration of extreme floods including the PMF. For these studies, TVA has attempted to (1) develop realistic and consistent procedures to determine the PMF using probability concepts, (2) define flood frequencies up to the PMF level, and (3) estimate dam failure probabilities from hydrologic causes and use this information in a risk assessment for setting priorities for dam safety evaluations and for design.

PMF Determinations

TVA finds distinct advantages to the concept of the PMF and its use in defining a practical upper boundary to the flood potential at a site. It provides an assessment of the maximum flood potential which considers both climatic and watershed variables. When used to define maximum reservoir levels as needed for dam design or safety evaluation, it permits a rational consideration of the sequence of events which could create maximum reservoir levels. While there is basic agreement within the hydrologic engineering community on the events or conditions to postulate in a PMF determination for dam design, there are significant differences of opinion on the appropriate magnitude in which they are to be combined. TVA has found that PMF estimates based upon a series of conservative assumptions produces a result which at best is unrealistically conservative and at worst, physically impossible. Our efforts have been directed toward finding an objective means using probability concepts for selecting a meteorologically and hydrologically reasonable sequence of events to compute the PMF.

The major components to a PMF determination are the principal storm, antecedent and/or subsequent storms, time and areal distribution of rainfall, loss rates, and the hydrograph determination. For dam design or evaluation, given the design flood inflow hydrograph, computed reservoir headwater levels depend upon initial pool levels, operating assumptions, and windwave effects. TVA has evaluated both flood and storm experience to determine what storm conditions antecedent to a PMP are likely and meteorologically reasonable. Studies by Newton and Lee (1969) showed that 75 percent of the major floods in the Tennessee Valley are caused by a pair of storms with a 3-day interval between them. An envelope of the antecedent storms, expressed as a percentage of the principal storm, decreased with principal storm magnitude as would be expected from meteorological analysis. Riedel, Schwarz, and Miller (1976) defined storm conditions antecedent to the PMP based upon studies of major storms. These studies and additional studies described in the various hydrometeorological reports applicable to the Tennessee Valley (Schwarz, 1965, 1973; Schwarz and Helfert, 1973) propose antecedent storm depths varying from 15 to 50 percent of PMP depending on storm duration, location and size of watershed, and season of occurrence.

The storm analysis showed that antecedent storms are likely and provided an envelope of maximum observed antecedent storm depths. The remaining decision is selection of the antecedent storm depth to be used in the PMF determination. A two-step procedure was proposed by Newton (1983) for selecting the combination of events to be used in a PMF determination. The first step is to define what is meant by "virtually no chance of being exceeded" or the "most severe flood that is reasonably possible," (as the PMF is defined) in terms which can be objectively evaluated. This requires attempting to define the PMF in probabilistic terms. Given a risk objective, the next step is to select the components of the PMF computation and reservoir design that achieve this objective. The data requirements and procedures for estimating the probability of a storm couplet, a storm antecedent to a storm producing PMP, and the PMP-producing storm have been discussed by Newton (1980, 1983). A realistic evaluation of windwaves expected at the time of reservoir cresting in a PMF requires wind velocity data by direction and probability. This wind information for the Tennessee Valley has been developed by the National Climatic Center (1975, 1979).

Flood Frequency for Extreme Events

TVA began exploring means to approximate the probability of the PMF in the late 1960s. A variety of approaches have been used (Newton and Cripe, 1973; Newton, 1980; Newton and Herrin, 1981) including both rain and flood-frequency analysis. The potentially most useful approach appears to be that proposed by Alexander (1963) which is based upon an analysis of all storms that have occurred in a meteorologically homogenous region which can be transposed to the watershed.

A basic approach (Newton, 1971) which can be used to define flood probabilities up to the PMF includes (1) developing a standardized procedure for computing the PMF, (2) selecting a probability to assign to the PMF, and (3) developing a standardized procedure for interpolating between flood probabilities defined by observed data and the PMF. This approach was used by Newton and Herrin (1981) to define flood frequencies as needed for a probabilistic risk assessment for TVA's nuclear plants.

All of the approaches to estimate PMP or PMF probabilities have their shortcomings. For use in nuclear plant and dam safety evaluations, TVA has assigned a probability to the PMF of 10^{-6} to 10^{-8} based upon a subjective evaluation of probability estimates of PMP and PMF made at various sites using a variety of estimating procedures.

Risk Assessment

In the safety review of its dams as called for by the Federal Guidelines for Dam Safety (Ad Hoc Interagency Committee on Dam Safety, 1979), TVA developed a priority ranking for future study of those dams judged deficient or possibly deficient by today's standards based upon a preliminary assessment of the relative risks and consequences of failure. The relative risks or probabilities of failure from hydrologic events were assigned using procedures described earlier and for seismic events, based upon the seismic history of the region. The floodwave from postulated dam failure was computed by unsteady flow routing procedures (Garrison, Granju, and Price, 1969; Price and Garrison, 1973).

An order-of-magnitude estimate of the economic and social consequences of failure was made based upon differences between estimated damages produced by downstream flood conditions just before failure and the maximum flood depth reached after failure. Social impacts were approximated based upon the resident population exposed as determined from a house count from maps. To assess the threat to life, the resident population was grouped according to warning times of 0 to 6 hours and 6 to 12 hours. Order-of-magnitude estimates of economic impacts were based upon the number of residences, commercial and industrial buildings flooded, highway and railroad bridges affected, and downstream dam failure(s). These failure consequences along with the broad assessment of failure probabilities were subjectively evaluated to select a priority schedule for detailed safety evaluation.

Selection of the safety design flood for those projects requiring design modifications has been largely based on a direct comparison of the social and economic consequence attributable to dam failure, the probability of failure, and the cost of modifications to permit the dam to safely pass the PMF or any other candidate safety design flood. The social consequences attributable to dam failure which are quantified include the resident population exposed and the estimated loss of life. The economic impacts that are quantified include the loss of the structure, the downstream physical damage attributable to dam failure, the loss of project benefits, and the increased costs for water supply and waste treatment services until repairs can be completed.

To date TVA has not limited itself to an economic analysis based on average annual damages when selecting a safety design flood for a dam because these tend to mask the catastrophic impact of dam failure. For most of the dams evaluated to date the economic, social, and environmental failure consequences are so great that it has been judged that they could not be tolerated. Although the estimated probability of failure, either annual or for a 100-year period, is usually small, the small chance of failure is overshadowed by the economic and social consequences of failure.

This rather simple approach has proven adequate for most of the dams evaluated to date. When the decision is less obvious it is anticipated that a more thorough identification and quantification of the social, economic, and environmental impacts of failure will be required as well as a more sophisticated economic analysis including consideration of expected values.

RECOMMENDATIONS

The field of hydrologic engineering has come a long way toward developing improved procedures for dam design or evaluation since the introduction of the unit hydrograph and PMP concepts in the late 1930s and the introduction of computers in the 1960s. Additional procedures are needed to enhance the engineer's capability for improved design and analysis in the future. The major needs TVA identifies in the field of flood frequency and risk analysis as related to the design and safety evaluation of dams and for community flood studies include:

- A nationally consistent standard for determining the PMF. The standard should cover regions where snowmelt causes floods or contributes to flooding. It should preserve the concept of maximum events without compounding one unlikely event with another to unrealistic extremes.
- An agreed-to procedure to evaluate flood probabilities up to and including the PMF. Because of the short flood records and limited information about extreme events, currently such a procedure can only approximate flood probabilities of extreme events. Order-of-magnitude

estimates following a consistent procedure which recognizes national variations in extreme storm probability are needed to improve design decisions.

- Procedures for incorporating into the design decision the economic, social, and environmental impacts of dam failure in addition to simple economic benefit/cost analysis. Constructing dams or other structures whose failure would impose severe economic and social consequences on society requires consideration of the economic, social, and environmental impacts of failure which cannot adequately be measured by simple cost/benefit analysis alone. Improved procedures are needed for identifying and quantifying dam failure impacts and incorporating this information into the safety design flood selection process.
- Improved procedures for estimating flood frequencies at ungaged locations as needed for land-use planning. May of the procedures in general use have been shown to be inaccurate and to give different results at the same location when applied by different people. An objective nationwide evaluation of alternate procedures is needed to provide a basis for identifying the most promising procedures so that efforts can be focused on these for improved implementation. Further, based upon TVA's experience, a more systematic nationwide effort is needed to document unusual floods that occur in communities where flood damage abatement programs are anticipated. This documentation would include marking flood profiles, making indirect flow measurements, and collecting rain data which would provide data necessary for hydraulic model calibration and to improve flow-frequency estimates.

REFERENCES

Ad Hoc Interagency Committee on Dam Safety. 1979. Federal Guidelines for Dam Safety. Federal Coordinating Council for Science, Engineering, and Technology, June 25.

Alexander, G. N. 1963. 'Using the Probability of Storm Transposition for Estimating the Frequency of Rare Floods.' Journal of Hydrology, Vol. 1, No. 1, March, pages 46-57.

Garrison, J. M., J. P. Granju, and J. T. Price. 1969. 'Unsteady Flow Simulation in Rivers and Reservoirs.' Journal of the Hydraulics Division, American Society of Civil Engineers, Vol. 95, No. HY5, Proceedings Paper 6771, September, pages 1559-1576.

National Climatic Center. 1979. "Tennessee Valley Extreme Wind Climatology." Prepared under Contract No. TV-36522A, Asheville, North Carolina, November.

National Climatic Center. 1975. "Extreme Wind Study for Selected Stations in the Tennessee Valley." Prepared under Contract No. TV-46522A, Asheville, North Carolina, August.

Newton, Donald W. 1983. 'Realistic Assessment of Maximum Flood Potential.' Journal of Hydraulic Engineering, American Society of Civil Engineers, Vol. 109, No. 6, pages 905-918.

Newton, Donald W. 1980. "Improving Probable Maximum Flood Estimates." Proceedings of the Symposium on Surface-Water Impoundments, University of Minnesota, American Society of Civil Engineers, American Water Resources Association, American Geophysical Union, Vol. 1, June, pages 273-290.

Newton, Donald W. 1971. "Suggested Procedure for Improving Consistency of Flow Frequency Analysis as Required for Floodplain Management." A proposal to the Interagency Work Group on Flow Frequency Analysis, Hydrology Committee, Water Resources Council.

Newton, Donald W. and C. D. Black. 1983. "Integrating Flood Damage Prevention Into City Planning at Huntsville, Alabama." Paper presented at the 19th Annual American Water Resources Association Conference, San Antonio, Texas, October.

Newton, Donald W. and M. W. Cripe. 1973. "Flood Studies for Safety of TVA Nuclear Plants, Hydrologic Embankment and Breaching Analysis." Tennessee Valley Authority, Knoxville, Tennessee.

Newton, Donald W. and Janet C. Herrin. 1983. "Estimating Flood Frequency at Ungaged Locations." Proceedings of the Conference on Frontiers in Hydraulic Engineering, Hydraulics Division, American Society of Civil Engineers, Massachusetts Institute of Technology, Boston Society of Civil Engineers, August.

Newton, Donald W. and Janet C. Herrin. 1982. 'Assessment of Commonly Used Flood Frequency Methods.' Transportation Research Record, Series TRR896, Washington, D.C.

Newton, Donald W. and Janet C. Herrin. 1981. Unpublished reports. Tennessee Valley Authority, Knoxville, Tennessee.

Newton, Donald W. and Ramon G. Lee. 1969. "Storms Antecedent to Major Floods." Paper presented at 50th Annual Meeting of the American Geophysical Union, Washington, D.C., April, With addendum of 1976.

Newton, D. W., M. W. Cripe., and R. C. Hughes. 1984. "Analysis for Planning Mitigation of Urban Flood Problems at Huntsville, Alabama." Proceedings of the 1984 International Symposium on Urban Hydrology, Hydraulics and Sediment Control, University of Kentucky, pages 195-205.

Newton, D. W., J. C. Herrin, S. D. Roser, D. F. Kibler, G. Aron, and D. J. Wall. 1984. Use of Limited Site Specific Flood Information in Estimating Flood Peaks. Tennessee Valley Authority, Knoxville, Tennessee, July.

Price, J. T. and J. M. Garrison. 1973. "Flood Studies for Safety of TVA Nuclear Plants - Flood Waves From Hydrologic and Seismic Dam Failures." Paper presented at American Society of Civil Engineers National Water Resources Engineering Meeting, January.

Reidel, John T., F. K. Schwarz, and J. F. Miller. 1976. "Rainfall Antecedent to the Probable Maximum Precipitation - Drainages of 100 to 3,000 Square Miles in the Tennessee Valley." File Report, National Weather Service, September.

Schwarz, Francis K. 1973. Meteorological Criteria for Extreme Floods for Four Basins in the Tennessee and Cumberland River Watersheds. Hydrometeorological Report No. 47, National Weather Service.

Schwarz, Francis K. 1965. Probable Maximum and TVA Precipitation Over the Tennessee River Basin Above Chattanooga. Hydrometeorological Report No. 41, U.S. Weather Bureau.

Schwarz, Francis K. and Norbert F. Helfert. 1973. Probable Maximum and TVA Precipitation for Tennessee River Basins up to 3,000 Square Miles in Area and Durations to 72 Hours. Hydrometeorological Report No. 45, National Weather Service, With addendum of June 1973.

Tennessee Valley Authority. 1961. Floods and Flood Control. Technical Report No. 26, Knoxville, Tennessee, Chapters 3 and 4.

U.S. Water Resources Council. 1981a. Guidelines for Determining Flood Flow Frequency. Bulletin 17B of the Hydrology Committee, Washington, D.C.

U.S. Water Resources Council. 1981b. Estimating Peak Flow Frequencies for Natural Ungaged Watersheds - A Proposed Nationwide Test. Hydrology Committee, Washington, D.C., 346 pages.

NUCLEAR REGULATORY COMMISSION USE OF FLOOD RISK ANALYSES

Rex G. Wescott
Hydraulic Engineer
U.S. Nuclear Regulatory Commission
Washington, D.C. 20555

The Nuclear Regulatory Commission (NRC) (formerly the Atomic Energy Commission) has recognized the potential hazards resulting from the flooding of a commercial nuclear power plant. Indeed, the siting regulation (10 CFR 100) recognized floods as a natural hazard. Flooding of a great enough magnitude in some cases could cause loss of electrical controls, emergency cooling water pumps, diesel generators and other safety systems through mechanisms of inundation, impact forces, and erosion. In order to independently evaluate the potential for flooding at proposed reactor sites, the Atomic Energy Commission in 1968 contracted the U.S. Army Corps of Engineers and the U.S. Geological Survey to evaluate flooding potential at coastal sites and river sites, respectively. For river sites, the USGS often used statistical extrapolation of historical flood peaks to determine a flood with a recurrence interval of thousands of years to be used as a design basis.

In 1970 the AEC hired Dwight Nunn and, in 1971, L. G. (Jerry) Hulman from the Corps of Engineers to develop and apply consistent guidance for use in determining flood protection requirements for all plant sites, and review flood protection aspects of proposed sites. Nunn and Hulman borrowed the concept of the "Probable Maximum Flood" that the Corps of Engineers used for spillway designs and applied it to power plant sites along streams and rivers. Guidance for determining the Probable Maximum Hurricane Surge, Probable Maximum Seiche and Probable Maximum Tsunami flooding was also developed and applied for plant sites along lakes and oceans.

In 1977 the Systematic Evaluation Program (SEP) was initiated by the U.S. Nuclear Regulatory Commission (NRC) to review the designs of older operating nuclear power plants in order to confirm and document their safety using present criteria. As

V. P. Singh (ed.), Application of Frequency and Risk in Water Resources, 439–441.

a result of SEP reviews it was discovered that some of the older plants could not safely withstand the Design Basis Flood (DBF) as determined using current criteria. In some cases the DBF resulted from runoff produced flooding in an adjacent stream, in others from upstream dam failure as a result of insufficient spillway capacity. In these cases the feasibility of using engineered measures to meet current flood criteria was always examined as the preferred alternative. Where meeting the current criteria was judged to be infeasible, a probabilistic evaluation of flooding potential was attempted in order to evaluate the risk involved with various alternative fixes.

One of the primary reasons the PMF has been used as a design basis by the Corps and others for structures where failure could result in catastrophic loss of human life is the uncertainty inherent in flood frequency calculations with recurrence intervals significantly greater then the period of record. With this in mind, probabilistic evaluations of flooding were undertaken with caution. This "cautious" approach to flood PRA was guided by two major principles:

1) To use as many independent procedures and/or data as possible to develop an understanding of the likely probability distribution of extreme flooding at the site.
2) To use the data collected to establish an upper limit of probability associated with inundation under the preferred course of action rather than attempt to estimate the probability of the PMF.

For example, at the R.E Ginna Plant, near Rochester, N.Y. it was discovered that while the PMF peak discharge on a small creek adjacent to the site was almost 39,000 cfs, the plant could only be feasibly protected against a flood with a discharge up to 26,000 cfs. Using regional peak discharge data, a method developed by James R. Wallis of the IBM Watson Research Center was used to estimate a probability for the 26,000 cfs peak discharge. A straight line interpolation on log-probability paper using the 100 yr flood and PMF (after assigning the PMF a probability valve of 10-5) was also performed.

The Standard Project Flood (15,000 cfs) was also computed for use as a benchmark. From all of this, the staff concluded that the peak discharge of 26,000 cfs has an annual probability of exceedance of no greater than 5 x 10-5. Hence, protection of the plant from flows up to 26,000 cfs was considered acceptable.

It should be recognized that the procedures described may not be adequate for other sites. The possibilities of non-hydrologic failures of upstream dams, ice jams, landslides, cutoffs and etc. may completely invalidate flood PRA based on historical stream flow data.

In conclusion, NRC has not established any standardized procedures for flood PRA's. Flood PRA's are only being used as one tool (i.e., in conjunction with probable maximum estimates) to evaluate older plants where it is not feasible to meet present flood protection criteria. The primary goal has been to conservatively quantify the risk associated with feasible alternatives for decreasing the flood vulnerability of such plants. Research is actively being pursued to meet this goal.

FLOOD RISK ANALYSIS FOR THE NATIONAL FLOOD INSURANCE PROGRAM

Brian R. Mrazik, Ph.D.
Assistant Administrator
Office of Risk Assessment
Federal Insurance Administration
Federal Emergency Management Agency
Washington, D.C. 20472

ABSTRACT. Flood frequency and risk analysis play a fundamental role in the insurance, floodplain management and hazard mapping aspects of the National Flood Insurance Program. This paper describes the analytic procedures commonly utilized in flood insurance studies of riverine and coastal floodplains in the United States. Consistency in flood frequency estimates over both time and space is needed to effectively administer the Program at national and local levels. Enhanced capability in risk analysis for flood-related hazards such as ice jams, erosion, levee failure, shifting channels, long term lake level fluctuations, and mudflows is also needed to support Program operations in these specialized hazard environments.

ROLE OF FLOOD FREQUENCY AND RISK ANALYSIS

Statutory Requirements

The role of flood frequency and risk analysis under the National Flood Insurance Program (NFIP) is defined in the Program's legislation. Section 1307 of the National Flood Insurance Act of 1968 (as amended) authorizes the Federal Emergency Management Agency (FEMA) to carry out studies necessary to estimate risk premium rates for flood insurance. Section 1360 of the Act also authorizes the Agency to identify and publish information with respect to all floodplain areas, including coastal areas, and establish flood risk zones for such areas. Finally, Section 1361 provides that on the basis of such studies and investigations, the Agency shall develop comprehensive criteria designed to encourage the adoption of adequate State and local measures which, to the maximum extent feasible, will:

1. constrict the development of land which is exposed to flood damage;
2. guide development of proposed construction away from locations which are threatened by flood hazards;
3. assist in reducing damage caused by floods; and

V. P. Singh (ed.), Application of Frequency and Risk in Water Resources, 443–453.

4. otherwise improve the long-range land management and use of flood-prone areas.

Thus, the program legislation provides for a dual approach toward implementation of the NFIP: Federal flood insurance and local floodplain management; both based upon a sound program of flood risk assessment.

In fulfillment of these statutory requirements FEMA conducts risk studies for individual communities participating in the Program. These studies typically yield the following hazard and risk information:

1. Flood Insurance Rate Maps containing:
   a. 100-year and 500-year floodplain delineations
   b. 100-year base flood elevations
   c. insurance risk zones based on the slope of the flood elevation - frequency relationship at various sites
   d. coastal high hazard areas subject to additional risks due to wave impact
2. Flood Boundary and Floodway Maps containing:
   a. 100-year and 500-year flood boundary delineations
   b. delineation of the regulatory floodway
3. Flood Insurance Study reports containing:
   a. flood elevation profiles for the 10-, 50-, 100-, and 500-year flood events for each reach of stream
   b. tabulated floodway data, including velocity information

Program Status

Currently, some 18,700 communities have had flood hazard areas mapped. Detailed risk studies have been completed for some 8,600 developed or developing communities and some 2,000 studies are underway. FEMA plans to initiate approximately 500 additional flood risk studies before entering into a maintenance level study and map revision effort in Fiscal Year 1988. Flood risk studies are prepared by other Federal agencies as well as private engineering firms under contract to FEMA. The risk data and mapping developed in these studies are utilized by lending institutions to establish the requirement to purchase flood insurance; by insurance agents to properly rate flood insurance policies; by local communities to regulate floodplain land use and enforce building standards; and by the other Federal agencies to administer their floodplain management requirements under Executive Order 11988.

Insurance Requirements

In implementation of the insurance aspects of the Program, risk information is utilized to compute expected annual damages and corresponding annual premiums for insured structures. Generalized depth-percent damage curves for various structure types are used in combination with elevation-frequency relationships established in the flood risk studies to develop flood insurance rate tables for various classes of structures. These rate tables also include adjustments for various program administrative costs. Individual structures are then rated by using the elevation difference between their lowest floor and the established base (100-year) flood elevation. Thus, insurance rates vary with the type of structure, the risk zone in which it is located, and the elevation of the structure's lowest floor.

Floodplain Management Requirements

In implementation of loss reduction aspects of the Program, risk information is utilized by participating communities as a basis for adoption and enforcement of floodplain building requirements and land use criteria. As a condition of participation in the NFIP, communities must adopt and enforce ordinances meeting minimum Federal floodplain management standards established by FEMA. The key elements of those standards are: the elevation (or floodproofing in the case of non-residential structures) of new construction to or above the base flood elevation; elevation of new construction on piles, piers, posts, or columns in identified coastal high hazard areas; and the prohibition of development in identified floodways which would cause any increase in flood levels.

PROCEDURES UTILIZED

Riverine Methods

Flood frequency analyses for gaged streams. Following extensive investigation of all the procedures in common usage by Federal and state water resource agencies, the U.S. Water Resources Council and the follow-on Interagency Advisory Committee on Water Data, U.S. Geological Survey, published a series of guidelines for determining flood flow frequency for gaged streams.

Procedures set forth in this continuing series of guidelines have been widely adopted by water resource investigators and planners for use in analyzing available systematic records by statistical methods. These guidelines have largely standardized the use of the log-Pearson Type III distribution as a base method for flood flow frequency studies.

Frequency analyses performed in accordance with the latest guidelines, presented in Bulletin 17B of the Hydrology Subcommittee (U.S. Geological Survey, 1981) are generally adopted for flood risk studies performed for the NFIP. The guidelines also set out consistent procedures for statistical treatment of previously difficult problems such as, estimating generalized skew coefficients by weighting factors, handling statistical outliers, adjusting for broken or incomplete systematic records, zero flood years, mixed populations and historic flood data.

Flood frequency analyses for ungaged streams. The most widely used flood insurance study procedures for estimating T-year flood discharges for ungaged areas are: regional equations derived from regression analyses using gaging station systematic records; single-event watershed models coupled with hypothetical T-year storm rainfalls; and simplified transfer methods which relate flood peak magnitude to one or, at most, a few watershed parameters.

Regression equations are generally utilized where only peak flows are needed to adequately predict flood elevations. Such regression equations have been developed by the U.S. Geological Survey for most of the contiguous 48 States. These equations predict peak discharges for floods having recurrence intervals typically ranging from 2 to 100 years. These regionalized equations define relationships between peak flows and one or more watershed parameters for basins which are not significantly affected by regulations, and are often rural in character. Watershed parameters which are frequently used include: drainage area size; main stream slope; percentage of basin covered by lakes and swamps; and average annual precipitation. Occasionally, other parameters, such as mean annual snowfall, have been successfully correlated with flood peaks.

Where a full flood hydrograph is needed, for flood routing or other purposes, single-event watershed models using synthetic T-year storm rainfall is an alternative procedure commonly used for deriving discharge-frequency relationships for flood insurance studies. The watershed model requires development of a natural or synthetic unit hydrograph based on watershed runoff characteristics. A natural unit hydrograph can be derived from observed flood hydrographs and corresponding storm rainfall. Calibration and verification of the unit hydrograph parameters can be obtained by testing the adopted unit hydrograph through application to observed storm rainfall and recorded flood hydrographs from other storms.

As in other methods of deriving discharge-frequency relationships, procedures based on watershed modeling with T-year storm rainfall require considerable judgment in selecting appropriate unit hydrograph and storm parameters. Synthetic unit hydrograph derivation methods are commonly required for watershed locations where only flood peaks or portions of flood hydrographs have been observed and recorded. In those basins for which there are no observed data, calibration of runoff models may be feasible through correlation of hydrologic parameters with nearby basins assumed to have similar rainfall-runoff characteristics.

The specific procedures typically utilized in flood insurance studies are the HEC-1 Flood Hydrograph Package developed by the Corps of Engineers (U.S. Army Corps of Engineers, 1981) and the TR-20 model developed by the Soil Conservation Service (Soil Conservation Service, 1969). Both procedures are capable of developing T-year flood hydrographs using derived and user input data. The HEC-1 Flood Hydrograph Package includes several options by which observed storms and concurrent flood hydrographs can be analyzed and optimal values of storm and watershed parameters can be derived. The TR-20 program derives synthetic unit hydrographs from prescribed user input and applies synthetic storm rainfall to produce T-year flood hydrographs. The resulting T-year hydrograph can be routed and combined to derive discharge-frequency relationships for various locations with a basin.

Other simplified methods of deriving peack discharge-frequency relationships may be used for very small watersheds where homogeneity of watershed runoff characteristics can be assumed. Methods such as the Soil Conservation Service TR-55 technique (Soil Conservation

Service, 1975) have been successfully used in small watersheds. Other simple computational transfer methods are frequently applied in flood insurance studies to derive peak flow estimates at various locations along the same stream, or where homogeneity of small runoff units can be assumed.

Stage-frequency determinations. Once satisfactory estimates of the discharge-frequency relationships have been completed, T-year flood discharges are utilized in conjunction with hydraulic models to compute water surface elevations. Conventional step-backwater procedures, assuming one-dimensional gradually varied flow conditions, are usually used to derive elevation-discharge relationships. Computerized computation methods, such as the Corps of Engineers HEC-2 "Water-Surface Profiles" model, the Soil Conservation Service WSP-2 model, and the U.S.G.S. WSPRO model, use stream cross-sections and bridge data, roughness and other energy loss coefficients in solving the energy equation. Two-dimensional unstead flow modeling is also used in certain situations when circumstances warrant.

Coastal Methods

In the Atlantic and Gulf coast areas of the United States, flooding is primarily due to storm surges from hurricanes and northeaster storms and associated wave action. Coastal flood insurance studies require the determination of elevation-frequency relationships from storm surges and wave heights. Numerical simulation is the most frequently used technique in evaluating these phenomena. A numerical storm surge model was developed for FEMA in 1976 and has been continually updated to incorporate improved analytic techniques (Federal Emergency Management Agency, 1985). A numerical wave height model based on the methodology recommended by the National Academy of Sciences (1977) was developed by FEMA and is used to calculate wave heights associated with the one percent annual chance (100-year) storm surge in the vicinity of the open coast. The following is an overview of the methodology and procedures used in preparing a coastal FIS.

Synthetic storm surge modeling. The storm surge methodology used by FEMA consists of a hydrodynamic model and a hurricane wind model. The hydrodynamic model simulates the response of a water body to hurricanes and calculates the resulting surge heights. This surge height is commonly referred to as the stillwater elevation. Hurricane characteristics, such as atmospheric pressure distribution and wind field, are modeled by the hurricane wind model using hurricane parameters such as the radius to maximum winds, the central pressure depression, and the forward velocity of the hurricane. The wind model provides the driving forces for the hydrodynamic model.

The hydrodynamic model is based on the principles of conservation of momentum and mass. The momentum equation used in the model represents inertial and convective forces balanced by gravity forces, wind stress, atmospheric pressure gradient forces, reactive bottom friction forces, and Coriolis forces. The continuity equation is used to calculate the water-surface elevation. These equations are solved by using an explicit, two dimensional (horizontal) finite difference scheme with appropriate boundary conditions. Input data to the model includes bathymetry, coastline configuration, and flow resistance. The surface wind stress and atmospheric pressure distributions obtained from the wind model are also used in the hydrodynamic model. One important aspect of the hydrodynamic model is that it is capable of resolving a moving water-land boundary. This allows the model to accurately define the hydrodynamics of coastal plains.

The Atlantic and Gulf coasts commonly have thin barrier islands with inlets connecting the back-bay areas and the open coast. Rivers and intracoastal waterways meandering to the estuaries are also common geographic features of these coastal areas. These physical configurations complicate the normal solution of the momentum equations and have to be treated by separate procedures in order to stimulate surges accurately. The hydrodynamic model accounts for the effects of barrier islands and inlets my means of flux continuity for overtopping discharge and flow through the inlet. The estuary rivers and other channels are represented by one-dimensional channel routines to account for the flow conveyed inland in channels.

<u>Statistical analyses</u>. A FIS must associate frequencies with the flood hazards and elevations depicted. This is accomplished in coastal studies using a statistical procedure called the Joint Probability Method (Myers, 1970). This method, based on the assumption that the surge elevation produced by a storm at any given location on the shoreline has the same frequency of occurrence as the storm, consists of three basic steps described as follows:

(1) The frequency rate of storms affecting the study area is determined from historical records. The probability distributions of the meteorologic parameters such as the central pressure depression, the radius of maximum wind, the forward velocity, and the hurricane track angle are also extracted from historical records;

(2) Surges for a set of synthetic storms constructed from various combinations of the meteorological parameters analyzed in step (1) are then simulated using the storm surge model; and

(3) Probabilities are established for the synthetic surge levels by the combination of individual storm parameter probabilities and probabilities of amplitude and phasing of astronomical tides.

<u>Wave analysis</u>. The wave height analysis methodology adopted by FEMA, and revised several times since 1977, accounts for the wind-generated wave effects on peak water levels associated with hurricane storm surges on the open coast.

Wave height calculations are performed along beach profiles that originate in open water off the shoreline and extend inland in a direction perpendicular to the shoreline. Along each profile in the study area, subdivided segments are designated as fetches or obstructions based on their physical features. The computations along each profile consider the combined effects of open fetch areas, changes in ground elevations, vegetation, obstructions, and stillwater flood elevations on wave generation and decay.

Combining the data from the storm surge modeling and wave analysis yields the 100-year base flood elevation which is a representation of peak water levels expected considering both storm surge and wave effects. This elevation is then depicted on the Flood Insurance Rate Maps for insurance and floodplain management purposes.

CAPABILITY REQUIREMENTS

NFIP needs for enhanced capability in flood frequency and risk assessment fall into two broad categories. First, improved consistency in flood frequency determinations would significantly facilitate administration of the program at both local and national levels. Secondly, the capability to analyze, in a probabilistic manner, several unique flood situations and flood-related hazards would facilitate the establishment of actuarial insurance rates and floodplain management standards.

Consistency in Flood Frequency Estimates

The NFIP is the first nationwide program to establish regulatory flood risk data to be applied by virtually every flood-prone community. Flood risk studies for these jurisdictions are usually conducted by different agencies or contractors at different times, utilizing different methods and data sets. Variations in flood frequency information caused by different methodologies, data bases, and analyst judgment result in development of inconsistent flood risk information between jurisdictions. These inconsistencies require time-consuming and expensive actions to resolve even though such expense is frequently not warranted when the uncertainty in each of the conflicting estimates is considered.

Similarly, consistency of estimates is an issue when flood risk data must be updated. FEMA is often called upon to revise floodplain mapping based upon the latest flood frequency estimates. Such updated analyses are usually based on additional years of record, or more sophisticated analytical techniques, rather than significantly changed physical conditions. Determining whether such changes are warranted requires careful consideration of the relative uncertainty inherent in estimates. This problem has been expensive to deal with, both in terms of costs associated with revising floodplain mapping and credibility problems at the community level created by frequent flood map revisions.

This problem was envisioned as early as 1966 by the President's Task Force on Federal Flood Control Policy in House Document No. 465, some two years prior to the enactment of the National Flood Insurance Act of 1968. This Task Force concluded that "Techniques for determining and reporting the frequency of floods used by several Federal agencies are not now in consistent form." This Task Force further recommended that "A uniform technique of determining flood frequency should be developed by a panel of the Water Resources Council." Out of this recommendation, a single guidance document has been issued and periodically updated: "Guidelines for Determining Flood Flow

Frequency," Bulletin 17B of the Hydrology Subcommittee, Interagency Advisory Committee on Water Data (U.S. Geological Survey, 1981). This document addresses the problem of consistent peak flow frequency estimates for gaged streams where sufficient length of record to warrant statistical analysis is available. Although this document has been of benefit in facilitating the coordination of flood flow frequency information among the Federal agencies, it has not totally eliminated the problem. In fact, criticism of this Federal standard methodology and an evolving state-of-the-art may result in its falling into disuse in the future.

No similar document exists for evaluating flood flow frequencies for ungaged sites, which constitute the vast majority of floodplain situations in the United States. The U.S. Water Resources Council's Hydrology Committee published a partial evaluation of several frequently-used procedures for estimating peak flow frequencies for natural ungaged watersheds as a preliminary step toward developing consistent national guidelines (U.S. Water Resources Council, 1981). Although this report provided further evidence of the great variability in peak flow frequency estimates between procedures, and among analysts applying the same procedures, the work group was reluctant to make any recommendations or establish any guidelines without a significant nationwide testing program that would permit discrimination between procedures at different levels of hydrologic and statistical significance.

Since 1981, work on establishing consistent approaches toward flood flow frequency estimation within the Federal sector has languished. Instead, attention seems to have been shifted toward evaluating probabilities associated with the Probable Maximum Flood and the variability in such procedures. It would seem that significant progress toward more consistent estimates of much more frequent events, such as the one percent annual chance flood, would be a prerequisite to any consistent approach toward risk assessments associated with flood events in the Probable Maximum Flood range.

Risk Assessment for Flood-Related Hazards

Ice Jams. In northern regions of the United States, where rivers can develop relatively thick ice covers during the winter, ice jamming can contribute significantly to flood hazards. The establishment of stage-frequency information for river reaches subject to ice jam flooding is essential for implementation of insurance rating and loss reduction activities under the NFIP.

Although the mechanisms of ice jam formation have been investigated for some time, and hydraulics of flow under ice covers have been modeled with success, little work has been done to establish flood elevation-frequency relationships for ice jams. Attempts to assign frequencies to stages from ice james must consider the joint probability of various discharge events occurring during periods of significant ice cover, and the probability of ice jam formation. Currently, FEMA utilizes historical ice jam stage data where possible to develop ice jam stage-frequency relationships. However, where insufficient stage data is available from the historical record, the probability of ice jam occurrence in the historical record is utilized, in conjunction with discharge frequency data and hydraulic modeling, to develop ice jam stage-frequency relationships.

Flood related erosion. The National Flood Insurance Act of 1968, as amended, provides that flood insurance be made available for "loss resulting from the erosion and undermining of shorelines by waves or currents in lakes and other bodies of water exceeding anticipated cyclical levels," i.e., flood-related erosion. The situations envisioned by the Act consider storm-induced erosion such as that commonly experienced on the Pacific Coast, the Great Lakes, and other coastal areas of the United States. The language of the Act implies some probabilistic means to distinguish extreme erosion events from more frequent events, even though the integral effects of both lead to long-term coastline rescission. Evaluation of the frequency of occurrence of various erosion rates along a coast could involve the consideration of numerous factors such as surge or lake level frequency, duration of wave attack, and the frequency of occurrence of wave spectrums of various energies. A long-term and detailed record of coastline rescission would allow the development of frequency relationships, but little work has been done to evaluate storm-induced erosion in a probabilistic manner.

Reliability of flood control structures. Although non-structural in its approach to flood loss mitigation, the National Flood Insurance Program is faced with the problem of evaluating the degree of flood protection offered by local flood protection systems; in particular, levees. Currently, the NFIP applies the 100-year flood standard to determine whether levee-protected areas should be identified as special flood hazard areas, thus triggering insurance purchase and floodplain management requirements. To make such determinations, consideration must be given to probabilities of various loading conditions on the levee system, the uncertainties involved in establishing those probabilities and the risk of failure under those various loading conditions. Because of the impracticality of performing such analyses for thousands of structures nationwide, FEMA must rely on standardized design safety factors in making such determinations. Thus, adequately maintained levee systems with a freeboard of three feet or greater above the one percent annual chance flood are generally credited with 100-year flood protection on NFIP maps. Enhanced capability in the areas of structural reliability analysis would benefit mapping, loss reduction, and insurance rating decisions under the NFIP.

Alluvial systems. Flood risk analyses for alluvial streams systems pose additional probabilistic considerations. In these situations, channel and floodplain configurations change during major and minor flow events. Attempts to map flood risks based on "todays" floodplain configuration can be misleading if fixed-bed assumptions are critical to the risk data which is finally produced. The classic example of this problem occurs with alluvial fan flooding where debris flows from mountain canyons emerge onto unconsolidated deposits formed by previous flood events. In this situation, channels are often formed by the flow itself and former channels on the fan surface may or may not influence the behavior of a major flood event.

Analysis of alluvial fan floods over geologic time, such as through paleo-flood records, would suggest that the probability of flooding is fairly uniform at any point along an elevation contour on the fan surface, and that the probability of flooding increases toward the fan apex and decreases toward the toe of the fan. The formation of the alluvial cone structure suggests such a geomorphologic process. Thus, FEMA considers both the flood flow frequency relationship at the canyon mouth, as well as a random channel formation, when establishing the one percent annual chance flow depths and velocities on the fan surface. Recent studies suggest that the random flow path assumption may not be applicable when relatively short-term risks (such as over the life of structures) are considered. Likewise, this approach has been questioned where man-made alteration of alluvial fan surfaces (e.g., development) may alter the probabilities that future floods will follow certain paths.

Closed basin flooding. Flood frequency and risk analysis in closed basin systems have proven to be troublesome for the NFIP. Such systems (the Great Salt Lake being a classic example), respond to long-term changes in the climatic balance of a precipation and evapotranspiration. Evaluation of lake stage-frequency based on short term records often proves to be highly misleading. Analysis of paleo-flood records can be helpful, but raises questions of climatic and watershed stationarity. Frequency analyses of synthetic lake levels developed from runoff models based upon long-term precipitation and evapotranspiration data appear to hold the most promise. Modeling using stochastic precipitation and evaporation data appears to be a relatively unexplored area.

Mudflows. In many arid and semi-arid regions of the United States, mudflows and mud-flood events are equally or more important in causing flood-related losses than "clearwater" flood events. Since mudflows and mud floods are covered under the definition of floods in the National Flood Insurance Program, probabilistic assessment of these hazards is important to loss reduction and insurance operations of the Program.

In these floodplain areas, mudflow volumes tend to be more critical than peak flow rates in determining areas subject to risk. Probabilistic estimate of mudflow volumes have only rarely been attempted and much additional research is required. The Los Angeles County Department of Public Works (1986) has utilized volume measurements of debris deposits collected behind debris dams to develop relationships between debris volumes and drainage areas for various event severity classes. However, no probabilities were assigned to the volumes. In considering probabilities of mudflow events, probabilities of related events such as rainfall, snowmelt, antecedent moisture or watershed burns may be appropriate. In flood insurance studies, 100-year peak flows or hydrographs are often "bulked" by a fixed sediment concentration in an attempt to consider the solid component. However, a true probabilistic approach to mudflow analysis has not been developed.

REFERENCES

Los Angeles County Department of Public Works. 1986. An Enginnering Methodology for Delineating Debris Flow Hazards in Los Angeles County. Los Angeles, CA

Federal Emergency Management Agency. 1985. Coastal Flooding Hurricane Storm Surge Model. Volume I - Methodology. Washington, D.C.

Myers, Vance A. 1970. Joint Probability Method of Tide Frequency Analysis Applied to Atlantic City and Long Beach Island, New Jersey U.S. Department of Commerce, Environmental Sciences Services Administration. Technical Memorandum WBTM, Hydro 11. Washington, D.C.

National Academy of Sciences. 1977. Methodology for Calculating Wave Action Effects Associated With Storm Surges. Washington, D.C.

U.S. Army Corps of Engineers. 1981. HEC-1 Flood Hydrograph Package Users Manual. Hydrologic Engineering Center. Davis, CA

U.S. Geological Survey. 1981. Guidelines for Determining Flood Flow Frequency. Bulletin 17B of the Hydrology Subcommittee, Interagency Advisory Committee on Water Data. Reston, VA

U.S. Water Resources Council. 1981. Estimating Peak Flow Frequencies for Natural Ungaged Watersheds. Hydrology Committee, Washington, D.C.

Soil Conservation Service. 1969. Computer Program for Project Formulation - Hydrology. Technical Release No. 20. Washington, D.C.

Soil Conservation Service. 1975. Urban Hydrology for Small Watersheds. Technical Release No. 55, Washington, D.C.

THE NATIONAL WEATHER SERVICE ROLE IN FLOOD FREQUENCY AND RISK ANALYSIS

E. Marshall Hansen
Chief, Water Management Information Division
National Weather Service
8060 13th Street
Silver Spring, Maryland 20910

ABSTRACT. The National Weather Service does not have a direct role in studies concerning flood-flow frequency and risk analysis; however, it does take part in a number of areas that are of interest to such studies. NWS publications, as the primary source of probable maximum precipitation estimates and precipitation frequency studies in the United States, have long been recognized throughout the world. A brief summary of these subjects and NWS involvement in work groups assembled to resolve aspects of the flood frequency problem is presented here.

1. INTRODUCTION

The National Weather Service (NWS), under the National Oceanic and Atmospheric Administration (NOAA) has responsibility for observation, collection and processing of meteorological data and issuing forecasts and warnings to the general public and other users. The NWS Office of Hydrology has broad control over 13 River Forecast Centers that issue stage forecasts, assist in collecting hydrologic data, and disseminate flood warnings. The Office of Hydrology also has the responsibility for overseeing the river stage forecasts and for research to improve flood warnings in an effort to reduce property damage and save lives. In these capacities the NWS does not have a role that requires direct concern for flood frequency and risk analysis.

There is, however, a part of the Office of Hydrology, namely the Water Management Information Division, that participates in studies of primary interest to the subjects being discussed at this conference. These responsibilities are in the area of precipitation frequency studies, deterministic estimates of probable maximum precipitation, and interagency study group participation.

In the middle 1930's, it became apparent to hydrologic design engineers that design estimates of maximum precipitation based on design practices of that period did not guarantee non-exceedance. Over a period of some 30 years prior to this time, numerous attempts had been made to achieve standards that met certain levels of expected safety. Thus, the U.S. Army Corps of Engineers in seeking to find a better

V. P. Singh (ed.), Application of Frequency and Risk in Water Resources, 455–461.

approach to design safety turned to a deterministic concept of maximization of observed storm precipitation. Because it was desired to maintain a certain level of credibility for development of procedures and results, it was agreed that such responsibilities should be placed in and controlled by an independent agency. The Weather Bureau (now NWS) was the logical agency for such responsibility because of their meteorological experience and access to precipitation data. The NWS offices that grew into the Water Management Information Division thus evolved from about 1940 with a mission to provide estimates of the maximum precipitation that could be expected over selected drainages. It should be noted that throughout this period, support for these studies has been bourne by federal agencies other than NWS.

2. PROBABLE MAXIMUM PRECIPITATION

The early methods to estimate maximum design levels of precipitation eventually developed some formalism and evolved into what in recent years has been referred to as probable maximum precipitation (PMP) estimates. The current definition of PMP is, "Theoretically the greatest depth of precipitation, physically possible for a given size storm area and duration at a particular location and time of the year." Over the course of some 45 years of making PMP estimates, the WMID has amassed information on more than 600 extreme storms that have occurred throughout the United States. Through this period, there has been a shift from primarily site-specific estimates to recent studies resulting in generalized estimates covering large portions of the country. The advantages of generalized studies are the regional consistency they provide, as well as a means to obtain results in data-poor regions.

The PMP concept, as perpetuated over these many years, is that the record of observed extreme precipitation events, when adjusted to account for non-optimum atmospheric conditions, provides a meteorologically-sound basis for determining a level of extreme precipitation that is essentially reliable, and will virtually not be exceeded.

The disadvantage in the PMP concept, as has been discussed at this conference and elsewhere is that no probability can be obtained for the PMP estimates. This problem has brought about considerable interest in ways to assign a probability to the PMP in an effort to enable a risk analysis approach to be made on lesser design precipitation levels. I will briefly discuss some of these efforts later in this paper under participation.

Because of the international scope of this conference, I have summarized in table 1 the current generalized PMP studies applicable to regions of the United States. These have been developed over a period of many years and, although all adhere to the basic concept of storm maximization, there are somewhat different techniques employed to obtain the PMP. Many of these techniques are already known and being applied in other countries.

Table 1. Generalized probable maximum precipitation studies currently applicable to portions of the United States

Hydrometeorological Report Number*	Applicable Region	Applicable Range of Drainage Areas (km^2)	Applicable Range of Durations (hr)	Remarks
36 (1961)	California	26 to 12,590	6 to 72	General-storm PMP
39 (1963)	Hawaii	2.6 to 518	30 min. to 24	
43 (1966)	Columbia and Snake River basins of Pacific Northwest	26 to 12,590 west of Cascade Mt. 26 to 2,590 east of Cascade Mt.	1 to 72	General-storm PMP plus local-storm (2.6-1259 km^2 1-6 hrs.)
49 (1977)	Colorado River and Great Basin Region of Southwest	26 to 12,590	6 to 72	General storm, plus CA local storm PMP
51 (1978)	U.S. east of the 105th meridian	26 to 51,800	6 to 72	General-storm PMP
52 (1982)	same	2.6 to 51,800	1 to 72	Application study plus PMP for < 26 km^2, < 6 hrs.
53 (1980)	same	26	24	Seasonal variation of general-storm PMP
54 (1983)	Southeastern Alaska	26 to 1036	6 to 72	General-storm PMP
55 (1984)	Continental Divide to 103rd meridian	26 to 12,590 in orographic region 26 to 51,800 otherwise	1 to 72	General-storm PMP plus local-storm
56 (1986)	Tennessee River Valley (revision to HMR No. 45)	2.6 to 7770	1 to 72	PMP < 260 km^2 and PMP 260 - 7770 km^2
Tech. Paper 42 (1961)	Puerto Rico & Virgin Islands	2.6 to 1036	1 to 24	
Tech. Paper 47 (1963)	Alaska	2.6 to 1036	1 to 24	

* A listing of complete titles for all reports may be obtained from the author

The World Meteorological Organization (1973) summarized procedures for estimating PMP for hydrological forecasting and design purposes. This Manual was prepared to provide insight to these procedures to users in other parts of the world and was based on examples from actual studies by the National Weather Service. Currently, this manual is being updated and should be available in early 1987.

3. PRECIPITATION FREQUENCIES

Along the way to developing PMP estimates came requests for estimates of precipitation amounts of selected frequencies, primarily out to the 1-percent exceedance probability. Over the past 40 years, WMID has had a major role in a number of precipitation frequency studies, some of which are still current (table 2). Foremost of these is Technical Paper No. 40 (Hershfield 1961) that combined much of the earlier work, and which in turn has been in part superseded by subsequent studies. Roughly 20 years of recording gage data supplemented by non-recording data (records up to 50 years) were expressed as a partial duration series by converting the annual series of maximum annual amounts. The Gumbel procedure was used to fit data for the Fisher-Tippett Type I distribution. Generalized maps giving precipitation frequencies for various return periods from 1 to 100 years are presented for the entire United States.

In the Western United States, Technical Paper No. 40 provided only a broad representation of the estimated distribution of frequencies. In 1973, WMID completed a reanalysis of western data and published its results in 11 volumes for the 11 Western United States (Miller et al. 1973). These superseded Technical Paper No. 40 for that region and attempted to account for the complex topographic influence. The method followed in developing NOAA Atlas 2 is essentially that of its predecessor studies, but made use of multiple-linear regression relations between precipitation frequencies and climatologic and topographic factors at observing sites to improve interpolation between stations. Much greater spatial detail is provided than was available in Technical Paper No. 40.

In the east, another study (Frederick et al. 1977) superseded Technical paper No. 40 for precipitation frequencies less than 1 hour. The improvement over T.P. 40 came as a result of using additional short-duration data to produce isohyetal maps of geographic variation in these data, as opposed to the nationwide average ratios previously applied to obtain short-duration values in T.P. 40.

Technical Paper No. 40 is still generally valid over the eastern United States for durations greater than 1 hour. Some 25 years of additional data have become available since that report was published and numerous inquiries have been received concerning the update of this report. To date there has been a general lack of financial support for such an update, however, some recent interest in the geographic effects of the Appalachian Mountains has resulted in WMID plans for a pilot

Table 2. Currently applicable rainfall frequency publications for the United States

Technical Paper Number*	Applicable Region	Applicable Range of Return Periods (yr)	Applicable Range of Drainage Areas (km^2)	Applicable Range of Durations (hr)
40 (1961)	Contiguous U.S. (Portions superseded)	1 to 100	2.6 to 1036	30 min to 24
42 (1961)	Puerto Rico & Virgin Islands	1 to 100	2.6 to 1036	30 min to 24
43 (1962)	Hawaii	1 to 100	2.6 to 518	1 to 24
47 (1963)	Alaska	1 to 100	2.6 to 1036	30 min to 24
49 (1964)	Contiguous U.S.	2 to 100	2.6 to 1036	48 to 10-day
51 (1965)	Hawaii	2 to 100	2.6 to 1036	48 to 10-day
52 (1965)	Alaska	2 to 100	2.6 to 1036	48 to 10-day
53 (1965)	Puerto Rico & Virgin Islands	2 to 100	2.6 to 1036	48 to 10-day
NOAA Atlas 2 (1973)	11 Western States	1 to 100	2.6 to 1036	30 min to 24
HYDRO-35 (1977)	U.S. east of the 105th meridian	2 to 100	2.6 to 26	5 min to 1

* A listing of complete titles for all reports may be obtained from the author

study to revise T.P. 40 for the States of Pennsylvania and West Virginia, beginning in late 1986.

Most recently, NWS has been involved in an evaluation study to determine the 1000-yr precipitation-frequency estimates for selected sites in the Eastern United States. The procedure and the reservations attached to it have been described in a paper by Richards and Wescott presented earlier in this conference.

4. STUDY GROUP PARTICIPATION

Another aspect of the problems being discussed at this conference are the numerous interagency work groups that have been assembled over the years to address state-of-the-art issues. Among those in the Federal establishment, the NWS has been a long-time participant. The various joint study groups are too numerous to recall in this brief overview, but a few examples may be useful to establish the NWS role. In the late 1960's a pioneering effort was that of the Hydrology Subcommittee of the Water Resources Council to develop a consistent approach to determination of flood-flow frequencies. The Hydrology Subcommittee is a panel of representatives from various Federal Agencies having hydrologic interests (currently 22 agencies are members). Some of the members of that committee are on this panel. The product of the subcommittee was the 1967 publication of Bulletin No. 15, (Water Resources Council 1967). Briefly stated, the Pearson type III distribution using log-transformed data was recommended for use in flood frequency studies.

Interest in improving and further extending the results in Bulletin 15 continued in the Water Resources Council and its follow-on organization, the Interagency Advisory Committee for Water Data, which lead to the publication of Bulletin No. 17, (Water Resources Council 1976). This study provided more complete guidance than earlier versions but was limited to locations where systematic records of peak discharge were available. Further modifications were to come in 1978 as Bulletin 17A and in 1982 with Bulletin 17B, which provide methods for weighting station skews according to generalized skews, treatment of outliers, and computation of confidence limits about the frequency curve.

Still another work group was formed to address the problem of guidelines for defining flood flow frequencies for ungaged sites. The results of this effort were published in 1981, "Estimating Peak Flow Frequencies for National Ungaged Watersheds." The report is important in laying out procedures for a nationwide testing program that would better aid selection of guidelines for ungaged sites.

At present, the NWS is represented on two work groups organized to establish the state-of-the-art regarding the feasibility of assigning a probability to the PMF and to determining which are the most promising approaches for research in determining the probability of rare floods. Although the second group has just begun their tasks, the former work group has completed their study. The conclusions reached were based on an extensive survey of current literature. The work group concluded that it is not within the state-of-the-art to calculate the probability of PMF-scale floods within definable confidence or error bands and that none of the probabilistic approaches presently available is capable of assigning probabilities to extreme floods in a reliable, consistent or credible manner. This report will be published later this year.

5. SUMMARY

Although the NWS has no direct responsibility for studies directed at flood flow frequency and risk analysis, there are a number of areas in which the Office of Hydrology, NWS, has been involved that are of general interest to the topics being discussed at this conference. Long-term studies in development of PMP and precipitation frequency estimates has made the NWS the acknowledged authority in these areas. Interest in applications of these estimates has led to NWS participation in other studies more directly related to flood frequency and risk analysis problems.

6. REFERENCES

Frederick, R.H., V.A. Myers and E.P. Auciello, 1977: 'Five-to 60-Minute Precipitation Frequency for the Eastern and Central United States,' NOAA Technical Memorandum NWS-HYDRO-35, National Weather Service, National Oceanic and Atmospheric Administration, Silver Spring, MD, 36 pp.

Hershfield, D.M., 1961: 'Rainfall Frequency Atlas of the United States,' Technical Paper No. 40, Weather Bureau, U.S. Department of Commerce, Washington, DC, 61 pp.

Miller, J.F., R.H. Frederick and R.J. Tracey, 1973: 'Precipitation-Frequency Atlas of the Western United States,' NOAA Atlas 2 (11 Vols.), National Weather Service, National Oceanic and Atmospheric Administration, Silver Spring, MD.

Water Resources Council, Hydrology Committee, 1967: 'A Uniform Technique for Determining Flood Flow Frequencies,' Bulletin No. 15, Washington, DC.

____________________, 1976: 'Guidelines for Determining Flood Flow Frequency,' Bulletin No. 17, Washington, DC (Revised 1978, 1982).

World Meteorological Organization, 1973: 'Manual for Estimation of Probable Maximum Precipitation,' Operational Hydrology Report No. 1, Geneva, 190 pp. (Update of this report to be released by WMO in 1987.)

THE ROLE OF FLOOD-FREQUENCY ANALYSIS IN THE U.S. GEOLOGICAL SURVEY

Wilbert O. Thomas, Jr.
U.S. Geological Survey
Water Resources Division
Mail Stop 415, National Center
Reston, Virginia 22092
USA

ABSTRACT. The U.S. Geological Survey is involved in several flood-related activities as part of its mission to monitor the Nation's water resources. This paper describes techniques and approaches utilized in (1) archiving flood data, (2) analyzing flood data at gaging stations, (3) documenting extreme floods, (4) estimating flood-peak discharges at ungaged sites, (5) estimating flood-peak depths at ungaged sites, and (6) developing flood-hazard maps. A brief description of the future direction of flood-frequency analysis within the U.S. Geological Survey is also given.

INTRODUCTION

The mission of the Water Resources Division (WRD) of the U.S. Geological Survey (USGS) is to provide the hydrologic information and understanding needed for the best use and management of the Nation's water resources for the benefit of the people of the United States. In support of this mission, WRD offices across the country collect, analyze, and interpret basic water-resources data, conduct problem-oriented research in hydrology to better understand hydrologic systems, disseminate water data and results of investigations and research, and maintain a water-resources data base. The collection and analysis of flood data are a major part of the surface-water activities of the WRD, USGS. The purpose of the paper is to discuss the various flood-related activities and studies of the USGS and provide a brief overview of the analytical techniques used in these stuides.

The USGS is involved in the following types of flood-related activities:

1. collection, archiving, and dissemination of flood-peak data;
2. analysis and dissemination of flood-frequency information at gaging stations;
3. documentation of extreme floods;
4. development of techniques for estimating flood-peak discharges at ungaged sites;

V. P. Singh (ed.), Application of Frequency and Risk in Water Resources, 463–484.

5. development of techniques for estimating flood-peak depths at ungaged sites; and
6. flood-plain delineation studies.

A brief description of each of these activities and the technique utilized is given in the following sections.

FLOOD-PEAK DATA

The USGS maintains a Peak Flow File as part of the National Water Data Storage and Retrieval System known as WATSTORE (Lepkin and DeLapp, 1979). This file contains data on annual instantaneous peak discharges and partial-duration peaks (peaks above a base value) for both active and discontinued stations. As of March 1986, this file contained about one-half million flood peaks at about 22,000 stations with an average record length of 22 years. A station must have a minimum of 5 years of annual peak data before it is added to the file. At about 8,300 stations partial-duration peaks are also available. There are approximately three partial-duration peaks per year of record for these stations. Those flood peaks affected by regulation, backwater, dam breaks, etc., are coded so that the analyst will be aware they do not represent natural flow conditions. The flood-peak data can be retrieved from the Peak Flow File using procedures documented by Lepkin and DeLapp (1979).

The annual flood peaks and partial-duration peaks are published each year in the annual data reports for each State (e.g. Water Resources Data, West Virginia, Water Year 1984). The partial-duration peaks are computed and published at all unregulated continuous-record stations. Annual peak discharges are published for all stations including crest-stage stations. Crest-stage stations are those stations where only peak stages greater than some predetermined level are recorded. Only the annual maximum peak discharge is published for these stations.

ANALYSIS OF FLOOD-PEAK DATA AT GAGING STATIONS

Like all Federal agencies in the United States, the USGS uses Bulletin 17B, Guidelines for Determining Flood-Flow Frequency (Hydrology Subcommittee, Interagency Advisory Committee on Water Data, 1982), for analyzing flood-peak data at gaging stations. Bulletin 17B guidelines recommend fitting the logarithms of the annual maximum peak discharges to a Pearson Type III distribution using the method of moments approach for parameter estimation as the standard method. Deviations from the standard method are allowed with proper documentation of the analysis. The objective of the analysis is to determine those flood discharges that have, for example, a 50-, 10-, 2-, or 1-percent chance of being exceeded in any given year. Bulletin 17B guidelines include outlier tests for identifying high and low outliers, procedures for including historic information in the analysis and procedures for weighting at-site skew with a regional value of skew. Thomas (1985) provides a description of the historical development of Bulletin 17B and discusses the current methodology used in Bulletin 17B.

Often the annual flood-peak discharges are caused by two different hydrologic events such as summer thunderstorms or snowmelt runoff. In these instances it may not be possible to adequately fit the annual peak discharges with a three-parameter distribution such as the Pearson Type III. For these mixed population problems, the USGS utilizes the combination of probabilities approach as described by Crippen (1978).

The USGS has developed a computer program for analyzing flood-peak data according to Bulletin 17B guidelines (Kirby, 1979). Many copies of this program (J407) have been provided to consulting engineers, universities, other Federal and State agencies, etc. In addition, flood-frequency estimates determined using Bulletin 17B guidelines are available from the various WRD District Offices of the USGS. These analyses generally are based on weighted skew using the skew map in Bulletin 17B and the available information in the USGS Peak Flow File. If alternate analyses are required using different data/assumptions, then these analyses must be made at the user's expense. As an example of the demand for this type of information, the USGS office in Jackson, Mississippi, answers about 100 requests per year for flood-frequency estimates from engineers and planners. The requests come from both public and private agencies for purposes ranging from highway and culvert design to flood-plain management planning.

The most up-to-date estimates of flood frequency at gaging stations can be obtained by contacting the local USGS office. A Water Resources Division Information Guide (U.S. Geological Survey, 1984) gives a listing of addresses, telephone numbers, and office hours for the USGS offices. However, the desired flood-frequency estimates can also be obtained from (1) published reports of the USGS describing flood-peak estimation techniques for ungaged sites (discussed later in this paper) and (2) the USGS Streamflow and Basin Characteristics File described by Dempster (1983).

DOCUMENTATION OF FLOODS

The documentation and description of extreme floods that occur from year to year is important since this information can be used in hydraulic design and flood-plain management. The USGS has documented these historical floods in several types and series of book reports and hydrologic atlases. The book reports include circulars, professional papers, water-supply papers, water-resources investigations, and open-file reports. The hydrologic atlases are map reports with a small text describing the flood and showing the area innudated by the particular flood on a topographic map. Table I is a summary of all such flood reports listed by year. Water-Resources Investigations Reports (WRIR's) and Open-File Reports (OFR's) are shown in Table I only if a more formal publication is in preparation. At least 50 WRIR's and OFR's documenting extreme floods have been prepared by USGS District Offices since 1980. These reports are not shown in Table I and must be obtained from the appropriate District Office. The reader can easily identify the year a major flood occurred in a given area by simply scanning Table I. These reports provide information on the frequency of the particular flood using Bulletin 17B guidelines or the appropriate guidelines at the time the report was written. Copies of these reports may be obtained by writing to U.S. Geological Survey, Books and Open-File Reports, Federal Center, Building 41, Box 25425, Denver, Colorado 80225.

The water-supply papers shown in Table I under the heading "Annual Flood Summary" are summaries of floods for the entire country for that year. Not every year has an annual summary and this practice was discontinued in 1969. The reports shown under "Area Affected by Flooding" describe a particular flood for that area. Therefore, for some years, there is an annual summary report for the entire country and one or more other reports for a specific area.

ESTIMATING FLOOD-PEAK DISCHARGES AT UNGAGED SITES

A major effort of USGS related to flood frequency is the development of techniques for estimating flood discharges (such as the 1-percent chance flood) for both rural and urban ungaged sites. These estimating techniques are generally multiple-regression equations relating a flood characteristic, such as the 1-percent chance flood, to watershed and climatic characteristics such as watershed size, slope of the main channel, an index of precipitation, percent of the watershed forested, mean elevation of the watershed, etc. Benson (1962, 1964) and Thomas and Benson (1970) describe some of the earlier applications of multiple-regression techniques for estimating flood-flow characteristics. Since 1973 every USGS District Office has published a similar report for their State. Table II is a listing by State of reports for estimating flood-peak discharges for rural watersheds. Table III is a similar listing for watersheds affected by urbanization. The PB numbers shown in these tables for some references are the numbers of the reports as contained in the National Technical Information Service (NTIS) of the Department of Commerce, Springfield, Virginia. Tables II and III contain the most recent reports (December 1986) for estimating flood-peak discharges. These reports are constantly being updated so the lists in Tables II and III will naturally change with time. For instance, regional flood-frequency studies are underway in Illinois, Alaska, Minnesota, Kentucky, Kansas, New Jersey, and undoubtedly other States. Therefore the reports for those States will be updated in the near future. General guidelines are that these reports should be updated every 5 to 10 years.

In the reports in Table II, flood-peak discharges are estimated as a function of watershed and climatic characteristics. The most frequent watershed and/or climatic characteristics utilized are watershed size, an index of watershed topography such as slope of the main channel, mean basin slope, or mean watershed elevation, and an index of precipitation such as the mean annual precipitation or the 2-year, 24-hour precipitation intensity. Most analysts divided a given State into hydrologic regions to improve the accuracy of the estimating technique.

ESTIMATING FLOOD-PEAK DEPTHS AT UNGAGED SITES

With the growth of the flood-insurance program in the United States in the early 1970's, there was a need for a quick and easy method for estimating flood depths for flood-plain management. One such method employed by USGS is to utilize multiple-regression techniques to estimate flood depths using watershed and climatic characteristics. Flood depths are computed as the difference between the flood-peak elevation and the elevation of the streambed or some low-water elevation such as the median flow (50 percent flow duration). Flood depths estimated by the regression equations can be added to streambed (or water-surface) contour elevations on topographic maps to obtain a flood elevation. By repeating this exercise at several locations on the same stream, a water-surface profile for a given percent chance flood can be determined for defining the flood-plain boundaries.

Table IV is a listing of reports for estimating flood-peak depths for ungaged sites. The techniques may be applicable for both urban and rural watersheds provided the stream channel has not been modified. For the studies shown in Table IV, the flood depths for various percent chance floods were generally computed by (1) estimating the required flood-peak discharge using Bulletin 17B guidelines (or the appropriate guidelines at the time of the study) and (2) using the stage-discharge relation for the site, determine the flood depth for a given percent chance flood as the difference between the elevation of the flood-peak discharge and the streambed elevation or elevation of median flow. The flood depths so computed are applicable to riverine flooding and not flooding due to backwater from ice, high tides, etc. If flood depths or elevations are required for these backwater conditions, then the frequency analysis should utilize the flood depths or elevations directly. The probability of being flooded due to these backwater conditions can then be combined with the probability of riverine flooding using techniques described by Crippen (1978).

In most of the studies shown in Table IV, watershed size was used to estimate the desired flood depth and the State was divided into regions based on the topography and channel characteristics. However, a few studies have used other watershed and climatic characteristics either in addition to or in place of watershed size. Examples of such studies include Colorado (streambed slope in vicinity of the site), Illinois (main channel slope), and Oklahoma (2-year, 24-hour rainfall).

The flood-depth frequency approach has been used to develop USGS flood-prone area maps and to define flood boundaries for less-detailed flood insurance studies for the Federal Emergency Management Agency (FEMA). Burkham (1978) has shown that the accuracy of flood profiles developed by the flood-depth frequency method is not significantly less than those developed by standard step-backwater techniques.

Copies of reports in Tables II-IV can be obtained from the local USGS office or from the Books and Open-File Reports Section in Denver, Colorado (address given earlier).

FLOOD-PLAIN DELINEATION STUDIES

Since 1969, the USGS has identified flood-prone areas throughout the Nation on more than 13,000 topographic maps. These flood-prone area maps show the approximate areas innundated by the 1-percent chance flood. The flood boundaries were determined by historical flood records and flood-depth frequency methods (as described in the previous section) rather than by detailed hydraulic field surveys.

In addition to the Hydrologic Atlases shown in Table I, there are approximately 170 more atlases that define flood-prone areas for either hypothetical floods of specified frequencies (using step-backwater techniques) or use information on historical floods to define the flood-prone areas. These atlases define the flood boundaries more accurately than the flood-prone area maps described above.

The USGS is also involved in developing flood profiles and defining flood boundaries for FEMA as part of the national flood-insurance program. As of January 1986, flood profiles and boundaries are being defined by standard step-backwater techniques in 38 communities in 11 States while less-detailed studies are underway in 515 communities in 35 States. The less-detailed studies use a variety of techniques requiring less effort than standard step-backwater techniques including the flood-depth frequency methods discussed earlier.

FUTURE DIRECTIONS

A work group of the Hydrology Subcommittee of the Interagency Advisory Committee on Water Data recently conducted a survey of all Federal agencies relative to problems that users may be having with Bulletin 17B guidelines. The general consensus of the replies was that there are no significant operational problems with the present methodology. The USGS participated in this survey so the general conclusions about the applicability of Bulletin 17B guidelines obviously pertain to the USGS as well. The general feeling among Federal agencies is that Bulletin 17B can be utilized to obtain reasonable flood-frequency estimates. However, Bulletin 17B techniques could be extended or enhanced to include topics not covered in the present bulletin. Certain enhancements and extensions were suggested in Bulletin 17B and Thomas (1985) provides additional details on possible improvements to Bulletin 17B. These improvements would not change the present Bulletin 17B but would extend and enhance the present guidelines making them more applicable for at-site frequency analysis.

In the area of estimating flood discharge for ungaged sites, the USGS is advocating the use of the generalized least squares technique to estimate the parameters of the multiple-regression equations (Stedinger and Tasker, 1985). Traditionally ordinary least squares regression has been used by the USGS in developing the multiple-regression equations. In this approach all stations are given equal weight regardless of record length and the possible correlation between flood estimates at nearby sites. However, the generalized least squares approach takes into consideration the variance of flood estimates at a given site (function of record length and at-site variability) and the correlation structure of flood estimates at nearby sites when estimating the parameters of the multiple-regression equation. In addition, the variability about the regression equation is partitioned into model error and sampling error. The model error results from assuming the wrong model form and this error cannot be reduced by collecting more data. The sampling error includes both time- and space-sampling errors and can be reduced by obtaining longer records at the present stations or possibly installing additional stations. This attribute of the generalized least squares technique makes it a natural for hydrologic network analysis and the evaluation of the worth of additional data collection. Therefore the USGS will be advocating an analysis of the regional hydrology network as a second step in the regionalization of flood-flow characteristics. This network analysis will identify how the stream-gaging network should be modified to achieve the maximum regional information on flood characteristics. Tasker (1986) and Medina (1985) provide details and examples of how this analysis should be performed.

SUMMARY

As part of its mission to monitor the Nation's water resources, the USGS is very involved in estimating flood discharges for gaged and ungaged sites, documenting extreme floods, archiving and disseminating flood-peak data, and preparing flood-hazard maps. A major portion of surface-water data collection and analysis activities relate to floods and this is not likely to change in the future.

REFERENCES

Benson, M.A. 1962. 'Factors Influencing the Occurrence of Floods in a Humid Region of Diverse Terrain.' U.S. Geological Survey Water-Supply Paper 1580-B, 62 p.

__________ 1964. 'Factors Affecting the Occurrence of Floods in the Southwest.' U.S. Geological Survey Water-Supply Paper 1580-D, 72 p.

Burkham, D.E. 1978. 'Accuracy of Flood Mapping.' U.S. Geological Survey Journal of Research, Vol. 6, No. 4, p. 515-527.

Crippen, J.R. 1978. 'Composite Log-Pearson Type III Frequency-Magnitude Curve of Annual Floods.' U.S. Geological Survey Open-File Report 78-352, 4 p.

Dempster, G.R., Jr. 1983. 'Streamflow/Basin Characteristics Retrieval (Program E796).' U.S. Geological Survey WATSTORE User's Guide, Vol. 4, Ch. II, Sec. B, 31 p.

Hydrology Subcommittee of the Interagency Advisory Committee on Water Data. 1982. Bulletin 17B, Guidelines for Determining Flood-Flow Frequency, U.S. Geological Survey, Office of Water Data Coordination, Reston, Virginia.

Kirby, W.H. 1979. 'Annual Flood Frequency Analysis Using U.S. Water Resources Council Guidelines (Program J407).' U.S. Geological Survey Open-File Report 79-1336-I, WATSTORE User's Guide, Vol. 4, Ch. I, Sec. C, 56 p.

Lepkin, W.D. and DeLapp, M.M. 1979. 'Peak Flow File Retrieval (Program J980).' U.S. Geological Survey Open-File Report 79-1336-I, WATSTORE User's Guide, Vol. 4, Ch. I, Sec. B, 64 p.

Medina, K.D. 1985. 'Analysis of Surface-Water Data Network in Kansas for Effectiveness in Providing Regional Streamflow Information--with a Section on Theory and Application of Generalized Least Squares.' U.S. Geological Survey Open-File Report 85-680, 39 p.

Stedinger, J.R. and Tasker, G.D. 1985. 'Regional Hydrologic Analysis--Ordinary, Weighted and Generalized Least Squares Compared.' Water Resources Research, Vol. 21, No. 9, pp. 1421-1432.

Tasker, G.D. 1986. 'Generating Efficient Gaging Plans for Regional Information.' Second Scientific Assembly of International Association of Hydrological Sciences, Budapest, Hungary, July 1986.

Thomas, D.M. and Benson, M.A. 1970. 'Generalization of Streamflow Characteristics from Drainage-Basin Characteristics.' U.S. Geological Survey Water-Supply Paper 1975, 55 p.

Thomas, W.O., Jr. 1985. 'A Uniform Technique for Flood Frequency Analysis.' American Society of Civil Engineers, Journal of Water Resources Planning and Management, Vol. 111, No. 3, July 1985, pp. 321-337.

U.S. Geological Survey. 1984. Water Resources Division (WRD) Information Guide, November 1984, 15 p.

TABLE I. List of reports documenting extreme floods

Year of flood	Annual flood summary	Area affected by flooding
1902		Passaic, New Jersey, WSP 88
1903	WSP 96	Passaic, New Jersey, WSP 92
1904	WSP 147	
1905	WSP 162	
1913		Ohio Valley, WSP 334
1916		Southern California, WSP 426
1921		Arkansas River, WSP 487; Central Texas, WSP 488
1922		Colorado, WSP 520-G
1923		Wyoming, WSP 520-G
1927		New England, WSP 636-C
1931	WSP 847	
1934	WSP 771	California, WSP 796-C
1935		New York, WSP 773-E; Kansas-Nebraska, WSP 796-B; Texas, WSP 796-G; and Ohio, WSP 869
1936		New England Rivers, WSP 798; Hudson River to Susquehanna River, WSP 799; Potomac River and Upper Ohio River, WSP 800; Texas, WSP 816
1937		Ohio and Mississippi Rivers, January-February, WSP 838; New Mexico, WSP 842; Northern California, WSP 843
1938		Connecticut, WSP 836-A; Northeastern States, WSP 867; Utah, WSP 994; Southern California, WSP 844; Texas, WSP 914; North Atlantic States, WSP 966
1939		Colorado River below Boulder Dam, WSP 967-A; Kentucky, WSP 967-B; Maine, WSP 967-C; Texas, WSP 914
1940		Texas, WSP 1046; Southeastern States, WSP 1066
1942		Pennsylvania, WSP 1134-B
1943		West Virginia, WSP 1134-B
1946		San Antonio, Texas, C 32
1947		North Atlantic States, WSP 966; Oregon, WSP 968-A; Washington, WSP 968-B
1948		Colorado, WSP 997; Colorado River, Washington, WSP 1080
1949		New England, C 155
1950	WSP 1137-I	Missouri, WSP 1137-A; Red River, WSP 1137-B; Hawaii, WSP 1137-C; Nebraska, WSP 1137-D; Oregon, WSP 1137-E; California, WSP 1137-F; Minnesota, WSP 1137-G; Nevada, WSP 1137-H; and Wichita Falls, C 99

Footnotes at end of table.

TABLE I. List of reports documenting extreme floods (continued)

Year of flood	Annual flood summary	Area affected by flooding
1951	WSP 1227-D	Kansas-Missouri, WSP 1139; Alabama, WSP 1227-A; Oklahoma-Texas, WSP 1227-B; New York, WSP 1227-C
1952	WSP 1260-F	Texas, WSP 1260-A; Missouri River Basin, WSP 1260-B; Upper Mississippi River and Red River, WSP 1260-C; California, WSP 1260-D; Utah-Nevada, WSP 1260-E
1953	WSP 1320-E	Iowa, WSP 1320-A; Montana, WSP 1320-B; Louisiana, WSP 1320-C; Oregon-California, WSP 1320-D
1954	WSP 1370-C	Iowa, WSP 1370-A; Illinois-Indiana, WSP 1370-B
1955	WSP 1455-B	New England to North Carolina, WSP 1420; Colorado-New Mexico, WSP 1455-A; Northeastern States, C 377; Far Western States, C 380
1956	WSP 1530	Far Western States Part I, WSP 1650-A and Far Western States Part II, WSP 1650-B
1957	WSP 1652-C	Kentucky, WSP 1652-A; Texas, WSP 1652-B; and Indiana, C 407
1958	WSP 1660-B	Louisiana, WSP 1660-A
1959	WSP 1750-B	Ohio River, WSP 1750-A; Ohio River, C 418; Indiana, C 440
1960	WSP 1790-B	Nebraska, WSP 1790-A; Puerto Rico, C 451
1961	WSP 1810	Southeastern States, C 452; Mississippi, C 465 and HA-127; Skagit River, Washington, WSP 1527
1962	WSP 1820	Idaho-Nevada, C 467; Tampa Bay, Florida, HA-66; Texas, HA-190, HA-238, and HA-240; Louisiana, HA-126
1963	WSP 1830-B	California-Nevada, WSP 1830-A
1964	WSP 1840-C	Ohio River, WSP 1840-A; Montana, WSP 1840-B; Far Western States, WSP 1866-A and WSP 1866-B
1965	WSP 1850-E	Upper Mississippi River, WSP 1850-A; Colorado, WSP 1850-B; Arizona-New Mexico, WSP 1850-C; Colorado-New Mexico, WSP 1850-D; Hawaii HA-473
1966	WSP 1870-D	Utah, WSP 1870-A; Texas, WSP 1870-B; California, WSP 1870-C; New Mexico, HA-318; Nebraska, HA-188
1967	WSP 1880-C	Alaska, WSP 1880-A and HA-294; Texas, WSP 1880-B; Nebraska, HA-352
1968	WSP 1970-B	Arkansas, WSP 1970-A; Louisiana, HA-374; Massachusetts, HA-371, HA-419, HA-482, and HA-500
1969	WSP 2030	Mississippi-Alabama, HA-395 to HA-408; Virginia, HA-409 to HA-412; California, HA-422 to HA-425
1970		Arizona, Utah, Colorado, and New Mexico, WSP 2052

Footnotes at end of table.

TABLE I. List of reports documenting extreme floods (continued)

Year of flood	Annual flood summary	Area affected by flooding
1972		Virginia, HA-547, HA-505, and C 667; South Dakota, PP-877 and HA-511; Northeastern States, PP-924; New York, HA-518 to HA-519; West Virginia, HA-547; Pennsylvania, HA-523, HA-530, and HA-541
1973		Mississippi River, PP-937; Southeastern States, PP-998
1974		Nevada, PP-930
1976		Maine, PP-1087; Idaho, WSP 565 and HA-565 to HA-581; Colorado, WSP 1115
1977		Appalachian Flood, PP-1098 and HA-588; Kansas City, Missouri, PP-1169; Kelly Barnes Dam, HA-613; West Virginia, HA-588
1978		Montana-Wyoming, PP-1244; Central Texas, PP-1332; Arizona, WSP 2223
1979		Mississippi-Alabama, PP-1319; Alabama-Florida, 1979, HA-621 to HA-641; Jackson, Mississippi, HA-655; Arizona-New Mexico, WSP 2241
1980		California-Arizona, February 1980 (PP in preparation); Wisconsin, HA-53
1981		Austin, Texas, HA-656
1982		San Francisco Bay, California, January 1982 (PP in preparation); Indiana, Michigan, Illinois, and Ohio, March 1982 WRI 83-4201 (PP in preparation); Missouri, HA-689
1983		Central and Southern Mississippi River Basin, December 1982, January and April 1983 OFR 83-213 (WRI in preparation); Colorado River Basin, May-June 1983 and May 1984 (WSP in preparation); Southeastern Arizona, October 1983 (WSP in preparation)
1984		Great Basin, Utah and Nevada, March, May, and July 1984 (WSP in preparation); Tulsa, Oklahoma, May 1984, HA-707
1985		Cheyenne, Wyoming, August 1985, HA-699
1986		California and Nevada, February 1986 (report in preparation)

C - Circular
HA - Hydrologic Investigations Atlas
PP - Professional Paper
WSP - Water-Supply Paper
WRI - Water-Resources Investigations
OFR - Open-File Report

TABLE II. List of reports for estimating rural flood-peak discharges

Alabama:

Olin, D. A., 1984, Magnitude and frequency of floods in Alabama: U.S. Geological Survey Water-Resources Investigations 84-4191.

Alaska:

Lamke, R. D., 1978, Flood characteristics of Alaskan streams: U.S. Geological Survey Water-Resources Investigations 78-129.

Arizona:

Eychaner, J. H., 1984, Estimation of magnitude and frequency of floods in Pima County, Arizona, with comparisons of alternative methods: U.S. Geological Survey Water-Resources Investigations 84-4142.

Roeske, R. H., 1978, Methods for estimating the magnitude and frequency of floods in Arizona: U.S. Geological Survey Open-File Report 78-711.

Arkansas:

Neely, B. L., Jr., 1986, Magnitude and frequency of floods in Arkansas: U.S. Geological Survey Water-Resources Investigations Report 86-4335.

California:

Waananen, A. O., and Crippen, J. R., 1977, Magnitude and frequency of floods in California: U.S. Geological Survey Water-Resources Investigations 77-21 (PB-272 510/AS).

Colorado:

Kircher, J. E., Choquette, A. F., and Richter, B. D., 1985, Estimation of natural streamflow characteristics in Western Colorado: U.S. Geological Survey Water-Resources Investigations Report 85-4086.

McCain, J. R., and Jarrett, R. D., 1976, Manual for estimating flood characteristics of natural-flow streams in Colorado: Colorado Water Conservation Board, Technical Manual No. 1.

Connecticut:

Weiss, L. A., 1975, Flood flow formula for urbanized and non-urbanized areas of Connecticut: Watershed Management Symposium of ASCE Irrigation and Drainage Division, p. 658-675, August 11-13, 1975.

Delaware:

Simmons, R. H., and Carpenter, D. H., 1978, Technique for estimating the magnitude and frequency of floods in Delaware: U.S. Geological Survey Water-Resources Investigations Open-File Report 78-93.

Florida:

Bridges, W. C., 1982, Technique for estimating magnitude and frequency of floods on natural-flow streams in Florida: U.S. Geological Survey Water-Resources Investigations 82-4012.

Georgia:

Price, M., 1979, Floods in Georgia, magnitude and frequency: U.S. Geological Survey Water-Resources Investigvations 78-137 (PB-80 146 244).

TABLE II. List of reports for estimating rural flood-peak discharges (continued)

Hawaii:

Matsuoka, I., 1978, Flow characteristics of streams in Tutuila, American Samoa: U.S. Geological Survey Open-File Report 78-103.

Nakahara, R. H., 1980, An analysis of the magnitude and frequency of floods on Oahu, Hawaii: U.S. Geological Survey Water-Resources Investigations 80-45 (PB-81 109 902).

Idaho:

Thomas, C. A., Harenburg, W. A., and Anderson, J. M., 1973, Magnitude and frequency of floods in small drainage basins in Idaho: U.S. Geological Survey Water-Resources Investigations 7-73 (PB-222 409).

Illinois:

Curtis, G. W., 1977, Technique for estimating magnitude and frequency of floods in Illinois: U.S. Geological Survey Water-Resources Investigations 77-117 (PB-277 255/AS).

Indiana:

Glatfelter, D. R., 1984, Techniques for estimating magnitude and frequency of floods in Indiana: U.S. Geological Survey Water-Resources Investigations 84-4134.

Iowa:

Lara, O., 1973, Floods in Iowa: Techniques manual for estimating their magnitude and frequency: State of Iowa Natural Resources Council Bulletin no. 11.

Kansas:

Jordan, P. R., and Irza, T. J., 1975, Magnitude and frequency of floods in Kansas, unregulated streams: Kansas Water Resources Board Technical Report no. 11.

Kentucky:

Hannum, C. H., 1976, Technique for estimating magnitude and frequency of floods in Kentucky: U.S. Geological Survey Water-Resources Investigations 76-62 (PB-263 762/AS).

Louisiana:

Lee, F. N., 1985, Floods in Louisiana, Magnitude and frequency, Fourth edition: Department of Transportation and Development, Water Resources Technical Report No. 36.

Lowe, A. S., 1979, Magnitude and frequency of floods for small watersheds in Louisiana: Louisiana Department of Transportation and Development, Office of Highways, Research Study No. 65-2H.

Maine:

Morrill, R. A., 1975, A technique for estimating the magnitude and frequency of floods in Maine: U.S. Geological Survey open-file report.

Maryland:

Carpenter, D. H., 1980, Technique for estimating magnitude and frequency of floods in Maryland: U.S. Geological Survey Water-Resources Investigations Open-File Report 80-1016.

TABLE II. List of reports for estimating rural flood-peak discharges (continued)

Massachusetts:

Wandle, S. W., 1983, Estimating peak discharges of small rural streams in Massachusetts: U.S. Geological Survey Water-Supply Paper 2214.

Michigan:

Holtschlag, D. J., and Croskey, H. M., 1984, Statistical models for estimating flow characteristics of Michigan streams: U.S. Geological Survey Water Resources Investigations 84-4207.

Minnesota:

Guetzkow, L. C., 1977, Techniques for estimating magnitude and frequency of floods in Minnesota: U.S. Geological Survey Water-Resources Investigations 77-31 (PB-272 509/AS).

Mississippi:

Colson, B. E., and Hudson, J. W., 1976, Flood frequency of Mississippi streams: Mississippi State Highway Department.

Missouri:

Hauth, L. D., 1974, A technique for estimating the magnitude and frequency of Missouri floods: U.S. Geological Survey open-file report.

Montana:

Omang, R. J., Parrett, C., and Hull, J. A., 1986, Methods of estimating magnitude and frequency of floods in Montana based on data through 1983: U.S. Geological Survey Water-Resources Investigations Report 86-4027.

Nebraska:

Beckman, E. W., 1976, Magnitude and frequency of floods in Nebraska: U.S. Geological Survey Water-Resources Investigations 76-109 (PB-260 842/AS).

Nevada:

Moore, D. O., 1976, Estimating peak discharges from small drainages in Nevada according to basin areas within elevation zones: Nevada State Highway Department Hydrologic Report no. 3.

New Hampshire:

LeBlanc, D. R., 1978, Progress report on hydrologic investigations of small drainage areas in New Hampshire--Preliminary relations for estimating peak discharges on rural, unregulated streams: U.S. Geological Survey Water-Resources Investigations 78-47 (PB-284 127/AS).

New Jersey:

Stankowski, S. J., 1974, Magnitude and frequency of floods in New Jersey with effects of urbanization: New Jersey Department of Environmental Protection Special Report 38.

New Mexico:

Waltmeyer, S. D., 1986, Techniques for estimating flood-flow frequency for unregulated streams in New Mexico: U.S. Geological Survey Water-Resources Investigations Report 86-4104.

TABLE II. List of reports for estimating rural flood-peak discharges (continued)

New York:

Zembrzuski, T. J., and Dunn, B., 1979, Techniques for estimating magnitude and frequency of floods on rural unregulated streams in New York, excluding Long Island: U.S. Geological Survey Water-Resources Investigations 79-83 (PB-80 201 148).

North Carolina:

Jackson, N. M., Jr., 1976, Magnitude and frequency of floods in North Carolina: U.S. Geological Survey Water-Resources Investigations 76-17 (PB-254 411/AS).

North Dakota:

Crosby, O. A., 1975, Magnitude and frequency of floods in small drainage basins in North Dakota: U.S. Geological Survey Water-Resources Investigations 19-75 (PB-248 480/AS).

Ohio:

Webber, E. E., and Bartlett, W. P., Jr., 1977, Floods in Ohio magnitude and frequency: State of Ohio, Department of Natural Resources, Division of Water, Bulletin 45.

Oklahoma:

Tortorelli, R. L., and Bergman, D. L., 1984, Techniques for estimating flood peak discharge for unregulated streams and streams regulated by small floodwater retarding structures in Oklahoma: U.S. Geological Survey Water-Resources Investigations 84-4358.

Oregon:

Harris, D. D., and Hubbard, L. E., 1982, Magnitude and frequency of floods in eastern Oregon: U.S. Geological Survey Water-Resources Investigations 82-4078.

Harris, D. D., Hubbard, L. L., and Hubbard, L. E., 1979, Magnitude and frequency of floods in western Oregon: U.S. Geological Survey Open-File Report 79-553.

Pennsylvania:

Flippo, H. N., Jr., 1977, Floods in Pennsylvania: A manual for estimation of their magnitude and frequency: Pennsylvania Department of Environmental Resources Bulletin no. 13.

Puerto Rico:

Lopez, M. A., Colon-Dieppa, E., and Cobb, E. D., 1978, Floods in Puerto Rico: magnitude and frequency: U.S. Geological Survey Water-Resources Investigations 78-141 (PB-300 855/AS).

Rhode Island:

Johnson, C. G., and Laraway, G. A., 1976, Flood magnitude and frequency of small Rhode Island streams--Preliminary estimating relations: U.S. Geological Survey open-file report.

South Carolina:

Whetstone, B. H., 1982, Floods in South Carolina--Techniques for estimating magnitude and frequency of floods with compilation of flood data: U.S. Geological Survey Water-Resources Investigations 82-1.

TABLE II. List of reports for estimating rural flood-peak discharges (continued)

South Dakota:

Becker, L.D., 1974, A method for estimating the magnitude and frequency of floods in South Dakota: U.S. Geological Survey Water-Resources Investigations 35-74 (PB-239 831/AS).

__________, 1980, Techniques for estimating flood peaks, volumes, and hydrographs on small streams in South Dakota: U.S. Geological Survey Water-Resources Investigations 80-80 (PB-81 136 145).

Tennessee:

Randolph, W. J., and Gamble, C. R., 1976, A technique for estimating magnitude and frequency of floods in Tennessee: Tennessee Department of Transportation.

Texas:

Schroeder, E. E., and Massey, B. C., 1977, Techniques for estimating the magnitude and frequency of floods in Texas: U.S. Geological Survey Water-Resources Investigations Open-File Report 77-110.

Utah:

Thomas, B. E., and Lindskov, K. L., 1983, Methods for estimating peak discharges and flood boundaries of streams in Utah: U.S. Geological Survey Water-Resources Investigations 83-4129.

Vermont:

Johnson, C. G., and Tasker, G. D., 1974, Flood magnitude and frequency of Vermont streams: U.S. Geological Survey Open-File Report 74-130.

Virginia:

Miller, E. M., 1978, Technique for estimating the magnitude and frequency of floods in Virginia: U.S. Geological Survey Water-Resources Investigations Open-File Report 78-5.

Washington:

Cummans, J. E., Collins, M. R., and Nassar, E. G., 1974, Magnitude and frequency of floods in Washington: U.S. Geological Survey Open-File Report 74-336.

Haushild, W. L., 1978, Estimation of floods of various frequencies for the small ephemeral streams in eastern Washington: U.S. Geological Survey Water-Resources Investigations 79-81.

West Virginia:

Runner, G. S., 1980, Technique for estimating magnitude and frequency of floods in West Virginia: U.S. Geological Survey Open-File Report 80-1218.

Wisconsin:

Conger, D. H., 1980, Techniques for estimating magnitude and frequency of floods for Wisconsin streams: U.S. Geological Survey Water-Resources Investigations Open-File Report 80-1214.

Wyoming:

Craig, G. S., Jr., and Rankl, J. G., 1977, Analysis of runoff from small drainage basins in Wyoming: U.S. Geological Survey Water-Supply Paper 2056.

TABLE II. List of reports for estimating rural flood-peak discharges (continued)

Lowham, H. W., 1976, Techniques for estimating flow characteristics of Wyoming streams: U.S. Geological Survey Water-Resources Investigations 76-112 (PB-264 224/AS).

Regional Reports:

Landers, M. N., 1985, Floodflow frequency of streams in the alluvial plain of the Lower Mississippi River in Mississippi, Arkansas and Louisiana: U.S. Geological Survey Water-Resources Investigations 85-4150.

Wetzel, K. L., and Bettandorff, J. M., 1986, Techniques for estimating streamflow characteristics in the Eastern and Interior Coal Provinces of the United States: U.S. Geological Survey Water-Supply Paper 2226.

TABLE III.--List of reports for estimating urban flood-peak discharges

Alabama:

Olin, D. A., and Bingham, R. H., 1982, Synthesized flood frequency of urban streams in Alabama: U.S. Geological Survey Water-Resources Investigations 82-683.

California:

Waananen, A. O., and Crippen, J. R., 1977, Magnitude and frequency of floods in California: U.S. Geological Survey Water-Resources Investigations 77-21 (PB-272 510/AS).

Connecticut:

Weiss, L. A., 1975, Flood flow formula for urbanized and non-urbanized areas of Connecticut: Watershed Management Symposium of ASCE Irrigation and Drainage Division, p. 658-675, August 11-13, 1975.

Florida:

Franklin, M. A., 1984, Magnitude and frequency of floods from urban streams in Leon County, Florida: U.S. Geological Survey Water-Resources Investigations 84-4004.

Lopez, M. A., and Woodham, W. M., 1982, Magnitude and frequency of flooding on small urban watersheds in the Tampa Bay area, west-central Florida: U.S. Geological Survey Water-Resources Investigations 82-42.

Georgia:

Inman, E. J., 1983, Flood-frequency relations for urban streams in metropolitan Atlanta, Georgia: U.S. Geological Survey Water-Resources Investigations 83-4203.

Illinois:

Allen, H. E., Jr., and Bejcek, R. M., 1979, Effects of urbanization on the magnitude and frequency of floods in northeastern Illinois: U.S. Geological Survey Water-Resources Investigations 79-36 (PB-299 065/AS).

Iowa:

Lara, O., 1978, Effects of urban development on the flood-flow characteristics of Walnut Creek basin, Des Moines metropolitan area, Iowa: U.S. Geological Survey Water-Resources Investigations 78-11 (PB-284 093/AS).

Kansas:

Peek, C. O., and Jordan, P. R., 1978, Determination of peak discharge from rainfall relations for urbanized basins, Wichita, Kansas: U.S. Geological Survey Open-File Report 78-974.

Missouri:

Becker, L. D., 1986, Techniques for estimating flood-peak discharges for urban streams in Missouri: U.S. Geological Survey Water-Resources Investigations Report 86-4322.

Spencer, D. W., and Alexander, T. W., 1978, Technique for estimating the magnitude and frequency of floods in St. Louis County, Missouri: U.S. Geological Survey Water-Resources Investigations 78-139 (PB-298 245/AS).

TABLE III. List of reports for estimating urban flood-peak discharges (continued)

New Jersey:

Stankowski, S. J., 1974, Magnitude and frequency of floods in New Jersey with effects of urbanization: New Jersey Department of Environmental Protection Special Report 38.

North Carolina:

Martens, L. A., 1968, Flood inundation and effects of urbanization in metropolitan Charlotte, North Carolina: U.S. Geological Survey Water-Supply Paper 1591-C.

Putnam, A. L., 1972, Effect of urban development on floods in the Piedmont province of North Carolina: U.S. Geological Survey open-file report.

Ohio:

Sherwood, J. M., 1986, Estimating peak discharges, flood volumes, and hydrograph stages of small urban streams in Ohio: U.S. Geological Survey Water-Resources Investigations Report 86-4197.

Oklahoma:

Sauer, V. B., 1974, An approach to estimating flood frequency for urban areas in Oklahoma: U.S. Geological Survey Water-Resources Investigations 23-74 (PB-235 307/AS).

Oregon:

Laenen, Antonius, 1980, Storm runoff as related to urbanization in the Portland, Oregon-Vancouver, Washington, area: U.S. Geological Survey Water-Resources Investigations Open-File Report 80-689.

Tennessee:

Neely, B. L., Jr., 1984, Flood frequency and storm runoff of urban areas of Memphis and Shelby County, Tennessee: U.S. Geological Survey Water-Resources Investigations 84-4110.

Robbins, C. H., 1984, Synthesized flood frequency for small urban streams in Tennessee: U.S. Geological Survey Water-Resources Investigations 84-4182.

Wibben, H. C., 1976, Effects of urbanization on flood characteristics in Nashville-Davidson County, Tennessee: U.S. Geological Survey Water-Resources Investigations 76-121 (PB-266 654/AS).

Texas:

Land, L. F., Schroeder, E. E., and Hampton, B. B., 1982, Techniques for estimating the magnitude and frequency of floods in the Dallas-Fort Worth Metropolitan Area, Texas: U.S. Geological Survey Water-Resources Investigations 82-18.

Liscum, F., and Massey, B. C., 1980, Technique for estimating the magnitude and frequency of floods in the Houston, Texas, metropolitan area: U.S. Geological Survey Water-Resources Investigations 80-17 (ADA-089 495).

Veenhuis, J. E., and Garrett, D. G., 1986, The effects of urbanization on floods in the Austin metropolitan area, Texas: U.S. Geological Survey Water-Resources Investigations Report 86-4069.

TABLE III. List of reports for estimating urban flood-peak discharges (continued)

Virginia:

Anderson, D. G., 1970, Effects of urban development on floods in Northern Virginia: U.S. Geological Survey Water-Supply Paper 2001-C.

Wisconsin:

Conger, D. H., 1986, Estimating magnitude and frequency of floods for ungaged urban streams in Wisconsin: U.S. Geological Survey Water-Resources Investigations Report 86-4005.

Regional Reports:

Sauer, V. B., Thomas, W. O., Jr., Stricker, V. A., and Wilson, K. V., 1983, Flood characteristics of urban watersheds in the United States--Techniques for estimating magnitude and frequency of urban floods: U.S. Geological Survey Water-Supply Paper 2207.

TABLE IV. List of reports for estimating flood-peak depths

Alabama:

Olin, D. A., 1985, Flood depth frequency relations for streams in Alabama: U.S. Geological Survey Water-Resources Investigations Report 85-4296.

Arkansas:

Hines, M. S., 1977, Graphs for determining the approximate elevation of the 50-year flood in Arkansas: Arkansas Geological Commission, Water Resources Summary No. 12.

__________, 1978, Graphs for determining the approximate evaluation of the 100-year flood in Arkansas: Arkansas Geological Commission, Water Resources Summary No. 13.

Colorado

McCain, J. R., and Jarrett, R. D., 1976, Manual for estimating flood characteristics of natural-flow streams in Colorado: Colorado Water Conservation Board, Technical Manual 1.

Georgia:

Price, M., 1977, Techniques for estimating flood-depth frequency relations in natural streams in Georgia: U.S. Geological Survey Water-Resources Investigations 77-90 (PB-275 381/AS).

Illinois:

Prugh, B. J., Jr., 1976, Depth and frequency of floods in Illinois: State of Illinois Department of Transportation, Division of Water Resources.

Louisiana:

Lowe, A. S., 1980, Flood depth-frequency relations for Louisiana: Louisiana Department of Transportation and Development, Office of Public Works, Water Resources Technical Report No. 23.

Maryland:

Herb, W. J., 1978, Exceedance probability-depth relationships of floods for Maryland streams west of Chesapeake Bay: U.S. Geological Survey Open-File Report 78-171.

Massachusetts:

Wandle, S. W., 1976, Relationship of flood height to drainage area for streams in Massachusetts: U.S. Geological Survey, Journal of Research, Professional Paper 1000.

Missouri

Gann, E. E., 1968, Flood height-frequency relations for the plains area in Missouri: U.S. Geological Survey Professional Paper 600-D, p. D52-D53.

New Jersey:

Velnick, A. J., and Laskowski, S. L., 1979, Technique for estimating depth of 100-year floods in New Jersey: U.S. Geological Survey Open-File Report 79-419.

Thomas, D. M., 1964, Height-frequency relations for New Jersey floods: Article 167 _in_ U.S. Geological Survey Professional Paper 475-D, p. D202-203.

TABLE IV. List of reports for estimating flood-peak depths (continued)

New York:

Archer, R. J., 1978, Discharge, gage-height, and elevation of 100-year floods in the Hudson River basin, New York: U.S. Geological Survey Open-File Report 78-332.

North Carolina:

Coble, R. W., 1979, A technique for estimating heights reached by the 100-year flood on unregulated non-tidal streams in North Carolina: U.S. Geological Survey Water-Resources Investigations 79-69 (PB-301 372).

Eddin, H. A., and Jackson, N. M., Jr., 1980, A technique for estimating flood heights on small streams in the city of Charlotte and Mecklenburg County, North Carolina: U.S. Geological Survey Water-Resources Investigations 80-106.

Oklahoma:

Thomas, W. O., Jr., 1976, Techniques for estimating flood depths for Oklahoma streams: U.S. Geological Survey Water-Resources Investigations 2-76 (PB-253 310/AS).

Pennsylvania:

Flippo, H. N., Jr., 1986, Technique for estimating depths of 100-year floods in Pennsylvania: U.S. Geological Survey Water-Resources Investigations Report 86-4195.

Tennessee:

Gamble, C. R., 1983, Technique for estimating depth of floods in Tennessee: U.S. Geological Survey Water-Resources Investigations 83-4060.

Gamble, C. R., and Lewis, J. G., 1977, Technique for estimating depth of 100-year floods in Tennessee: U.S. Geological Survey Open-File Report 77-668.

Utah:

Thomas, B. E., and Lindskov, K. L., 1983, Methods for estimating peak discharges and flood boundaries of streams in Utah: U.S. Geological Survey Water-Resources Investigations 83-4129.

Virginia:

Miller, E. M., 1977, Equation for estimating regional flood depth-frequency relation for Virginia: U.S. Geological Survey open-file report.

SOME REMARKS ON PRACTICE OF FLOOD FREQUENCY AND RISK ANALYSIS BY THE FEDERAL ENERGY REGULATORY COMMISSION

Cornelius L. Cooper
Deputy Chief, Design Review Branch
Federal Energy Regulatory Commission
Washington, D.C. 20426
United States of America

ABSTRACT. The Federal Energy Regulatory Commission, formerly the Federal Power Commission, has been in dam safety regulation since 1920. As of October 1, 1985, there were about 2,050 dams under our jurisdiction. Almost all of our spillway inflow design floods creep higher and higher with passing years. Mathematical probability methods provide no guidance to the shape and magnitude of the hydrographs of rare floods which would be substantially above the measured data. Risk analysis should include the probability of engineering, administrative, or other human error. Therefore, we require that the adequacy of both new and existing dams be evaluated by considering the hazard potential which would result from failure of the project works during flood flows. If structural failure would present a hazard to human life or cause significant property damage, the dam must withstand the loading or overtopping which may occur from a flood up to the probable maximum (risk analysis that involves loss of life is not permitted). We do in most cases, allow the dam owner to use risk analysis, if the failure will not cause loss of life or significant property damage.

INTRODUCTION

My agency, the Federal Energy Regulatory Commission, has not issued any opinions specifically on flood frequency and risk analysis. The opinions I express here are my own and do not necessarily reflect the opinions of the Commission. I will discuss the practices and procedures used by our staff in carrying out our responsibilities for dam safety and how they relate to flood frequency and risk analysis. These procedures were included in our June 12, 1984, letter to the National Research Council describing our inflow design flow criteria for dams. They were also submitted to the Federal Emergency Management Agency by prior Chairman O'Connor on December 3, 1985, in his report describing our agency's progress in implementing the Federal Guidelines for Dam Safety.

The Commission is authorized by Part I of the Federal Power Act

V. P. Singh (ed.), Application of Frequency and Risk in Water Resources, 485–489.

to license non-Federal hydroelectric projects. The Federal Power Act, which I will refer to as "the Act," authorizes the Commission to issue licenses for the purpose of constructing, operating, and maintaining facilities necessary for the development of non-Federal hydroelectric projects located on navigable streams, or public lands of the United States, utilizing surplus water or water power from any Federal government dam, or located on streams over which the Congress has jurisdiction under the Commerce Clause of the Constitution.

The Energy Security Act (ESA) of 1980, Public Law 96-294, authorizes the Commission to exempt certain small hydroelectric power projects from all or part of Part I of the Act, including licensing. This exemption authority was provided to encourage small hydropower development. Pursuant to its authority under the ESA, the Commission exempts small hydroelectric projects from all licensing provisions of the Act, except as they relate to dam safety.

As of October 1, 1985, there were about 2,050 dams under Commission jurisdiction. In addition, there are several hundred applications for licenses or exemptions pending before the Commission.

COMMISSION PRACTICES

The Commission has developed and implemented policies and procedures and committed resources to ensure the safe design, construction, and operation of each dam under its jurisdiction. The Director of the Office of Hydropower Licensing has been delegated the responsibility for administering the Commission's Dam Safety Program.

The procedures and policies of the Commission's Dam Safety Program include the review and monitoring of all phases of the project development to ensure that the licensee's responsibilities are fulfilled. Therefore, our staff independently reviews and evaluates the safety of dams under the Commission's jurisdiction during the design and construction stages. We ensure that the dams are properly operated and maintained through an annual on-site inspection program. Between October 1983 and October 1985, our staff independently reviewed the safety and adequacy of 188 projects at the design stage. This included both unconstructed projects and existing projects undergoing structural modifications. The reviews included evaluation of site geology and subsurface investigations; hydrologic and hydraulic studies; and stability and stress analysis of all major structures under all probable loading conditions.

To supplement the review of our staff, the terms and conditions of the licenses for major unconstructed projects require the licensees to engage an independent qualified Board of Consultants, approved by our staff, to review the design and conduct periodic inspections during construction of the project. There is only one exception to this requirement. When a non-Federal hydropower development is licensed at a Federal dam, the design and construction of the licensed hydropower facility (that will be an integral part of or that could affect the structural integrity or operation of the Federal project) is subject to the review and approval of the appropriate Federal agency that owns the dam. Fourteen of the projects under construction have been subject to the review by independent Boards of Consultants, five (located on Corps of Engineers' dams) have been monitored by Corps' personnel, and three have been monitored by the Bureau of Reclamation.

In addition to the dam safety reviews by our staff, the Commission's regulations require an independent consultant to inspect and evaluate every five years each licensed or exempted dam that exceeds 32.8 feet (10 meters) in height, has a reservoir with a gross storage capacity of 2,000 acre-feet (2,500,000 m^3) or more, or has a high downstream hazard potential as determined by our staff. The consultant is required to evaluate the adequacy of spillways and the stability and structural adequacy of all structures under all credible loading conditions to determine if the structures meet current accepted engineering design criteria and practices. The consultant's report is reviewed by our staff to ensure that the evaluation conforms to the requirements of the regulations and that the consultant's recommendations are implemented by the licensee. Independent consultants performed 268 formal inspections between October 1983 and October 1985.

As of October 1, 1985, there were 64 existing projects under license undergoing modifications to insure their safety. The recommendations for modification of 40 of the projects were based on a combination of licensee, independent consultants and our staff reviews. The recommendations for the remaining 24 were based on our staff's reviews.

The criteria for flood design which follows apply to the reviews of designs by our staff, the Boards of Consultants and the independent consultants.

The adequacy of both new and existing projects for extreme flood conditions is evaluated by considering the hazard potential which would result from failure of the project works during flood flows. If structural failure would present a hazard to human life or cause significant property damage, the project is evaluated as to its ability to withstand the loading or overtopping which may occur from a flood up to the probable maximum (risk analysis that involve loss of life is not permitted). We do, in most cases, allow the dam owner to assume the risk of the loss of a dam, if we can be absolutely assured that the failure will not cause loss of life or significant property damage. Therefore, if structural failure would not present a hazard to human life or cause significant property damage, a spillway design flood of lesser magnitude than the probable maximum flood would be acceptable provided that there is an adequate basis for showing that there would not be a hazard to human life and such basis can be documented.

The 1984 Federal Guidelines for Selecting and Accommodating Inflow Design Floods for Dams pointed out that risk analysis procedures are still evolving and were beyond the scope of the guidelines. Our staff has not developed any analytical techniques or methodologies for risk analysis. We, however, are keeping abreast of the research and development in this area through training and education courses. Pending further development of the risk techniques and methodologies, we utilize case-by-case analysis of the potential consequences of a failure or accident to a dam as the basis for establishing emergency action planning, the priority of inspections, and the urgency for any required remedial work. At this point I should note that the National Research Council's 1985 report on Safety of Dams, Flood and Earthquake Criteria, did not advocate risk-based analysis for non-Federal dams and concluded that until the risk analysis approach can be extended to non-Federal dams, the potential overall savings to society that it appears to offer will not be realized. The Council considered that regulatory agencies may be reluctant to accept risk-based analysis because of non-Federal dam owners' financial capabilities to sustain a major loss and the trend toward owners being held for full liability for damage as demonstrated by recent court decisions.

The conclusions drawn in the draft report "Feasibility of Assigning a Proability to the Probable Maximum Flood" prepared by the Interagency Committee on Water Data's Work Group on "PMF Risk Assessment", support my experiences at the Federal Energy Regulatory Commission that an exceedance probability cannot be assigned to rare floods or the probable maximum flood with any degree of reliability or credibility.

CONCLUSION

Finally, based on my Agency's jurisdiction over more than 2,000 dams and my work as a member of the work group on "PMF Risk Assessment" I will give a personal opinion on flood frequency and risk analysis. Mathematical probability methods provide no guidance to the shape and magnitude of the hydrographs of rare floods which would be substantially above the measured data. Risk analysis should include the probability of engineering, administrative, or other human error. If we accepted the National Research Council's guideline that the probable maximum flood may be selected as a 10,000-year flood, this probability of occurence is less than the probability of a major error occurring in the risk analysis or the data used in the risk analysis. Most of our new spillway hydrology design techniques, such as dambreak analysis, do not include factors of safety as do our structural design techniques for dams.

The Federal Energy Regulatory Commission, formerly the Federal Power Commission, has been in dam safety regulation since 1920. Our files are full of examples where competent engineering hydrologic studies have been modified over time at the same damsite to reflect a constantly lengthening period of hydrologic records. It should be no surprise that almost all spillway inflow design floods creep higher and higher with passing years. This very practical observation serves to temper any tendency to adopt analytical techniques overly dependent on relatively short periods of record.

The use of flood frequency and risk analysis, therefore, must exclude those situations where failure of structures cannot be tolerated. This class of structures would include, among others, structures located in areas where failure could jeopardize critical national defense facilities. Many of these facilities are secret and design engineers are unaware of their existence. This class also would include structures, where failure would cause irreparable economic, environmental, and social damages, such as the City of Vidalia's proposed hydropower project, FERC No. 2754, located about 80 miles north of here (at the Corps of Engineer's Old River Project) on the Mississippi River. Failure of this structure could cause the Mississippi River to change its course by diverting into the Atchafalaya River. Last year the Public Broadcasting System televised a simulated time-lapsed destruction scenario which could be caused by failure of the Old River Project as envisioned by one of the professors at this university. The simulation clearly demonstrates that the application of flood frequency and risk analysis should be limited to only specific classes of structures where a prior determination has been made by the highest level of management officials that failure can be tolerated.

SUBJECT INDEX

AUTHOR INDEX